Polymer Interface and Adhesion

Polymer Interface and Adhesion

SOUHENG WU

E. I. du Pont de Nemours & Company
Wilmington, Delaware

MARCEL DEKKER, INC. New York and Basel

LIBRARY OF CONGRESS CATALOGING IN PUBLICATION DATA

Wu, Souheng, [date]
  Polymer interface and adhesion.

  Includes bibliographical references and index.
  1. Polymers and polymerization--Surfaces. I. Title.
QD381.8.W87   547.7'045453   81-22215
ISBN 0-8247-1533-0   AACR2

The following figures are adapted or reprinted by permission of John Wiley & Sons, Inc.
9.5, 9.6, 9.7, 9.8, 9.10, 9.14, 9.15, 9.16, 9.17, 9.18, 9.19, 9.21, 9.22, 9.23, 9.26, 9.28, 11.2, 11.26, 11.30, 11.31, 14.9, 14.17, 14.19, 14.20, 14.21, 14.22, 14.23, 14.49, 14.50, 14.64, 14.71, 15.1, 15.2

COPYRIGHT © 1982 by MARCEL DEKKER, INC.   ALL RIGHTS RESERVED

Neither this book nor any part may be reproduced or transmitted in any form or by any means, electronic or mechanical, including photocopying, microfilming, and recording, or by any information storage and retrieval system, without permission in writing from the publisher.

MARCEL DEKKER, INC.

270 Madison Avenue, New York, New York 10016

Current printing (last digit):
10 9 8 7 6

PRINTED IN THE UNITED STATES OF AMERICA

AN IMPORTANT MESSAGE TO READERS

A Marcel Dekker, Inc., Facsimile Edition contains the identical content of the original MDI publication.

Reprinting scholarly works in an economical format ensures that they will remain accessible. This contemporary printing process allows us to meet the demand for limited quantities of books that would otherwise go out of print.

We are pleased to offer this specialized service to our readers in the academic and scientific communities.

TO ASTER AND LAWREN

Preface

Interfacial energy, structure, and adhesion of polymers are important in the science and technology of plastics, elastomers, films, fibers, coatings, and adhesives. Many polymer materials, such as polymer blends, composites, laminates, coatings and adhesive joints, have multiple-phase structures containing one or more interfaces. The physical and mechanical properties of these multiple-phase materials are significantly affected by the structure and strength of the interfaces. To understand this, one must know how the interface is formed, its physical and chemical structures, and how the interface modifies the stress field and fracture process of a multiple-phase material. The problem is thus quite complicated, requiring an interdisciplinary approach combining interfacial science, rheology, stress analysis, and fracture mechanics to adequately analyze it. Although several basic questions have not been completely answered, significant advances have been made during the last ten years. The objective of this book is to develop and present a comprehensive, interdisciplinary treatment of polymer interface and adhesion in the light of recent advances.

The significant recent advances include (1) direct and accurate measurements of interfacial tensions between polymer melts, (2) theoretical and experimental analyses of interfacial structure between polymers, (3) surface chemical analysis by electron spectroscopy, and (4) application of flaw theory to adhesive (interfacial) fracture. These advances provide a new insight into the energy and structure of polymer interfaces, and a logical analysis of the fracture strength of adhesive bonds consistent with the views of cohesive fracture mechanics.

In the present approach, the interface is explicitly recognized as a region of finite thickness, formed by interfacial flow of contiguous phases, interdiffusion of macromolecular segments, and/or interfacial chemical bonding. The diffuseness or sharpness of interfacial structure, chemical composition and magnitude of intermolecular forces all combine to determine the mechanical strength of the interfacial zone. Thus, wetting, diffusion, chemical bonding, and mechanical interlocking are all important in determining the fracture process and strength of adhesive bonds. This is in contrast to several previous schools of thought; each suggests that only a single factor is important in adhesion. This has created controversy and has left the gap between the theory and practice unbridged . The present approach gives a more unified view, and hopes to unify theory and practice.

In this book, physical concepts and mechanisms are emphasized, using appropriate mathematical details. Extensive illustrations, tabulations, and references are given so that the book can serve as a source book for scientists, engineers, and students interested in polymer interface and adhesion. It is hoped that this book provides an up-to-date scientific basis for the technology of polymer interface and adhesion, and will promote further advances in polymer blends, composites, coatings and adhesives.

Souheng Wu

Acknowledgments

The permissions to reuse copyrighted materials from the following copyright holders are gratefully acknowledged.

Academic Press, Inc., New York

American Ceramic Society, Columbus, Ohio

American Chemical Society, Washington, D.C.

American Institute of Physics, New York

American Society for Testing and Materials, Philadelphia

American Society of Mechanical Engineers, New York

Chemical Institute of Canada, Ottawa

Communication Channels, Inc., Atlanta

Elsevier Publishing Company, New York

Gordon and Breach Science Publishers, Inc., New York

Huthig and Wepf Publishers, New York

Institute of Metal Finishing, London

IPC Science and Technology Press Ltd, Guildford Surrey, England

John Wiley and Sons, Inc., New York

Lutzmann International Consulting, Cleveland, Ohio

Macmillan Journals Ltd, London

Malaysian Natural Rubber Producers' Research Association, Hertford, England

McGraw-Hill Publications Company, New York

Naval Research Laboratory, Washington, D.C.

Oxford University Press, London

Pergamon Press, Elmsford, New York

Plenum Publishing Corporation, New York

Royal Society, London

Rubber Division-American Chemical Society, Inc., Akron, Ohio

Society of Plastics Engineers, Brookfield Center, Connecticut

Society of Rheology, New York

Technomic Publishing Co., Inc., Westport, Connecticut

Contents

Preface
Acknowledgments

1. INTERFACIAL THERMODYNAMICS 1

    1.1 Formulation of Interfacial Thermodynamics 1
    1.2 Work of Adhesion and Work of Cohesion 4
    1.3 Interfacial and Hydrostatic Equilibrium 5
    1.4 Effect of Curvature on Vapor Pressure and Surface Tension 7
    1.5 Spreading Pressure and Spreading Coefficient 8
    1.6 Contact Angle Equilibrium: Young Equation 11
    1.7 Contact Angle Hysteresis 15
        References 26

2. MOLECULAR INTERPRETATIONS 29

    2.1 Microscopic Theories of van der Waals Forces 30
    2.2 Additivity and Fractional Contributions of Various Types of Molecular Forces 40
    2.3 Approximations for Unlike Molecules 42
    2.4 Lennard-Jones Potential Energy Function 44
    2.5 Attraction Between Macroscopic Bodies 47
    2.6 Macroscopic Theory of van der Waals Forces: The Lifshitz Theory 54
        References 62

3. INTERFACIAL AND SURFACE TENSIONS OF POLYMER MELTS AND LIQUIDS 67

3.1 Surface Tensions of Polymer Melts and Liquids 67
3.2 Interfacial Tensions Between Polymers 96
References 129

4. CONTACT ANGLES OF LIQUIDS ON SOLID POLYMERS 133

4.1 Equilibrium Spreading Pressure 133
4.2 Spreading Coefficient 139
4.3 Effect of Temperature on Contact Angle 139
4.4 Effect of Primary and Secondary Transitions 147
4.5 Contact Angle and Heat of Wetting 148
4.6 Polarity of Liquids 148
4.7 Prediction of Contact Angles 152
4.8 Effect of Solute Adsorption on Contact Angle 152
4.9 Tabulation of Equilibrium Contact Angle 157
4.10 Distortion of Liquid Surfaces: Marangoni Effect 161
4.11 Spreading of Partially Submerged Drops 164
References 165

5. SURFACE TENSION AND POLARITY OF SOLID POLYMERS 169

5.1 Determination of Surface Tension and Polarity 169
5.2 Surface Tension and Polarity of Organic Pigments 198
5.3 Constitutive Effect on Surface Tension 201
5.4 Morphological Effect on Surface Tension 201
5.5 Effects of Additives, Polymer Blends, Copolymers, Conformation, and Tacticity on Surface Tension 209
References 211

6. WETTING OF HIGH-ENERGY SURFACES 215

6.1 Introduction 215
6.2 Spreading on High-Energy Surfaces 219
6.3 Effect of Water and Organic Contaminations 222
6.4 Kinetics of Surface-Energy Variation 229
References 231

7. DYNAMIC CONTACT ANGLES AND WETTING KINETICS 235

7.1 Introduction 235
7.2 Kinetics of Spontaneous Motion 236
7.3 Kinetics of Forced Motion 247
References 254

8. EXPERIMENTAL METHODS FOR CONTACT ANGLES AND INTERFACIAL TENSIONS 257

8.1 Measurements of Contact Angles 257
8.2 Measurements of Interfacial and Surface Tensions of Polymer Liquids and Melts 266
References 274

9. MODIFICATIONS OF POLYMER SURFACES: MECHANISMS OF WETTABILITY AND BONDABILITY IMPROVEMENTS 279

9.1 Chemical Treatments 280
9.2 Flame and Thermal Treatments 296
9.3 Plasma Treatments 298
9.4 Photochemical Treatments 322
9.5 Crystalline Modifications of Polymer Surfaces 323
9.6 Miscellaneous Modifications 327
References 328

10. ADHESION: BASIC CONCEPT AND LOCUS OF FAILURE 337

10.1 Definitions 337
10.2 Basic Concept of Adhesive Bond Strength 338
10.3 Locus of Failure 344
References 354

11. FORMATION OF ADHESIVE BOND 359

11.1 Elementary Processes in Adhesive Bond Formation 359
11.2 Wetting and Adhesion 360
11.3 Diffusion and Adhesion 380
11.4 Chemical Adhesion 406
11.5 Mechanical Adhesion 434
References

12. WEAK BOUNDARY LAYERS 449

    12.1 Weak-Boundary-Layer Theory 449
    12.2 Examples of Weak Boundary Layers 450
    12.3 Critique of Weak-Boundary-Layer Theory 456
    References 460

13. EFFECT OF INTERNAL STRESS ON BOND STRENGTH 465

    13.1 Shrinkage of a Coating Adhering to a Substrate 467
    13.2 Free Thickness Contraction in a Butt Joint 470
    13.3 No Thickness Contraction in a Sandwich Structure 472
    13.4 No Thickness Contraction in a Long, Narrow Adhesive Layer in an Annulus 472
    References 473

14. FRACTURE OF ADHESIVE BOND 475

    Part I. Fundamentals of Fracture Mechanics 475

    14.1 Linear Elastic Fracture Mechanics 476
    14.2 Local Plastic Flow 484
    14.3 Generalization of Flaw Theory to Viscoelastic Materials: Rivlin and Thomas Theory 486
    14.4 Properties of Fracture Energy 488

    Part II. Analysis and Testing of Adhesive Bonds 497

    14.5 Tensile Tests (Butt Joints) 497
    14.6 Shear Tests (Lap Joints) 510
    14.7 Peel Tests (Peel Joints) 530
    14.8 Cantilever Beam Tests 555
    14.9 Other Fracture Energy Tests 560
    References 563

15. CREEP, FATIGUE, AND ENVIRONMENTAL EFFECTS 571

    Part I. Creep and Fatigue of Adhesive Joints 571

    15.1 Creep Fracture 572
    15.2 Fatigue Fracture 579

Part II. Environmental Effects 586

15.3 Spontaneous Separation 589
15.4 Effect of Interfacial Chemical Bonds 593
15.5 Stress Corrosion Cracking and Fatigue in Wet Environment 603
References 609

Appendix I: Calculation of Surface Tension and Its Nonpolar and Polar Components from Contact Angles by the Harmonic-Mean and the Geometric-Mean Methods 613

Appendix II: Unit Conversion Tables 619

Index 621

Polymer Interface and Adhesion

1
Interfacial Thermodynamics

1.1. FORMULATION OF INTERFACIAL THERMODYNAMICS

The interface (surface) is a region of finite thickness (usually less than 0.1 μm) in which the composition and energy vary continuously from one bulk phase to the other. The pressure (force field) in the interfacial zone is therefore nonhomogeneous, having a gradient perpendicular to the interfacial boundary. In contrast, the pressure in a bulk phase is homogeneous and isotropic. Therefore, no net energy is expended in reversibly transporting the matter within a bulk phase. However, a net energy is required to create an interface by transporting the matter from the bulk phase to the interfacial zone. The reversible work required to create a unit interfacial (surface) area is the interfacial (surface) tension, that is,

$$\gamma = \left(\frac{\partial G}{\partial A}\right)_{T,P,n} \tag{1.1}$$

where γ is the interfacial (surface) tension, G the Gibbs free energy of the total system, A the interfacial area, T the temperature, P the pressure, and n the total number of moles of matter in the system.

Consider two homogeneous bulk phases α and β separated by an interfacial layer σ of thickness t (Figure 1.1). For a two-component (1 and 2) system at equilibrium, we have

$$\mu_1^\alpha = \mu_1^\beta = \mu_1^\sigma = \mu_1 \tag{1.2}$$

$$\mu_2^\alpha = \mu_2^\beta = \mu_2^\sigma = \mu_1 \tag{1.3}$$

where μ is the chemical potential. The general variations of the Gibbs free energy are thus given by

$$dG^j = -S^j \, dT + V^j \, dP + \mu_1 \, dn_1^j + \mu_2 \, dn_2^j \tag{1.4}$$

$$dG^\sigma = -S^\sigma \, dT + V^\sigma \, dP + \gamma \, dA + \mu_1 \, dn_1^\sigma + \mu_2 \, dn_2^\sigma \tag{1.5}$$

where S is the entropy, V the volume, n the number of moles of matter, and j = α, β (the two bulk phases). The integral forms of the above are

$$G^j = n_1^j \mu_1 + n_2^j \mu_2 \tag{1.6}$$

$$G^\sigma = \gamma A + n_1^\sigma \mu_1 + n_2^\sigma \mu_2 \tag{1.7}$$

Dividing Eq. (1.7) by A gives

$$f^\sigma = \gamma + \Gamma_1 \mu_1 + \Gamma_2 \mu_2 \tag{1.8}$$

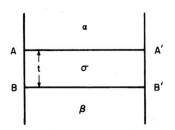

Figure 1.1. Guggenheim model for interface.

Formulation of Interfacial Thermodynamics

where $f^\sigma = G^\sigma/A$ is the specific surface free energy and $\Gamma = n^\sigma/A$ is the surface concentration. Thus, surface tension γ is the *excess* specific surface free energy, that is, $\gamma = f^\sigma - (\Gamma_1 \mu_1 + \Gamma_2 \mu_2)$. Equation (1.8) defines the distinction between the surface tension and the specific surface free energy.

Subtracting Eqs. (1.4) and (1.5) from the total differentials of Eqs. (1.6) and (1.7) gives the Gibbs-Duhem equation,

$$S^j \, dT - V^j \, dP + n_1^j \, d\mu_1 + n_2^j \, d\mu_2 = 0 \tag{1.9}$$

$$S^\sigma \, dT - V^\sigma \, dP + A \, d\gamma + n_1^\sigma \, d\mu_1 + n_2^\sigma \, d\mu_2 = 0 \tag{1.10}$$

Dividing Eq. (1.10) by the area A gives the well-known Gibbs adsorption equation,

$$\bar{S}^\sigma dT - t \, dP + d\gamma + \Gamma_1 \, d\mu_1 + \Gamma_2 \, d\mu_2 = 0 \tag{1.11}$$

where $\bar{S}^\sigma$ is the surface entropy per unit surface area. Rearrangement of the above gives the temperature coefficient of surface tension as

$$-\frac{d\gamma}{dT} = \bar{S}^\sigma - t \frac{dP}{dT} + \Gamma_1 \frac{d\mu_1}{dT} + \Gamma_2 \frac{d\mu_2}{dT} \tag{1.12}$$

Since $d\mu = \bar{V} dP - \bar{S} dT$, where $\bar{V}$ and $\bar{S}$ are partial molal quantities, Eq. (1.12) becomes

$$\frac{-d\gamma}{dT} = (\bar{S}^\sigma - \Gamma_1 \bar{S}_1 - \Gamma_2 \bar{S}_2) - (t - \Gamma_1 \bar{V}_1 - \Gamma_2 \bar{V}_2) \frac{dP}{dT} \tag{1.13}$$

where the first term on the right is the entropy of surface formation per unit surface area, and the second term arises from volume change of the process. The term dP/dT is the temperature coefficient of the equilibrium vapor pressure. Thus, at constant volume, the entropy of interfacial formation per unit area is given by

$$\Delta \bar{S}^\sigma = -\frac{d\gamma}{dT} \tag{1.14}$$

and the energy (or enthalpy) of interfacial formation per unit area is given by

$$\Delta h^\sigma = \gamma - T \frac{d\gamma}{dT} \tag{1.15}$$

The latent heat of interfacial formation per unit area is given by

$$\bar{L}^\sigma = -T \frac{d\gamma}{dT} \tag{1.16}$$

Interfacial thermodynamics was first formulated exactly by Gibbs [1]. A mathematical dividing surface of zero volume is invoked, and the extensive interfacial properties are defined as excess quantities. The formulation is exact, but the use of a dividing surface is unnatural. This formulation has been discussed in many standard texts [2]. An algebraic method that avoids specific reference to a dividing surface has been proposed by Goodrich [3]. Alternatively, Guggenheim [4] proposed a formulation that specifies the interface to have a definite volume. The present formulation is based on the Guggenheim's method.

1.2. WORK OF ADHESION AND WORK OF COHESION

The work required to separate reversibly the interface between two bulk phases α and β from their equilibrium separation to infinity is the work of adhesion,

$$W_a = \gamma_\alpha + \gamma_\beta - \gamma_{\alpha\beta} \tag{1.17}$$

where W_a is the work of adhesion, γ_α the surface tension of phase α, γ_β the surface tension of phase β, and $\gamma_{\alpha\beta}$ the interfacial tension between phases α and β (Figure 1.2). This was apparently first proposed by Dupré [5]. When the two phases are identical, the reversible work is the work of cohesion,

$$W_{cj} = 2\gamma_j \tag{1.18}$$

where W_{cj} is the work of cohesion for phase j.

The work of adhesion is the decrease of Gibbs free energy per unit area when an interface is formed from two individual surfaces. Thus,

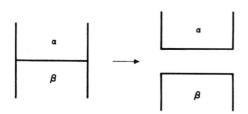

Figure 1.2. Work of adhesion.

Interfacial and Hydrostatic Equilibrium

the greater the interfacial attraction, the greater the work of adhesion will be. Rearrangement of Eq. (1.17) gives

$$\gamma_{\alpha\beta} = \gamma_\alpha + \gamma_\beta - W_a \qquad (1.19)$$

indicating that the greater the interfacial attraction, the smaller the interfacial tension will be. The work of adhesion can be related to the works of cohesion theoretically. Thereby, the interfacial tension can be linked to the properties of the two individual phases; see Chapters 2 and 3.

1.3. INTERFACIAL AND HYDROSTATIC EQUILIBRIUM

The configuration of a liquid meniscus is determined by the balance of interfacial and hydrostatic forces. Several useful equations are discussed below.

1.3.1. Laplace Equation

The Laplace equation governs the configurations of all macroscopic liquid interfaces, drops, and bubbles, and is the basis for all static measurements of interfacial and surface tensions. The pressure drop across a curved interface, ΔP, is balanced by the capillary force, giving [6–9]

$$\Delta P = P^\alpha - P^\beta = \gamma \left(\frac{1}{G_1} + \frac{1}{G_2} \right) \qquad (1.20)$$

where P^α is the pressure in phase α, P^β the pressure in phase β, γ the interfacial tension, and G_1 and G_2 are the two principal radii of curvature. For a plane interface, $\Delta P = 0$.

1.3.2. Equation of Bashforth and Adams

Bashforth and Adams [10] transformed the Laplace equation into convenient dimensionless forms. The shapes of liquid menisci were calculated and tested against carefully measured ones, thus verifying the Laplace equation [10].

Consider the meridional profiles of some axisymmetrical menisci (Figure 1.3). The shapes in Figure 1.3a and d are for a liquid of lower density than that of the surrounding medium. The shapes in Figure 1.3 b and c are for a liquid of higher density than that of the surrounding medium. The two principal radii of curvature at the apex of the meniscus (placed at the origin O) are equal, designated as b. Let P_0 be the pressure at the apex. Then, the pressure at any point $P(x,z)$ on the meridional profile is

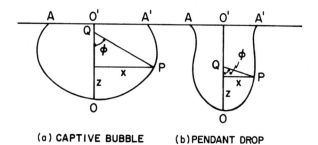
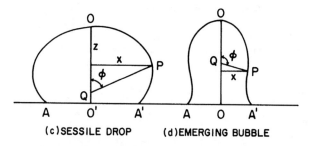

Figure 1.3. Configurations of fluid drops: (a) captive bubble; (b) pendant drop; (c) sessile drop; (d) emerging bubble. (After Ref. 11.)

$$P = P_0 + \rho g z \qquad (1.21)$$

where ρ is the liquid density, g the gravitational acceleration, and z the vertical height of the point $P(x,z)$ measured from the apex. At the apex, $z = 0$ and $G_1 = G_2 = b$. Thus, Eq. (1.20) gives

$$P_0 = \frac{2\gamma}{b} \qquad (1.22)$$

Let the two principal radii of curvature at the point $P(x,z)$ be R in the plane of the paper, and be $x/\sin \phi$ in the plane perpendicular to the plane of the paper, where ϕ is the angle between a tangent and the horizontal. The radius $x/\sin \phi$ is the length PQ which rotates about the vertical axis OO'. Using these in Eq. (1.20) transforms the Laplace equation to

$$2 + \beta\left(\frac{z}{b}\right) = \frac{1}{R/b} + \frac{\sin \phi}{x/b} \qquad (1.23)$$

Effect of Curvature on Vapor Pressure and Surface Tension

which is known as the equation of Bashforth and Adams. The quantity β is the shape factor, defined by

$$\beta = \frac{b^2 \rho g}{\gamma} \tag{1.24}$$

which is positive for the shapes in Figure 1.3a and c, and negative for the shapes in Figure 1.3b and d.

Alternatively, Cartesian coordinates may be used. The two principal radii of curvature are

$$\frac{1}{G_1} = z''(1 + z'^2)^{-3/2} \tag{1.25}$$

$$\frac{1}{G_2} = \frac{z'}{x}(1 + z'^2)^{-1/2} \tag{1.26}$$

where $z' = dz/dx$ and $z'' = d^2z/dx^2$. Using these in Eq. (1.20) gives

$$\hat{z}'' = (2 + \beta\hat{z})[1 + (\hat{z}')^2]^{3/2} - \frac{\hat{z}'}{\hat{x}}[1 + (\hat{z}')^2] \tag{1.27}$$

where $\hat{z} = z/b$ and $\hat{x} = x/b$. Equation (1.27) is an alternative form of the equation of Bashforth and Adams.

Applications of the equation of Bashforth and Adams for the calculation of many meniscus shapes have been reviewed extensively [11,12].

1.4. EFFECT OF CURVATURE ON VAPOR PRESSURE AND SURFACE TENSION

1.4.1. Effect of Curvature on Vapor Pressure

The Laplace equation indicates that there is a pressure drop across a curved interface. The reversible work associated with changing mechanical pressure at constant temperature is

$$G = \int v dP \tag{1.28}$$

where v is the molar volume of the liquid. Assuming the liquid to be incompressible (constant v) and using the Laplace equation for ΔP in Eq. (1.28) gives

$$G = \gamma v \left(\frac{1}{G_1} + \frac{1}{G_2}\right) \tag{1.29}$$

Further, assume ideal gas behavior for the vapor. Then, for a spherical liquid drop of radius R,

$$\ln \frac{P}{P_0} = \frac{v}{RT} \frac{2\gamma}{R} \qquad (1.30)$$

where P_0 is the normal vapor pressure of the liquid (across a flat interface) and P is the vapor pressure over the curved interface. Equation (1.30) is known as the Kelvin equation.

For liquid drops, ΔP is positive, and thus an increased vapor pressure with curvature. For water, P/P_0 is about 1.011 if R is 10^{-5} cm, and 1.114 if R is 10^{-6} cm. This has been verified experimentally for several liquids down to 0.1 μm [13]. A more rigorous derivation of the Kelvin equation has been given recently [14].

1.4.2. Effect of Curvature on Surface Tension

Thermodynamic consideration shows that the surface tension will decrease with increased curvature. Tolman [15] proposed

$$\frac{\gamma}{\gamma^0} = \left(1 + \frac{2t}{R}\right)^{-1} \qquad (1.31)$$

where γ^0 is the normal surface tension and t is the thickness of the surface layer (on the order of 10^{-8} cm). If $t/R = 0.1$, γ/γ^0 is 0.83. On the other hand, Benson and Shuttleworth [16] calculated that the surface tension of a water drop consisting of 13 molecules (radius of curvature 4.6 Å) should be 61 dyne/cm at room temperature, as compared with 72 dyne/cm for the macroscopically flat case. It is, however, uncertain whether macroscopic thermodynamics still applies in such a small phase. To summarize, the effect is quite small except when the phase size is comparable to the surface layer thickness, such as in nucleation and embryo formation.

1.5. SPREADING PRESSURE AND SPREADING COEFFICIENT

Spreading refers to movement of a phase front whereby the interfacial area is increased. Sometimes, it is used specifically to mean that the final contact angle is zero in such a process. Other times, it is used to mean liquid motion, regardless of what is the final contact angle. Either usage should, however, be obvious from the context.

1.5.1. Spreading Pressure

Adsorption of a vapor on a liquid or on a solid will change the surface tension of the substrate. Generally, adsorption will occur if the free energy of the system is thereby reduced. This will occur when the condensed vapor has a surface tension similar to or lower than that of the substrate, or in other words, when the condensed vapor exhibits a low contact angle (less then about 10°) on the substrate (see Chapter 4).

The equilibrium spreading pressure of the vapor on the substrate is defined as

$$\pi_e = \gamma_S - \gamma_{SV} \tag{1.32}$$

where π_e is the equilibrium spreading pressure, γ_S the surface tension of the substrate (solid or liquid) in vacuum (or in equilibrium with its own vapor), and γ_{SV} the surface tension of the substrate in equilibrium with the saturated vapor of the wetting liquid. π_e is therefore the decrease of surface tension due to vapor adsorption. The work of adhesion is thus given by

$$W_a = \gamma_S + \gamma_{LV} - \gamma_{SL}$$

$$= \gamma_{LV}(1 + \cos \theta) + \pi_e \tag{1.33}$$

where θ is the equilibrium contact angle of the liquid (condensed vapor) on the substrate; see Section 1.6. Alternatively, the work of adhesion may be defined as

$$W'_a = \gamma_{SV} + \gamma_{LV} - \gamma_{SL}$$

$$= \gamma_{LV}(1 + \cos \theta) \tag{1.34}$$

where W'_a is the work of adhesion between a liquid in equilibrium with its own saturated vapor and a solid in equilibrium with the saturated vapor of the liquid.

The equilibrium spreading pressure may be measured by vapor adsorption,

$$\pi_e = \int_0^{p^0} \Gamma \, d\mu = RT \int_0^{p^0} \Gamma \, d(\ln p) \tag{1.35}$$

where p^0 is the saturated vapor pressure, Γ the number of moles of the vapor adsorbed on a unit area of the substrate, μ the chemical potential of the vapor, and p the vapor pressure.

1.5.2. Spreading Coefficients

If a liquid has a zero contact angle on a solid, the liquid may spread to a monolayer, a multilayer, or a duplex film. When the spread film is sufficiently thick such that its surface tension is identical to that of the bulk liquid, it is called a duplex film [17].

When a duplex film spreads by one unit area, the decrease in free energy is

$$\lambda_{LS} = \lambda_S - \lambda_{LV} - \lambda_{SL} \tag{1.36}$$

where λ_{LS} is the spreading coefficient of a liquid (L) on a substrate (S). Applying Eqs. (1.17) and (1.18) gives

$$\lambda_{LS} = W_a - W_{cL} \tag{1.37}$$

where W_a is the work of adhesion and W_{cL} is the work of cohesion of the liquid. Spontaneous spreading (zero contact angle) requires that

$$\lambda_{LS} \geq 0 \tag{1.38}$$

that is,

$$W_a \geq W_{cL} \tag{1.39}$$

For a pair of partially miscible liquids, three stages of spreading can be distinguished. In the first stage, mutual saturation occurs only at the interfacial layer, while the two bulk phases are not changed. The initial spreading coefficient is thus

$$\lambda_{LS}^o = \gamma_S - \gamma_{LV} - \gamma_{SL}^* \tag{1.40}$$

where the asterisk refers to mutual saturation. In the second stage, the thin duplex film attains saturation. The intermediate (semifinal) spreading coefficient is thus

$$\lambda_{LS}^m = \gamma_S - \gamma_{LS}^* - \gamma_{SL}^* \tag{1.41}$$

Contact Angle Equilibrium: Young Equation

In the final stage, all phases attain saturation. The final spreading coefficient is thus

$$\lambda_{LS}^{f} = \gamma_{S}^{*} - \gamma_{LV}^{*} - \gamma_{SL}^{*} \tag{1.42}$$

where, of course, $\gamma_{S}^{*} = \gamma_{SV}$, and $\lambda_{LS}^{m} - \lambda_{LS}^{f} = \gamma_{S} - \gamma_{S}^{*} = \pi_{e}$.

Spreading coefficients for some systems are tabulated in Chapters 3 and 5. Additional data for simple liquids can be found elsewhere [17,18]. In some cases, the initial and intermediate spreading coefficients are positive, but the final spreading coefficients are negative. Such liquids will first spread, and then retract to form lenses: an example is benzene one water (Table 4.4).

1.6. CONTACT ANGLE EQUILIBRIUM: YOUNG EQUATION

1.6.1. Contact Angle Phenomena

A liquid in contact with a solid will exhibit a contact angle (Figure 1.4). If the sytem is at rest, a static contact angle is obtained. If the system is in motion, a dynamic contact angle is obtained (Chapter 8). Here, static contact angles are discussed. A system at rest may be in stable equilibrium (the lowest energy state), or in metastable equilibrium (an energy trough separated from neighboring states by energy barriers).

Stable equilibrium will be obtained if the solid surface is ideally smooth, homogeneous, planar, and nondeformable; the angle formed is the equilibrium contact angle, θ_{e}.

On the other hand, if the solid surface is rough or compositionally heterogeneous, the system may reside in one of many metastable states; the angle formed is a metastable contact angle. The amount of mechanical energy in the liquid drop (such as vibrational energy) determines

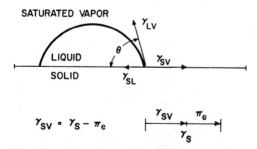

Figure 1.4. Contact angle equilibrium on a smooth, homogeneous, planar, and rigid surface.

which metastable state is to be occupied. Therefore, metastable contact angles vary with drop volume, external mechanical energy (such as vibration), and how the angle is formed (whether by advancing or receding the liquid front on the solid). The stable equilibrium contact angle may sometimes (but rarely) be observed on a rough or heterogeneous surface. This equilibrium angle corresponds to the lowest energy state, discussed further in Section 1.7.

The angle formed by advancing the liquid front on the solid is termed the *advancing contact angle*, θ_a. The angle formed by receding the liquid front on the solid is termed the *receding contact angle*, θ_r. Advancing contact angles are usually greater than receding contact angles when the system is in a metastable state. On the other hand, the advancing and the receding angles are identical when equilibrium contact angles are formed. Many real surfaces are rough or heterogeneous. Thus, variable contact angles are often observed. This has previously led to concern as to whether contact angle is a true thermodynamic quantity. The origin of variable contact angle has now been clearly established and the thermodynamic status of contact angle ascertained.

1.6.2. The Young Equation

The equilibrium contact angle (abbreviated θ here) for a liquid drop on an ideally smooth, homogeneous, planar, and nondeformable surface (Figure 1.4) is related to the various interfacial tensions by

$$\gamma_{LV} \cos \theta = \gamma_{SV} - \gamma_{SL} \qquad (1.43)$$

where γ_{LV} is the surface tension of the liquid in equilibrium with its saturated vapor, γ_{SV} the surface tension of the solid in equilibrium with the saturated vapor of the liquid, and γ_{SL} the interfacial tension between the solid and the liquid. This is known as the *Young equation*. Young [7] described the relation in words, and did not attempt to prove it. Several proofs were offered later by others.

A simple and descriptive proof considers the mechanical equilibrium of interfacial forces at the three-phase boundary (Figure 1.4). The horizontal force acting to the left is $\gamma_{SL} + \gamma_{LV} \cos \theta$; that acting to the right is $\gamma_{SV} = \gamma_S - \pi_e$. Equating the two gives the Young equation. The vertical component of the liquid surface tension is $\gamma_{LV} \sin \theta$, which is balanced by the elastic stress induced in the solid. On a rigid surface, the elastic strain induced is negligibly small. On a soft surface, however, a circular ridge may be raised at the drop periphery [19,20].

A simple thermodynamic proof which neglects the gravity effect was given by Poynting and Thompson [21]. The variation of Gibbs free energy as the meniscus slides on the solid reversibly is

Contact Angle Equilibrium: Young Equation

$$dG = \gamma_{LV} \, dA_{LV} + \gamma_{SV} \, dA_{SV} + \gamma_{SL} \, dA_{SL} = 0 \qquad (1.44)$$

where G is the Gibbs free energy, and A the surface (or interfacial) area. Geometrical consideration gives $dA_{SL} = -dA_{SV}$. Therefore,

$$dA_{LV} = \cos\theta \, dA_{SL} \qquad (1.45)$$

which combines with Eq. (1.44) to give the Young equation.

A rigorous proof based on the minimization of free energy in the absence of gravity was given by Gibbs [1], and in the presence of gravity was given by Johnson [22]. The total Helmholtz free energy of the system consisting of the liquid, solid, and solid-liquid interface in a gravity field is given by [22]

$$F = \int_V dF^V + \int_\sigma dF^\sigma + \int_V gz \, dm^V + \int_\sigma gz \, dm^\sigma \qquad (1.46)$$

where g is the gravitational acceleration, z the height of the element above a horizontal plane, dm^V the mass of an element dV in the bulk phase, dm^σ the mass of an element $d\sigma$ in the interfacial zone, and the symbols $\int_V$ and $\int_\sigma$ indicate integrations over all the volumes and interfaces, respectively. At constant temperature, volume, and mass, the condition for equilibrium is

$$(\delta F)_{T,V,n_i} = 0 \qquad (1.47)$$

Applying Eq. (1.47) in Eq. (1.46) with appropriate mathematical transformations gives the Young equation.

In order for the free energy of the system to be at the minimum, the drop must be a spherical cap having minimum surface area, in the absence of gravity. However, in the presence of gravity, minimum free energy does not correspond to minimum drop surface area. The gravity tends to flatten the drop, whose two principal radii of curvature are no longer identical or constant. The drop shape is described by the Laplace equation. Several other derivations of the Young equation have also been given [3,23,24].

The Young equation is valid for all contact angles (including zero angle) for a system *in equilibrium*. At zero contact angle, just enough liquid vapor will adsorb on the solid surface such that $\pi_e = \gamma_S - \gamma_{LV} - \gamma_{SL}$, and

$$\gamma_{SV} - \gamma_{LV} - \gamma_{SL} = 0 \quad (\theta = 0) \qquad (1.48)$$

so that Eq. (1.43) is satisfied. This has been verified experimentally [25]. On the other hand, if the vapor pressure of the liquid drop is very low (such as with most molten polymers), equilibrium adsorption of the vapor on the solid cannot occur within the time scale of the experiment. In this case, we will have

$$\gamma_S - \gamma_{LV} - \gamma_{SL} \geq 0 \qquad (1.49)$$

Equilibrium contact angle is independent of drop volume. However, contact angles are often found to vary with drop size, particularly with small drops. In such cases, the angles observed must not be the equilibrium contact angle. The variation of contact angle can arise from surface roughness or heterogeneity. However, recent experimental evidence seems to suggest the existence of a negative line tension at the drop periphery which could also cause the drop-size effect [26].

1.6.3. Neumann Triangle

A liquid drop resting on a liquid substrate will form a lens (Figure 1.5). Minimization of the total free energy of the system gives

$$\gamma_{LV} \cos \theta_1 = \gamma_{SV} \cos \theta_2 - \gamma_{SL} \cos \theta_3 \qquad (1.50)$$

where the subscript S now refers to the liquid substrate, and the contact angles θ_1, θ_2, and θ_3 are defined in Figure 1.5. The three vectors form a triangle, known as the *Neumann triangle* [27]. The relation has been verified experimentally [28]. Using the laws of cosines, the relation can be rewritten as

$$\gamma_{SL} = \gamma_{LV}^2 + \gamma_{SV}^2 - 2\gamma_{LV}\gamma_{SV} \cos \beta \qquad (1.51)$$

where β is the Neumann angle, defined in Figure 1.5. The relation can also be obtained by equating the horizontal components of the three interfacial tensions.

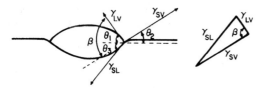

Figure 1.5. Neumann triangle: contact angle equilibrium on a liquid substrate.

1.7. CONTACT ANGLE HYSTERESIS

Many real surfaces are rough or heterogeneous. A liquid drop resting on such a surface may reside in the stable equilibrium (the lowest energy state), or in a metastable equilibrium (energy trough separated from neighboring states by energy barriers). The equilibrium contact angle θ_e corresponds to the lowest energy state for a system. On an ideally smooth and compositionally homogeneous surface, the equilibrium contact angle is the *Young's angle* θ_Y, which is also the microscopic local contact angle on any rough or heterogeneous surface, hence also known as the *intrinsic contact angle* θ_0. The fact that θ_0 equals θ_Y has been proved theoretically as the condition for minimization of system free energy (Section 1.7.1).

The equilibrium contact angle on a rough surface is Wenzel's angle θ_W. The equilibrium contact angle on a heterogeneous surface is *Cassie's angle* θ_C. These angles correspond to the lowest energy state, but are often not observed experimentally. Instead, the system often resides in a metastable state, exhibiting a metastable contact angle. In this case, advancing and receding angles are different, known as *hysteresis*. The difference $\theta_a - \theta_r$ is the extent of hysteresis.

Consider a liquid drop having a steady contact angle on a horizontal planar surface. If the surface is ideally smooth and homogeneous, addition of a small volume of the liquid to the drop will cause the drop front to advance, and the same contact angle will reestablish. Subtraction of a small volume of the liquid from the drop will cause the drop front to recede, but the same contact angle will again reestablish. On the other hand, if the surface is rough or heterogeneous, addition of the liquid will make the drop grow taller without moving its periphery, and the contact angle becomes larger. When enough liquid is added, the drop will suddenly advance with a jerk. The angle at the onset of this sudden advance is the maximum advancing contact angle. Removal of the liquid will make the drop become flatter without moving its periphery, and the contact angle will become smaller. When enough liquid is removed, the drop front will suddenly retract. The angle at the onset of this sudden retraction is the minimum receding contact angle.

Both the advancing and the receding angles can be seen in one drop when the substrate is tilted (Figure 1.6). When the drop width w is very large, the Laplace equation gives [29]

$$\sin \theta_a - \sin \theta_r = \frac{g \rho z_0^2}{2\gamma} \tag{1.52}$$

where g is the gravitational acceleration, ρ the liquid density, z_0 the vertical distance between the top and the bottom of the horizontal

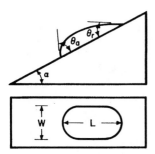

Figure 1.6. A liquid drop on a tilted plate.

liquid drop, and γ the liquid surface tension. The contact angles are related to the tilt angle by [30,31]

$$mg \sin \alpha = L\gamma(\cos \theta_a - \cos \theta_r) \tag{1.53}$$

where m is the drop mass, α the tilt angle, and L the drop length. The θ_a and θ_r at which the drop just starts to roll down the tilted plate are the maximum advancing angle and the minimum receding angle, respectively. On the other hand, if there is no hysteresis, the drop will roll down at the slightest tilt of the plate. This shows that contact angle hysteresis arises from the existence of energy barriers at the liquid front.

1.7.1. Hysteresis Due to Surface Roughness

Hysteresis on a rough surface results from existence of numerous closely spaced metastable states. Several analyses have been given [32–34]. Minimization of free energy requires that the microscopic local contact angle must be the Young's angle (or the intrinsic angle) [33].

Consider a drop on a tilted plate (Figure 1.7). Both the front and the rear edges meet the solid with the same intrinsic angle θ_0. The macroscopic angle observed is the angle with respect to the tilt plane. At the front is the advancing angle θ_a, and at the rear is the receding angle θ_r. The relationship between the local intrinsic angle and the macroscopic angle is illustrated further in Figure 1.8.

Wenzel's Equation

Wenzel's equation [35] corresponds to the lowest energy state on a rough surface. The ratio of the true surface area A (taking into account the peaks and valleys on the surface) to the apparent surface area A' is defined as the roughness factor r = A/A'. Thus, Eq. (1.45) becomes

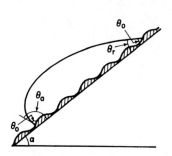

Figure 1.7. Contact angle hysteresis of a liquid drop on a tilted rough surface. Note that the microscopic local contact angles at the front and the rear are identical.

$$r\, dA_{LV} = \cos\theta\, dA_{SL} \qquad (1.54)$$

and the Wenzel's equation is obtained as

$$\cos\theta_W = r\cos\theta_0 \qquad (1.55)$$

where θ_W is Wenzel's angle and θ_0 is the intrinsic contact angle. Other derivations of the equation have been given [36–38]. Wenzel's angle is the equilibrium contact angle on a rough surface. It is, however, seldom observed practically, as the system is more likely to reside in a neighboring metastable state.

Analysis of a Roughness Model

Several analyses of roughness models have been given [32–34]. The analysis of Johnson and Dettre [38] is perhaps the most illuminating. Consider a liquid drop placed at the center of concentric grooves on an idealized rough surface (Figure 1.9). The roughness is large compared with molecular dimensions, but small compared with the resolution of observation apparatus. Gravity is neglected so that the drop is a spherical cap. This simplifies the mathematics but does not af-

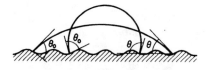

Figure 1.8. Two metastable configurations of a liquid drop on a rough surface.

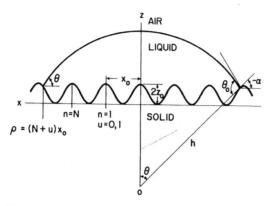

Figure 1.9. A liquid drop on a model rough surface. (After Ref. 38, Johnson and Dettre, *Contact Angle, Wettability, and Adhesion*.)

fect the general features of the results. The microscopic local contact angle is the intrinsic angle (or Young's angle) θ_0 as required for energy minimization. The observed macroscopic angle θ with respect to the horizontal is

$$\theta = \theta_0 + \alpha \tag{1.56}$$

where α is the angle of inclination of the surface at the point of liquid-solid contact. The grooves are assumed to be sinusoidal. These conditions place two geometrical limitations on the system. The first is that the maximum and the minimum macroscopic angles are

$$\theta_{max} = \theta_0 + \alpha_{max} \tag{1.57}$$

$$\theta_{min} = \theta_0 - \alpha_{max} \tag{1.58}$$

where α_{max} is the maximum inclination of the surface. The second is that not all values of θ are allowed. There are only two positions for the drop periphery in each groove such that the macroscopic angle is θ and the drop remains as a spherical cap with constant volume. One position is metastable with the drop edge near the crest; the other is unstable with the drop edge near the bottom of the valley. The number of metastable configurations allowed is the number of crests under the drop counted from the origin, N, that is,

$$\rho = (N + u)x_0 \tag{1.59}$$

Contact Angle Hysteresis

where ρ is the radius of the drop on the surface, u the fraction of horizontal distance between adjacent crests, and x_0 the horizontal distance between adjacent crests.

The difference in Helmholtz free energy between two configurations j and k is given by

$$\Delta F_{jk} = \sum (\gamma A)_j - \sum (\gamma A)_k$$

$$= \gamma_{LV}(F_j^{rel} - F_k^{rel}) \qquad (1.60)$$

where the summations are to be carried over all interfaces, and F^{rel} is a relative Helmholtz free energy for a given configuration and is given by

$$F^{rel} = A_{LV} - A_{SL} \cos \theta \qquad (1.61)$$

Geometrical consideration gives

$$F^{rel} = \frac{2\pi \rho^2}{1 + \cos \theta} - \pi r \rho^2 \cos \theta_0 \qquad (1.62)$$

where r is the Wenzel roughness factor.

The relation between ρ and θ is an implicit function, obtainable by geometrical consideration. That is, θ cannot be expressed explicitly as a function of ρ. Therefore, Eq. (1.62) is solved numerically. The results show the existence of numerous metastable states separated by energy barriers between adjacent states (Figure 1.10). The absolute minimum occurs at the Wenzel's angle (stable equilibrium). The energy barriers are higher nearer Wenzel's angle, and approach zero at the maximum advancing and the minimum receding angle. The energy barriers increase with increasing ridge height and ridge slope [33].

Observed macroscopic angles will therefore depend on the mechanical (vibrational) energy of the drop. For a drop to remain in a given metastable state, the energy barriers of that state must be higher than the vibrational energy of the drop. This vibrational energy is termed the drop energy ϵ. Increased vibration to a system tends to decrease the hysteresis [44,45]. The height of energy barrier for a model system is plotted as a function of contact angle in Figure 1.11. The two horizontal lines represent two different drop energy values. The intersections of the horizontal lines with the curves give the advancing and receding angles for the corresponding ϵ.

Families of hysteresis curves for $\theta_0 = 45°$ and $120°$ are given in Figure 1.12. Hysteresis increases with increasing roughness. For

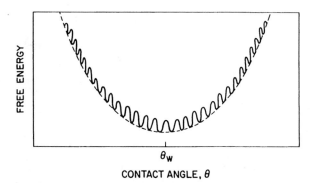

Figure 1.10. Free energy versus contact angle for a liquid drop on a rough surface, showing the existence of numerous metastable states. (After Ref. 32.)

$\theta_0 < 90°$, Wenzel's angle decreases with increasing roughness. For $\theta_0 > 90°$, on the other hand, Wenzel's angle increases with increasing roughness. Spontaneous wicking will occur at a critical roughness, r_c, where

$$r_c = \frac{1}{\cos \theta_0} \quad \text{for } \theta_0 < 90° \tag{1.63}$$

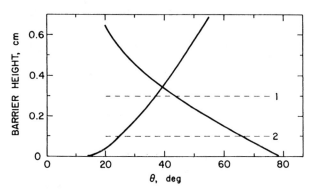

Figure 1.11. Barrier height versus contact angle on a rough surface with $\theta_0 = 45°$, $r = 1.09$, and $\theta_w = 39°$. (1) $\varepsilon = 0.3$ cm, $\theta_a = 44.0°$, $\theta_r = 36.5°$; (2) $\varepsilon = 0.1$ cm, $\theta_a = 64.5°$, $\theta_r = 24.5°$. (After Ref. 32.)

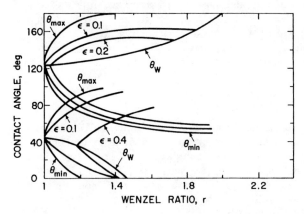

Figure 1.12. Contact angle hysteresis on a model rough surface with $\theta_0 = 120°$ and $\theta_0 = 45°$, respectively. (After Ref. 32.)

Contact angle behavior consistent with Figure 1.12 has been observed experimentally [39–42]. Examples are water on roughened paraffin wax [39]; glycerol and Methyl Cellosolve on paraffin wax embossed with arrays of pyramids [40]; and water, methylene iodide, and hexadecane on randomly roughened fluorocarbon wax [41]. All confirm the qualitative features of Figure 1.12. Wenzel's equation was verified experimentally with mercury drops on silica [43]. The condition for spontaneous wicking was verified with methanol on roughened paraffin wax [40].

1.7.2. Hysteresis Due to Surface Heterogeneity

Hysteresis can also occur on a compositionally heterogeneous surface. Energy barriers exist at the boundaries between regions of different intrinsic contact angles. The liquid front tends to stop at the phase boundaries. Advancing angles tend to reflect the higher intrinsic angle region (or lower surface energy region); receding angles tend to reflect the lower intrinsic angle region (or higher surface energy region) of a heterogeneous surface.

Cassie's Equation

The equilibrium contact angle (corresponding to the lowest energy state) on a heterogeneous surface is Cassie's angle θ_C. Minimization of free energy of the system for a heterogeneous surface consisting of two types of regions of intrinsic angle θ_{01} and θ_{02} gives

$$\cos \theta_C = f_1 \cos \theta_{01} + f_2 \cos \theta_{02} \tag{1.64}$$

where f_j is the fraction of surface area with intrinsic angle θ_{0j} and $f_1 = 1 - f_2$. This is known as Cassie's equation, and can easily be obtained by applying Eq. (1.44) to a heterogeneous surface [46]. Cassie's angle is the equilibrium contact angle on a heterogeneous surface. It is seldom observed, however, as a system is more likely to reside in a metastable state.

Analysis of a Heterogeneous Model

Several analyses of heterogeneous models are available [47,48]. The analysis of Johnson and Dettre is perhaps the most illuminating. Consider a model heterogeneous surface, consisting of concentric circular bands of alternating intrinsic angles θ_{01} and θ_{02} (Figure 1.13). The heterogeneity is assumed to be much smaller than the drop dimensions. Gravity is neglected to simplify the mathematics. This does not affect the general features of the results.

The Helmholtz free energy difference between two states j and k is given by

$$\Delta F_{jk} = \gamma_{LV}(F_j^{rel} - F_k^{rel}) \tag{1.65}$$

where

$$F^{rel} = A_{LV} - A_{SL}\cos\theta_{01} - A_{SL}\cos\theta_{02} \tag{1.66}$$

$$= \frac{2\pi\rho^2}{1+\cos\theta} - \pi x_0^2 \cos\theta_{01}[Nu_1^2 + N(N+1)u_1]$$

$$- \pi x_0^2 \cos\theta_{02}[M(1-u_1^2) + M(M+1)(1-u_1)] \tag{1.67}$$

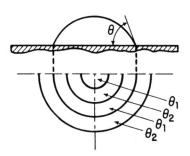

Figure 1.13. A liquid drop on a model heterogeneous surface.

where ρ is the radius of the drop on the surface, θ the macroscopic angle, N the number of θ_{01} strips under the drop, M the number of θ_{02} strips, u_1 the fraction of surface with θ_{01}, and x_0 the width of a circular band.

The results of calculation show that as with the roughness case, a large number of metastable states are accessible to the system. Free energy versus contact angle plots are similar to Figure 1.10 for rough surfaces. The absolute minimum in the curve occurs at Cassie's angle, θ_C. The energy barriers are maximum near θ_C and approach zero at θ_{01} and θ_{02}.

The observed macroscopic angles will, again, depend on the vibrational energy of the liquid drop, ϵ. Hysteresis curves can be constructed using ϵ as for rough surfaces. A family of hysteresis curves showing the variation of contact angle with the fraction of surface covered by the low-contact-angle (θ_{02}) phase is given in Figure 1.14. The center line is calculated from Cassie's equation. The curves above Cassie's curve are possible advancing angles for various drop energies; those below are possible receding angles. As the vibrational energy of the drop increases or as the size of the surface heterogeneity decreases, the observed contact angles tend to approach Cassie's curve. As the drop vibration decreases or the surface heterogeneity increases, the advancing contact angle will approach θ_{01} and the receding contact angle will approach θ_{02}.

Several predictions of the model are discussed below. Advancing angles are more reproducible on largely low energy surfaces. Receding angles are more reproducible on largely high energy surfaces. Advancing angles tend to be close to the intrinsic angles of the lower-energy region. Receding angles tend to be close to the intrinsic angles of the higher-energy region.

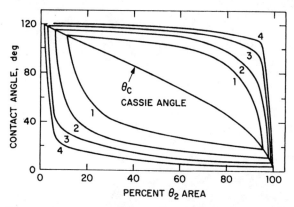

Figure 1.14. Contact angle hysteresis on a model heterogeneous surface. (1) $\epsilon = 0.2$ cm; (2) $\epsilon = 0.1$ cm; (3) $\epsilon = 0.05$ cm; (4) $\epsilon = 0.025$ cm. (After Ref. 47.)

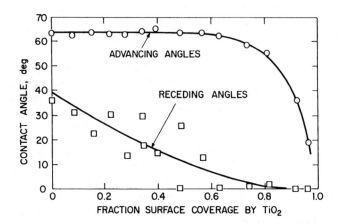

Figure 1.15. Water contact angle on partial monolayers of $H(CH_2)_{18}N(CH_3)_3$ as a function of surface coverage by hydrophilic TiO_2. (After Ref. 32.)

Since $f_2 = 1 - f_1$, Cassie's equation can be rewritten as

$$f_1 = \frac{\cos \theta_C - \cos \theta_{02}}{\cos \theta_{01} - \cos \theta_{02}} \tag{1.68}$$

Cassie's angle may be approximated by

$$\cos \theta_C \simeq \frac{1}{2}(\cos \theta_a + \cos \theta_r) \tag{1.69}$$

An application of Eq. (1.69) has been discussed [49].

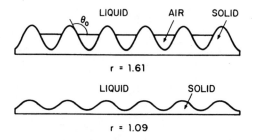

Figure 1.16. Composite interfaces on rough surfaces.

Contact Angle Hysteresis

A small amount of low-energy contaminant on a high-energy surface will greatly increase the advancing contact angle but will have little effect on the receding contact angle. On the contrary, a small amount of high-energy contaminant on a low-energy surface will greatly reduce the receding contact angle, but will have little effect on the advancing contact angle.

These predicted features have been observed experimentally [50–57]. For instance, contact angles of water and methylene iodide on heterogeneous surfaces consisting of monolayer of $CH_3(CH_2)_{17}N(CH_3)_3$ with patches of TiO_2 supported on glass have been investigated [50]. The advancing and receding angles of water versus TiO_2 coverage are shown in Figure 1.15. The similarity between Figures 1.14 and 1.15 is evident.

1.7.3. Comparison of Roughness and Heterogeneity

Both roughness and heterogeneity can cause contact angle hysteresis. Generally, the hysteresis will be negligible when the roughness is below 0.5–0.1 µm, or when the heterogeneous phase is smaller than 0.1 µm.

The barrier heights for roughness and heterogeneity are about the same. On heterogeneous surfaces, the maximum hysteresis $\theta_a - \theta_r$ may be as much as 110°. But on rough surfaces, the maximum hysteresis will be $2\alpha_{max}$, which is usually less than 10° for polished surfaces. Thus, any hysteresis on an optically smooth surface must arise from surface heterogeneity.

1.7.4. Composite Interfaces

On a very rough surface, a liquid with high intrinsic angle may not completely wet the crevices (Figure 1.16). Such an incompletely wetted surface is called a *composite interface* [32]. Composite interfaces are possible only when the roughness has a slope α such that

$$|\alpha| = 180° - \theta_0 \tag{1.70}$$

Let θ_{01} be the intrinsic angle for region 1 (the wetted region) and let $\theta_{02} = 180°$ for region 2 (the unwetted region); then

$$\cos \theta_C = f_1 \cos \theta_{01} - f_2 \tag{1.71}$$

which is known as the *equation of Cassie and Baxter* [58,59].

Composite surfaces have lower energy barriers than do noncomposite surfaces. Both advancing and receding angles tend to approach Cas-

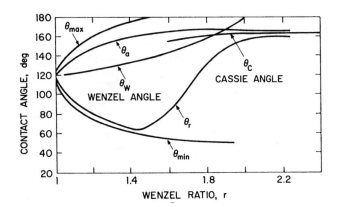

Figure 1.17. Contact angle hysteresis on a model rough surface with $\theta_0 = 120°$, showing transition from noncomposite to composite interface. (After Ref. 32.)

sie's angle, when a noncomposite surface becomes a composite surface, such as when the Wenzel ratio r increases (Figure 1.17). The intersection of Wenzel's angle and Cassie's angle is the point of transition from a noncomposite to a composite interface. This has been observed, for instance, for water on roughened paraffin wax, fluorocarbon wax, porous polyolefins, and a fluoropolymer [41,58,59]. Hysteresis curves similar to Figure 1.17 are found [60].

REFERENCES

1. J. W. Gibbs, *The Collected Works of J. W. Gibbs*, Vol. 1, Longmans, Green, New York, 1931; also, *The Scientific Papers of J. Willard Gibbs*, Vol. 1, *Thermodynamics*, Dover, New York, 1961.
2. G. N. Lewis and M. Randall, *Thermodynamics*, revised by K. S. Pitzer and L. Brewer, 2nd ed., McGraw-Hill, New York, 1961.
3. F. C. Goodrich, in *Surface and Colloid Science*, Vol. 1, E. Matijevic, ed., Wiley-Interscience, New York, 1969, pp. 1–38.
4. E. A. Guggenheim, Trans. Faraday Soc., 36, 397 (1940).
5. A. Dupré, *Théorie mécanique de la chaleur*, Paris, 1869, p. 368.
6. P. S. de Laplace, *Mécanique celeste*, Suppl. au Xième Livre, Courier, Paris, 1805.
7. T. Young, *Proc. Roy. Soc. (Lond.)*, December 1804, Collected Works, D. Peacock, ed.; Royal Society of London, London; also, Phil. Trans., 95, 65, 84 (1805).

References

8. K. F. Gauss, *Principae Generalia Theoreticae Figurae Fluidorum in Statu Aequilibrii*, Göttingen, 1830.
9. S. D. Poisson, *Nouvelle théorie de l'action capillaire*, Paris, 1831.
10. S. Bashforth and J. C. Adams, *An Attempt to Test the Theory of Capillary Action*, Cambridge University Press and Deighton, Bell & Co., London, 1892.
11. J. F. Padday, in *Surface and Colloid Science*, Vol. 1, E. Matijevic, ed., Wiley-Interscience, New York, 1969, pp. 39—252.
12. H. M. Princen, in *Surface and Colloid Science*, Vol. 2. E. Matijevic, ed., Wiley-Interscience, New York, 1969, pp. 1—84.
13. V. K. La Mer and R. Gruen, Trans. Faraday Soc., *48*, 410 (1952).
14. P. J. McElroy, J. Colloid Interface Sci., *72*, 147 (1979).
15. R. C. Tolman, J. Chem. Phys., *17*, 333 (1949).
16. G. C. Benson and R. Shuttleworth, J. Chem. Phys., *19*, 130 (1951).
17. W. D. Harkins, *The Physical Chemistry of Surface Films*, Reinhold, New York, 1952.
18. D. J. Donahue and F. E. Bartell, J. Phys. Chem., *56*, 480 (1952).
19. A. S. Michaels and S. W. Dean, Jr., J. Phys. Chem., *66*, 34 (1962).
20. G. L. J. Bailey, Proc. 2nd Int. Congr. Surf. Act., *3*, 189 (1957).
21. J. H. Poynting and J. J. Thompson, *A Textbook of Physics: Properties of Matter*, 8th ed., Charles Griffin, London, 1920.
22. R. E. Johnson, Jr., J. Phys. Chem., *63*, 1655 (1959).
23. F. P. Buff, in *Handbuch der Physik*, Vol. 10, S. Flugge, ed., Springer-Verlag, Berlin, 1960, pp. 279—304.
24. G. Bakker, in *Wien Harm's Handbuch der Experimental Physik*, Vol. 6, Akademische Verlag, Leipzig, 1928.
25. R. E. Johnson, Jr., and R. H. Dettre, J. Colloid Interface Sci., *21*, 610 (1966).
26. R. J. Good and M. N. Koo, J. Colloid Interface Sci., *71*, 283 (1979).
27. F. Neumann, *Vorlesungen über die Theorie der Capillariat*, B. G. Teubner, Leipzig, 1894.
28. W. Fox, J. Am. Chem. Soc., *67*, 700 (1945).
29. J. J. Bikerman, *Physical Surfaces*, Academic Press, New York, 1970.
30. J. L. Rosano, Mem. Serv. Chim. Etat (Paris), *36*, 437 (1951).
31. C. G. Furmidge, J. Colloid Interface Sci., *17*, 309 (1962).
32. R. E. Johnson, Jr., and R. H. Dettre, in *Surface and Colloid Science*, Vol. 2, E. Matijevic, ed., Wiley-Interscience, New York, 1969, pp. 85—153.
33. J. D. Eick, R. J. Good, and A. W. Neumann, J. Colloid Interface Sci., *53*, 235 (1975).

34. C. Huh and S. G. Mason, J. Colloid Interface Sci., 60, 11 (1977).
35. R. N. Wenzel, Ind. Eng. Chem., 28, 988 (1936); J. Phys. Chem., 53, 1466 (1949).
36. R. Shuttleworth and G. L. J. Bailey, Discuss. Faraday Soc., 3, 16 (1948).
37. R. J. Good, J. Am. Chem. Soc., 74, 5041 (1952).
38. R. E. Johnson, Jr., and R. H. Dettre, in *Contact Angle, Wettability, and Adhesion* (Frederick M. Fowkes, ed.), Adv. Chem. Ser., 43, American Chemical Society, 1964, p. 112.
39. B. R. Ray and F. E. Bartell, J. Colloid Sci., 8, 214 (1953).
40. F. E. Bartell and J. W. Shepard, J. Phys. Chem., 57, 211, 455, 458 (1953).
41. R. H. Dettre and R. E. Johnson, Jr., in *Contact Angle, Wettability, and Adhesion* (Frederick M. Fowkes, ed.), Adv. Chem. Ser., 43, American Chemical Society, 1964, p. 136.
42. A. J. G. Allan and R. Roberts, J. Polym. Sci., 39, 1 (1959).
43. Y. Tamai and K. Aratani, J. Phys. Chem., 76, 3267 (1972).
44. W. Phillipoff, S. R. B. Cooke, and D. E. Caldwell, Mining Eng., 4, 283 (1952).
45. G. R. M. del Giudice, Eng. Mining J., 137, 291 (1936).
46. A. B. D. Cassie, Discuss. Faraday Soc., 3, 11 (1948).
47. R. E. Johnson, Jr., and R. H. Dettre, J. Phys. Chem., 68, 1744 (1964).
48. A. W. Neumann and R. J. Good, J. Colloid Interface Sci., 38, 341 (1972).
49. R. E. Johnson, Jr., and R. H. Dettre, J. Adhes., 2, 3 (1970).
50. R. H. Dettre and R. E. Johnson, Jr., J. Phys. Chem., 69, 1507 (1965).
51. F. E. Bartell and K. E. Bristol, J. Phys. Chem., 44, 86 (1940).
52. J. W. Shepard and J. P. Ryan, J. Phys. Chem., 63, 1729 (1959).
53. L. O. Brockway and R. L. Jones, Adv. Chem. Ser., 43, 275 (1964).
54. G. L. Gaines, Jr., J. Colloid Interface Sci., 15, 321 (1960).
55. F. E. Bartell and A. D. Wooley, J. Am. Chem. Soc., 55, 778 (1952).
56. F. E. Bartell and B. R. Ray, J. Am. Chem. Soc., 74, 778 (1952).
57. B. R. Ray, J. R. Anderson, and J. J. Scholz, J. Phys. Chem., 62, 1220, 1227 (1958).
58. A. B. D. Cassie and S. Baxter, Trans. Faraday Soc., 40, 546 (1944).
59. S. Baxter and A. B. D. Cassie, J. Text. Inst., 36, T67 (1945).
60. R. H. Dettre and R. E. Johnson, Jr., in *Wetting*, Society of Chemical Industry, London, Monograph No. 25, Gordon and Breach, London, 1967, p. 144.

2
Molecular Interpretations

Van der Waals recognized the existence of intermolecular force in 1879, and introduced an attractive energy term and an excluded volume term into the ideal gas law. Subsequent workers have found that there are various types of intermolecular forces, including dispersion (nonpolar) force, dipole (dipole-dipole) force, induction (dipole-induced dipole) force, and hydrogen bonding. These intermolecular forces are commonly known as the *van der Waals forces*. The van der Waals force between two molecules is a short-range force, varying with r^{-7}, where r is the intermolecular distance. On the other hand, the van der Waals force between two macroscopic bodies is a long-range force, varying with z^{-3}, where z is the distance between two flat plates. It should be noted here that the variation of van der Waals force between two macroscopic bodies depends on the shapes of the bodies. For instance, it varies with z^{-2} between two spheres (see Section 2.5.5). Various molecular forces are discussed and used to analyze the interfacial energies in this chapter. Several reviews of intermolecular forces are available elsewhere [1–8].

2.1. MICROSCOPIC THEORIES OF VAN DER WAALS FORCES

2.1.1. Dispersion (Nonpolar) Energy

Dispersion force exists between any pair of molecules. Its magnitude depends on the electronic frequency of the molecule, which is measurable from the dispersion of refractive index, hence the name *dispersion force*. However, the term *nonpolar force* should be preferrable. In 1927, Wang [9] showed that neutral and nonpolar molecules attract each other. In 1930, London [10] made the first quantum mechanical treatment; hence the force is also known as the *London dispersion force* or *London force*. London's treatment has since been refined [8, 10–19].

A molecule, with or without a permanent dipole moment, has an instantaneous dipole moment as its electrons fluctuate. This instantaneous dipole will induce a dipole in another molecule. The interaction between these two dipoles averaged over all instantaneous electronic configurations is the dispersion force. Note that the interaction between instantaneous dipoles of two molecules averaged over all electronic configurations is small compared with the interaction between an instantaneous dipole and its induced dipole, because two instantaneous dipoles in two molecules lack synchronization.

The attractive dispersion energy between two like molecules of type 1 is given by

$$U_{11}^d = -\frac{3}{4}\frac{\alpha_1^2 C_1}{r_{11}^6} \tag{2.1}$$

where U^d is the dispersion energy, α the polarizability, r the intermolecular distance, and C a constant. The attractive dispersion energy between a molecule of type 1 and a molecule of type 2 is given by

$$U_{12}^d = -\frac{3}{4}\frac{\alpha_1 \alpha_2}{r_{12}^6}\frac{2C_1 C_2}{C_1 + C_2} \tag{2.2}$$

The constant C is difficult to express completely in terms of molecular constants. For a hydrogenlike atom (having only one orbital electron), London [10] gave

$$C = h\nu_0 \tag{2.3}$$

where h is Planck's constant and ν_0 is the natural frequency of the electron in the hydrogenlike atom, which is related to the polarizability by [20]

Microscopic Theories of van der Waals Forces

$$\nu_0 = \frac{e}{2\pi (m_e \alpha)^{1/2}} \tag{2.4}$$

where e is the electronic charge (4.803×10^{-10} esu) and m_e is the electronic mass (9.109×10^{-28} g). Experimentally, ν_0 is obtained from the dispersion of refractive index, hence the name *dispersion energy*. For atoms (or molecules) having many electrons, all electrons are assumed to have identical characteristic frequency ν_v, but contribute different amounts to the total polarizability. Equation (2.3) is then written as

$$C = h\nu_v \tag{2.5}$$

If S is the effective number of dispersion electrons, then ν_0 should be replaced by ν_v and α by α/S in Eq. (2.4), giving

$$\frac{\nu_v}{\nu_0} = S^{1/2} \tag{2.6}$$

Therefore,

$$C = h\nu_0 S^{1/2} \tag{2.7}$$

which is similar to the Slater-Kirkwood relation [21],

$$C = h\nu_0 Z^{1/2} \tag{2.8}$$

where Z is the number of outer shell electrons; S is usually less than Z, and may be taken as the number of valency electrons. An improved relation is obtained by replacing ν_0 with ν_v in Eq. (2.8) [22], giving

$$C = h\nu_0 (SZ)^{1/2} \tag{2.9}$$

Applying Eq. (2.4) in Eq. (2.9) gives

$$C = \frac{he}{2\pi} (m_e \alpha)^{-1/2} (SZ)^{1/2} \tag{2.10}$$

which gives excellent results when used to calculate the surface tension of alkanes [23].

On the other hand, London [10] noted that $h\nu_0$ is similar to the ionization potential; that is, $h\nu_0 \simeq I$. Thus,

$$C \sim I \tag{2.11}$$

which has become popular because of its simplicity. Using Eq. (2.11) in Eqs. (2.1) and (2.2) gives

$$U_{11}^d = -\frac{3}{4}\frac{\alpha_1^2 I_1}{r_{11}^6} \tag{2.12}$$

$$U_{12}^d = -\frac{3}{4}\frac{\alpha_1 \alpha_2}{r_{12}^6}\frac{2 I_1 I_2}{I_1 + I_2} \tag{2.13}$$

Finally, Neugebauer [24] proposed

$$C = -\frac{8}{3} m_e c^2 \frac{\chi}{\alpha} \tag{2.14}$$

where c is the speed of light and χ is the diamagnetic susceptibility.

Retardation Effect

The r^{-6} dependence of the dispersion energy is valid only at small distances. At distances greater than about 100 Å, Casimir and Polder [16] showed that the dispersion energy is much smaller and drops off even more rapidly, that is,

$$U_{12}^d = -\frac{23}{8\pi^2}\frac{hc\alpha_1\alpha_2}{r_{12}^7} \tag{2.15}$$

This retardation effect arises from the phase lag between the instantaneous dipole and the induced dipole, as a finite time is required for the electromagnetic wave to travel (at the speed of light) from one molecule to another. The intermolecular attraction at 100 Å is already negligible as compared with that at, say, 10 Å, so the retardation effect is unimportant. However, it explains why the attractive force between two large flat plates at large distances decays with z^{-4} (where z is the distance of separation between two macroscopic plates) rather than with z^{-3} as predicted by the London theory.

Experimental confirmations of the London theory have been reported [25-29] and reviewed [4,6].

2.1.2. Dipole-Dipole Energy (Dipole Energy)

Debye [30,31] gave the interaction energy between two dipoles of fixed orientation as

$$U_{12}^p = \frac{\mu_1 \mu_2}{r^3}[2\cos\theta_1 \cos\theta_2 - \sin\theta_1 \sin\theta_2 \cos(\phi_1 - \phi_2)]$$

(2.16)

where U^p is the energy of interaction, μ the dipole moment, r the center-to-center distance between the two dipoles, and θ and ϕ are as shown in Figure 2.1. The subscripts 1 and 2 refer to the dipoles 1 and 2, respectively. The lowest-energy configuration is the head-to-tail orientation, whose energy is given by

$$U_{12}^p = -\frac{2\mu_1 \mu_2}{r^3}$$

(2.17)

When the dipoles are freely rotating, such as when the dipole energy is smaller than the thermal energy, that is, $2\mu_1\mu_2/r^3 < kT$, any pair of dipoles will not persist in a particular orientation for more than a short time. Keesom [32] gave the dipole energy averaged over all possible orientations as

$$U_{12}^p = -\frac{2\mu_1^2 \mu_2^2}{3kTr^6}$$

(2.18)

which should be more appropriate for polymers, since the dipoles in polymers are highly restricted and hence should be randomly oriented. Table 2.1 compares the head-to-tail and the random dipole energies.

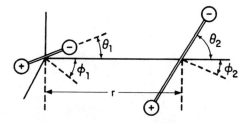

Figure 2.1. Dipole orientation.

Table 2.1. Comparison of Biomolecular Dipole Energies for Heat-to-Tail and Random Orientations at 25°C [a]

Dipole	μ_1, debyes	μ_1', debyes	Assumed intermolecular distance Å	Dipole energy, ergs	
				Random Eq. (2.18)	Head-to-tail Eq. (2.17)
CHI_3-CHI_3	0.9	0.9	5.0	7×10^{-16}	1.3×10^{-14}
$CHCl_3$-$CHCl_3$	1.1	1.1	4.5	6×10^{-15}	3.8×10^{-14}
CH_3Cl-CH_3Cl	1.94	1.94	4.0	6.2×10^{-14}	1.2×10^{-13}
CH_3Cl-H_2O	1.94	1.85	3.5	7.8×10^{-14}	1.7×10^{-13}
CH_3NO_2-CH_3NO_2	3.1	3.1	4.5	1.8×10^{-13}	2.1×10^{-13}
H_2O-H_2O	1.85	1.85	3.0	2.6×10^{-13}	2.5×10^{-13}
CH_3NO_2-CH_3Cl	3.1	1.94	4.2	1.1×10^{-14}	3.1×10^{-13}

[a] In calculating the dipole energy, only the orientation polarization is considered; induction and atom polarizations are ignored.
Source: From Ref. 65.

Microscopic Theories of van der Waals Forces

2.1.3. Induction Energy (Dipole-Induced Dipole Energy)

A permanent dipole will induce a dipole in another molecule. The magnitude of the induced dipole is given by

$$\mu_{induced} = \alpha E \tag{2.19}$$

where α is the polarizability and E is the electric field strength. Debye [33] gave the attractive energy between a dipole and its induced dipole as

$$U_{12}^i = -\frac{(\mu_1^2 \alpha_2 + \mu_2^2 \alpha_1)}{r^6} \tag{2.20}$$

where μ is the permanent dipole moment and α is the polarizability.

2.1.4. Hydrogen Bonding

A hydrogen bond is formed between a proton acceptor and a hydrogen atom attached to a highly electronegative atom or group. The proton acceptor may be a highly electronegative atom (O, Cl, F, N) or group ($-CCl_3$, $-CN$, etc.). Dipole-dipole attraction accounts for most of the properties of a hydrogen bond. For shorter bonds (2.5 Å or less), covalent character may be as much as 25%. For longer bonds (2.8 Å or greater), however, covalent character is negligible. Hydrogen bonds were first recognized by Latimer and Rhodebush [34] in 1920, and have been discussed and reviewed extensively [35–38].

Hydrogen bonds are formed only when both the donor and the acceptor atoms (or groups) are present. Ketones, esters, and nitriles have acceptor atoms (or groups) but no donor atoms (or groups); therefore, they have few or no hydrogen bonds. However, they will form strong hydrogen bonds in mixtures with other compounds having hydrogen bond donors, such as between an alcohol and a ketone, or between an alcohol and a nitrile [39].

A hydrogen bond is a short-range interaction, arising from head-to-tail interaction of dipoles. It is not a primary chemical bond. The intermolecular energy for a hydrogen bond may formally be given as

$$U_{12}^h = -\frac{(H_1^a H_2^d + H_1^d H_2^a)}{r^6} \tag{2.21}$$

$$U_{11}^h = -\frac{2H_1^a H_1^d}{r^6} \qquad (2.22)$$

where H_j^a is the acceptor attraction constant for the hydrogen bond of molecule j, H_j^d the donor attraction constant for the hydrogen bond, and r the hydrogen-bond length. If molecule 1 has only acceptor, then $H_1^d = 0$ and $U_{11}^h = 0$. On the other hand, if molecule 1 has only acceptor and molecule 2 has only donor, then $H_1^d = H_2^a = 0$ and $U_{12}^h = -H_1^a H_2^d/r^6$.

Hydrogen-bond energies have been listed extensively [35]. A method for estimating from group contributions has been proposed [40]. Table 2.2 lists some hydrogen-bond energies.

2.1.5. Numerical Values of Molecular Constants

Polarizability

Experimental polarizability values have been tabulated [41,43]. It may also be calculated from the refractive index by the Clausius-Mosotti equation [31,42],

Table 2.2. Typical hydrogen-bond Energies

Compound	Hydrogen-bond energy, kcal/mole
Inorganics	
HF-HF	6.3-7.0
HCN-HCN	3.3-4.4
H_2O-H_2O	3.4-5.8
Organics	
Alcohols, ROH-ROH	3.2-6.2
Amines, RNH_2-RNH_2	3.1-4.5
Amides, $RCONH_2-CONH_2$	3.5-5.2
Carboxylic acids, RCOOH-RCOOH	4-11.5
Phenols, $\phi OH-\phi OH$	3.5-4.4
Phenol-DMF [a]	6.4
Alcohol-DMF	4.5
$CHCl_3$-ketone	2.5

[a] DMF = N,N-dimethyl formamide.
Source: From Ref. 35.

$$\alpha = \frac{3}{4\pi} \frac{M}{\rho N} \frac{n^2 - 1}{n^2 + 2} \tag{2.23}$$

where M is the molecular weight, ρ the density, N Avogadro's number, and n the refractive index. The refractive index of polymers has been tabulated [43]. The refractive index may also be calculated from the molar refraction [44],

$$Mr = \frac{M}{\rho} \frac{n^2 - 1}{n^2 + 2} \tag{2.24}$$

where Mr is the molar refraction, which may be estimated from group contributions [43]. Polarizabilities for a number of polymers are listed in Table 2.3.

The polarizability of a molecule may not be isotropic. The mean polarizability of a bond, averaged over all possible orientations, is

$$\alpha_{bond} = \frac{1}{3}(\alpha_{\parallel} + 2\alpha_{\perp}) \tag{2.25}$$

where $\alpha_{\parallel}$ is the polarizability in the direction of the bond and $\alpha_{\perp}$ that in the direction perpendicular to the bond. For a C—C bond, $\alpha_{\parallel}$ is 1.88×10^{-24} cm^3, and $\alpha_{\perp}$ is only 0.02×10^{-24} cm^3 [45]. For a molecule, the mean polarizability is

$$\alpha_{mean} = \frac{1}{3}(\alpha_1 + \alpha_2 + \alpha_3) \tag{2.26}$$

where α_1, α_2, and α_3 are the three principal components of α.

Ionization Potential

Experimental values of ionization potential for various molecules have been tabulated [46–49]. Some values are given in Table 2.4.

Dipole Moment

Experimental values have been tabulated [30,31,43,50–52]. Some values are given in Table 2.5. The Debye equation [30] relates the dipole moment to the dielectric constant (500 to 5000 kHz),

$$\frac{\varepsilon - 1}{\varepsilon + 2} \frac{M}{\rho} = \frac{4\pi N}{3}\left(\alpha + \frac{\mu^2}{3kT}\right) \tag{2.27}$$

where ε is the dielectric constant, k the Boltzmann constant, and T the temperature. The left-hand side of Eq. (2.27) is known as the

Table 2.3. Segmental Polarizability, Molar Diamagnetic Susceptibility, and Lennard-Jones 6-12 Potential Parameters for Some Polymer Repeat Units

Repeat unit	$\alpha \times 10^{24}$ cm^3	$-\chi_M \times 10^6$ cgs units	r_e, Å	U_0/k, K
—CH$_2$—CH$_2$—	3.68	22.8	4.19	225
—CH(CH$_3$)—CH$_2$—	5.49	34.1	4.70	275
—C(CH$_3$)$_2$—CH$_2$—	7.32	45.4	5.13	306
—CFH—CH$_2$—	3.60	26.9	4.16	274
—CHCl—CH$_2$—	5.59	36.9	4.72	297
—CF$_2$—CH$_2$—	3.66	30.9	4.18	316
—CCl$_2$—CH$_2$—	7.52	50.9	5.18	340
—CF$_2$—CFH—	3.60	34.9	4.16	365
—CClF—CF$_2$—	5.59	49.3	4.72	410
—CF$_2$—CF$_2$—	3.66	38.8	4.18	408
—CH(OH)—CH$_2$—	4.27	27.9	4.36	259
—CH(C$_6$H$_5$)—CH$_2$—	13.4	75.9	6.27	321
—CH(OCCH$_3$)—CH$_2$— ‖ O	8.01	48.5	5.29	308
—C(COCH$_3$)—CH$_2$— ‖ | O CH$_3$	9.81	58.5	5.65	317
—C(COC$_4$H$_9$)—CH$_2$— ‖ | O CH$_3$	15.3	92.6	6.57	351
—CH$_2$—CH$_2$—O—	4.34	27.3	4.38	251
—CH$_2$—CH(CH$_3$)—O—	6.16	40.0	4.86	305
—Si(CH$_3$)$_2$—O—	7.39	47.0	5.15	316

Source: Collected from Refs. 43, 63, and 64.

Table 2.4. Ionization Potentials

Compound	I, eV	Compound	I, eV
Water	12.59	Methyl acetate	10.27
Ammonia	10.15	n-Butylamine	8.70
Methane	12.98	N,N-Dimethylformamide	9.12
n-Hexane	10.18	1-Butene	9.58
2-Chlorobutane	10.65	Vinyl chloride	9.99
Ethanol	10.48	Vinyl acetate	9.19
Methyl ethyl ketone	9.53	Benzene	9.24
Acetic acid	10.37	$C_5F_{11}CF{=}CF_2$	10.48

Source: From Ref. 47.

Table 2.5. Dipole Moments

Compound	μ, debyes
Small molecules	
Water	1.85
Ethanol	1.7
Toluene	0.4
Chlorobenzene	1.6
Fluorobenzene	1.35
Nitrobenzene	3.9
Acetone	2.85
Acetonitrile	3.4
Formamide	3.4
Ethyl acetate	1.85
Polymers (per monomer unit)	
Polyisobutylene	0
Polystyrene	0.26–0.44
Poly(vinyl acetate)	1.61–1.70
Poly(methyl methacrylate)	1.27–1.43
Poly(butyl methacrylate)	1.39–1.65
Poly(vinyl chloride)	1.67–1.75
Polychloroprene	1.45–1.62
Polychlorotrifluoroethylene	0.54–0.60
Polyoxyethylene	1.41–1.68
Polyacrylonitrile	3.4

Source: Collected from Refs. 50–52.

total polarization. The first term on the right-hand side is the *distortion polarization* (due to distortion of electrons) and the second term the *orientation polarization* (due to dipole orientation).

At optical frequencies, permanent dipoles cannot orient themselves in response to the field. In this case, the Maxwell relation holds [53],

$$\varepsilon = n^2 \tag{2.28}$$

and the orinetation polarization drops out, giving the Clausius-Mosotti equation.

Dielectric constants of polymers have been tabulated [43]. The dipole moment may also be estimated from group contributions [41–43] or by a modified Bottcher equation [41,54].

Diamagnetic Susceptibility

The molar diamagnetic susceptibility, $\chi_M = M\chi$, can be estimated from additive atomic and group contributions, whose values have been tabulated [43]. Values for some polymers are listed in Table 2.3.

2.2. ADDITIVITY AND FRACTIONAL CONTRIBUTIONS OF VARIOUS TYPES OF MOLECULAR FORCES

Summation of various molecular energies gives the total attractive energy

$$U_{12} = U_{12}^d + U_{12}^p + U_{12}^i + U_{12}^h = \frac{-A_{12}}{r^6} \tag{2.29}$$

Here, A_{12} is the total attraction constant, given by

$$A_{12} = A_{12}^d + A_{12}^p + A_{12}^i + A_{12}^h \tag{2.30}$$

where A_{12}^d is the attraction constant for dispersion interaction, A_{12}^p the attraction constant for dipole interaction, A_{12}^i the attraction constant for induction interaction, and A_{12}^h the attraction constant for hydrogen bonding. These are given by

$$A_{12}^d = \frac{3}{4} \alpha_1 \alpha_2 \frac{2 I_1 I_2}{I_1 + I_2} \tag{2.31}$$

$$A_{12}^p = \frac{2 \mu_1^2 \mu_2^2}{3kT} \tag{2.32}$$

Table 2.6. Comparison of Various Intermolecular Energies

				Various intermolecular energies, 10^{-6} erg-cm^6			
Compound	μ, debyes	α, Å^3	I, eV	Dipole-dipole energy, $2\mu^4/3kT$	Dipole-induced dipole energy, $2\alpha\mu^2$	London dispersion energy, $(3/4)\alpha^2 I$	London energy / Dipole energy, $2\alpha^2 I/(\mu^4/kT)$
Ar	0	1.63	15.8	–	–	50	–
CH$_4$	0	2.58	13.1	–	–	97	–
Na	0	29.7	5.15	–	–	5340	–
HI	0.78	3.85	10.3	6.2	4.0	191	31
HCl	1.03	2.63	12.78	18.6	5.4	106	5.7
CH$_3$Cl	1.6	4.56	11.35	109	23	284	2.6
H$_2$O	1.85	1.48	12.67	190	10	33	0.18
CH$_3$NO$_2$	3.1	3.01	11.34	1530	58	124	0.08

Source: From Ref. 65.

$$A_{12}^i = \mu_1^2 \alpha_2 + \mu_2^2 \alpha_1 \tag{2.33}$$

$$A_{12}^h = H_1^a H_2^d + H_1^d H_2^a \tag{2.34}$$

Various molecular energies for some molecules are compared in Table 2.6. The induction energy is usually small compared with the dispersion energy and the dipole energy in small molecules. However, in large molecules wherein polar groups are far apart, the induction energy may be greater than the dipole energy [55].

The fractional contributions of various molecular energies can be defined as [56–59]

$$x^k = \frac{U^k}{U} = \frac{A^k}{A} \tag{2.35}$$

where $k = d, p, i,$ or h and $x^d + x^p + x^i + x^h = 1$. For simplicity, the various polar energies may be combined into one term (see Chapter 3).

2.3. APPROXIMATIONS FOR UNLIKE MOLECULES

The attraction constant between two unlike molecules A_{12} may be approximated by those for like molecules A_{11} and A_{22}. This provides a basis for expressing adhesive energy in terms of cohesive energies of the two phases.

2.3.1. Approximations for Dispersion Energy

Using Eq. (2.12) in Eq. (2.13) to eliminate the polarizability gives

$$A_{12}^d = \left[\frac{2(I_1 I_2)^{0.5}}{I_1 + I_2}\right](A_{11}^d A_{22}^d)^{0.5} \tag{2.36}$$

If, $I_1 \simeq I_2$, which is true in many cases, then [56]

$$A_{12}^d \simeq (A_{11}^d A_{22}^d)^{0.5} \tag{2.37}$$

which is the geometric-mena approximation for the dispersion energy between unlike molecules.

On the other hand, if the ionization potential is eliminated, instead, then [56]

Approximations for Unlike Molecules

$$A_{12}{}^d = \frac{2A_{11}{}^d A_{22}{}^d}{A_{11}{}^d (\alpha_2/\alpha_1) + A_{22}{}^d (\alpha_1/\alpha_2)} \qquad (2.38)$$

If, $\alpha_1 \sim \alpha_2$, which is true in many other cases, then

$$A_{12}{}^d \sim \frac{2A_{11}{}^d A_{22}{}^d}{A_{11}{}^d + A_{22}{}^d} \qquad (2.39)$$

which is the harmonic-mean approximation for the dispersion energy between unlike molecules.

2.3.2. Approximation for Dipole Energy

From Eq. (2.33), valid for random orientation, it is readily shown that

$$A_{12}{}^p = (A_{11}{}^p A_{22}{}^p)^{1/2} \qquad (2.40)$$

which is the geometric-mean relation for dipole-dipole energy between unlike molecules.

2.3.3. Approximation for Induction Energy

From Eq. (2.34), it is readily shown that

$$A_{12}{}^i = \frac{1}{2}\left[\left(\frac{\mu_1}{\mu_2}\right)^2 A_{11}{}^i + \left(\frac{\mu_2}{\mu_1}\right)^2 A_{22}{}^i\right] \qquad (2.41)$$

$$= \frac{1}{2}\left(\frac{\alpha_2}{\alpha_1} A_{11}{}^i + \frac{\alpha_1}{\alpha_2} A_{22}{}^i\right) \qquad (2.42)$$

If $\alpha_1 \sim \alpha_2$, true in many cases, then

$$A_{12}{}^i \sim \frac{1}{2}(A_{11}{}^i + A_{22}{}^i) \qquad (2.43)$$

which is the arithmetic-mean approximation for the induction energy between unlike molecules.

2.3.4. Approximation for Hydrogen-Bond Energy

The hydrogen-bond attraction constant A_{12}^h is generally a cross-product function [Eq. (2.34)]. If $H_1^a/H_1^d \simeq H_2^a/H_2^d$, then

$$A_{12}^h \simeq (A_{11}^h A_{22}^h)^{0.5} \tag{2.44}$$

which is the geometric-mean approximation for hydrogen-bond energy between unlike molecules.

2.3.5. Total Attraction Constant

The total attraction constant is the sum of all types of attraction constants, given by using Eqs. (2.37)–(2.44) in Eq. (2.30). For simplicity, it is conveneint to combine various polar energies into one term empirically. Two particularly useful relations are [56–59]

$$A_{12} = (A_{11}^d A_{22}^d)^{1/2} + (A_{11}^p A_{22}^p)^{1/2} \tag{2.45}$$

which is the geometric-mean approximation preferred for interactions between high-energy phases or between a low-energy phase and a high-energy phase, and

$$A_{12} = \frac{2A_{11}^d A_{22}^d}{A_{11}^d + A_{22}^d} + \frac{2A_{11}^p A_{22}^p}{A_{11}^p + A_{22}^p} \tag{2.46}$$

which is the harmonic-mean approximation preferred for interactions between low-energy phases [56–59]. In Eqs. (2.45) and (2.46), the terms with a superscript p refer to combined polar energies (including dipole, induction, and hydrogen-bond energies).

2.4. LENNARD-JONES POTENTIAL ENERGY FUNCTION

Two neutral molecules will repel each other at close distances when molecular orbitals overlap. At equilibrium separation, the attractive force equals the repulsive force. The net force is zero and the energy is minimal. Various attractive energies have been discussed in Section 2.1. Several repulsive-energy functions have been proposed [1, 8,22,60]. Particularly useful is the Lennard-Jones 6-12 potential energy function for two isolated neutral molecules,

Lennard-Jones Potential Energy Function

$$U = -\frac{A}{r^6} + \frac{B}{r^{12}} \tag{2.47}$$

where A is the attractive constant and B is the repulsive constant. The first term is the attractive energy, which follows from Section 2.1. The second term is the repulsive energy, which may be adequately expressed by any inverse power of 8 to 14. The choice of r^{-12} is made for mathematical convenience. Kihara [61] refined the function by displacing the force center by a core radius. Kihara potential has been used to treat interfacial energies [62].

The attraction constant and the repulsion constant are related to each other. The force F between two molecules is given by $F = -(dU/dr)$, that is,

$$F = -\frac{6A}{r^7} + \frac{12B}{r^{13}} \tag{2.48}$$

At equilibrium separation r_0, the net force is zero, and the energy is at minimum. Letting $F(r_0) = 0$ gives

$$r_0 = \left(\frac{2B}{A}\right)^{1/6} \tag{2.49}$$

and the equilibrium energy U_0 is obtained as

$$U_0 = -\frac{A^2}{4B} = -\frac{A}{2r_0^6} \tag{2.50}$$

Thus, the equilibrium energy is half the attractive energy in the absence of repulsion.

If the molecules approach closer than r_0, the potential energy will rise sharply. The distance r_e at which the energy is zero is given by

$$r_e = \left(\frac{1}{2}\right)^{1/6} r_0 \tag{2.51}$$

The maximum attractive force occurs at r_m, when $dF/dr = 0$ and $-d^2U/dr^2 = 0$. Therefore,

$$r_m = \left(\frac{13}{7}\right)^{1/6} r_0 \tag{2.52}$$

and the maximum attractive force is given by

$$F_m = -1.345 \frac{A}{r_0^7} = -2.690 \frac{U_0}{r_0} \tag{2.53}$$

An alternative form of Lennard-Jones potential energy function can thus be written as

$$U = U_0\left[2\left(\frac{r_0}{r}\right)^6 - \left(\frac{r_0}{r}\right)^{12}\right] = 4U_0\left[\left(\frac{r_e}{r}\right)^6 - \left(\frac{r_e}{r}\right)^{12}\right] \tag{2.54}$$

A Lennard-Jones 6-12 potential energy function is illustrated in Figure 2.2.

An empirical method for estimating the Lennard-Jones potential parameters for polyatomic molecules has been proposed [63]:

$$\ln(r_e^2 - 5.4) = 1.456 + 0.797 \ln(\alpha \times 10^{24}) \tag{2.55}$$

$$\ln\left(\frac{U_0}{k} \frac{r_e^6}{10^4}\right) = -0.1445 + 1.1148 \ln(-\alpha\chi_M \times 10^{30}) \tag{2.56}$$

where α is the polarizability (in cm^3), χ_M the molar diamagnetic susceptibility (in cgs units), and k the Boltzmann constant. Thus, the parameters U_0/k (in kelvin) and r_e (in Å) can be calculated from α and χ_M, which can be estimated from atomic and group contributions. Values for some polymers are given in Table 2.3.

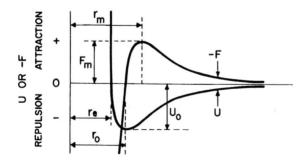

Figure 2.2. Intermolecular energy and force based on the Lennard-Jones 6-12 potential energy function.

2.5. ATTRACTION BETWEEN MACROSCOPIC BODIES

Attractive force and energy between two macroscopic bodies can be obtained by summation of all pairwise interactions. Various summation and integration methods have been proposed [65–70]. The method of Good [65] based on Fowler and Guggenheim integral [18] is discussed below.

2.5.1. Ideal Cohesive and Adhesive Strengths

The attractive force per unit area between two semiinfinite parallel plates (Figure 2.3) can be obtained by integrating the force between a molecule at position x in phase 1 and a molecule at a distance r from the first molecule in phase 2 over the whole of the two semiinfinite volumes,

$$\sigma_{12}(z) = 2\pi \int_z^\infty q_1 \, dj \int_j^\infty f \, df \int_f^\infty q_2 \left(\frac{\partial U}{\partial r}\right) dr \qquad (2.57)$$

$$\sigma_{11}(z) = 2\pi \int_z^\infty q_1 \, dj \int_j^\infty f \, df \int_f^\infty q_1 \left(\frac{\partial U}{\partial r}\right) dr \qquad (2.58)$$

where $\sigma(z)$ is the attractive force per unit area between two phases separated by a distance of z and q is the number of molecules per unit volume. The subscripts 1 and 2 refer to phases 1 and 2, respectively. The q is assumed to be continuous (quasi-continuum model).

Using the Lennard-Jones potential energy function in Eqs. (2.57) and (2.58) gives

$$\sigma_{12}(z) = \frac{2\pi q_1 q_2}{z^3} \left(\frac{A_{12}}{12} - \frac{B_{12}}{90z^6}\right) \qquad (2.59)$$

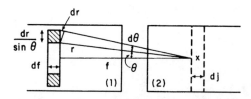

Figure 2.3. Integration of intermolecular attractions between two semi-infinite parallel plates. (After Ref. 65.)

and

$$\sigma_{11}(z) = \frac{2\pi q_1^2}{z^3}\left(\frac{A_{11}}{12} - \frac{B_{11}}{90z^6}\right) \tag{2.60}$$

When $z = z_0$ (the equilibrium separation), $\sigma = 0$. Therefore,

$$B = \frac{15}{2} A z_0^6 \tag{2.61}$$

Thus, the attractive forces become

$$\sigma_{12}(z) = \frac{2\pi q_1 q_2 A_{12}}{6z^3}\left[1 - \left(\frac{z_{0,12}}{z}\right)^6\right] \tag{2.62}$$

$$\sigma_{11}(z) = \frac{2\pi q_1^2 A_{11}}{6z^3}\left[1 - \left(\frac{z_{0,11}}{z}\right)^6\right] \tag{2.63}$$

which are illustrated in Figure 2.4.

The maximum force occurs at z_m, where

$$z_m = (3)^{1/6} z_0 = 1.20 z_0 \tag{2.64}$$

and the maximum force is given by [65]

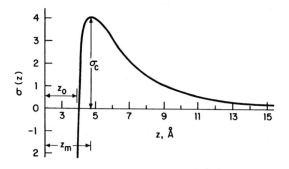

Figure 2.4. Attractive force between two macroscopic bodies (thick parallel plates) in vacuum according to Eq. (2.63), where $A = 4 \times 10^{-58}$ erg/cm^6 and $q = 6 \times 10^{23}$. (After Ref. 65.)

Attraction Between Macroscopic Bodies

$$\sigma_{12}{}^a = \frac{2\pi q_1 q_2 A_{12}}{9(3)^{1/2} z_{0,12}^3} \qquad (2.65)$$

$$\sigma_{11}{}^c = \frac{2\pi q_1{}^2 A_{11}}{9(3)^{1/2} z_{0,11}^3} \qquad (2.66)$$

where $\sigma_{12}{}^a$ is the ideal adhesive strength and $\sigma_{11}{}^c$ is the ideal cohesive strength, that is, the maximum force per unit area required to pull apart two semi-infinite parallel plates at equilibrium separation reversibly without viscoelastic deformations.

Note that the attractive force between two molecules varies with z^{-7}, whereas the attractive force between two large flat plates varies with z^{-3}. Thus, the former is a short-range force, whereas the latter is a long-range force.

The Young's modulus is defined as $E = z_0 (d\sigma/dz)$ at $z = z_0$. Therefore [65],

$$E = \frac{2\pi q_1{}^2 A_{11}}{z_0^3} = 9(3)^{1/2} \sigma_{11}{}^c \qquad (2.67)$$

which is the theoretical Young's modulus.

2.5.2. Interfacial and Surface Tensions

The free energy of attraction per unit area, ΔF, between two semi-infinite parallel plates separated by a distance z is the reversible work required to separate the two bodies from the distance z to infinity, that is,

$$\Delta F(z) = \int_z^\infty \sigma_{12}(z)\,dz \qquad (2.68)$$

Using Eqs. (2.62) and (2.63) in the above gives

$$\Delta F_{12}(z) = \frac{\pi q_1 q_2 A_{12}}{6z^2}\left[1 - \frac{1}{4}\left(\frac{z_0}{z}\right)^6\right] \qquad (2.69)$$

$$\Delta F_{11}(z) = \frac{\pi q_1{}^2 A_{11}}{6z^6}\left[1 - \frac{1}{4}\left(\frac{z_0}{z}\right)^6\right] \qquad (2.70)$$

The work of adhesion W_a is the reversible work required to separate phases 1 and 2 from equilibrium separation to infinity, that is, $\Delta F_{12}(z_{0,12})$; the work of cohesion W_{c11} is the reversible work required to separate phase 1 along an imaginary line in the bulk to infinity, that is, $\Delta F_{c,11}(z_{0,11})$. Thus, from Eqs. (2.69) and (2.70), we have

$$W_a = \frac{\pi q_1 q_2 A_{12}}{8 z_{0,12}^2} \tag{2.71}$$

$$W_{c11} = \frac{\pi q_1^2 A_{11}}{8 z_{0,11}^2} \tag{2.72}$$

Applying Eqs. (1.17) and (1.18) in Eqs. (2.71) and (2.72) gives [65]

$$\gamma_1 + \gamma_2 - \gamma_{12} = \frac{\pi q_1 q_2 A_{12}}{8 z_{0,12}^2} \tag{2.73}$$

$$\gamma_1 = \frac{\pi q_1^2 A_{11}}{16 z_{0,11}^2} \tag{2.74}$$

Note that the attractive energies vary with the inverse square of the distance, that is, z^{-2}. Equations (2.73) and (2.74) express the interfacial and surface tensions in terms of molecular constants. Using these equations in Eqs. (2.65) and (2.66) gives

$$\sigma_{12}^a = \frac{16}{9(3)^{1/2}} \frac{W_a}{z_{0,12}}$$

$$= \frac{16}{9(3)^{1/2}} \frac{\gamma_1 + \gamma_2 - \gamma_{12}}{z_{0,12}} \tag{2.75}$$

and

$$\sigma_{11}^c = \frac{16}{9(3)^{1/2}} \frac{W_{c11}}{z_{0,11}}$$

$$= \frac{16}{9(3)^{1/2}} \frac{2\gamma_1}{z_{0,11}} \tag{2.76}$$

Attraction Between Macroscopic Bodies

A relation between surface tension and modulus is obtained by using Eq. (2.67) in Eq. (2.76),

$$\gamma = \frac{E z_0}{32} \tag{2.77}$$

On the other hand, from Eq. (3.41), $\gamma = (1/4)\rho^{-1/3}\delta^2$, where γ is in dyne/cm, ρ in g/ml, and δ in $(cal/ml)^{1/2}$. Letting $z_0 = (6v/\pi N)^{1/3}$, where $v = M_i/\rho$ is the molar volume of an interacting element (M_i = 46.8; see Chapter 3), Eq. (2.76) becomes

$$\sigma_{11}^{\ c} = \delta^2 \tag{2.78}$$

where $\sigma_{11}^{\ c}$ is in MPa and δ in $(cal/ml)^{1/2}$. Thus, the ideal cohesive strength numerically equals the cohesive energy density for the units used. A linear correlation between fracture strength and cohesive energy density has been demonstrated for polymers and metals [66].

2.5.3. Contact Angles

The contact angle θ of a liquid 1 on a solid 2 is obtained by combining Eqs. (2.73), (2.74), and (1.34),

$$1 + \cos \theta = 2 \frac{q_2}{q_1} \frac{A_{12}}{A_{11}} \left(\frac{z_{0,12}}{z_{0,11}}\right)^2$$

$$= 2 \left(\frac{A_{22}}{A_{11}}\right)^{1/2} \left(\frac{v_1}{v_2}\right)^{5/3} \tag{2.79}$$

where v is the molar volume. From Eq. (2.74), the surface tension of phase j is obtained as

$$\gamma_j = \frac{9}{8} \left(\frac{\pi N}{6 v_j}\right)^{8/3} A_{jj} \tag{2.80}$$

If θ_1 is the contact angle of liquid 1 on solid 2, and θ_3 that of liquid 3 on solid 2, then

$$\frac{1 + \cos \theta_1}{1 + \cos \theta_3} = \left(\frac{A_{33}}{A_{11}}\right)^{1/2} \left(\frac{v_1}{v_3}\right)^{5/3} \tag{2.81}$$

2.5.4. Comparison with Hamaker Equation

If only the attractive force is considered, and the repulsive force is neglected, the force between two semi-infinite parallel plates is, from Eq. (2.60), given by

$$\sigma_{11}(z) = \frac{\pi q_1^2 A_{11}}{6z^3} \tag{2.82}$$

The work of cohesion is thus given by

$$W_{c11} = \int_{z_0}^{\infty} \frac{\pi q_1^2 A_{11}}{6z^3} dz = \frac{H_{11}}{12\pi z_0^2} \tag{2.83}$$

which is known as the Hamaker equation. The constant H_{11} is the Hamaker constant, given by

$$H_{11} = \pi^2 q_1^2 A_{11} \tag{2.84}$$

Comparison with the complete equation, that is, Eq. (2.72), shows that the Hamaker constant is smaller than the complete constant by a factor of 2/3.

2.5.5. Effect of Body Shapes on Attractive Energy

The van der Waals *energy* between two macroscopic bodies depends on the shapes of the two large bodies. For instance, it varies with z^{-2} between two flat plates, but varies with z^{-1} between two spheres. Nonretarded van der Waals energies between some pairs of large bodies of different shapes are given in Figure 2.5. In these expressions, only the attractive energies are considered; the repulsive energies are omitted in the integrations.

2.5.6. Repulsive van der Waals Forces

Van der Waals forces between two identical bodies are always attractive, but those between unlike bodies separated with an intervening third phase may be attractive or repulsive. Consider two bodies i and j separated with a medium m. The attraction constant is given by [68,70]

$$A_{imj} = A_{ij} + A_{mm} - A_{im} - A_{jm} \tag{2.85}$$

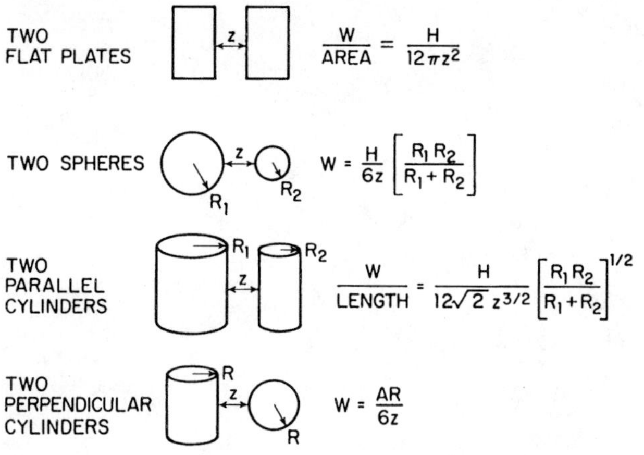

Attraction Between Macroscopic Bodies

TWO FLAT PLATES $\dfrac{W}{AREA} = \dfrac{H}{12\pi z^2}$

TWO SPHERES $W = \dfrac{H}{6z}\left[\dfrac{R_1 R_2}{R_1 + R_2}\right]$

TWO PARALLEL CYLINDERS $\dfrac{W}{LENGTH} = \dfrac{H}{12\sqrt{2}\, z^{3/2}}\left[\dfrac{R_1 R_2}{R_1 + R_2}\right]^{1/2}$

TWO PERPENDICULAR CYLINDERS $W = \dfrac{AR}{6z}$

Figure 2.5. Nonretarded van der Waals attractive energies W between large bodies of various shapes and size (R >> z) in terms of the Hamaker constant H.

Using the geometric-mean approximation for A_{ij}, A_{im}, and A_{jm} in the above gives

$$A_{imj} = \tfrac{1}{2}[(A_{ii}^{1/2} - A_{mm}^{1/2})^2 + (A_{jj}^{1/2} - A_{mm}^{1/2})^2 - (A_{ii}^{1/2} - A_{jj}^{1/2})^2] \qquad (2.86)$$

which is plotted schematically in Figure 2.6. Thus, the force between like bodies is always attractive, but the force between unlike bodies separated in a medium will be repulsive when A_{mm} is intermediate between A_{ii} and A_{jj}. The maximum repulsive force occurs when

$$A_{mm} = \tfrac{1}{4}(A_{ii}^{1/2} + A_{jj}^{1/2})^2 \qquad (2.87)$$

and the attraction constant is given by

$$A_{imj,min} = -\tfrac{1}{4}(A_{ii}^{1/2} - A_{jj}^{1/2})^2 \quad \text{(maximum repulsion)} \qquad (2.88)$$

For nonpolar bodies, only the dispersion force operates. Then, it can be shown, by using Eqs. (2.23) and (2.31), that

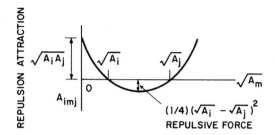

Figure 2.6. Repulsive van der Waals force between two macroscopic bodies i and j separated by a medium m.

$$A_{imj} = \frac{27}{128}\left(\frac{M}{\pi\rho N}\right)^2 \left(\frac{I}{(n^2+2)^2}\right)[(n_i^2 - n_m^2)^2 + (n_j^2 - n_m^2)^2 - (n_i^2 - n_j^2)^2] \quad (2.89)$$

where n is the refractive index, and since for organic molecules, I and $(n^2 + 2)$ are nearly constant. Thus, the force will be repulsive, when n_m is between n_i and n_j. The modern Lifshitz theory, discussed later, confirms this.

2.6. MACROSCOPIC THEORY OF VAN DER WAALS FORCES: THE LIFSHITZ THEORY

In the classical (microscopic) theory, the van der Waals force between two macroscopic bodies is obtained by summation of pairwise interactions over the whole bodies. This is strictly valid only in rarefied gases. In condensed bodies, the pair potentials are distorted by neighboring molecules. Such effects are not accounted for in the classical theory.

In 1955, Lifshitz [71,72] proposed a macroscopic theory, which considers the interactions of resultant electromagnetic waves (arising from motions of electrons) emanating from macroscopic bodies, instead of considering the interactions between isolated individual molecules. These electromagnetic fields are completely described in the complex dielectric spectrum, which is therefore the only material property needed to completely specify the attractive force between macroscopic bodies. The theory applies to distances larger than interatomic distances (10–15 Å), covering contributions from all frequencies and thus encompassing dispersion, dipole, and induction energies. Strictly, however, the theory is not applicable to hydrogen bonding, since its

Macroscopic Theory of van der Waals Forces

bond length is very short. The theory confirms that the van der Waals force between two large bodies can be repulsive in certain cases. Several applications of the theory have been reported, including adsorption of hydrocarbons on water [73], surface tensions and contact angles of liquids [74], attraction between polymers, metals, and inorganic materials [75], biological systems [76–81], and colloidal particles [82]. Reviews and refinements of the theory have been given elsewhere [83–92].

2.6.1. Van der Waals Forces Between Two Flat Plates

The general expression, the Lifshitz equation, for the force σ per unit area between two semi-infinite parallel plates 1 and 2 separated by a dielectric medium 3 of thickness z (Figure 2.7) is given by

$$\sigma(z,T) = \frac{kT}{\pi c^3} \sum_{n=0}^{\infty}{}' \varepsilon_3^{3/2} \beta_n^2 \int_1^{\infty} p^2 (\Delta + \overline{\Delta})\, dp \qquad (2.90)$$

where

$$\Delta = \left[\frac{(s_1 + p)(s_2 + p)}{(s_1 - p)(s_2 - p)} e^{rzp} - 1\right]^{-1} \qquad (2.91)$$

$$\overline{\Delta} = \left[\frac{(s_1 + p\varepsilon_1/\varepsilon_3)(s_2 + p\varepsilon_2/\varepsilon_3)}{(s_1 - p\varepsilon_1/\varepsilon_3)(s_2 - p\varepsilon_2/\varepsilon_3)} e^{rzp} - 1\right]^{-1} \qquad (2.92)$$

$$r = \frac{2}{c} \varepsilon_3^{1/2} \beta_n \qquad (2.93)$$

$$\beta_n = \frac{(2\pi)^2 nkT}{h} \qquad (2.94)$$

$$s_1 = \left(\frac{\varepsilon_1}{\varepsilon_3} - 1 - p^2\right)^{1/2} \qquad (2.95)$$

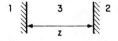

Figure 2.7. Schematics of the system analyzed in the Lifshitz theory.

$$s_2 = \left(\frac{\varepsilon_2}{\varepsilon_3} - 1 - p^2\right)^{1/2} \tag{2.96}$$

k the Boltzmann constant, T the temperature, c the speed of light, h the Planck constant, and ε_1, ε_2, and ε_3 are dielectric constants taken on the imaginary frequency $i\beta$, that is,

$$\varepsilon_j = \varepsilon_j(i\beta) \tag{2.97}$$

where $i = (-1)^{1/2}$ and ε_1, ε_2, and ε_3 are all real numbers. The summation is to be taken over all integral values of n and the prime on the summation denotes that the term for n = 0 is to be multiplied by 1/2.

2.6.2. Complex Dielectric Spectrum

Complex dielectric constants of the bodies and the medium over the whole spectral range, that is, from $\beta = 0$ to ∞, must be known to calculate the forces by the Lifshitz equation. The dielectric constant as a function of real frequency ω is given by

$$\varepsilon(\omega) = \varepsilon'(\omega) + i\varepsilon''(\omega) \tag{2.98}$$

$$= [\mu(\omega) + i\kappa(\omega)]^2 \tag{2.99}$$

where $\varepsilon'(\omega)$ is the measured (real part of dielectric constant, $\varepsilon''(\omega)$ the loss component which measures the dissipation of energy as heat when the electromagnetic wave propagates in the medium, μ the real part of the refractive index, and κ the absorption coefficient. Separation of the real and the imaginary parts gives

$$\varepsilon'(\omega) = \mu^2(\omega) - \kappa^2(\omega) \tag{2.100}$$

$$\varepsilon''(\omega) = 2\mu(\omega)\kappa(\omega) \tag{2.101}$$

Thus, measurements of $\mu(\omega)$ and $\kappa(\omega)$ will give $\varepsilon'(\omega)$ and $\varepsilon''(\omega)$.

The quantity that determines the van der Waals force is the function $\varepsilon(\omega)$ evaluated on the imaginary frequency $\omega = i\beta$, that is, $\varepsilon(i\beta)$, which is related to $\varepsilon''(\omega)$ through the Kramers-Kronig relation [93],

$$\varepsilon(i\beta) = 1 + \frac{2}{\pi} \int_0^\infty \frac{\varepsilon''(\omega)\omega \, d\omega}{\omega^2 + \beta^2} \tag{2.102}$$

Macroscopic Theory of van der Waals Forces

Thus, only $\epsilon''(\omega)$ is needed. Since $\epsilon''(\omega)$ is real and positive, $\epsilon(i\beta)$ is real, positive, and a monotonically decreasing function of β. At $\beta = 0$, $\epsilon(i\beta) = \epsilon_0$, the electrostatic dielectric constant; at $\beta \to \infty$, $\epsilon(i\beta) = 1$ at very high frequencies (far UV and above). Since $\epsilon''(\omega)$ reaches peak values only near absorption peaks [42], the strengths and positions of the absorption peaks will determine the van der Waals forces.

The function $\epsilon'(\omega)$ for a hypothetical material containing polar molecules capable of orientation is shown schematically in Figure 2.8. At very high frequencies (far UV and above), that is, above ω_e, both the dielectric constant and the refractive index approach unity. At UV and optical frequencies, that is, between ω_e and ω_r, electronic polarization occurs, and the dielectric constant equals the square of the refractive index, $\epsilon = n^2$. At infrared frequencies, that is, between ω_r and ω_m, displacement polarization occurs. In the microwave region, that is, below ω_m, dipole orientation makes a contribution.

Ninham and Parsegian [79] proposed that a suitable expression for $\epsilon(\omega)$ satisfying the Kramers-Kronig relation is

$$\epsilon(\omega) = 1 + \frac{c_m}{1 - i(\omega/\omega_m)} + \sum_j \frac{c_j}{1 - (\omega/\omega_j)^2 + i\gamma_j(\omega/\omega_j)} \qquad (2.103)$$

and the function $\epsilon(i\beta)$ is thus given by

$$\epsilon(i\beta) = 1 + \frac{c_m}{1 + (\beta/\omega_m)} + \sum_j \frac{c_j}{1 + (\beta/\omega_j)^2} \qquad (2.104)$$

where the damping term $\gamma_j(\omega/\omega_j)$ is taken to be negligible. For materials capable of optical polarization only, such as nonpolar materials, the dielectric loss is negligible, and the function $\epsilon(i\beta)$ becomes

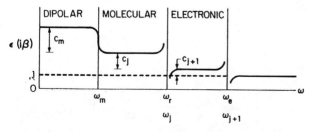

Figure 2.8. Schematics of dielectric constant spectrum.

$$\varepsilon(i\beta) = 1 + \frac{n^2 - 1}{1 + (\beta/\omega_e)^2} \qquad (2.105)$$

where the refractive index n is independent of frequency lower than ω_e. The functions $\varepsilon(i\beta)$ for water and hydrocarbon are shown in Figure 2.9. The constants for water are $c_m = 75.2$, $\omega_m = 1.06 \times 10^{11}$, $c_j = c_r = 3.42$, $\omega_j = \omega_r = 5.66 \times 10^{14}$, $c_{j+1} = c_e = 0.78$, and $\omega_{j+1} = \omega_e = 1.906 \times 10^{16}$.

2.6.3. Lifshitz Equation for Small-Distance Regimes (Nonretarded Limit)

When z is small, or when c is assumed to be infinite, the Lifshitz equation simplifies to

$$\sigma = \frac{\hbar\bar{\omega}}{8\pi^2 z^3} \qquad (2.106)$$

where

$$\bar{\omega} = \int_0^\infty \frac{[\varepsilon_1(i\beta) - \varepsilon_3(i\beta)][\varepsilon_2(i\beta) - \varepsilon_3(i\beta)]}{[\varepsilon_1(i\beta) + \varepsilon_3(i\beta)][\varepsilon_2(i\beta) + \varepsilon_3(i\beta)]} d\beta \qquad (2.107)$$

in which $\hbar = h/2\pi$. The quantity $\hbar\bar{\omega}$ has been termed the Lifshitz–van der Waals constant, whose values for some systems are given in Table 2.7.

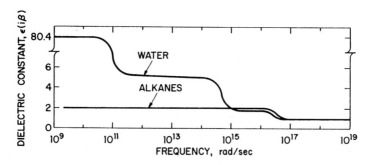

Figure 2.9. Dielectric constant spectra for water and alkanes. (After Ref. 5.)

Table 2.7. Lifshitz-van der Waals Constant for Some Systems

	Lifshitz-van der Waals constant ($\hbar\bar{\omega}$, eV [a])	
	Theoretical [b]	Experimental [c]
Water/vacuum/water	0.94-1.65	-
Polystyrene/vacuum/polystyrene	1.60-2.07	-
Silver/vacuum/silver	10.5	-
Gold/vacuum/gold	4.78-11.77	-
Hydrocarbon/vacuum/hydrocarbon	1.31 (liquid)	-
	2.35 (solid)	2.61 (solid)
Mica/vacuum/mica	2.61	3.53
Silica/vacuum/silica	2.09-2.61	-
Polystyrene/water/polystyrene	0.34	-
Gold/water/gold	3.24-8.63	-
Hydrocarbon/water/hydrocarbon	0.089-0.18	-
Silica/water/silica	0.26	-
Gold/water/polystyrene	0.60	-
Water/hydrocarbon/water	0.089-0.18	0.15

[a] Hamaker constant $H = (3/4\pi)\hbar\bar{\omega}$.
[b] Theoretical values from Refs. 74, 75, 77, and 80.
[c] Experimental values from Refs. 26 and 27.

The equations above confirm the classical prediction that the forces between two large flat plates decay with the inverse third power of distance, and the forces will be repulsive when the dielectric constant of the medium is intermediate between those of the two bodies over sufficiently wide frequency ranges, that is, $\varepsilon_1 < \varepsilon_3 < \varepsilon_2$.

The reversible work required to separate the two parallel plates from a distance z to infinity is then given by

$$\Delta F_{12}(z) = \frac{\hbar\bar{\omega}}{16\pi^2 z^2} \qquad (2.108)$$

which confirms the classical result that the van der Waals energy between two large flat plates decays with the inverse square of the distance. Comparison with Eq. (2.88) gives

$$H_{12} = \frac{3}{4\pi} \hbar\bar{\omega} \qquad (2.109)$$

which relates the Hamaker constant to the Lifshitz-van der Waals constant.

Table 2.8. Surface Tension of Some Hydrocarbons and Water Calculated by Lifshitz Theory [a]

Compound	L	M	ρ, g/ml	z_0, Å	n	H,[b] 10^{-13} erg	γ, dyne/cm Calculated	γ, dyne/cm Measured
n-Hexane	20	86.17	0.6594	2.029	1.3749	5.48	17.7	18.4
2-Methylpentane	20	86.17	0.6532	2.035	1.3715	5.40	17.3	17.4
n-Octane	26	114.22	0.7025	1.999	1.3974	6.08	20.2	21.8
2-Methylheptane	26	114.22	0.6979	2.004	1.3949	6.02	19.9	20.6
2,3-Dimethylhexane	26	114.22	0.7121	1.990	1.4011	6.18	20.7	21.0
n-Dodecane	38	170.33	0.7487	1.971	1.4216	6.75	23.1	25.4
n-Eicosane	62	282.54	0.7887	1.947	1.44	7.35	25.7	29.0
Water	3	18.01	1.0000	1.973	—	5.1	17.4	72.8
						6.3	21.5	

[a] Calculated by $\gamma = \hbar \bar{\omega}_0 / 32\pi^2 z_0^2$.

[b] H = Hamaker constant, z_0 = mean separation between close-packed planes defined by z_0 = $(2/3)^{1/2}(2)^{1/6}(M/NL\rho)^{1/3}$ (where M = molecular weight, L = number of atoms per molecule, N = Avogadro's number), ρ = density, n = refractive index, $\bar{\omega}_0$ = absorption frequency (2.43×10^{16} rad/sec). All data for 20°C.

Source: From Ref. 74.

Macroscopic Theory of van der Waals Forces

If we let $z = z_0$, then for identical bodies in vacuum, then the surface tension is obtained as

$$\gamma = \frac{\hbar \bar{\omega}_0}{32\pi^2 z_0^2} \qquad (2.110)$$

where

$$\bar{\omega}_0 = \int_0^\infty \left[\frac{\epsilon(i\beta) - 1}{\epsilon(i\beta) + 1}\right]^2 d\beta \qquad (2.111)$$

which has been used to calculate the surface tension of water and hydrocarbons [74]. Good agreement between the calculated and the measured values is obtained for hydrocarbons (Table 2.8). The calculated value for water is about 20 dyne/cm, much smaller than the measured value of 72 dyne/cm. The calculated value is, however, practically identical to the dispersion component (21–22 dyne/cm). Thus, it appears that the calculation failed to account for hydrogen-bond interaction. This may be because the dielectric spectrum used does not cover the hydrogen-bond absorption at very low frequencies, or the Lifshitz theory is simply not applicable at hydrogen-bond (very short) distances.

For two bodies separated by vacuum and absorbing strongly only at one frequency in the UV (such as for saturated hydrocarbons), the nonretarded dispersion energy per unit area can be given explicitly by

$$\Delta F_{12}(z) = \left[\frac{\hbar}{32\pi(2)^{1/2} z^2}\right] \frac{(n_1^2 - 1)(n_2^2 - 1)}{(n_1^2 + 1)^{1/2}(n_2^2 + 1)^{1/2}}$$

$$\times \frac{\omega_1 \omega_2}{(n_1^2 + 1)^{1/2} \omega_1 + (n_2^2 + 1)^{1/2} \omega_2} \qquad (2.112)$$

where n is the refractive index and ω is the frequency at which strong absorption occurs. For identical bodies, Eq. (2.112) becomes

$$\Delta F_{11}(z) = \frac{\hbar \omega (n^2 - 1)^2}{64\pi(2)^{1/2} z^2 (n^2 + 1)^{3/2}} \qquad (2.113)$$

The corresponding relation based on Hamaker equation is given by

$$\Delta F_{11}(z) = \frac{9\hbar\nu}{128z^2}\left(\frac{n^2-1}{n^2+2}\right)^2 \qquad (2.114)$$

where ν is the characteristic electronic frequency.

REFERENCES

1. H. Margenau and N. Kestner, *Theory of Intermolecular Forces*, 3rd ed., Pergamon Press, London, 1971.
2. J. O. Hirschfelder, ed., *Intermolecular Forces*, Interscience, New York, 1967.
3. *Intermolecular Forces*, Discuss. Faraday Soc., *40* (1965).
4. J. N. Israelachvilli and D. Tabor, Prog. Surf. Membr. Sci., *7*, 1 (1973).
5. J. N. Israelachvilli, Qt. Rev. Biophys., *6*(4), 341 (1974).
6. J. N. Israelachvilli, in *Yearbook of Science and Technology*, McGraw-Hill, New York, 1976, pp. 23–31.
7. H. Krupp, Adv. Colloid Interface Sci., *1*, 111 (1967).
8. J. O. Hirschfelder, C. F. Curtiss, and R. B. Bird, *Molecular Theory of Gases and Liquids*, Wiley, New York, 1954.
9. S. C. Wang, Phys. Z., *28*, 663 (1927).
10. F. London, Z. Phys., *63*, 245 (1930); Z. Phys. Chem. (Leipzig), *B11*, 222 (1930); Trans. Faraday Soc., *33*, 57 (1937); R. Eisenshitz and F. London, Ann. Phys., *60*, 491 (1930).
11. J. G. Kirkwood, Phys. Z., *33*, 57 (1932).
12. A. Muller, Proc. R. Soc. Lond., *A154*, 624 (1936).
13. K. S. Pitzer, in *Advances in Chemical Physics*, Vol. 2, I. Prigogine, ed., Wiley-Interscience, New York, 1959.
14. L. Salem, J. Chem. Phys., *37*, 2100 (1962).
15. L. Salem, Can. J. Biochem. Physiol., *40*, 1287 (1962).
16. H. B. G. Casimir and D. Polder, Phys. Rev., *73*, 360 (1948).
17. O. Sinanoglu and K. S. Pitzer, J. Chem. Phys., *32*, 1279 (1960).
18. R. H. Fowler and E. A. Guggenheim, *Statistical Thermodynamics*, Oxford University Press, New York, 1939.
19. J. Mahanty and B. W. Ninham, *Dispersion Forces*, Academic Press, New York, 1976.
20. P. K. L. Drude, *Theory of Optics*, Longmans, Green, London, 1933.
21. J. C. Slater and J. G. Kirkwood, Phys. Rev., *37*, 682 (1931).
22. E. A. Moelwyn-Hughes, *Physical Chemistry*, 2nd ed., Pergamon Press, London, 1961.
23. J. F. Padday and N. D. Uffindell, J. Phys. Chem., *72*, 1407 (1968).
24. T. Neugebauer, Z. Phys., *107*, 785 (1937).

References

25. D. Tabor and R. H. S. Winterton, Nature, *219*, 1120 (1968); Proc. R. Soc. Lond., *A 312*, 43 (1969).
26. J. N. Israelachvilli and D. Tabor, Proc. R. Soc. Lond., *A 331*, 19 (1972).
27. D. A. Haydon and J. L. Taylor, Nature, *217*, 739 (1968).
28. B. V. Derjaguin and I. I. Abrikosova, Zh. Eksp. Teor. Fiz., *21*, 945 (1951).
29. J. T. G. Overbeek and M. J. Sparnaay, J. Colloid Sci., *7*, 343 (1952).
30. P. J. W. Debye, *Polar Molecules*, Chemical Catalog, New York, 1929.
31. C. P. Smyth, *Dielectric Behavior and Structure*, McGraw-Hill, New York, 1955.
32. W. H. Keesom, Phys. Z., *22*, 126, 643 (1922); *23*, 225 (1923).
33. P. J. W. Debye, Phys. Z., *21*, 178 (1920); *22*, 302 (1921).
34. W. M. Latimer and W. H. Rhodebush, J. Am. Chem. Soc., *42*, 1419 (1920).
35. G. C. Pimentel and A. L. McClellan, *The Hydrogen Bond*, W. H. Freeman, San Francisco, 1960.
36. L. Pauling, *The Nature of Chemical Bonds*, 3rd ed., Cornell University Press, Ithaca, N.Y., 1960.
37. D. Hadzi and H. W. Thompson, eds., *Hydrogen Bonding*, Pergamon Press, London, 1959.
38. C. A. Coulson, *Valence*, 2nd ed., Oxford University Press, New York, 1961.
39. R. C. Nelson, R. W. Hemwall, and G. D. Edwards, J. Paint Technol., *42*(550), 636 (1970).
40. C. M. Hansen and A. Beerbower, in *Kirk-Othmer's Encyclopedia of Chemical Technology*, 2nd ed., Suppl. Vol., Wiley, New York, 1971, pp. 889–910.
41. H. H. Landolt and R. Bornstein, *Zahlenwerte und Funktionen*, Vol. 1, Part 3, Springer-Verlag, Berlin, 1951.
42. A. R. von Hippel, *Dielectrics and Waves*, Wiley, New York, 1954; A. R. von Hippel, ed., *Dielectric Materials and Applications*, Wiley, New York, 1954.
43. D. W. van Krevelen, *Properties of Polymers—Correlations with Chemical Structure*, 2nd ed., Elsevier, Amsterdam, 1976.
44. H. A. Lorentz, Wied. Ann. Phys., *9*, 64 (1880); L. V. Lorenz, ibid., *11*, 70 (1880).
45. K. G. Denbigh, Trans. Faraday Soc., *36*, 936 (1940).
46. H. Watanabe, T. Takayama, and S. Mettle, *Final Report on Ionization Potential of Molecules by Photoionization Method*, Pamphlet 158, U.S. Dept. of Commerce, Office of Technical Service, 1959, p. 317.
47. H. Watanabe, T. Takayama, and S. Mettle, J. Quant. Spectrosc. Radiat. Transfer, *2*, 369 (1962).
48. W. C. Price, P. V. Harris, and T. R. Passmore, J. Quant. Spectrosc. Radiat. Transfer, *2*, 327 (1962).

49. P. G. Wilkinson, Astrophys. J., *138*, 778 (1963).
50. A. L. McClellan, *Tables of Experimental Dipole Moments*, Vol. 1, W. H. Freeman, San Francisco, 1963; Vol. 2, Rahara Enterprises, El Cerrito, Calif., 1974.
51. *Digest of Literature on Dielectrics*, National Academy of Science, Vols. 35, 36, 37, Washington, D.C., 1971, 1972, 1975.
52. W. K. Krigbaum and J. V. Dawkins, in *Polymer Handbook*, 2nd ed., J. Brandrup and E. H. Immergut, eds., Wiley, New York, 1975, pp. IV319–322.
53. J. C. Maxwell, *A Treatise on Electricity and Magnetism*, Clarendon, Oxford, 1892.
54. C. J. F. Bottcher, *Theory of Electric Polarization*, Elsevier, New York, 1952.
55. E. F. Meyer and R. E. Wagner, J. Phys. Chem., *70*, 3166 (1966); E. F. Meyer, T. A. Renner and K. S. Stec, ibid., *75*, 642 (1971).
56. S. Wu, J. Adhes., *5*, 39 (1973), also in *Recent Advances in Adhesion*, L. H. Lee, ed., Gordon and Breach, New York, 1973, pp. 45–63.
57. S. Wu, J. Polym. Sci., *C34*, 19 (1971).
58. S. Wu, J. Macromol. Sci., *C10*, 1 (1974).
59. S. Wu, in *Polymer Blends*, Vol. 1, D. R. Paul and S. Newman, eds., Academic Press, New York, 1978, pp. 243–293.
60. J. M. Morris, Aust. J. Chem., *26*, 649 (1973).
61. T. Kihara, Adv. Chem. Phys., *5*, 147 (1963); Rev. Mod. Phys., *25*, 831 (1953).
62. R. J. Good and E. Ebling, Ind. Eng. Chem., *62*(3), 54 (1970).
63. B. W. Davis, J. Colloid Interface Sci., *59*, 420 (1977).
64. Y. Oh and M. S. Jhon, J. Colloid Interface Sci., *73*, 467 (1980).
65. R. J. Good, in *Treatise on Adhesion and Adhesives*, Vol. 1, R. L. Partick, ed., Marcel Dekker, New York, 1967. pp. 9–68.
66. J. L. Gardon, in *Treatise on Adhesion and Adhesives*, Vol. 1, R. L. Partick, ed., Marcel Dekker, New York, 1967, pp. 269–324.
67. H. C. Hamaker, Physica, *4*, 1058 (1937).
68. H. T. G. Overbeek, in *Colloid Science*, Vol. 1, R. H. Kruyt, ed., Elsevier, Amsterdam, 1952, pp. 245–277; also, E. J. W. Verwey and J. T. G. Overbeek, *Theory of the Stability of Lyophobic Colloids*, Elsevier, New York, 1948.
69. J. Padday, in *Surface and Colloid Science*, Vol. 1, E. Matijevic, ed., Wiley-Interscience, New York, 1969, pp. 39–100.
70. W. F. Jaep, private communication.
71. E. M. Lifshitz, Zh. Eksp. Teor. Fiz., *29*, 94 (1955); Sov. Phys., *2*, 73 (1956).
72. I. E. Dzyaloshinskii, E. M. Lifhsitz, and L. P. Pitaevskii, Adv. Phys., *10*, 165 (1961).

References

73. P. Richmond and B. W. Ninham, J. Colloid Interface Sci., 45, 69 (1973).
74. J. N. Israelachvilli, J. Chem. Soc., Faraday Trans. II, 69, 1729 (1973).
75. H. Krupp, W. Schnabel, and G. Walter, J. Colloid Interface Sci., 39, 421 (1972).
76. V. A. Parsegian and B. W. Ninaham, Nature, 224, 1197 (1969).
77. B. W. Ninham and V. A. Parsegian, Biophys. J., 10, 646 (1970).
78. V. A. Parsegian and B. W. Ninham, Biophys. J., 10, 664 (1970).
79. B. W. Ninham and V. A. Parsegian, J. Chem. Phys., 52, 4578 (1970).
80. V. A. Parsegian and B. W. Ninham, J. Colloid Interface Sci., 37, 332 (1971).
81. V. A. Parsegian and B. W. Ninham, J. Adhes., 4, 283 (1972).
82. J. E. Kiefer, V. A. Parsegian, and G. H. Weiss, J. Colloid Interface Sci., 51, 543 (1975).
83. A. A. Abrikosov, L. P. Gorkov, and I. E. Dzyaloshinskii, Zh. Eksp. Teor. Fiz., 36, 900 (1959); Sov. Phys., 9, 636 (1959); *Methods of Quantum Field Theory in Statistical Physics*, Prentice-Hall, Englewood Cliffs, N.J., 1963.
84. D. Langbein, Solid State Commun., 12, 853 (1973).
85. A. D. McLachlan, Proc. R. Soc. Lond., $A271$, 387 (1963); $A274$, 80 (1963).
86. J. N. Israelachvilli, Proc. R. Soc. Lond, $A331$, 39 (1972).
87. D. Langbein, J. Adhes., 1, 237 (1969); 3, 213 (1972).
88. D. Langbein, J. Phys. Chem. Solids, 32, 133, 1657 (1971).
89. D. Langbein, Phys. Rev., $B2$, 3371 (1970).
90. N. G. van Kempen, B. R. A. Nijboer, and K. Schram, Phys. Lett., 26A, 307 (1968).
91. W. L. Blade, J. Chem. Phys., 27, 1280 (1957).
92. B. R. A. Nijboer and M. J. Renne, Chem. Phys. Lett., 1, 317 (1967); 2, 35 (1968).
93. L. C. Landau and E. M. Lifshitz, *Statistical Physics*, Pergamon Press, London, 1959.

3
Interfacial and Surface Tensions of Polymer Melts and Liquids

Most polymers are mutually immiscible and can form stable interfaces, unlike small-molecule organic liquids, which are often mutually miscible. Despite their importance, direct and accurate measurements of interfacial tensions between polymer melts and liquids were not reported until 1969, because of experimental difficulty in handling and ensuring equilibrium of highly viscous melts [1,2]. Suitable experimental methods are discussed in Chapter 9.

The pendant drop method has been used the most extensively [1,2]. A pendant drop of poly(vinyl acetate) melt in equilibrium with surrounding molten polyethylene is shown in Figure 3.1. Extensive tabulations of interfacial and surface tensions of polymer melts, liquids, and solids have become available [3,4] and are used to test theoretical and semiempirical relations [3–13]. Here, the surface tension is discussed first and the interfacial tension next.

3.1. SURFACE TENSIONS OF POLYMER MELTS AND LIQUIDS

3.1.1. Temperature Dependence

Surface tensions of small-molecule liquids vary linearly with temperature with $-(d\gamma/dT)$ about 0.1 dyne/cm at ordinary temperatures far

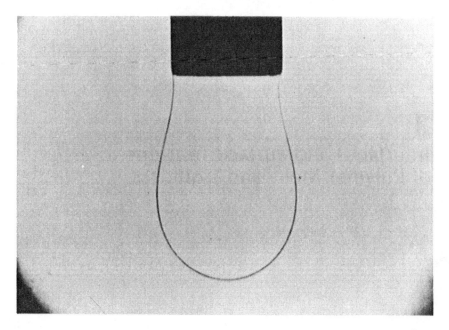

Figure 3.1. A pendant drop of poly(vinyl acetate) melt in equilibrium with molten polyethylene as the surrounding medium at 140°C.

below the critical temperature [14–16]. Similarly, surface tensions of polymers also vary linearly with temperature, with $-(d\gamma/dT)$ about 0.05 dyne/cm [3,4]. Since $-(d\gamma/dT)$ is the surface entropy, the smaller $-(d\gamma/dT)$ for a polymer is attributed to conformational restrictions of long-chain molecules. The γ versus T plots for some polymers are shown in Figures 3.2 and 3.3 [1,17–19]. Linearity is found for all polymers shown. One notable exception is perfluoroalkanes, whose γ versus T plot fits better with a quadratic equation [20].

The Guggenheim equation [14], developed for small-molecule liquids, has also been found to apply to polymers,

$$\gamma = \gamma_0 \left(1 - \frac{T}{T_c}\right)^{11/9} \tag{3.1}$$

where γ_0 is the surface tension at T = 0 K and T_c is the critical temperature. The γ_0 and T_c values found by least-squares regression of γ versus T for polymers are given in Table 3.1. The T_c values for polymers thus obtained agree well with those extrapolated from small-molecule liquids. Differentiation of Eq. (3.1) with respect to temperature gives

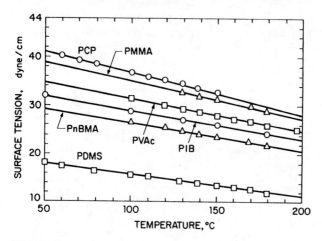

Figure 3.2. Surface tension versus temperature for some polymer melts and liquids. (After Refs. 3 and 4.)

$$-\frac{d\gamma}{dT} = \frac{11}{9} \frac{\gamma_0}{T_c} \left(1 - \frac{T}{T_c}\right)^{2/9} \tag{3.2}$$

which is nearly constant at temperatures far below T_c, confirming the linearly of γ versus T at ordinary temperature ranges. The critical temperature of a polymer is typically 600–900°C. Therefore, $-(d\gamma/dT)$ is practically constant at ordinary temperatures such as 0–200°C.

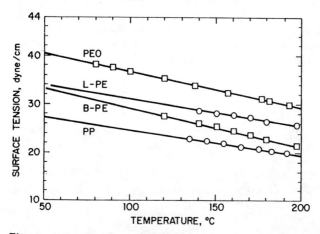

Figure 3.3. Surface tension versus temperature for some polymer melts. (After Refs. 3 and 4.)

Table 3.1. Values of γ_0 and T_c of Guggenheim Equation, Obtained by Least-Squares Fit of Surface Tension Versus Temperature Data

Polymer	γ_0 dyne/cm	T_c,[a] K
Polyethylene (linear)	53.71	1032 (1028)
Polyethylene (branched)	56.38	921
Polypropylene	47.16	914
Polyisobutylene	53.74	918
Polystyrene	63.31	967
Polytetrafluoroethylene	43.96	823 (828)
Polychloroprene	70.95	892
Poly(vinyl acetate)	57.37	948
Poly(methyl methacrylate)	65.09	935
Polydimethylsiloxane	35.31	776

[a] The T_c values in parentheses are obtained by extrapolation of values for gaseous and liquid homologs.

Macleod's equation [21], developed for small-molecule liquids, has also been shown to apply to polymers [1,4,19],

$$\gamma = \gamma^\circ \rho^n \tag{3.3}$$

where ρ is the density, and γ° and n are both positive constants and independent of temperature. This relation is accurately obeyed with n = 4 for unassociated small-molecule liquids [21–23]. For polymers, n varies from 3.0 to 4.4, and may be approximated with 4.0 just as for small-molecule liquids. Macleod's plots for polyehtylene and poly(vinyl acetate) are shown in Figure 3.4. The n values for some polymers are listed in Table 3.2. Differentiation of the above gives

$$-\frac{d\gamma}{dT} = n\alpha\gamma \tag{3.4}$$

where α is the isobaric thermal expansion coefficient, $\alpha = (1/v)(dv/dT)_p$. Note that the Macleod's equation can be derived from Eq. (2.74), however, with n = 8/3.

Eötvös equation [24,25], although applicable to small-molecule liquids, is not applicable to polymers, as it diverges at infinite molecular weight.

Surface Tensions of Polymer Melts and Liquids

Table 3.2. Macleod's Exponent for Some Polymers

Polymer	Macleod's exponent, n	Reference
Polychloroprene	4.2	18
Poly(methyl methacrylate)	4.2	17
Poly(n-butyl methacrylate)	4.2	17
Polystyrene	4.4	17
Poly(vinyl acetate)	3.4	1
Poly(ethylene oxide)	3.0	19
Polyisobutylene	4.1	1
Polypropylene	3.2	19
Polyethylene, linear	3.2	1
Polyethylene, branched	3.3	19
Polydimethylsiloxane	3.5	19

3.1.2. Effect of Chemical Composition

For small-molecule liquids, Sugden [26] showed that $\gamma°$ is determined by chemical constitution alone. For polymers, Sugden's relation becomes [1]

$$\gamma° = \left(\frac{P}{M}\right)^n \quad (3.5)$$

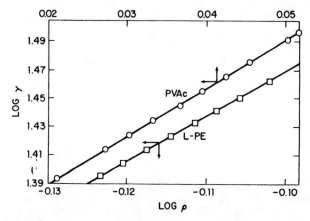

Figure 3.4. Macleod plots for poly(vinyl acetate) (PVAc, M_w = 45,000) and a linear polythylene (L-PE, M_w = 67,000). Arrows indicate which scales apply to each curve. (After Refs. 1 and 4.)

where P is the parachor for a repeat unit and M is the molecular weight of a repeat unit. As P is independent of temperature, the variation of surface tension with temperature arises solely from the variation of density with temperature. Values of parachor for various atoms and groups have been listed elsewhere [27,28]. Combining Eqs. (3.3) and (3.5) gives

$$\gamma = \left(\frac{P}{M}\right)^n \rho^n \tag{3.6}$$

which can predict the surface tension of polymers within 1–2 dyne/cm over wide temperature ranges in most cases (Table 3.3).

3.1.3. Molecular-Weight Dependence

The surface tension of homologous series tends to increase, while the surface entropy tends to decrease, with increasing molecular weight. At infinite molecular weight, both the surface tension and the surface entropy are, however, finite.

The bulk properties of homologous series vary linearly with the reciprocal of molecular weight [29–34],

$$X_b = X_{b\infty} - \frac{k_b}{M_n} \tag{3.7}$$

where X_b is a bulk property such as glass transition temperature, heat capacity, specific volume, thermal expansion coefficient, refractive index, tensile strength, and the like; $X_{b\infty}$ the bulk property at infinite molecular weight; k_b a constant; and M_n the number-average molecular weight. This relation is, however, not applicable to surface properties. Plots of γ versus M^{-1} show various degrees of curvature [35–38].

Instead, the surface tension of a homologous series varies linearly with $M_n^{-2/3}$ [39,40],

$$\gamma = \gamma_\infty - \frac{k_e}{M_n^{2/3}} \tag{3.8}$$

where γ_∞ is the surface tension at infinite molecular weight and k_e is a constant. This equation fits the data for n-alkanes with standard deviations in γ of about 0.05 dyne/cm, and for perfluoroalkanes, polyisobutylenes, polydimethylsiloxanes, and polystyrenes with standard deviations in γ about 0.2 dyne/cm (Table 3.4). A plot of γ versus $M^{-2/3}$ for n-alkanes is given in Figure 3.5. A simple lattice analysis gives [3]

Table 3.3. Comparison Between Measured and Parachor-Calculated Surface Tensions

Polymer	Surface tension,[a] dyne/cm						Maximum difference
	Measured			Calculated			
	20°C	140°C	180°C	20°C	140°C	180°C	
Polychloroprene	43.6	33.2	29.8	44.2	32.4	29.0	0.8
Polystyrene	40.7	32.1	29.2	37.5	30.9	28.5	1.2 (3.2)
Poly(methyl methacrylate)	41.1	32.0	28.9	41.1	32.0	38.9	0
Poly(n-butyl methacrylate)	31.2	24.1	21.7	38.1	28.6	26.0	4.5 (6.9)
Poly(vinyl acetate)	36.5	28.6	25.9	38.8	27.7	24.8	1.1 (2.3)
Polyethylene, linear	35.7	28.8	26.5	35.3	25.0	22.4	4.1

[a] The measured surface tension values at 20°C are obtained by linear extrapolation of γ-T plots of the melts. The parenthetical maximum differences refer to the case when the extrapolated values are also considered.

Source: From Ref. 4.

Table 3.4. Numerical Constants for Molecular-Weight Dependence of Surface Tension

Polymer	Temperature, °C	Eq. (3.12), $\gamma^{1/4}$ vs. M^{-1}			Eq. (3.8), γ vs. $M^{-2/3}$		
		γ_∞, dyne/cm	k_s	σ,[a] dyne/cm	γ_∞, dyne/cm	k_e	σ,[a] dyne/cm
n-Alkanes	20	34.75	30.69	0.04	37.81	385.9	0.03
Polyisobutylenes	24	34.50	46.15	0.19	35.62	382.7	0.34
Polydimethylsiloxanes	20	20.33	22.22	0.15	21.26	166.1	0.09
Perfluoroalkanes	20	23.94	12.21	0.29	25.85	682.8	0.30
Polystyrenes	176	29.50	75.42	0.01	29.97	372.7	0.08
Poly(ethylene oxide)-dimethyl ether	24	41.50	28.13	0.38	44.35	342.8	0.44

[a] σ is the standard deviation in γ.

Surface Tensions of Polymer Melts and Liquids

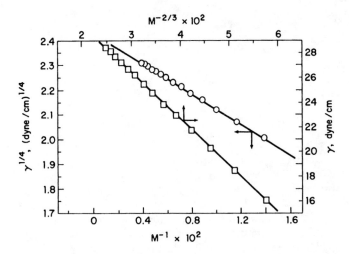

Figure 3.5. Variation of surface tension with molecular weight for n-alkanes (orthobaric data at 20°C). (After Ref. 3.)

$$k_e \sim (\gamma_\infty - \gamma_e)(2m)^{2/3} \tag{3.9}$$

where γ_e is the surface tension of the end groups and m is the formula weight of a repeat unit.

On the other hand, Macleod's equation provides an alternative relation. The density of a homologous series is given by [31]

$$\frac{1}{\rho} = \frac{V_r}{m} + \frac{V_e}{M} \tag{3.10}$$

where V_r is the molar volume of a repeat unit and V_e is that of the two chain ends. Combining with the Macleod equation gives [3]

$$\gamma^{1/n} = \frac{1}{m}\left[\frac{P_r M + (mP_e - m_e P_r)}{(V_r/m)M + V_e}\right] \tag{3.11}$$

where $M = rm + m_e$, $P = rP_r + P_e$, r the number of repeat units, m_e the formula weight of the two chain ends, P_r the parachor of a repeat unit, and P_e the parachor of the two chain ends. Equation (3.11) can be expanded in several different power series. A particularly useful relation is obtained by letting $n = 4$ and truncating all terms higher than $1/M$, that is,

$$\gamma^{1/4} = \gamma_\infty^{1/4} - \frac{k_s}{M_n} \qquad (3.12)$$

where

$$k_s = \frac{m_e P_r - mP_e}{V_r} + \frac{mV_e P_r}{V_r^2} \qquad (3.13)$$

Equation (3.12) accurately fits the data for n-alkanes with standard deviation in γ about 0.05 dyne/cm, and for perfluoroalkanes, polyisobutylenes, polydimethylsiloxanes and polystyrenes with standard deviations in γ about 0.2 dyne/cm (Table 3.4). A plot of $\gamma^{1/4}$ versus M^{-1} for n-alkanes is given in Figure 3.5. Thus, Eqs. (3.8) and (3.12) are equivalent. However, the two relations give somewhat different γ_∞ values. The $\gamma^{1/4}$ versus M^{-1} relation is, however, more consistent with the γ versus T relation, discussed in Section 3.1.5.

The extent of surface tension variation decreases with increasing molecular weight. The k_e and k_s values in Table 3.4 indicate that γ will be smaller than γ_∞ by less than 1 dyne/cm when the molecular weight is greater than about 3000. Accordingly, for instance, the surface tensions of poly(vinyl acetate) melts having molecular weights 11,000–120,000 are found to be practically independent of molecular weight [1].

On the other hand, the surface tensions of poly(ethylene glycols) have been found to be independent of molecular weight even in the oligomeric ranges [41,42]. This appears to arise from strong hydrogen bonding (between the terminal hydroxyl groups), which effectively makes the oligomers behave as having infinite molecular weight. If the formation of hydrogen bond is prevented by converting the terminal hydroxyl groups into nonhydrogen-bonding methyl ether or acetate groups, the surface tensions become dependent on molecular weight as expected (Figure 3.6).

3.1.4. Effects of Glass and Phase Transitions

Macleod's equation can be used to analyze the effects of glass (secondary) and phase (primary) transitions on surface tension [4]. The temperature coefficient of surface tension in the glassy region (below T_g) is related to that in the rubber region (above T_g) by

$$\left(\frac{d\gamma}{dT}\right)_g = \frac{\alpha_g}{\alpha_r} \left(\frac{d\gamma}{dT}\right)_r \qquad (3.14)$$

Surface Tension of Polymer Melts and Liquids

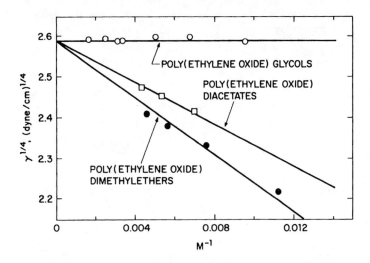

Figure 3.6. Variation of surface tension with molecular weight for hydrogen-bonding poly(ethylene oxide)-glycols, and non-hydrogen-bonding poly(ethylene oxide)-diacetates and -dimethylethers. (Data from Refs. 41 and 42.)

where the subscripts g and r refer to glassy and rubbery regions, respectively. Thus, a γ versus T plot is continuous but changes the slope at T_g. Since $\alpha_g < \alpha_r$, $-(d\gamma/dT)_g < -(d\gamma/dT)_r$.

On the other hand, at the crystal-melt transition, the surface tension of the crystalline phase γ^c is related to that of the amorphous phase γ^a by

$$\gamma^c = \left(\frac{\rho_c}{\rho_a}\right)^n \gamma^a \tag{3.15}$$

where ρ_c is the crystalline density and ρ_a the amorphous density. Thus, at the crystal-melt transition, the surface tension changes discontinuously, since the density is discontinuous. As ρ_c is usually greater than ρ_a, the crystalline phase will have higher surface tension than the amorphous phase. For instance, polyethylene has n = 3.2, γ^a = 35.7 dyne/cm, and ρ_a = 0.855 g/ml at 20°C. The crystalline density ρ_c is 1.000 g/ml. Thus γ^c is calculated by Eq. (3.15) to be 58.9 dyne/cm, which compares rather well with an experimental value of 53.6 dyne/cm [43].

Thermodynamic analysis [4] confirms that the surface tension should change discontinuously at the crystal-melt transition, and continuously at the glass transition with discontinuous $d\gamma/dT$. Let the change in the Gibbs free energy for the entire system for the crystal-melt transition be given by

$$\Delta G_{cm} = G_c - G_m \tag{3.16}$$

where G_c is the total Gibbs free energy of the crystal phase and G_m that of the melt phase. Since the crystal-melt transition is a first-order transition, the stability condition requires that [4]

$$\Delta G_{cm} = 0 \quad \text{at constant T, p} \tag{3.17}$$

and $(\partial \Delta G_{cm}/\partial A)_{T,p} \neq 0$, that is,

$$\left(\frac{\partial \Delta G_{cm}}{\partial A}\right)_{T,p} = \left(\frac{\partial G_c}{\partial A}\right)_{T,p} - \left(\frac{\partial G_m}{\partial A}\right)_{T,p}$$

$$= \gamma^c - \gamma^a \neq 0 \tag{3.18}$$

where A is the surface area. Since $\rho_c > \rho_a$, $\gamma^c > \gamma^a$ at the crystal-melt transition.

On the other hand, let the change in the Gibbs free energy for the entire system for glass-rubber transition be given by

$$\Delta G_{gr} = G_g - G_r \tag{3.19}$$

where G_g is the total Gibbs free energy for the glassy state and G_r is that for the rubbery state. Since glass-rubber transition is a second-order transition, the stability condition requires that [4]

$$\Delta G_{gr} = 0 \quad \text{at constant T, p} \tag{3.20}$$

and

$$\left(\frac{\partial \Delta G_{gr}}{\partial A}\right)_{T,p} = \gamma^g - \gamma^r = 0 \tag{3.21}$$

$$\left(\frac{\partial^2 \Delta G_{gr}}{\partial A \, \partial T}\right) = \left(\frac{\partial \gamma^g}{\partial T}\right) - \left(\frac{\partial \gamma^r}{\partial T}\right) \neq 0 \tag{3.22}$$

where γ^g is the surface tension for the glassy state and γ^r is that for the rubbery state. Therefore, $\gamma^g = \gamma^r$, but $(d\gamma^g/dT) \neq (d\gamma^r/dT)$ at the glass-rubber transition.

Semicrystalline polymers tend to be covered with an amorphous surface layer. As the amorphous phase has lower surface tension, it tends to migrate to the surface. Various degrees of surface crystallinity can, however, be induced on the surface by nucleating the polymers against certain mold surfaces. The effects of surface nucleation and crystallinity on surface tension of solid polymers are discussed in Sections 5.4 and 9.5.

3.1.5. Application to Solid Polymers

Surface tension of solid polymers may be obtained by extrapolation of melt data (γ versus T relation) or by extrapolation of liquid homolog data (γ versus M relation). When extrapolating from the melt data, the effects of primary and secondary transitions should be included. However, linear extrapolations are usually adequate, as the effect of glass transition is small and semicrystalline polymers usually have amorphous surfaces when prepared by cooling from the melts.

On the other hand, the surface tensions of solid polymers may also be obtained by extrapolation from liquid homologs by Eq. (3.8) or (3.12). Such extrapolation will give amorphous values. The γ_s (obtained from γ versus T melt data) and γ_∞ (obtained from γ versus M of liquid homologs) for some polymers are compared in Table 3.5. The $\gamma^{1/4}$ versus M^{-1} relation gives results more consistent with the γ versus T relation than does the γ versus $M^{-2/3}$ relation.

The surface tensions of solid polymers may also be determined from wettability data. The values obtained by the two methods described above are in excellent agreement with those obtained from wettability, discussed in Chapter 5.

3.1.6. Copolymers, Blends, and Additives

Low-energy components in copolymers or blends tend to preferentially adsorb on the surfaces, just as in small-molecule liquids, as this will lower the free energy of the system.

Random Copolymers

The surface tension of a random copolymer usually follows the linear relation [44]

$$\gamma = x_1\gamma_1 + x_2\gamma_2 \qquad (3.23)$$

Table 3.5. Comparison of γ_s (from γ-T Plot) and γ_∞ (from Liquid Homologs)

Polymer	Temperature °C	γ_s from γ-T plot	γ_∞ from liquid homologs	
			Eq. (3.12) $\gamma^{1/4}$ vs. M^{-1}	Eq. (3.8) γ vs. $M^{-2/3}$
Polyethylene	20	35.7	34.75	37.81
Polyisobutylene	24	34.0	34.50	35.62
Poly(dimethylsiloxane)	20	19.8	20.33	21.26
Polytetrafluoroethylene	20	23.9	23.94	25.85
Poly(ethylene glycol)	24	42.5	41.50	44.35
Polystyrene	176	29.5	29.50	29.97

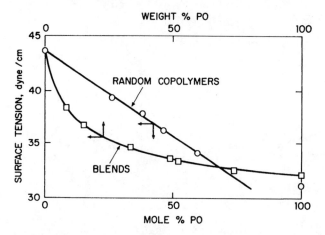

Figure 3.7. Linear additivity of surface tension of random copolymers of ethylene oxide and propylene oxide (PO), and surface-active behavior of blends of poly(ethylene oxide) (PEG 300) and poly(propylene oxide) (PPG 425). (After Ref. 44.)

where γ is the surface tension and x is the mole fraction. The subscripts 1 and 2 refer to the components 1 and 2, respectively. Such behavior is shown for random copolymers of ethylene oxide and propylene oxide in Figure 3.7. The lack of surface activity is probably due to restrictions on chain conformation, which precludes preferential arrangement of short segments of lower-energy component on the surface.

Block and Graft Copolymers

However, block and graft copolymers show considerable surface activity of the lower-energy component, when the lower-energy blocks or grafts are sufficiently long that they can accumulate and orient on the surfaces independently of the rest of the molecule. For instance, pronounced surface activity is observed for ABA block copolymers of ethylene oxide (A block, higher surface tension) and propylene oxide (B block, lower surface tension) (Figure 3.8). When the degree of polymerization (DP) of the B block is greater than about 56, the surface tension of the block copolymer is practically equal to that of the B block homopolymer, even when the B block is only 20% by weight of the block copolymer. On the other hand, a DP of about 20 of the lower-energy block is sufficient to reduce the surface tension of an ABA block copolymer of polyether (A block, higher surface tension) and dimethylsiloxane (B block, lower surface tension) to that of polydimethylsiloxane [45].

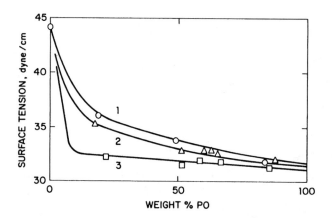

Figure 3.8. Surface tension versus composition for ABA block copolymers of ethylene oxide (A block) and propylene oxide (B block) at 25°C. Degrees of polymerization are (a) DP = 16; (2) DP = 30; (3) DP = 56. (After Ref. 44.)

Blends of Polymers

Blends of both compatible and incompatible polymers show pronounced surface activity, incompatible blends being more pronounced than compatible blends. The surface activity of an incompatible blend is further complicated by heterogeneous phase structure.

Surface activities of compatible blends of poly(ethylene oxide) and poly(propylene oxide) and of poly(propylene oxide) and polyepichlorohydrin are shown in Figure 3.9. The surface activity increases with increasing molecular weight, apparently because of increased incompatibility.

The equation of Belton and Evans [46] fits the surface tensions of the above two compatible blends rather well [44],

$$\gamma = \gamma_1^0 + \frac{kT}{a} \ln \frac{c}{1 + (c-1)x_1} \tag{3.24}$$

where

$$c = \exp[a(\gamma_2^0 - \gamma_1^0)] \tag{3.25}$$

where x is the mole fraction, γ_j^0 the surface tension of the pure polymer j, a the surface area occupied by a molecule, k the Boltzmann constant, and T the temperature. The a value may be calculated from the γ value at x = 0.5.

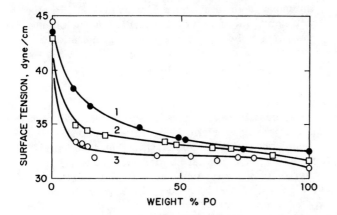

Figure 3.9. Surface tensions of blends of compatible homopolymers at 25°C. (1) Poly(propylene glycol) (PPG 425) + poly(ethylene glycol) (PEG 300); (2) PPG 2025 + polyepichlorohydrin (PECH 1500); (3) PPG 400 + PECH 2000. (After Ref. 44.)

For polymer solutions, the equation of Gaines [47] which combines the Prigogine-Marechal parallel layer model [48] with the Flory-Huggins lattice model has been found to give excellent results:

$$\gamma = \gamma_1^0 + \frac{kT}{a}\left[\ln\frac{\phi_1^s}{\phi_1} + \frac{r-1}{r}(\phi_2^s - \phi_2)\right] - \left(\frac{\chi}{aN}\right)\phi_2^2 \quad (3.26)$$

$$= \gamma_2^0 + \frac{kT}{ra}\left[\ln\frac{\phi_2^s}{\phi_2} + (r-1)(\phi_2^s - \phi_2)\right] - \left(\frac{\chi}{aN}\right)\phi_1^2 \quad (3.27)$$

and

$$\frac{\phi_1^s}{\phi_1} = \left(\frac{\phi_2^s}{\phi_2}\right)^{1/r} \exp\left[\frac{(\gamma_2^0 - \gamma_1^0)a}{kT}\right]\exp\left[\frac{\chi}{NkT}(1-2\phi_1)\right] \quad (3.28)$$

where ϕ_j is the volume fraction of component j in the bulk, ϕ_j^s the volume fraction of component j in the surface layer, r the number of segments in the r-mer (polymer) molecule, a the surface lattice parameter, χ the Flory-Huggins interaction parameter and N is the Avogadro's number. The parameter a may be taken to be the surface area of a solvent molecule, $a = (V_1/N)^{2/3}$, where V_1 is the molar volume of the solvent and N is the Avogadro's number. The r may be taken to be

V_2/V_1, where V_2 is the molar volume of the polymer. Excellent results have been obtained with solutions of polydimethylsiloxane in toluene and tetrachloroethylene [47], polyisobutylene in n-heptane and tetralin [49], three component solutions of the above [50], and polydimethylsiloxane oligomers and polyisobutylene oligomers [51].

Additives

Low-energy additives can greatly lower the surface tension of polymers, for instance, fatty amides in polyethylene [52], fluorocarbons in poly(methyl methacrylate) and poly(vinyl chloride) [53], and dimethylsiloxane block copolymers in polycarbonate and polystyrene [54].

ABA block copolymer of polyether (A block) and polydimethylsiloxane (B block) at 0.1-1% by weight reduces the surface tension of a liquid polyol (γ = 32 dyne/cm at 25°C) to about 21 dyne/cm, resembling the surface tension of pure polydimethylsiloxane [45]. A block copolymer of styrene and dimethylsiloxane reduces the surface tension of polystyrene melt (27.7 dyne/cm at 200°C) to 14.7 dyne/cm at 0.5% by weight additive, and to 11.9 dyne/cm at 5% by weight additive, as compared with the surface tension of 11.2 dyne/cm (at 200°C) for pure polydimethylsiloxane [55].

The rate at which the additive migrates to the surface is diffusion-controlled. The rate of change of surface tension is given by [55]

$$\gamma = \gamma_0 - 2RTc_0 \left(\frac{Dt}{\pi}\right)^{1/2} \qquad (3.29)$$

where γ is the surface tension at time t, γ_0 that at zero time, c_0 the bulk concentration of the additive, R the gas constant, T the temperature, and D the diffusion coefficient.

3.1.7. Theories of Surface Tension

Quasi-Continuum Theory

Summation of pairwise potential gives $\gamma = (\pi/16)q^2 A/z_0^2$; see Eq. (2.74), which can be rewritten as

$$\gamma = \frac{\pi}{16} A \left(\frac{\rho}{m}\right)^{8/3} \qquad (3.30)$$

where A is the attraction constant, ρ the density, and m the mass of an interacting element. For nonpolar molecules, only the dispersion force needs to be considered, and the attraction constant is given by

$$A = \frac{3he\alpha^{3/2}(SZ)^{1/2}}{8\pi(m_e)^{1/2}} \qquad (3.31)$$

which follows from Eq. (2.10). The relationships above predict the surface tensions of n-alkanes (up to C_{20} and triacontane) quite well. In the calculation, CH_2 and CH_3 are chosen to be the interacting elements [56].

Lifshitz Theory

The surface tension is given by $\gamma = h\bar{\omega}_0/64\pi^3 z_0^2$; see Eq. (2.110). For a nonpolar liquid which absorbs strongly at one UV frequency ω, the relation becomes

$$\gamma = \frac{(n^2 - 1)^2 h\omega}{256(2)^{1/2}\pi^2(n^2 + 1)^{3/2} z_0^2} \qquad (3.32)$$

which follows from Eq. (2.113). The relation predicts the surface tensions of n-alkanes (up to C_{20}) remarkably well, but gives only the dispersion component when applied to water, as expected; see Table 1.12 [57].

Corresponding States Theory

The reduced surface tension is a universal function of the reduced temperature, given by [10,11]

$$\tilde{\gamma}(\tilde{T}) = \frac{\gamma}{k^{1/3} P^{*2/3} T^{*1/3}} \qquad (3.33)$$

where $\tilde{\gamma}$ is the reduced surface tension, $\tilde{\gamma} = \gamma(r,T)/\gamma^*(r)$, $\tilde{T}$ is the reduced temperature, $\tilde{T} = T/T^*$, k is the Boltzmann constant, r the number of repeat units in the macromolecule, and P^*, T^* and γ^* are Prigogine's reduction parameters for pressure, temperature, and surface tension, respectively.

Various models can be used to calculate the reduction parameters. The cell model gives

$$\alpha T = \frac{1 - \tilde{V}^{-1/3}}{(4/3)\tilde{V}^{-1/3} - 1} \qquad (3.34)$$

where α is the isobaric thermal expansion coefficient and $\tilde{V}$ is the reduced volume, which can thus be calculated from α. The reduced temperature $\tilde{T}$ is related to the reduced volume by

$$\tilde{T} = \tilde{V}^{-1}(1 - \tilde{V}^{-1/3}) \quad (3.35)$$

and P^* is related to the reduced volume by

$$P^* = \frac{\alpha}{\beta} T\tilde{V}^2 \quad (3.36)$$

where β is the isothermal compressibility. The reduced surface tension $\tilde{\gamma}$ is related to the reduced volume by

$$\tilde{\gamma}\tilde{V}^{5/3} = 0.29 - (1 - \tilde{V}^{-1/3}) \ln \frac{\tilde{V}^{1/3} - 0.5}{\tilde{V}^{1/3} - 1} \quad (3.37)$$

The values of $\tilde{\gamma}$, P^*, and T^* are than used in Eq. (3.33) to obtain the surface tension γ. The calculation requires the data for α and β.

Alternatively, since $\tilde{\alpha} = \alpha T^*$ and $\tilde{\beta} = \beta P^*$, it follows from Eq. (3.33) that $\gamma k^{-1/3} \alpha^{1/3} \beta^{2/3} = \tilde{\gamma}\tilde{\alpha}^{1/3}\tilde{\beta}^{2/3}$, which is a universal function of αT, that is,

$$\gamma k^{-1/3} \alpha^{1/3} \beta^{2/3} = \Omega(\alpha T) \quad (3.38)$$

which has been plotted for small-molecule and macromolecular liquids (Figure 3.10). The liquids include n-alkanes from methane to poly-

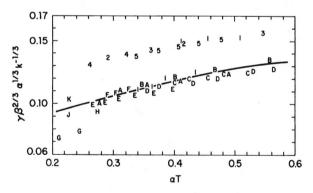

Figure 3.10. Corresponding states plot. (1) Ar; (2) O_2; (3) N_2; (4) methane; (5) propane; (A) butane; (B) pentane; (C) hexane; (D) octane; (E) hexadecane; (F) polymethylene; (G) polyisobutylene; (H) polydimethylsiloxane; (I) dimer, trimer, tetramer, and pentamer of dimethylsiloxane; (J) poly(propylene glycol); (K) poly(ethylene glycol). The solid line is drawn to the theoretical cell model. (After Ref. 10.)

Table 3.6. Prediction of Surface Tension for Some Polymers from Cohesive Energy Density at 20°C [a]

Polymer	Amorphous density, g/ml	Amorphous δ cal$^{1/2}$/cm$^{3/2}$	Surface tension, dyne/cm	
			Calc.	Exp.
Polyethylene	0.85	10.87	32.1	35.4
Polypropylene	0.85	10.25	28.5	29.3
Polyisobutylene	0.84	10.62	30.8	33.8
Polystyrene	1.05	12.36	38.7	40.7
Poly(vinyl acetate)	1.19	12.68	39.1	36.5
Poly(methyl methacrylate)	1.17	12.29	36.9	41.1
Poly(n-butyl methacrylate)	1.053	11.58	34.0	31.2
Poly(propylene oxide)	1.00	12.23	31.6	30.9
Poly(tetramethylene oxide)	0.98	11.39	33.7	31.9
Poly(ethylene terephthalate)	1.33	13.26	41.1	42.1
Poly(hexamethylene adipamide)	1.07	13.41	45.2	47.9
Polydimethylsiloxane	1.14	9.64	20.3	19.8
Poly(vinyl chloride)	1.385	12.82	38.0	41.9

[a] Calculated by Eq. (3.41).

Table 3.7. Surface Tension of Polymers by Direct Measurement on Polymer Melts and Liquids

Polymer	Molecular weight	Surface 20°C
Hydrocarbon polymers		
Polyethylene, linear (L-PE)	$M_w = 67,000$	35.7
Polyethylene, branched (B-PE)	$M_n = 7000$	35.3
	$M_n = 2000$	33.7
Polyethylene, ideal (n-alkane)	$M = \infty$	36.8
Polypropylene, atactic (PP)	-	29.4
	$M_n = 3000$	28.3
Polypropylene, isotatic and atacic	-	30.1
Polyisobutylene (PIB)	$M_n = 2700$	33.6
Polystyrene (PS)	$M_V = 44,000$	40.7
	$M_n = 9290$	39.4
	$M_n = 1680$	39.3
Poly(α-methyl styrene) (PMS)	$M_n = 3000$	38.7
Halogenated hydrocarbon polymers		
Polychloroprene (PCP)	$M_V = 30,000$	43.6
Polychlorotrifluoro-ethylene (PCTFE)	$M_n = 1280$	30.9
Polytetrafluoroethylene (PTFE)	$M = \infty$	23.9
$C_{21}F_{44}$	$M = 1088$	21.5
Vinyl polymers		
Poly(vinyl acetate) (PVAc)	$M_W = 11,000$	36.5
Methacrylate polymers		
Poly(methyl methacrylate) (PMMA)	$M_V = 3000$	41.1
Poly(ethyl methacrylate) (PEMA)	$M_V = 5200$	35.9
Poly(n-propyl methacrylate) (PnPMA)	$M_V = 8500$	33.2
Poly(n-butyl methacrylate) (PnBMA)	$M_V = 37,000$	31.2

tension, dyne/cm		$-(d\gamma/dT)$ dyne/cm-deg	Polarity,[a] x^p	References[b]
140°C	180°C			
28.8	26.5	0.057	0	1, 19, 63
27.3	24.6	0.067	0	18, 19
26.5	24.1	0.060	0	63
30.0	27.7	0.056	0	4
22.7	20.4	0.056	-	19
23.5	21.9	0.040	-	64
23.1	20.8	0.058	0.020	65, 66
25.9	23.4	0.064	-	1, 19, 38, 39, 65
32.1	29.2	0.072	0.168	17, 65-69
31.6	29.0	0.065	-	40
30.0	26.9	0.077	-	40
31.7	29.4	0.058	-	37, 65
33.2	29.8	0.086	0.108	18, 70
22.9	20.2	0.067	0.282	71
16.9	14.6	0.058	0.089	4
13.7	11.1	0.065	-	20
28.6	25.9	0.066	0.329	1
32.0	28.9	0.076	0.281	17
27.5	24.7	0.070	-	72
25.4	22.8	0.065	-	72
24.1	21.7	0.059	0.158	17

Table 3.7. (Continued)

Polymer	Molecular weight	Surface 20°C
Poly(i-butyl methacrylate) (PiBMA)	M_v = 35,000	30.9
Poly(t-butyl methacrylate) (PtBMA)	M_v = 6000	30.4
Poly(n-hexyl methacrylate) (PnHMA)	M_v = 52,000	30.0
Poly(2-ethylhexyl) methacrylate) (PEHMA)	M_v = 64,000	28.8
Acrylate polymers		
Poly(methyl acrylate) (PMA)	M_n = 25,000	41.0
Poly(ethyl acrylate) (PEA)	M_n = 28,000	37.0
Poly(n-butyl acrylate) (PnBA)	M_n = 32,000	33.7
Poly(2-ethylhexyl acrylate) (PEHA)	M_n = 34,000	30.2
Polyethers		
Poly(ethylene oxide)-diol (PEO-DO)	M = 6000	42.9
	M = 86-17,000	42.9
Poly(ethylene oxide)-dimethyl ether (PEO-DME)	M = 114	28.1 (25°C)
	M = 148	30.6 (25°C)
	M = 182	32.4 (25°C)
	M = 600	37.1 (25°C)
Poly(propylene oxide)-diol (PPO-DO)	M_n = 3000	30.9
	M = 400-4100	30.4
Poly(propylene oxide)-dimethyl ether (PPO-DME)	M = 3000	30.2 (25°C)
Poly(tetramethylene oxide)-diol (PTMO-DO)	M_n = 43,000	31.9
Polyepichlorohydrin (PECH)	M = 1500	43.2 (25°C)
Polyesters		
Poly(ethylene terephthalate) (PET)	M_n = 25,000	44.6

tension, dyne/cm		$-(d\gamma/dT)$ dyne/cm-deg	Polarity,[a] x^p	References [b]
140°C	180°C			
23.7	21.3	0.060	0.139	18
23.3	21.0	0.059	0.120	18
22.6	20.1	0.062	-	72
21.4	18.9	0.062	-	72
31.8	28.7	0.077	0.248	72
27.8	24.7	0.077	0.174	72
25.3	22.5	0.070	0.098	72
21.8	19.0	0.070	0.028	72
33.8	30.7	0.076	0.284	19, 41, 42, 44
31.1	27.2	0.098	-	41, 42, 44
-	-	-	-	42, 41
-	-	-	-	42, 41
-	-	-	-	42, 41
-	-	-	-	42, 41
21.4	18.3	0.079	-	42, 44
20.8	17.6	0.080	-	42, 44
-	-	-	-	42, 44
24.6	22.2	0.061	0.142	18, 5
-	-	-	-	44
28.3 (270°C)	26.4 (300°C)	0.065	0.221	72, 73

Table 3.7. (Continued)

Polymer	Molecular weight	Surface 20°C
Polyamides		
Poly(hexamethylene adipamide), nylon 66	M_n = 19,000	46.5
Polycaproamide, nylon 6	-	36.1 (265°C)
Poly(hexamethylene sebacamide), nylon 610	-	37.0 (265°C)
Polyundecylamide, nylon 11	-	22.6 (225°C)
Polyorganosiloxane		
Polydimethylsiloxane (PDMS)	60,000 cS	19.8
	M_n = 1274	19.9
	M_n = 607	18.8
	M_n = 310	17.6
	M_n = 162	15.7
Polydiethylsiloxane	158 cS	25.7
Polymethylphenylsiloxane	102 cS	26.1
Random copolymers		
Poly(ethylene/vinyl acetate)		
E/VAc weight ratio 75/25	M_n = 17,000	35.5
E/VAc weight ratio 82.3/17.7	-	34.1
E/VAc weight ratio 73.4/26.6	-	31.3
E/VAc weight ratio 69.1/30.9	-	30.7
E/VAc weight ratio 61.3/38.7	-	33.0
Poly(ethylene oxide/ propylene oxide)	-	
Block copolymers		
ABA block copolymers of ethylene oxide (A block) and propylene oxide (B block)		

tension, dyne/cm		$-(d\gamma/dT)$ dyne/cm-deg	Polarity,[a] x^p	References [b]
140°C	180°C			
30.2 (270°C)	28.3 (300°C)	0.065	0.344	72, 73, 74
-	-	-	-	74
-	-	-	-	74
-	-	-	-	74
14.0	12.1	0.048	0.042	18, 19, 48, 75
-	-	-	-	51
-	-	-	-	51
-	-	-	-	51
-	-	-	-	51
16.9	14.0	0.073	-	75
12.9	8.5	0.11	-	75
27.5	24.8	0.067	-	2, 37, 76
27.6	25.5	0.054	-	37, 76
26.9	25.4	0.037	-	37, 76
26.7	25.4	0.033	-	37, 76
27.4	25.5	0.047	-	37, 76
Linear additivity with mole fraction of comonomer; no surface excess (see Figure 3.7)			-	44
Pronounced surface excess behavior when the lower-energy block is sufficiently long (degree of polymerization greater than 20 to 50); see Figure 3.8			-	44

Table 3.7. (Continued)

Polymer	Molecular weight	Surface 20°C
ABA block copolymers of polyether (A block) and dimethylsiloxane (B block)		
Mixtures of homopolymers		
Mixtures of poly(ethylene oxide) and poly(propylene glycol)		

[a] The polarity ($x^p = \gamma^p/\gamma$) is calculated from the interfacial tension against polyethylene by the harmonic-mena equation, and is independent of temperature.

[b] The first reference is the source from which the data are taken. The following numbers are additional references.

ethylene over a 100°C temperature range, for polyisobutylenes over a 100°C temperature range, for polydimethylsiloxanes from oligomers to polymers at 30°C, for poly(ethylene oxide) and poly(propylene oxide) at 20°C, and for argon, nitrogen, and oxygen over a 25°C temperature range. The data separate into two distinct curves. The lower curve holds for large molecules such as n-alkanes from butane to polymethylenes and other polymers to within ±5%; the upper curve (about 30% higher) holds for small molecules such as methane, propane, argon, nitrogen, and oxygen [10,11].

Significant Structure Theory

Several analyses based on the significant structure of liquids have been given [13,58]. Amorphous polymers are considered to have gaseous and solidlike regions. Consideration of the partition function gives the surface tension as [13]

$$\gamma = \frac{2\lambda(r_e)^3}{(V/N)^2} \left(\frac{r_e}{\lambda} U_0 - 0.4465kT \right) \tag{3.39}$$

where λ is the distance between two neighboring repeat units in the same chain, r_e the collision diameter [that is, the intermolecular distance at which the intermolecular energy is zero (see Section 2.4)], V the molar volume of a repeat unit, N is Avogadro's number, and U_0 the equilibrium intermolecular energy (see Section 2.4). This equa-

tension, dyne/cm		$-(d\gamma/dT)$ dyne/cm-deg	Polarity,[a] x^p	References[b]
140°C	180°C			
Same as above; see Section 3.1.6			-	44
Pronounced surface excess behavior; see Figure 3.7.				44, 55

tion reasonably predicts the surface tensions of polymer melts over wide temperature ranges [13].

Poser and Sanchez Theory

The Helmholtz free energy of a system having a density gradient, as given by the Cahn-Hilliard theory of inhomogeneous system, is used to express the surface tension. The partition function is evaluated by a lattice fluid model. The theory reasonably predicts the surface tension and its molecular-weight dependence on small-molecule liquids and polymers [12].

Relation to Cohesive Energy Density

The surface tension of small-molecule liquids is related to the cohesive energy density by [59,60]

$$\gamma = 0.07147 \delta^2 V^{1/3} \tag{3.40}$$

where δ is the solubility parameter and V is the molar volume. Theoretical derivations have been given [61].

For polymers, V should not be identified as the molar volume of the polymer molecule as a whole, nor as that of a repeat unit. Rather, V should be the molar volume of an interacting element, which is, however, unknown a priori. An early approach is to take V as the average atomic volume [62]. A recent improvement is to let $V = M_i/\rho$, where M_i is the molecular weight of an interacting unit, which should vary with chemical structure. Regression of experimental data, however, gives an average value $M_i = 46.8$. Using this in Eq. (3.40) gives [3]

$$\gamma = 0.2575 \delta^2 \rho^{-1/3} \tag{3.41}$$

applicable to polymers (Table 3.6), where δ is in $(cal/ml)^{1/2}$ and ρ in g/ml.

3.1.8. Tabulation of Surface Tensions

Surface tensions of polymers obtained by direct measurements on melts and liquids are tabulated in Table 3.7. The values for solid polymers are obtained by linear extrapolation from the melts. The polarity, $x^p = \gamma^p/\gamma$, is calculated from the interfacial tension against polyethylene (a nonpolar polymer) by using the harmonic-mean equation, where $\gamma = \gamma^d + \gamma^p$ (see Section 3.2.4).

3.2. INTERFACIAL TENSIONS BETWEEN POLYMERS

3.2.1. Antonoff's Rule

Antonoff's rule [77] is the oldest relation linking interfacial tension to surface tensions of the two phases,

$$\gamma_{12} = \gamma_2 - \gamma_1 \quad (\gamma_2 \geq \gamma_1) \tag{3.42}$$

which is valid only when phase 1 has a zero contact angle on phase 2, and the two phases are in equilibrium vapor adsorption and mutual saturation [78,79]. The relation follows directly from the Young equation (Section 1.6.2).

The relation is, however, generally invalid to polymers. As most polymers have negligible vapor pressure, equilibrium vapor adsorption cannot occur within the time scale of experiment. Thus, for polymers, when the contact angle is zero, the Young equation becomes [3,4]

$$\gamma_{12} \leq \gamma_2 - \gamma_1 \tag{3.43}$$

which has been verified experimentally [3,4].

3.2.2. Theory of Good and Girifalco

In terms of work of adhesion W_a, the interfacial tension is given by

$$\gamma_{12} = \gamma_1 + \gamma_2 - W_a \tag{3.44}$$

Interfacial Tensions Between Polymers

which follows from Eq. (1.19). Good and Girifalco [80–83] introduced an interaction parameter ϕ, defined by

$$\phi = \frac{W_a}{(W_{c1}W_{c2})^{1/2}} \tag{3.45}$$

Combining Eqs. (3.44) and (3.45) gives

$$\gamma_{12} = \gamma_1 + \gamma_2 - 2\phi(\gamma_1\gamma_2)^{1/2} \tag{3.46}$$

which is the equation of Good and Girifalco.

To express ϕ in terms of molecular parameters, the quasi-continuum model of Section 2.5.2 is used. Using Eqs. (2.71) and (2.72) in Eq. (3.45) gives [83]

$$\phi = \phi_V \phi_A \tag{3.47}$$

where

$$\phi_V = \frac{z_{01}z_{02}}{z_{012}^2} = \frac{4V_1^{1/3}V_2^{1/3}}{(V_1^{1/3}+V_2^{1/3})^2} \tag{3.48}$$

$$\phi_A = \frac{\frac{3}{4}\alpha_1\alpha_2\left(\frac{2I_1 I_2}{I_1+I_2}\right) + \alpha_1\mu_2^2 + \alpha_2\mu_1^2 + \frac{2}{3}\left(\frac{\mu_1^2\mu_2^2}{kT}\right)}{\left[\left(\frac{3}{4}\alpha_1^2 I_1 + 2\alpha_1\mu_1^2 + \frac{2}{3}\frac{\mu_1^4}{kT}\right)\left(\frac{3}{4}\alpha_2^2 I_2 + 2\alpha_2\mu_2^2 + \frac{2}{3}\frac{\mu_2^4}{kT}\right)\right]^{1/2}} \tag{3.49}$$

where V is the molar volume of an interacting unit, which is, however, unknown for a polymer. If the volumes of interacting units in both phases are similar, as expected in many cases, $\phi_V = 1$ and $\phi = \phi_A$.

It can be shown that when the polarities of the two phases are exactly the same, ϕ_A will have the maximum value of unity. Its value decreases with increasing disparity in the polarities of the two phases. If both phases are nonpolar, then $\mu_1 = \mu_2 = 0$, and

$$\phi_A = \frac{2(I_1 I_2)^{1/2}}{I_1 + I_2} \tag{3.50}$$

which has the maximum value of unity when $I_1 = I_2$. The ϕ values between many small-molecule organic liquids and water have been calculated by Eq. (3.49), and give reasonable predictions of the interfacial tensions [80–83].

The ϕ values have also been calculated from surface and interfacial tensions, between small-molecule liquids (Table 3.8), and between polymer melts and liquids (Table 3.9). For instance, the ϕ value ranges from about 0.55 between alkane and water to 1.09 between alcohol and water (Table 3.8), and from 0.80 between polyethylene and poly(vinyl acetate) to 0.97 between poly(methyl methacrylate) and poly(n-butyl methacrulate) (Table 3.9). The ϕ has been shown to be independent of temperature [17],

$$\frac{d\phi}{dT} = 0 \tag{3.51}$$

which can readily be seen in Table 3.9.

3.2.3. Theory of Fractional Polarity

Various molecular forces are linearly additive (Sections 2.2 and 2.3). The total attraction constant consists of two terms [18,70],

$$A_{ij} = A_{ij}^d + A_{ij}^p \tag{3.52}$$

where the superscripts d and p refer to dispersion (nonpolar) and polar components, respectively. Here, the various polar interactions (including dipole energy, induction energy, and hydrogen bonding) are combined into one polar term for simplicity.

The quasi-continuum theory gives the surface tension as $\gamma = \pi q^2 A / 16 z_0^2$, which follows from Eq. (2.72). Using Eq. (3.52) for the attraction constant separates the surface tension into two terms [18,70],

$$\gamma = \gamma^d + \gamma^p \tag{3.53}$$

Similarly, the work of adhesion is separated into two terms, giving

$$\gamma_{12} = \gamma_1 + \gamma_2 - W_a^d - W_a^p \tag{3.54}$$

which follows from Eq. (1.19), and

$$W_a^k = \frac{\pi q_1 q_2 A_{12}^k}{8 z_{0,12}^2} \quad (k = d, p) \tag{3.55}$$

Table 3.8. Interaction Parameter φ for Some Liquid Pairs at 20°C

Liquid pair	γ_1, dyne/cm	γ_{12}, dyne/cm	φ
Organic liquid (subscript 1) versus water (subscript 2)			
n-Butyl alcohol	24.6	1.8	1.13
Cyclohexanol	32.7	3.9	1.04
Oleic acid	32.5	15.7	0.92
Methyl n-butyl ketone	25.0	9.6	1.03
Ethyl isovalerate	23.7	18.4	0.94
Butyronitrile	28.1	10.4	1.00
Nitrobenzene	43.9	25.7	0.81
n-Hexane	18.4	51.1	0.55
n-Octane	21.8	50.8	0.55
n-Tetradecane	25.6	52.2	0.53
Cyclohexane	25.5	50.2	0.55
Decalin	29.9	51.4	0.55
Toluene	28.5	36.1	0.71
Carbon tetrachloride	27.0	45.0	0.61
α-Bromonaphthalene	44.6	42.1	0.66
Carbon disulfide	32.3	48.4	0.58
Perfluorodibutyl ether	12.2	51.9	0.55
Perfluorotributylamine	16.8	25.6	0.95
Fluorocarbon (subscript 1) versus benzene (subscript 2)			
Perfluorodibutyl ether	12.2	5.7	0.96
Perfluorotributylamine	16.8	6.4	0.89
Fluorocarbon (subscript 1) versus n-heptane (subscript 2)			
Perfluorodibutyl ether	12.2	3.6	0.95
Perfluorotributylamine	16.8	1.6	0.96
Miscellaneous liquid (subscript 1) versus mercury (subscript 2)			
n-Hexane	18.4	378	0.62
Toluene	28.5	361	0.61
Carbon tetrachloride	27.0	359	0.64
Nitrobenzene	43.9	350	0.59
n-Butyl acetate	25.0	375	0.58
Ethyl alcohol	22.8	389	0.53
n-Octyl alcohol	27.5	352	0.70
Oleic acid	32.5	322	0.75
Water	72.8	428	0.32

Source: From Ref. 80.

Table 3.9. Interaction Parameter ϕ for Some Polymer Pairs

Polymer pair [a]	ϕ Value		
	20°C	140°C	180°C
PVAc/L-PE	0.798	0.804	0.798
PVAc/PIB	0.860	0.864	0.865
PMMA/L-PE	0.845	0.841	0.838
PnBMA/L-PE	0.896	0.903	0.906
PS/L-PE	0.893	0.905	0.907
PMMA/PS	0.962	0.974	0.976
PnBMA/PVAc	0.941	0.950	0.950
PMMA/PnBMA	0.960	0.975	0.982
PCP/B-PE	0.947	0.943	0.942
PCP/PDMS	0.958	0.944	0.937
PVAc/PDMS	0.891	0.880	0.873

[a] See Table 3.7 for abbreviations of polymer names.
Source: From Refs. 1, 17, and 18.

which follows from Eqs. (2.71) and (2.72).

The dispersion component of the attraction constant, A_{12}^d, may be approximated with the harmonic-mean relation, valid when $\alpha_1 \simeq \alpha_2$, which is true for many low-energy materials, that is,

$$A_{12}^d \simeq \frac{2A_{11}^d A_{22}^d}{A_{11}^d + A_{22}^d} \quad (3.56)$$

which follows from Eq. (2.39). Assume the geometric-mean combining rule for intermolecular distances, that is,

$$z_{0,12} = (z_{0,11} z_{0,22})^{1/2} \quad (3.57)$$

and the relation

$$\frac{q_1}{z_{0,11}} \simeq \frac{q_2}{z_{0,22}} \quad (3.58)$$

which is physically reasonable. Then, from Eq. (3.55) the dispersion component of the work of adhesion is obtained as [18,70]

$$W_a^d \simeq \frac{2W_{c11}^d W_{c22}^d}{W_{c11}^d + W_{c22}^d} \tag{3.59}$$

$$= \frac{4\gamma_1^d \gamma_2^d}{\gamma_1^d + \gamma_2^d} \tag{3.60}$$

which is the harmonic-mean approximation, preferred between low-energy materials.

On the other hand, the geometric-mean relation, valid when $I_1 \simeq I_2$, which is true between a low-energy and a high-energy material, may be used, that is,

$$A_{12}^d \simeq (A_{11}^d A_{22}^d)^{1/2} \tag{3.61}$$

which follows from Eq. (2.37). Then, from Eq. (3.55) the dispersion component of the work of adhesion is obtained as [18,70]

$$W_a^d \simeq (W_{c11}^d W_{c22}^d)^{1/2} \tag{3.62}$$

$$= 2(\gamma_1^d \gamma_2^d)^{1/2} \tag{3.63}$$

which is the geometric-mean approximation, preferred between a low-energy material and a high-energy material.

The polar component consists of dipole (p*), induction (i) and hydrogen-bonding (h) interactions. Therefore, polar surface tension may be written as

$$\gamma^p = \gamma^{p*} + \gamma^i + \gamma^h \tag{3.64}$$

Here, p* is used to designate the dipole-dipole interaction, whereas p is used to designate the combined polar interactions. The hydrogen-bonding component γ^h may be separated into acceptor and donor components, that is,

$$\gamma^h = 2\gamma^{ha} \gamma^{hd} \tag{3.65}$$

which follows from Section 2.1.4, and where γ^{ha} is the hydrogen-bond acceptor component, and γ^{hd} the hydrogen-bond donor component. The polar component of surface tension now becomes

$$\gamma^p = \gamma^{p*} + \gamma^i + 2\gamma^{ha}\gamma^{hd} \tag{3.66}$$

Using the appropriate approximations for the various polar attraction constants (Section 2.3) gives

$$W_a^p = 2(\gamma_1^{p*}\gamma_2^{p*})^{1/2} + (\gamma_1^i + \gamma_2^i) + (\gamma_1^{ha}\gamma_2^{hd} + \gamma_1^{hd}\gamma_2^{ha}) \tag{3.67}$$

which is the complete polar component of work of adhesion.

Several special cases deserve attention here. If the dipole-dipole interaction is predominant, then

$$W_a^p = 2(\gamma_1^p \gamma_2^p)^{1/2} \tag{3.68}$$

If the dipole-induced dipole interaction is predominant, then

$$W_a^p = \gamma_1^p + \gamma_2^p \tag{3.69}$$

If the hydrogen-bond interaction is predominant, then

$$W_a^p = \gamma_1^{ha}\gamma_2^{hd} + \gamma_1^{hd}\gamma_2^{ha} \tag{3.70}$$

On the other hand, a particularly useful approximation is the harmonic-mean relation [18,70]

$$W_a^p = \frac{4(\gamma_1^p \gamma_2^p)}{\gamma_1^p + \gamma_2^p} \tag{3.71}$$

which adequately represents the combined polar interactions in most cases and is valid for low-energy materials. Two particularly useful equations can be obtained by using appropriate relations for W_a^d and W_a^p in Eq. (3.54), that is, the harmonic-mean equation and the geometric-mean equation.

The harmonic-mean equation is obtained by using Eqs. (3.60) and (3.71) in Eq. (3.54),

$$\gamma_{12} = \gamma_1 + \gamma_2 - \frac{4\gamma_1^d \gamma_2^d}{\gamma_1^d + \gamma_2^d} - \frac{4\gamma_1^p \gamma_2^p}{\gamma_1^p + \gamma_2^p} \tag{3.72}$$

which is valid between low-energy materials [18,70]. Comparison of Eq. (3.72) with Eq. (3.46) gives the interaction parameter of Good and Girifalco as [70]

$$\phi = 2 \left(\frac{x_1^d x_2^d}{g_1 x_1^d + g_2 x_2^d} + \frac{x_1^p x_2^p}{g_1 x_1^p + g_2 x_2^p} \right) \tag{3.73}$$

$$g_j = \left(\frac{\gamma_j}{\gamma_i} \right)^{1/2} \tag{3.74}$$

$$x_j^d = \frac{\gamma_j^d}{\gamma_j} \tag{3.75}$$

$$x_j^p = \frac{\gamma_j^p}{\gamma_j} \tag{3.76}$$

where x_j^p is defined as the polarity, and $x_j^d + x_j^p = 1$. The ϕ has a maximum value, when the polarities of the two phases are identical, given by

$$\phi_{max} = 2(g_1 + g_2)^{-1} \tag{3.77}$$

The harmonic-mean equation is shown to predict quite well the interfacial tension between polymers and low-energy materials in Section 3.2.5 and Chapter 5.

The geometric-mean equation is obtained by using Eqs. (3.63) and (3.68) in Eq. (3.54),

$$\gamma_{12} = \gamma_1 + \gamma_2 - 2(\gamma_1^d \gamma_2^d)^{1/2} - 2(\gamma_1^p \gamma_2^p)^{1/2} \tag{3.78}$$

which is valid between a low-energy material and a high-energy material [17,18,70]. Comparison of Eq. (3.78) with Eq. (3.46) gives the interaction parameter of Good and Girifalco as [17]

$$\phi = (x_1^d x_2^d)^{1/2} + (x_1^p x_2^p)^{1/2} \tag{3.79}$$

which has a maximum value $\phi_{max} = 1$ when the polarities of the two phases are identical. Similar equations were also proposed empirically [84,85]. When the polar term is neglected, Eq. (3.78) becomes

$$\gamma_{12} = \gamma_1 + \gamma_2 - 2(\gamma_1^d \gamma_2^d)^{1/2} \tag{3.80}$$

which is applicable only when polar interaction is absent. This is, however, not true between polar polymers. Equation (3.80) is known as the Fowkes equation [86]. The geometric-mean equation is shown to be very much inferior to the harmonic-mean equation in predicting the interfacial tensions of polymers and low-energy materials in Section 3.2.5 and Chapter 5. It is, however, preferred between a low-energy material and a high-energy material.

3.2.4. Polarity of Polymers

The polarity of a material can be determined from its interfacial tension against a nonpolar material. In this case, the polar interaction term vanishes. The dispersion component of the polar polymer γ^d is then calculated from the surface and the interfacial tensions by the harmonic-mean equation or the geometric-mean equation [18,70]. The polar component γ^p is obtained as $\gamma^p = \gamma - \gamma^d$. The harmonic-mena equation is preferred between low-energy materials (polymers, organic liquids, water, etc.), whereas the geometric-mean equation is preferred between a low-energy material and a high-energy material (mercury, silica, metal oxides, etc.). Polyethylene is a convenient choice as the nonpolar polymer for the determination of polymer polarities. The polarities of many polymers have thus been determined (Table 3.10).

The polarity is found to be independent of temperature [1,70],

$$\frac{dx^p}{dT} = 0 \tag{3.81}$$

which follows from Eq. (3.51) and can be verified readily experimentally.

The polarity may also be defined in terms of the cohesive energy density. The quasicontinuum theory gives the cohesive energy ΔE_c as [72]

$$\Delta E_c = \delta^2 V = \frac{NcA}{4z_0^6} \tag{3.82}$$

where δ is the solubility parameter, V the molar volume, N is Avogadro's number, c the coordination number of the interacting unit, and A the attraction constant. Equation (3.82) can be derived from Section 2.5. Linear additivity of molecular energies gives

$$\delta^2 = \delta_d^2 + \delta_p^2 \tag{3.83}$$

Table 3.10. Surface Tension Components and Polarity of Some Polymers from Interfacial Tension Against Polyethylene at 140°C

Polymer	γ, dyne/cm	Components of γ (dyne/cm) and polarity					
		Harmonic-mean eq.			Geometric-mean eq.		
		γ^d	γ^p	x^p	γ^d	γ^p	x^p
Polychloroprene	33.2	29.6	3.6	0.11	29.5	3.7	0.11
Polystyrene	32.1	26.7	5.4	0.17	26.3	5.8	0.18
Poly(methyl methacrylate)	32.0	23.0	9.0	0.28	22.7	9.3	0.29
Poly(n-butyl methacrylate)	24.1	20.3	3.8	0.16	19.7	4.4	0.18
Poly(i-butyl methacrylate)	23.7	20.4	3.3	0.14	20.0	3.7	0.16
Poly(t-butyl methacrylate)	23.3	20.5	2.8	0.12	19.2	4.1	0.18
Poly(vinyl acetate)	28.6	19.2	9.4	0.33	18.4	10.2	0.36
Poly(ethylene oxide)	33.8	24.2	9.6	0.28	24.5	9.3	0.28
Poly(tetramethylene oxide)	24.6	21.1	3.5	0.14	20.8	3.8	0.15
Polydimethylsiloxane	14.1	13.5	0.6	0.04	12.1	2.0	0.14
Poly(methyl acrylate)	31.8	23.9	7.9	0.25	23.6	8.2	0.26
Poly(ethyl acrylate)	27.8	23.0	4.8	0.17	22.6	5.2	0.19
Poly(n-butyl acrylate)	25.3	22.8	2.5	0.10	22.5	2.8	0.11
Poly(2-ethylhexyl acrylate)	21.8	21.2	0.6	0.03	20.6	1.2	0.05

which follows from Eq. (3.52). Since $x^d = A^d/A = \gamma^d/\gamma$ and $x^p = A^p/A = \gamma^p/\gamma$,

$$x^d = \frac{\Delta E_c^d}{\Delta E_c} = \left(\frac{\delta_d}{\delta}\right)^2 \tag{3.84}$$

$$x^p = \frac{\Delta E_c^p}{\Delta E_c} = \left(\frac{\delta_p}{\delta}\right)^2 \tag{3.85}$$

which have been proposed before by Hansen [87]. Values of δ_d and δ_p have been tabulated [28,87,88]. The polarity values obtained from the interfacial tension (by the harmonic-mean equation), from solubility parameter, and from wettability (by harmonic-mean equation) agree quite well with one another (Table 3.11).

3.2.5. Prediction of Interfacial Tension

The interfacial tension can be calculated from the surface tensions and their components of the two phases by the harmonic-mean equation or the geometric-mean equation. The calculated and measured values of interfacial tensions between polymer melts and liquids are compared in Table 3.12. The surface tensions and polarities for polymers as listed in Table 3.10 are used in the calculations. It can be seen that the harmonic-mean equation predicts the interfacial tensions remarkably well. In contrast, the geometric-mean equation is much inferior [3,4, 17,18,70]. This has been further confirmed recently [89]. The harmonic-mean equation also predicts the surface tensions of solid polymers from contact angles remarkably well, whereas the geometric-mean equation is inferior (discussed in Chapter 5).

3.2.6. Effect of Polarity on Interfacial Tension

The magnitude of interfacial tension is determined primarily by the disparity in the polarities of the two phases. The greater the polarity difference, the greater will be the interfacial tension. If polar interactions are neglected, the predicted interfacial tension will always be negligibly small. For instance, the interfacial tension between poly(vinyl acetate) and polyethylene at 140°C is measured to be as high as 11.3 dyne/cm [1]. The harmonic-mean equation shows that the high interfacial tension arises from the high polarity of poly(vinyl acetate), $x^p = 0.33$, and the low polarity of polyethylene, $x^p = 0$. If both polymers are assumed to be nonpolar, the calculated interfacial

Table 3.11. Comparison of Polarity Values Obtained from Interfacial Tension, Solubility Parameter, and Wettability

Polymer	From interfacial tension by harmonic-mean equation	Polarity value, x^p	
		From solubility parameter	From contact angles by harmonic-mean equation
Poly(vinyl acetate)	0.33	0.33	–
Poly(methyl methacrylate)	0.28	0.25	0.25
Poly(vinyl chloride)	–	0.11	0.15
Polystyrene	0.17	0.14	0.10
Poly(n-butyl methacrylate)	0.16	0.16	–

Table 3.12. Comparison of Measured and Calculated Interfacial Tensions Between Polymers at 140°C

	Interfacial tension, dyne/cm		
		Calculated by:	
Polymer pair [a]	Measured [b]	Harmonic-mean eq.	Geometric-mean eq.
PEO/PDMS	9.9	10.7	4.8
PTMO/PDMS	6.3	3.8	1.5
PVAc/PDMS	7.4	8.5	3.6
PCP/PDMS	6.5	8.2	4.1
PnBMA/PDMS	3.8	3.7	1.4
PtBMA/PDMS	3.3	2.9	1.2
PCP/PnBMA	1.6	1.7	0.9
PCP/PS	0.5	0.5	0.4
PVAc/PS	3.7	2.3	1.2
PVAc/PnBMA	2.9	2.4	1.2
PMMA/PnBMA	1.9	2.2	1.0
PMMA/PtBMA	2.3	3.4	1.6
PMMA/PS	1.7	1.2	0.5
PEO/PTMO	3.9	3.1	1.4
PVAc/PTMO	4.6	2.8	1.6
PMA/PEA	1.4	0.8	0.4
PMA/PnBA	3.1	2.9	1.6
PMA/PEHA	5.8	6.4	3.3
PEA/PnBA	1.4	0.8	0.5
PEA/PEHA	3.3	3.3	1.5
PnBA/PEHA	1.2	1.2	0.5

[a] See Table 3.7 for abbreviations of polymer names.
[b] The measured interfacial tensions are from Table 3.20.

tension is nearly zero. Thus, contrary to popular thinking, the dispersion interaction, in fact, plays a minor part in determining the magnitude of interfacial tension. The polar interactions are all important.

3.2.7. Contact Angles and Spreading Coefficients

Several interfacial energetic functions are tabulated in Tables 3.13 and 3.14. The contact angle θ_{ij} means phase i on phase j, calculated by

$$\cos \theta_{ij} = \frac{\gamma_j - \gamma_{ij}}{\gamma_i} \qquad (3.86)$$

Table 3.13. Interfacial Energetic Functions for Some Polymer Pairs at 140°C [a]

Polymer pairs	Interfacial tension, γ_{12}, dyne/cm	Work of adhesion, W_a, erg/cm^2	Work of cohesion W_{c1} erg/cm^2	Work of cohesion W_{c2} erg/cm^2	Contact angle θ_{12} degrees	Contact angle θ_{21} degrees	Spreading coefficient λ_{12} erg/cm^2	Spreading coefficient λ_{21} erg/cm^2
Polar/nonpolar pairs								
PCP/B-PE	3.7	56.8	66.4	54.6	44	0	-9.6	+2.2
PMMA/L-PE	9.7	51.1	64.0	57.6	53	39	-12.9	-6.5
PnBMA/L-PE	5.3	47.6	48.2	57.6	13	51	-0.6	-10.0
PS/L-PE	5.9	55.0	64.2	57.6	45	24	-9.2	-2.6
PVAc/L-PE	11.3	46.2	57.2	57.6	52	53	-11.1	-11.5
Polar/polar pairs								
PCP/PnBMA	1.6	55.7	66.4	48.2	47	0	-10.7	+7.5
PCP/PDMS	6.5	40.8	66.4	28.2	76	0	-25.6	+12.6
PCP/PS	0.5	64.8	66.4	64.2	18	0	-1.6	+0.6
PMMA/PnBMA	1.9	54.2	64.0	48.2	44	0	-9.8	+6.0
PMMA/PS	1.7	62.4	64.0	64.2	18	19	-1.6	-1.8
PVAc/PnBMA	2.9	49.8	57.2	48.2	42	0	-7.4	+1.6

[a] See Table 3.7 for abbreviations of polymer names. The subscript 12 means phase 1 on phase 2; the subscript 21 means phase 2 on phase 1.
Source: From Refs. 1, 17, 18, and 70.

Table 3.14. Interfacial Energetic Functions for Some Polymer Pairs at Various Temperatures

Temperature, °C	γ_{12}, dyne/cm	Work of cohesion erg/cm²		Work of adhesion, W_a, erg/cm²	Contact angle, deg		Spreading coefficient,[a] erg/cm²	
		W_{c1}	W_{c2}		θ_{12}	θ_{21}	λ_{12}	λ_{21}
Poly(methyl methacrylate)-polyethylene								
20	11.8	82.2	71.4	64.9	55	35.5	-17.3	-6.5
140	9.7	64.0	57.6	51.1	53	39	-12.9	-6.5
180	9.0	57.8	53.0	46.4	53	40	-11.4	-6.6
Poly(n-butyl methacrylate)-polyethylene								
20	7.1	62.4	71.4	59.8	22	46	-2.6	-11.6
140	5.3	48.2	57.6	47.6	14	49	-0.6	-10.0
180	4.7	43.4	53.0	43.5	0	49	+0.1	-9.5
Polystyrene-polyethylene								
20	8.3	81.4	71.4	68.1	47	24	-13.3	-3.3
140	5.9	64.2	57.6	55.0	45	24	-9.2	-2.6
180	5.1	58.4	53.0	50.6	43	24	-7.8	-2.4

Interfacial Tensions Between Polymers

Poly(methyl methacrylate)-polystyrene								
20	3.2	82.2	71.4	78.6	21	24	-3.6	-2.8
140	1.6	64.0	57.6	62.4	18	19	-1.8	-1.6
180	1.1	57.8	53.0	56.9	17	19	-1.5	-0.9
Poly(n-butyl methacrylate)-poly(vinyl acetate)								
20	4.2	62.4	73.0	63.5	0	42	+1.1	-9.5
140	2.9	48.2	57.2	49.8	0	42	+1.6	-7.4
180	2.5	43.4	51.8	45.1	0	42	+1.7	-6.7
Poly(methylacrylate)-poly(n-butyl methacrylate)								
20	3.4	82.2	62.4	68.9	48	0	-13.3	+6.5
140	1.9	64.0	48.2	54.2	46	0	-9.8	+6.0
180	1.4	57.8	43.4	49.2	45	0	-8.6	+5.8

[a] The subscript ij means phase i on phase j.

Source: From Refs. 1, 17, 18, and 70.

where the substrate phase j is assumed nondeformable. Similarly, the spreading coefficient λ_{ij} means phase i spreading on phase j, calculated by

$$\lambda_{ij} = \gamma_j - \gamma_i - \gamma_{ij} \tag{3.87}$$

The contact angle θ_{12} and the spreading coefficient λ_{12} for phase 1 on phase 2 are generally different from those for phase 2 on phase 1, that is, θ_{21} and λ_{21}. The polymer having lower surface tension may not necessarily exhibit a zero contact angle on the polymer having a higher-surface tension. If polymer 1 spreads (exhibits a zero contact angle) on polymer 2, polymer 2 will not spread on polymer 1.

Interfacial tension plays a key role in determining the wettability. For instance, neither poly(methyl methacrylate) nor polystyrene will spread on the other, because the interfacial tension between the two (1.7 dyne/cm at 140°C) is greater than the surface tension difference (0.1 dyne/cm at 140°C).

The polarity plays a key role in determining the interfacial tension, as pointed out in Section 3.2.6. When the polarities are very different, the work of adhesion can be smaller than the work of cohesion of either phase.

The contact angle can either decrease or increase with temperature. Using the equation of Good and Girifalco in the Young equation gives

$$\cos \theta_{12} = 2\phi \left(\frac{\gamma_2}{\gamma_1}\right)^{1/2} - 1 \tag{3.88}$$

Differentiating Eq. (3.88) with respect to temperature gives [17]

$$\frac{d \cos \theta_{12}}{dT} = \frac{\phi}{\gamma_1 \gamma_2}\left(\frac{d\gamma_2}{dT} - \frac{\gamma_2}{\gamma_1}\frac{d\gamma_1}{dT}\right) \tag{3.89}$$

which indicates that $d \cos \theta_{12}/dT$ can be either positive or negative, depending on the sign of the bracketed quantity on the right-hand side. Since $d\gamma/dT$ is usually small, $d \cos \theta/dT$ is accordingly small.

3.2.8. Temperature Dependence

The temperature coefficient of interfacial tension $(d\gamma_{12}/dT)$ between polymers is typically about -0.01 dyne/cm (Figure 3.11), much smaller than that of surface tension. Differentiation of the equation of Good and Girifalco gives [4]

Interfacial Tensions Between Polymers

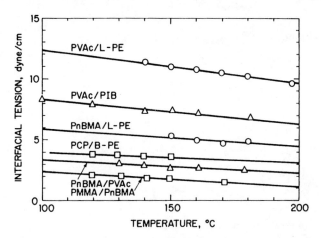

Figure 3.11. Interfacial tension between polymers versus temperature. (After Refs. 1, 3, 4, and 17.)

$$\frac{d\gamma_{12}}{dT} = \frac{d\gamma_1}{dT} + \frac{d\gamma_2}{dT} - \phi \left(g_1 \frac{d\gamma_2}{dT} + g_2 \frac{d\gamma_1}{dT} \right) \quad (3.90)$$

in which we have used $d\phi/dT = 0$, Eq. (3.51). $d\gamma_{12}/dT$ can then be calculated from $d\gamma_1/dT$ and $d\gamma_2/dT$. Good agreement between the calculated and the measured values can be seen in Table 3.15.

Table 3.15. Comparison of Measured and Calculated $-(d\gamma_{12}/dT)$ Values

	$-(d\gamma_{12}/dT)$, dyne/cm-deg	
Polymer pairs	Measured	Calc., Eq. (3.90)
Polyethylene-poly(vinyl acetate)	0.027	0.025
Polyethylene-poly(methyl methacrylate)	0.018	0.022
Polyisobutylene-poly(vinyl acetate)	0.020	0.018
Poly(n-butyl methacrylate)-poly(vinyl acetate)	0.010	0.006

Source: From Ref. 4.

3.2.9. Molecular-Weight Dependence

Using Eq. (3.8) in the equation of Good and Girifalco gives

$$\gamma_{12} = k_0 - \frac{k_1}{M_1^{2/3}} - \frac{k_2}{M_2^{2/3}} \qquad (3.91)$$

where

$$k_0 = \gamma_1 + \gamma_2 - 2\phi \left(\gamma_1 - \frac{k_1}{M_1^{2/3}}\right)^{1/2} \left(\gamma_2 - \frac{k_2}{M_2^{2/3}}\right)^{1/2} \qquad (3.92)$$

Thus, k_0 is strictly dependent on molecular weight. However, the functional property of k_0 is such that it is practically independent of molecular weight [90]. Thus, if the molecular weight of phase 1 is kept constant and that of phase 2 is varied, a plot of γ_{12} versus $M_2^{-2/3}$ will be linear. Such plots between n-alkanes and a fluorocarbon ($C_{12.5}F_{27}$) and between polydimethylsiloxanes and the same fluorocarbon are shown in Figure. 3.12.

Two polymers will become compatible when $\gamma_{12} \leq 0$. Thus, the condition for compatibility can be given by [90]

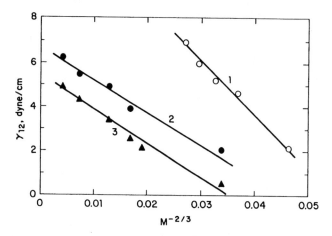

Figure 3.12. Interfacial tension (24°C) between n-alkanes and $C_{12.5}F_{27}$ (curve 1), and between polydimethylsiloxane and $C_{12.5}F_{27}$ (curve 2), or C_8F_{18} (curve 3) plotted against $M^{-2/3}$, where M is the molecular weight of the n-alkanes or the dimethylsiloxanes. (After Ref. 90.)

Interfacial Tensions Between Polymers

$$\frac{k_1}{M_1^{2/3}} + \frac{k_2}{M_2^{2/3}} \leq k_0 \tag{3.93}$$

The compatibility limits for homologous series can thus be estimated. Some examples are given in Table 3.16.

3.2.10. Mean-Field Theory of Polymer Interface

Helfand and coworkers [5–8] considered that polymer segments interdiffuse to lower the free energy and form the interface. The statistical segment length b_j for polymer j is defined as

$$r_j b_j^2 = \langle R^2 \rangle_j \tag{3.94}$$

where r_j is the degree of polymerization and $\langle R^2 \rangle_j$ is the mean-square end-to-end distance of component j. The reduced number density $\tilde{\rho}_j$ of component j is defined as

$$\tilde{\rho}_j = \frac{\rho_j}{\rho_{j0}} \tag{3.95}$$

where ρ_j is the number density and ρ_{j0} is that of pure component j. By solving a diffusion equation using mean-field approximation, the interfacial profile is obtained as

$$x = \frac{1}{2\alpha^{1/2}} \left(B_1 \ln\left\{ \frac{B_1 - [B_1^2(1-\tilde{\rho}_1) + B_2^2 \tilde{\rho}_1]^{1/2}}{B_1 + [B_1^2(1-\tilde{\rho}_1) + B_2^2 \tilde{\rho}_1]^{1/2}} \right. \right.$$

$$\left. \times \frac{B_1 + (B_1^2/2 + B_2^2/2)^{1/2}}{B_1 - (B_1^2/2 + B_2^2/2)^{1/2}} \right\}$$

$$- \left(\frac{1}{2}\right) B_2 \ln\left[\frac{[B_1^2(1-\tilde{\rho}_1) + B_2^2 \tilde{\rho}_1]^{1/2} - B_2}{[B_1^2(1-\tilde{\rho}_1) + B_2^2 \tilde{\rho}_1]^{1/2} + B_2} \right.$$

$$\left. \left. \times \frac{(B_1^2/2 + B_2^2/2)^{1/2} + B_2}{(B_1^2/2 + B_2^2/2)^{1/2} - B_2} \right] \right) \tag{3.96}$$

Table 3.16. Miscibility Limit Estimated by Eq. (3.93)

System	Estimated miscibility limit	k_0, dyne/cm
Perfluoroalkane-alkane	C_6F_{14}-$C_{6.5}H_{15}$	18.3
Perfluoroalkane-dimehtylsiloxane	C_7F_{16}-hexamethyldisiloxane	11.0
Perfluoroalkane-polyisobutylene	$C_{10.5}F_{23}$-isooctane	18.0
Polyisobutylene-dimethylsiloxane	($\overline{M}_n$ = 1250) Dodecamethylpentasiloxane	5.8

Source: From Ref. 90.

Interfacial Tensions Between Polymers

where x is the distance perpendicular to the interfacial boundary with $x = 0$ at $\tilde{\rho}_1 = \tilde{\rho}_2 = 1/2$, and the parameter B is defined as

$$B_j = \frac{1}{6} \rho_{0j} b_j^2 \tag{3.97}$$

The parameter α is related to the Flory-Huggins χ parameter by

$$\alpha = \rho_0 \chi \tag{3.98}$$

which can be estimated by using the Hildebrand-Scott relation

$$\alpha = \frac{1}{kT} (\delta_1 - \delta_2)^2 \tag{3.99}$$

The interfacial density profile between polyethylene and poly(n-butyl methacrylate) is calculated and shown in Figure 3.13.

A characteristic interfacial thickness is defined as

$$a_I = \left(\frac{dx}{d\tilde{\rho}_j}\right)_{\tilde{\rho} = 1/2}$$

$$= 2\left(\frac{B_1^2 + B_2^2}{2\alpha}\right)^{1/2} \tag{3.100}$$

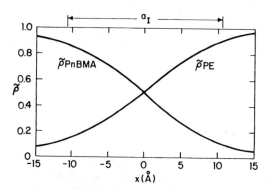

Figure 3.13. Density profile across the interface and interfacial thickness between polyethylene and poly(n-butyl methacrylate) at 140°C as calculated by Eq. (3.96) according to Helfand's mean field theory. (After Ref. 6.)

and the interfacial tension is given by

$$\gamma_{12} = \frac{2}{3} kT \alpha^{1/2} \frac{B_1^3 - B_2^3}{B_1^2 - B_2^2} \qquad (3.101)$$

which has been used to calculate the interfacial tension between some polymer pairs at 140°C. The parameters used in the calculation are listed in Table 3.17. The calculated and the measured interfacial tensions agree quite well, as listed in Table 3.18. Interfacial thicknesses are also calculated and listed, which are of the order of 10–100 Å.

Eliminating the α in Eqs. (3.100) and (3.101) gives a relation between the interfacial tension and the interfacial thickness, that is,

$$\gamma_{12} = \frac{(2)^{3/2}}{3} kT \frac{(B_1^3 - B_2^3)(B_1^2 + B_2^2)^{1/2}}{B_1^2 - B_2^2} a_I^{-1} \qquad (3.102)$$

Since the B parameters do not vary widely among polymers, the interfacial tension should be approximately inversely proportional to the interfacial thickness. Figure 3.14 shows a log-log plot of the inter-

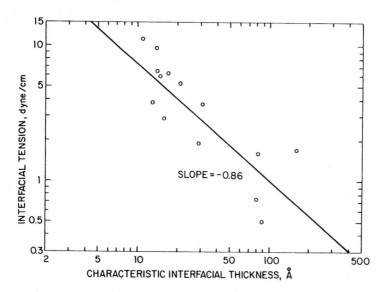

Figure 3.14. Relation between interfacial tension (measured) and interfacial thickness (calculated).

Table 3.17. Parameters Used in the Mean-Field Theory Calculation of Interfacial Tension Between Polymers

Polymer	δ, $(\text{cal/ml})^{1/2}$	Density, mole/ml	$\langle R^2 \rangle^{1/2}/M^{1/2} \times 10^9$, $\text{cm/g}^{1/2}$	b, Å	$B \times 10^9$, $\text{cm}^{-1/2}$
Polystyrene	9.0	0.0102	6.5	6.6	2.7
Poly(methyl methacrylate)	9.1	0.0116	6.4	6.4	2.8
Poly(n-butyl methacrylate)	8.6	0.0073	5.1	6.1	2.1
Poly(vinyl acetate)	9.55	0.0139	7.1	6.5	3.2
Polyethylene	7.8	0.0305	9.5	5.0	3.6
Polychloroprene	8.8	0.0140	7.5	7.1	3.4
Poly(tetramethylene oxide)	8.6	0.0135	9.6	8.1	3.9
Polydimethylsiloxane	7.5	0.0132	6.7	5.8	2.7

Source: From Ref. 6.

Table 3.18. Comparison Between Measured and Mean-Field Theory Calculated Interfacial Tensions Between Polymers at 140°C

Polymer pairs	Interfacial tension, dyne/cm		Interfacial thickness [b] a_I, Å
	Measured [a]	Calc., Eq. (3.101) [b]	
PE/PS	5.9	4.7	15
PE/PVAc	11.3	7.4	11
PE/PMMA	9.7	5.1	14
PE/PnBMA	5.3	2.8	21
PS/PCP	0.5	0.7	88
PS/PVAc	3.7	1.9	31
PS/PMMA	1.7	0.3	160
PDMS/PVAc	7.4	7.8	8
PDMS/PCP	6.5	4.9	14
PDMS/PnBMA	3.8	3.3	13
PDMS/PTMO	6.3	4.4	17
PnBMA/PVAc	2.9	3.1	16
PnBMA/PCP	1.6	0.7	82
PnBMA/PMMA	1.9	1.5	29

[a] Measured interfacial tensions from Refs. 1, 4, 17, 18, and 70.
[b] Calculated interfacial tensions and interfacial thicknesses from Ref. 6.

Interfacial Tensions Between Polymers

facial tension (measured at 140°C) versus the (calculated) interfacial thickness. Least squares give a straight line with a slope of -0.86 and the relation

$$\gamma_{12} = 55 a_I^{-0.86} \tag{3.103}$$

where the interfacial tension γ_{12} is in dyne/cm and the interfacial thickness a_I is in Å. Alternatively, a straight line with a slope of -1 may be drawn, giving $\gamma_{12} = 102 a_I^{-1}$, which represents the data about as well as does Eq. (3.103). Either of these equations may be used to estimate the interfacial tension from interfacial thickness which can be determined by electron microscopy.

3.2.11. Lattice Theory of Polymer Interface

Several lattice analyses of polymer interfaces have been given [8,9]. For infinite molecular weight, Helfand [8] obtained

$$\gamma_{12} = \frac{kT}{a} (m\chi)^{1/2} \tag{3.104}$$

and a characteristic interfacial thickness a_I as

$$a_I = 2 \left(\frac{m}{\chi} \right)^{1/2} \tag{3.105}$$

where a is the cross-sectional area of a lattice cell and m a lattice constant defined such that the number of nearest neighbors in the same layer parallel to the interface is $z(1 - 2m)$ and in each of the adjacent layers is zm, where z is the number of nearest neighbors of a lattice cell. This lattice theory is consistent with the mean field theory in that both predict $\gamma_{12} \propto \chi^{1/2}$ and $a_I \propto \chi^{-1/2}$.

On the other hand, for infinite molecular weight, Roe [9] obtained

$$\gamma_{12} = \frac{4}{3} (2)^{-1/4} \frac{kT}{a} m^{1/2} \chi^{3/4} \tag{3.106}$$

and

$$a_I = 2(2)^{3/4} m^{1/2} b \chi^{-1/4} \tag{3.107}$$

where b is the distance of separation between adjacent lattice layers. Equations (3.106) and (3.107) predict that $\gamma_{12} \propto \chi^{3/4}$ and $a_I \propto \chi^{-1/4}$, which are different from Helfand's results. Experimental verification of the lattice theories are difficult, since the parameters a, b, and m are unknown a priori.

The variations of interfacial tension and interfacial thickness with chain length r (degree of polymerization) are calculated for the case $r_1 = r_2$ in terms of Roe's analysis (Figure 3.15). At the critical chain length for phase separation ($r = 2/\chi$), the interfacial tension increases rapidly and reaches a plateau, while the interfacial thickness decreases rapidly and reaches a flat valley, consistent with expectation.

3.2.12. Classical Thermodynamic Analysis

Kammer [91] analyzed polymer interfaces by classical thermodynamics and also by statistical thermodynamics. The classical thermodynamic analysis gives the interfacial composition as

$$x_2 = \frac{d\gamma_{12}/dT + d\gamma_1/dT}{d\gamma_1/dT + (1/2)(d\gamma_2/dT)} \tag{3.108}$$

where x_2 is the mole fraction of component 2 in the interfacial zone. The interfacial thickness is obtained as

$$a_I = \frac{V}{RT}\left(\frac{\gamma_{12} - \gamma_2}{r\chi\phi_1^2 + \ln \phi_2}\right) \tag{3.109}$$

where ϕ_j is the volume fraction of component j in the interfacial zone. Calculated interfacial composition and thickness are reasonable.

3.2.13. Effect of Additives

Interfacial tensions between polymers can be reduced with additives. Block and graft copolymers having segments similar to those of the two bulk polymers are particularly effective [92]. About 1–2% by weight of the additive is sufficient to achieve the maximum reduction of interfacial tension (Figure 3.16). The effects of various block copolymers and homopolymers containing functional groups on the interfacial tensions between a polydimethylsiloxane and a poly(oxyethylene-oxypropylene) are given in Table 3.19 [93].

Interfacial Tensions Between Polymers

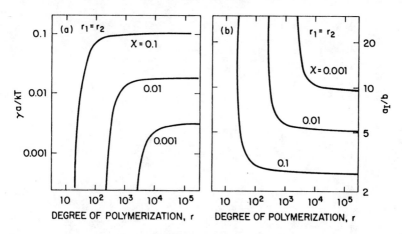

Figure 3.15. (a) Interfacial tension (in units of $\gamma a/kT$) versus degree of polymerization r. Arrows indicate the values of r ($= 2/\chi$) at incipient phase separation; (b) interfacial thickness (in units of b) versus degree of polymerization r. Note that incipient phase separation occurs at $r = 2/\chi$. (After Ref. 9.)

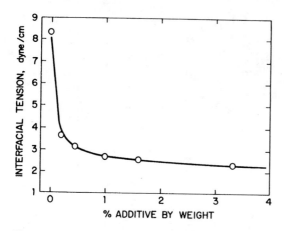

Figure 3.16. Effect of addition of poly(dimethylsiloxane-oxyethylene) on the interfacial tension between polydimethylsiloxane (10,300 cP) and a copolyester fluid (Ucon 75, 11,000 cP) at 25°C. (After Ref. 93.)

Table 3.19. Effect of Additives on Interfacial Tension Between Polydimethylsiloxane (DC-200, 100,000 cS) and Poly(oxyethylene-oxypropylene) (Ucon 75H-30,000)

Additives at 2% by weight	Interfacial tension at 25°C, dyne/cm	Percent reduction obtained with additive
None	8.3	–
60/40 Polydimethylsiloxane-polyoxyethylene copolymer	2.3	72
25/75 Polydimethylsiloxane-polyoxyethylene copolymer	3.0	64
25/75 Polydimethylsiloxane-polyoxypropylene copolymer	4.1	51
Polydimethylsiloxane with 10% carboxyl groups on alkyl side chains	3.1	63
Polydimethylsiloxane with 20% carboxyl groups on alkyl side chains	3.6	57
Polydimethylsiloxane with carboxyl end groups	4.2	49
Polydimethylsiloxane with hydroxyl end groups	8.1	0
Polydimethylsiloxane with 1% amino groups on alkyl side chains	6.0	28
Polydimethylsiloxane with 6% amino groups on alkyl side chains	6.8	18

Source: From Ref. 93.

3.2.14. Interfacial Tensions Between Demixed Polymer Solutions

Interfacial tensions between demixed polymer solutions are usually quite low, below about 10^{-2} dyne/cm [94,95]. The interfacial tensions between diethyl oxalate solutions of poly(vinyl acetate) and chlorinated poly(vinyl chloride) are given in Figure 3.17. The interfacial tensions between demixed solutions of monodisperse polystyrenes in methylcyclohexane have also been reported [95].

Several theoretical analyses have been given [9,96,97]. Only qualitative agreement with experimental result is achieved. Vrij [96] gave the interfacial tension as

$$\gamma_{12} = (6)^{-1/2} (\overline{s^2})^{1/2} Q \phi_p \sigma_r \tag{3.110}$$

where $(\overline{s^2})^{1/2}$ is the radius of gyration, ϕ_p the total volume fraction of the polymers, Q a heat of mixing parameter, and σ_r is a function of the critical solution temperature. On the other hand, Nose [97] gave

$$\gamma_{12} = \sigma_0 \left(1 - \frac{T}{T_c}\right)^{3/2} \tag{3.111}$$

where T_c is the critical solution temperature and σ_0 is given by

$$\sigma_0 = \frac{2}{3} \phi_c^{1/2} kT_c \langle R^2 \rangle_0^{1/2} r^{-1/2} v^{-1} \tag{3.112}$$

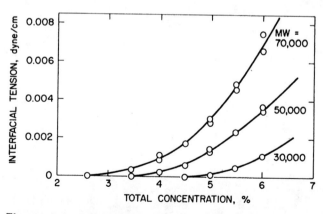

Figure 3.17. Interfacial tension between demixed solutions of poly-(vinyl acetate) and chlorinated poly(vinyl chloride) at a weight ratio of 1:2 in diethyl oxalate. (After Ref. 94.)

Table 3.20. Interfacial Tension Between Polymers (by Direct Measurements on Polymer Melts and Liquids)

	Interfacial tension,[b] dyne/cm				$-(d\gamma/dT)$ dyne/cm-deg	References[c]
Polymer pairs[a]	20°C	100°C	140°C	180°C		
Polyethylene vs. others						
PE/PP	–	–	1.1	–	–	65
L-PE/PS	8.3	6.7	5.9	5.1	0.020	17, 65
B-PE/PCP	4.6	4.0	3.7	3.4	0.0075	18
L-PE/PVAc	14.6	12.4	11.3	10.2	0.027	1, 5, 66
B-PE/PEVAc (25% VAc)	2.0	1.6	1.4	1.2	0.005	2, 66
L-PE/PMMA	11.8	10.4	9.7	9.0	0.018	17
L-PE/PnBMA	7.1	5.9	5.3	4.7	0.015	17
B-PE/PiBMA	5.5	4.7	4.3	3.9	0.010	18
B-PE/PtBMA	5.9	5.2	4.8	4.4	0.009	18
L-PE/PMA	10.6	9.1	8.4	7.7	0.018	72
L-PE/PEA	7.5	6.2	5.6	4.9	0.016	72
L-PE/PnBA	5.0	3.9	3.3	2.7	0.014	72
L-PE/PEHA	3.1	2.3	1.9	1.5	0.010	72
B-PE/PEO	11.6	10.3	9.7	9.1	0.016	2
B-PE/PTMO	5.1	4.5	4.2	3.9	0.007	18, 2
B-PE/PDMS	5.3	5.2	5.1	5.0	0.002	18, 2, 98
Polypropylene vs. others						
PP/PS	–	–	5.1	–	–	65
PP/PDMS	3.2	3.0	2.9	2.8	0.002	65
Polyisobutylene vs. others						
PIB/PVAc	9.9	8.3	7.5	6.7	0.020	1
PIB/PDMS	4.9	4.4	4.2	4.0	0.006	98

Interfacial Tensions Between Polymers

Polystyrene vs. others						
PS/PCP	0.7	0.6	0.5	0.4	0.0014	18
PS/PVAc	4.2	3.9	3.7	3.5	0.0044	18
PS/PMMA	3.2	2.2	1.6	1.1	0.013	17
PS/PEVAc (38.7% VAc)	–	–	5.6	–	–	66
PS/PDMS	6.1	6.1	6.1	6.1	0.000	98
Polychloroprene vs. others						
PCP/PnBMA	2.2	1.8	1.6	1.4	0.0047	18
PCP/PDMS	7.1	6.7	6.5	6.3	0.0050	18
Poly(vinyl acetate) vs. others						
PVAc/PnBMA	4.2	3.3	2.9	2.5	0.010	17
PVAc/PDMS	8.4	7.7	7.4	7.1	0.0081	18, 2, 98
PVAc/PTMO	5.5	4.9	4.6	4.3	0.0081	2
PVAc/PEVAc (25% VAc)	6.3	6.0	5.8	5.6	0.0043	2
Poly(methyl methacrylate) vs. others						
PMMA/PnBMA	3.4	2.4	2.0	1.5	0.012	17
PMMA/PtBMA	3.0	2.5	2.3	2.1	0.006	18
Polyacrylate vs. polyacrylate						
PMA/PEA	2.4	1.7	1.4	1.1	0.008	72
PMA/PnBA	4.0	3.4	3.1	2.8	0.008	72
PMA/PEHA	6.6	6.1	5.8	5.5	0.007	72
PEA/PnBA	2.1	1.6	1.4	1.1	0.006	72
PEA/PEHA	3.9	3.5	3.3	3.1	0.005	72
PnBA/PEHA	1.8	1.4	1.2	1.0	0.005	72

Table 3.20. (Continued)

Polymer pairs [a]	Interfacial tension,[b] dyne/cm				$-(d\gamma/dT)$ dyne/cm-deg	References [c]
	20°C	100°C	140°C	180°C		
Polydimethylsiloxane vs. others						
PDMS/PnBMA	4.2	3.9	3.8	3.6	0.0037	18
PDMS/PtBMA	3.6	3.4	3.3	3.2	0.0025	18
PDMS/PEO	10.8	10.2	9.9	9.6	0.0078	2
PDMS/PTMO	6.4	6.3	6.3	6.2	0.0012	18, 2
Polyether vs. others						
PEO/PEVAc (25% VAc)	6.4	6.0	5.8	5.6	0.0045	2
PEO/PTMO	4.5	4.1	3.9	3.7	0.0051	2
PTMO/PEVAc (25% VAc)	1.5	1.3	1.2	1.1	0.0023	2

[a] Abbreviations: PCP, polychloroprene; PDMS, polydimethylsiloxane; PE, polyethylene; B-PE, branched (low-density) polyethylene; L-PE, linear (high-density) polyethylene; PEO, poly(ethylene oxide); PEVAc, poly(ethylene-covinyl acetate); 25% VAc, 25% by weight of vinyl acetate as comonomer; 38.7% VAc, 38.7% by weight of vinyl acetate as comonomer; PIB, polyisobutylene; PMMA, poly(methyl methacrylate); PiBMA, poly(i-butyl methacrylate); PtBMA, poly(t-butyl methacrylate); PMA, poly(methyl acrylate); PEA, poly(ethyl acrylate); PnBA, poly(n-butyl acrylate); PEHA, poly(2-ethylhexylacrylate); PP, polypropylene; PS, polystyrene; PTMO, poly(tetramethylene oxide); PVAc, poly(vinyl acetate); PnBMA, poly(n-butyl methacrylate).

[b] Interfacial tensions are measured directly on polymer melts and liquids. The data at lower temperatures (if the polymers are solid) are obtained by linear extrapolation.

[c] The first reference is the source of the data. The following numbers are additional references.

where ϕ_c is the critical volume fraction of the polymer, $\langle R^2 \rangle_0$ the unperturbed mean-square end-to-end distance, r the number of segment in the polymer, and v the volume of a segment (or of a solvent molecule).

3.2.15. Tabulation of Interfacial Tensions

The interfacial tensions between polymers obtained by direct measurements on polymer melts and liquids are tabulated in Table 3.20.

REFERENCES

1. S. Wu, J. Colloid Interface Sci., *31*, 153 (1969).
2. R. J. Roe, J. Colloid Interface Sci., *31*, 228 (1969).
3. S. Wu, in *Polymer Blends*, Vol. 1, D. R. Paul and S. Newman, eds., Academic Press, New York, 1978, pp. 243–293.
4. S. Wu, J. Macromol. Sci., *C10*, 1 (1974).
5. E. Helfand and Y. Tagami, J. Polym. Sci., *B9*, 741 (1971); J. Chem. Phys., *56*, 3592 (1972); *57*, 1812 (1972).
6. E. Helfand and A. M. Sapse, J. Chem. Phys., *62*, 1327 (1975).
7. E. Helfand, Acc. Chem. Res., *8*, 295 (1975).
8. E. Helfand, J. Chem. Phys., *63*, 2192 (1975).
9. R. J. Roe, J. Chem. Phys., *62*, 490 (1975).
10. D. Patterson and A. K Rastogi, J. Phys. Chem., *74*, 1067 (1970).
11. K. S. Siow and D. Patterson, Macromolecules, *4*, 26 (1971).
12. C. I. Poser and I. C. Sanchez, J. Colloid Interface Sci., *69*, 539 (1979).
13. Y. Oh and M. S. Jhon, J. Colloid Interface Sci., *73*, 467 (1980).
14. E. A. Guggenheim, J. Chem. Phys., *13*, 253 (1945).
15. L. Rideal, Chem. Ing. Tech., *27*, 209 (1955).
16. J. F. Padday, in *Surface and Colloid Science*, Vol. 1, E. Matijevic, ed., Wiley, New York, 1969, pp. 39–99.
17. S. Wu. J. Phys. Chem., *74*, 632 (1970).
18. S. Wu, J. Polym. Sci., *C34*, 19 (1971).
19. R. J. Roe, J. Phys. Chem., *72*, 2013 (1968).
20. R. H. Dettre and R. E. Johnson, Jr., J. Colloid Interface Sci., *31*, 568 (1969); J. Phys. Chem., *71*, 1529 (1967).
21. D. B. Macleod, Trans. Faraday Soc., *19*, 38 (1923).
22. A. Furguson and S. J. Kennedy, Trans. Faraday Soc., *32*, 1474 (1936).
23. F. J. Wright, J. Appl. Chem., *11*, 193 (1961).
24. R. Eötvös, Ann. Phys., *27*, 448 (1886).
25. J. R. Partington, *Advanced Treatise on Physical Chemistry*, Vol. 2, Longmans, Green, New York, 1951, pp. 140–141.

26. S. Sugden, J. Chem. Soc., *125*, 32 (1924).
27. O. R. Quayle, Chem. Rev., *53*, 439 (1953).
28. D. W. Van Krevelen, *Properties of Polymers*, Elsevier, Amsterdam, 1976.
29. T. G. Fox and P. J. Flory, J. Appl. Phys., *21*, 581 (1959).
30. T. G. Fox and P. J. Flory, J. Polym. Sci., *14*, 315 (1954).
31. T. G. Fox and S. Loshaek, J. Polym. Sci., *15*, 371 (1955).
32. K. Ueberreiter and G. Kanig, Z. Naturforsch., *A6*, 551 (1955).
33. R. Boyer, Rubber Chem. Technol., *36*, 1303 (1963).
34. J. R. Martin, J. F. Johnson, and A. R. Cooper, J. Macromol. Sci., *C8*, 57 (1972).
35. M. C. Phillips and A. C. Riddiford, J. Colloid Interface Sci., *22*, 149 (1966).
36. R. H. Dettre and R. E. Johnson, Jr., J. Phys. Chem., *71*, 1529 (1967).
37. T. Hata, Kobunshi, *17*, 594 (1968).
38. H. Edwards, J. Appl. Polym. Sci., *12*, 2213 (1968).
39. D. G. LeGrand and G. L. Gaines, Jr., J. Colloid Interface Sci., *31*, 162 (1969).
40. D. G. LeGrand and G. L. Gaines, Jr., J. Colloid Interface Sci., *42*, 181 (1973).
41. G. W. Bender, D. G. LeGrand, and G. L. Gaines, Jr., Macromolecules, *2*, 681 (1969).
42. A. K. Rastogi and L. E. St. Pierre, J. Colloid Interface Sci., *35*, 16 (1971).
43. H. Schonhorn and F. W. Ryan, J. Phys. Chem., *70*, 3811 (1966).
44. A. K. Rastogi and L. E. St. Pierre, J. Colloid Interface Sci., *31*, 168 (1969).
45. T. C. Kendrick, B. M. Kingston, N. C. Lloyd, and M. J. Owen, J. Colloid Interface Sci., *24*, 135 (1967).
46. J. W. Belton and M. G. Evans, Trans. Faraday Soc., *41*, 1 (1945). See also R. S. Hansen and L. Sogar, J. Colloid Interface Sci., *40*, 424 (1972).
47. G. L. Gaines, Jr., J. Phys. Chem., *73*, 3143 (1969).
48. I. Prigogine and J. Marechal, J. Colloid Sci., *7*, 122 (1952).
49. G. L. Gaines, Jr., J. Polym. Sci., A-2, *7*, 1379 (1969).
50. G. L. Gaines, Jr., J. Polym. Sci., A-2, *9*, 1333 (1971); *10*, 1529 (1972).
51. D. G. LeGrand and G. L. Gaines, Jr., J. Polym. Sci., *C34*, 45 (1971).
52. A. J. G. Allan, J. Colloid Sci., *14*, 206 (1959).
53. N. L. Jarvis, R. B. Fox, and W. A. Zisman, Adv. Chem. Ser., *43*, 317 (1964). R. C. Bowers, N. L. Jarvis, and W. A. Zisman, Ind. Eng. Chem., Prod. Res. Dev., *4*, 86 (1965).
54. D. G. LeGrand and G. L. Gaines, Jr., Polym. Prepr., Am. Chem. Soc., *11*(2), 442 (1970).

References

55. G. L. Gaines, Jr., and G. W. Bender, Macromolecules, 5, 82 (1972).
56. J. F. Padday and N. D. Uffindell, J. Phys. Chem., 72, 1407 (1968).
57. J. N. Israelachvilli, J. Chem. Soc., Faraday Trans. II, 69, 1729 (1973).
58. H. Kammer, Z. Phys. Chem. (Leipzig), 255(3), S607 (1974).
59. J. H. Hildebrand and R. L. Scott, *Solubility of Nonelectrolytes*, D. Van Nostrand, Princeton, N.J., 1950.
60. A. Beerbower, J. Colloid Interface Sci., 35, 126 (1971).
61. P. Becher, J. Colloid Interface Sci., 38, 291 (1972).
62. S. Wu, J. Phys. Chem., 72, 3332 (1968).
63. R. H. Dettre and R. E. Johnson, Jr., J. Colloid Interface Sci., 21, 367 (1966).
64. H. Schonhorn and L. H. Sharpe, J. Polym. Sci., A3, 569 (1965).
65. Y. Oda and T. Hata, Prepr., 17th Annu. Meet. High Polymer Soc., Japan, 1968, p. 267.
66. Y. Oda and T. Hata, Prepr., 6th Symp. Adhes. Adhes., p. 69 (1968).
67. W. Y. Lau and C. M. Burns, Surf. Sci., 30, 478 (1972).
68. W. Y. Lau and C. M. Burns, J. Colloid Interface Sci., 45, 295 (1973).
69. W. Y. Lau and C. M. Burns, J. Polym. Sci., Polym. Phys. Ed., 12, 431 (1974).
70. S. Wu, J. Adhes., 5, 39 (1973); also in *Recent Advances in Adhesion*, L. H. Lee, ed., Gordon and Breach, New York, 1973, pp. 45–63.
71. H. Schonhorn, F. W. Ryan, and L. H. Sharpe, J. Polym. Sci., A-2, 4, 538 (1966).
72. S. Wu, Org. Coat. Plast. Chem., 31(2), 27 (1971).
73. H. T. Patterson, K. H. Hu, and T. H. Grindstaff, J. Polym. Sci., C34, 31 (1971).
74. F. J. Hybart and T. R. White, J. Appl. Polym. Sci., 3, 118 (1960).
75. H. W. Fox, P. W. Taylor, and W. A. Zisman, Ind. Eng. Chem., 39, 1401 (1947).
76. T. Hata, Hyoman [Surface, Japan], 6, 659 (1968).
77. G. Antonoff, J. Chim. Phys., 5, 371 (1907); Ann. Phys., 35, 5(1939); J. Phys. Chem., 46, 497 (1942).
78. D. J. Donahue and F. B. Bartell, J. Phys. Chem., 56, 480 (1952).
79. R. E. Johnson, Jr., and R. H. Dettre, J. Colloid Interface Sci., 21, 610 (1966).
80. L. A. Girifalco and R. J. Good, J. Phys. Chem., 61, 904 (1957).
81. R. J. Good and Elbing, Ind. Eng. Chem., 62(3), 55 (1970).
82. R. J. Good, Adv. Chem. Ser., 43, 74 (1964).

83. R. J. Good, in *Treatise on Adhesion and Adhesives*, Vol. 1, R. L. Patrick, ed., Marcel Dekker, New York, 1967, pp. 9–68.
84. D. K. Owens and R. C. Wendt, J. Appl. Polym. Sci., *13*, 1741 (1969).
85. D. H. Kaelble, *Physical Chemistry of Adhesion*, Wiley-Interscience, New York, 1970.
86. F. M. Fowkes, in *Chemistry and Physics of Interfaces*, American Chemical Society, Washington, D.C., 1965, pp. 1–12.
87. C. M. Hansen and A. Beerbower, in *Kirk-Othmer's Encyclopedia of Chemical Technology*, 2nd ed., Suppl. Vol., Wiley, New York, 1971, pp. 889–910.
88. C. M. Hansen, J. Paint Technol., *39*(505), 104 (1967).
89. T. Hata and T. Kasemura, in *Adhesion and Adsorption of Polymers*, Part A, L. H. Lee, ed., Plenum Press, New York, 1980, pp. 15–41.
90. D. G. LeGrand and G. L. Gaines, Jr., J. Colloid Interface Sci., *50*, 272 (1975).
91. H. W. Kammer, Z. Phys. Chem. (Leipzig), *258*(6), S1149 (1977).
92. N. G. Gaylord, Adv. Chem. Ser., *142*, 76 (1975).
93. H. T. Patterson, K. H. Hu, and T. H. Grindstaff, J. Polym. Sci., *C34*, 31 (1971).
94. G. Langhammer and L. Nester, Makromol. Chem., *88*, 179 (1965).
95. T. Nose and T. V. Tan, J. Polym. Sci., Polym. Lett. Ed., *14*, 705 (1976).
96. A. Vrij, J. Polym. Sci., A-2, *6*, 1919 (1968).
97. T. Nose, Polym. J. (Japan), *8*, 96 (1976).
98. Y. Kitazaki and T. Hata, 18th Annu. Meet. High Polymer Soc. (Japan), 1969, p. 478.

4
Contact Angles of Liquids on Solid Polymers

4.1. EQUILIBRIUM SPREADING PRESSURE

Adsorption of liquid vapor on a solid tends to lower the surface tension of the solid and increase the contact angle. The surface tension decrement at equilibrium adsorption is the equilibrium spreading pressure, $\pi_e = \gamma_S - \gamma_{SV}$, where γ_S is the solid surface tension in vacuum and γ_{SV} that in equilibrium with the saturated vapor of the liquid [Eq. (1.32)]. The π_e is usually negligible when the contact angle is larger than about 10°, but becomes significant when the contact angle is smaller than about 10°. When the contact angle is zero, just enough vapor will adsorb on the solid such that $\pi_e = \gamma_S - \gamma_{LV} - \gamma_{SL}$ (Section 1.6.2).

The π_e can be determined by several methods: (1) vapor adsorption isotherms, (2) surface tensions of liquid substrates, and (3) contact angles on solids. A molecular model has also been used to estimate it. All the results show that π_e is negligible when the contact angle is large, and can be quite significant when the contact angle approaches zero.

Integration of vapor adsorption isotherm gives the π_e, (Section 1.5.1). Some typical results are listed in Table 4.1 [1-7]. On the other hand, some anomalous behavior was also reported, where high π_e was found with large contact angles. For instance, π_e for water

Table 4.1. Equilibrium Spreading Pressure Obtained by Vapor Adsorption Measurement

Solids	Liquids	Temperature	Contact angle, deg	π_e, dyne/cm	Reference
Polypropylene	Nitrogen	78 K	0	12	1
	Argon	90 K	0	13	1
	CF_4	142 K	0	9	1
	Ethane	187 K	0	14	1
Polytetrafluoro-ethylene	Ethane	182 K	0	5.9	2
	n-Butane	-5°C	0	4.9	2
	n-Hexane	25°C	12	3.28	3
	n-Octane	25°C	26	4.9	2
				2.95	3
TiO_2	Water	25°C	0	300	3
				196	4
	1-Propanol	25°C	0	114	5
				108	4
	Benzene	25°C	0	85	5
	n-Heptane	25°C	0	58	5
				46	4
SiO_2	Water	25°C	0	316	5
	1-Propanol	25°C	0	134	5
	Acetone	25°C	0	109	5
	Benzene	25°C	0	81	5
	n-Heptane	25°C	0	59	5
$BaSO_4$	Water	25°C	0	318	5
	1-Propanol	25°C	0	101	5
	n-Heptane	25°C	0	58	5

Equilibrium Spreading Pressure

Fe₂O₃	n-Heptane	25°C		54	5
SnO₂	Water	25°C	0	292	5
	1-Propanol	25°C	0	220	4
	Propyl acetate	25°C	0	104	5
	n-Heptane	25°C	0	117	4
			0	104	5
				54	5
Graphite	Water	25°C	—	64	5
	1-Propanol	25°C	0	95	5
	Benzene	25°C	—	76	5
	n-Heptane	25°C	0	57	6
Mercury	Water	25°C	63	101	5
	1-Propanol	25°C	—	108	5
	Acetone	25°C	0	86	5
	Benzene	25°C	0	119	5
	n-Octane	25°C	0	101	5
Copper	n-Heptane	25°C	0	29	7
Silver	n-Heptane	25°C	0	37	7
Lead	n-Heptane	25°C	0	49	7
Iron	n-Heptane	25°C	0	53	7
Tin	Water	25°C	—	168	4
	n-Heptane	25°C	0	50	4
	1-Propanol	25°C	0	83	4

on polyethylene was 14.4 dyne/cm and the contact angle was 88° (at 22°C); π_e for water on carbon was 142 dyne/cm and the contact angle was 72° [8]. Such anomalous behavior may be due to the presence of porosity or hydrophilic sites on the solid.

The π_e can be determined accurately by measuring the surface tension when the substrate is a liquid. The vapor should have negligible solubility in the substrate liquid. This is true for n-alkanes on fluorocarbons and on water. Their π_e values have thus been measured accurately [9,10] (Table 4.2). When the contact angle approaches zero, π_e becomes quite large (Figure 4.1).

The π_e may also be determined from contact angles. Young's equation is written as

$$\gamma_S = \gamma_{LV} \cos\theta + \gamma_{SL} + \pi_e \tag{4.1}$$

Using the equation of Good and Girifalco in Eq. (4.1) gives

$$\pi_e = \gamma_{LV}\left[2\phi\left(\frac{\gamma_S}{\gamma_{LV}}\right)^{1/2} - (1 + \cos\theta)\right] \tag{4.2}$$

where the interaction parameter ϕ can be determined by the harmonic-mean equation [Eq. (3.73)]. Some results are listed in Table 4.3. The γ_S values for the solid polymers are obtained by extrapolation from the melts. The results are consistent with normal behavior.

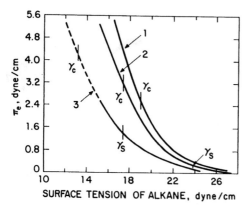

Figure 4.1. Equilibrium spreading pressure π_e of n-alkanes on various low-energy liquid substrates. (1) Water; (2) $H(CF_2)_4CH_2OH$; (3) $F(CF_2)_2CH_2OH$. γ_c and γ_S are critical surface tension and surface tension of the substrate, respectively. The γ_S for water is 72.8 dyne/cm (not shown). (After Ref. 9.)

Table 4.2. Equilibrium Spreading Pressure of Liquids on Liquid Substrates Determined by Direct Measurement of the Surface Tension of the Liquid Substrates in Equilibrium with the Saturated Vapor of the Wetting Liquids at 24.5°C

Substrate	Pentane	Hexane	Heptane	Octane	Decane	Dodecane
Water	(γ_S = 72.1 dyne/cm, γ_C = 19.1 dyne/cm)					
γ_{LV}, dyne/cm	15.7	18.0	19.7	21.4	23.5	25.1
γ_{SL}, dyne/cm	50.3	50.7	51.2	51.5	52.0	52.8
π_e, dyne/cm	6.6	3.9	1.9	1.1	0.5	0.3
θ, degree	0	0	0	24.3	33.5	40.8
$F(CF_2)_2CH_2OH$	(γ_S = 17.4 dyne/cm, γ_C = 13.2 dyne/cm)					
γ_{SL}, dyne/cm	1.7	2.5	3.5	4.4	5.6	6.7
π_e, dyne/cm	2.3	1.2	0.7	0.6	0.2	0.1
θ, degree	31.4	40.4	47.9	54.6	60.4	65.0
$H(CF_2)_2CH_2OH$	(γ_S = 26.0 dyne/cm, γ_C = 17.8 dyne/cm)					
γ_{SL}, dyne/cm	5.1	5.9	6.5	7.3	8.3	9.4
π_e, dyne/cm	5.0	2.3	1.4	0.8	0.4	0.2
θ, degree	0	0	23.2	33.2	42.6	49.2
$H(CF_2)_4CH_2OH$	(γ_S = 23.8 dyne/cm, γ_C = 17.5 dyne/cm)					
γ_{SL}, dyne/cm	3.4	3.8	4.9	5.4	6.7	7.6
π_e, dyne/cm	5.0	2.3	1.4	0.8	0.4	0.2
θ, degree	0	0	27.3	34.7	44.7	50.4

Source: From Ref. 9.

Table 4.3. Equilibrium Spreading Pressure Calculated from the Contact Angle at 20°C

Solid[a]	Liquid	γ_{LV}, dyne/cm	x_{LV}^d	ϕ	θ deg	π_e dyne/cm
Polyethylene	Water	72.8	0.30	0.548	94	0
	Methylene iodide	50.8	0.95	0.977	52	0
	Tricresyl phosphate	40.9	0.96	0.979	34	0
	Hexadecane	27.6	1.00	1.000	0	7.6
	Hexane	18.4	1.00	1.000	0	14.5
Polytetrafluoroethylene	Octane	21.8	1.00	0.954	26	0.04
	Hexane	18.4	1.00	0.954	12	0.13
Poly(methyl methacrylate)	Water	72.8	0.30	0.907	76	9.2
	Methylene iodide	50.8	0.96	0.941	41	0
	Tricresyl phosphate	40.9	0.96	0.939	19	0
	Hexane	18.4	1.00	0.848	0	10.0

[a] For polyethylene, $\gamma_s = 35.7$ dyne/cm, $X_s^d = 1.0$; for polytetrafluoroethylene, $\gamma_s = 22.5$ dyne/cm, $X_s^d = 0.91$; for poly(methyl methacrylate), $\gamma_s = 41.1$ dyne/cm, $X_s^d = 0.72$; see chapters 3 and 5.
[b] Calculated by Eq. (4.2).

Contact angles of several liquids (alkanes, ethers, and silicones) on polytetrafluoroethylene are found to be identical, either measured in open air or in saturated vapors, except when the contact angles are near zero [11].

A molecular model is used to estimate the π_e by considering the free energy of transferring vapor molecules from the liquid to the adsorbed monolayer on the solid [12]. The π_e increases rapidly as the contact angle approaches zero. However, the estimated π_e values are rather low.

4.2. SPREADING COEFFICIENT

The spreading coefficient is the reduction of free energy in forming a duplex film of unit interfacial area (Section 1.5.2). Three stages of spreading are recognized: initial, intermediate, and final. For a volatile liquid, just enough vapor will adsorb on the substrate such that $\gamma_{SV}^* - \gamma_{LV}^* - \gamma_{SL}^* = 0$ at a zero contact angle. Therefore, for a volatile liquid, the final spreading coefficient can only be zero or negative. On the other hand, the final spreading coefficient can be positive, zero, or negative for a nonvolatile liquid such as a polymer melt. Examples of spreading coefficients for volatile liquids are given in Table 4.4. Additional examples are available [9,13,14]. The examples for polymer melts are given in Chapter 3.

4.3. EFFECT OF TEMPERATURE ON CONTACT ANGLE

Contact angles may either decrease or increase with temperature, depending on the relative magnitudes of the surface entropies of the two phases (Section 3.2.7). The $-(d\theta/dT)$ values are quite small, about 0.05 deg/°C, at ordinary temperatures. However, at temperatures near the boiling point, the liquid surface tension decreases rapdily with temperature, and the contact angle will rapidly approach zero.

Contact angles of water on poly(ethylene terephthalate) versus temperature are plotted in Figure 4.2. The data above 100°C are measured in a pressurized cell [15]. Below the boiling point, the advancing angle decreases and the receding angle increases slowly. Above the boiling point, both angles decrease rapidly toward zero. Similar behavior is observed for hexadecane and octane on perfluorinated ethylene-propylene copolymer (FEP) [16] (Figure 4.3). Such behavior arises from the rapid reduction of the liquid surface tension near the boiling point [17]. Neglecting π_e, Eq. (4.2) becomes

$$\cos \theta = 2\phi \left(\frac{\gamma_S}{\gamma_{LV}} \right)^{1/2} - 1 \qquad (4.3)$$

Table 4.4. Spreading Coefficient, Contact Angle, and Equilibrium Spreading Pressure of Some Liquids on Water at 20°C

	Methylene iodide	Carbon disulfide	Benzene	n-Heptyl alcohol	Heptane
γ_{LV}, dyne/cm	50.68	32.35	28.88	26.97	19.7
γ_{LV}^*, dyne/cm	50.52	31.81	28.82	26.48	19.7
γ_S^*, dyne/cm [a]	72.20	70.49	62.36	28.53	70.2
π_e, dyne/cm	0.55	2.26	10.39	44.22	1.9
γ_{SL}^*, dyne/cm	45.87	48.63	35.03	7.95	50.5
$\theta°$, deg	57.9	41.8	0	0	0
θ^f, deg	58.6	46.6	18.5	39.0	0
$\lambda°$, erg/cm^2	-23.80	-8.23	8.84	37.83	1.2
λ^m, erg/cm^2	-23.64	-7.69	8.90	38.32	1.2
λ^f, erg/cm^2	-24.19	-9.95	-1.49	-5.90	0

[a] γ_S for clean water = 72.75 dyne/cm at 20°C.
Source: From Ref. 9 and 13.

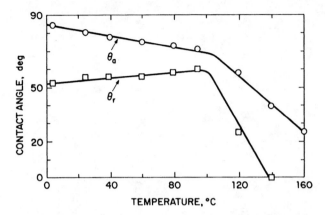

Figure 4.2. Advancing and receding contact angles of water on poly(ethylene terephthalate). (After Ref. 15.)

which can be used to calculate the contact angles at various temperatures by fitting both γ_S and γ_{LV} to the Guggenheim equation [Eq. (3.1)]. The calculated results (solid lines) adequately describe the experimental data (circles and squares) (Figure 4.4). Contact angles and their temperature coefficients for some liquid on solid polymers are collected in Table 4.5 [15,16,18]. Those for polymer melts are given in Tables 3.13 and 3.14.

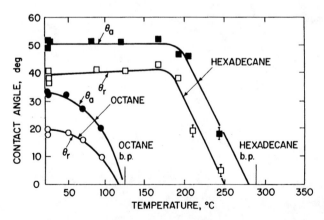

Figure 4.3. Advancing and receding contact angles of n-hexadecane and n-octane on perfluorinated ethylene-propylene copolymer versus temperature. (After Ref. 16.)

Table 4.5. Temperature Coefficient of Contact Angle for Liquids on Solid Polymers

Solid polymer	Liquid	Contact angle,[a] θ, deg		$d\theta/dT$, deg/°C	
Fluorinated ethylene-propylene copolymer (FEP)	n-Hexadecane	θ_a	53.0 (20°C)	0	(20-200°C)
		θ_r	31.2 (20°C)	+0.022	(20-200°C)
		θ_a	39 (225°C)	-0.65	(200-275°C)
		θ_r	20 (225°C)	-0.65	(200-275°C)
	n-Octane	θ_a	33.2 (20°C)	-0.025	(20-50°C)
		θ_r	20.0 (20°C)	-0.050	(20-50°C)
		θ_a	20.0 (100°C)	-0.332	(50-148°C)
		θ_r	9.5 (100°C)	-0.271	(50-118°C)
Polytetra-fluoroethylene	Water	θ_a	109 (20°C)	-0.05	(20-100°C)
		θ_r	106 (20°C)	-0.04	(20-100°C)
		θ_a	97 (120°C)	-0.33	(100-160°C)
		θ_r	93 (120°C)	-1.95	(100-160°C)
	Glycerol	θ_a	101 (20°C)	-0.04	(0-160°C)
		θ_r	96 (20°C)	-0.03	(0-160°C)
	Formamide	θ_a	92 (20°C)	-0.06	(0-160°C)
		θ_r	88 (20°C)	-0.05	(0-160°C)

Effect of Temperature on Contact Angle

	Ethylene glycol	Θ_a	86	(20°C)	-0.06 (0-160°C)
		Θ_r	82	(20°C)	-0.04 (0-160°C)
	n-Decane	Θ_Y	40.1	(20°C)	-0.106 (20-70°C)
	n-Undecane	Θ_Y	42.1	(20°C)	-0.057 (20-70°C)
	n-Dodecane	Θ_Y	42.7	(20°C)	-0.012 (20-70°C)
	n-Tridecane	Θ_Y	43.4	(20°C)	0 (20-70°C)
	n-Tetradecane	Θ_Y	44.8	(20°C)	0 (20-70°C)
	n-Pentadecane	Θ_Y	46.0	(20°C)	0 (20-70°C)
	n-Hexadecane	Θ_Y	46.0	(20°C)	0 (20-70°C)
Polystyrene	Water	Θ_a	91	(20°C)	-0.04 (20-100°C)
		Θ_r	84	(20°C)	-0.01 (20-100°C)
		Θ_a	69	(120°C)	-1.25 (100-160°C)
		Θ_r	48	(120°C)	-1.25 (100-160°C)
	Glycerol	Θ_a	73	(20°C)	+0.05 (20-160°C)
		Θ_r	64	(20°C)	+0.07 (20-160°C)
	Formamide	Θ_a	69	(20°C)	+0.13 (20-80°C)
		Θ_r	65	(20°C)	+0.10 (20-40°C)
	Ethylene glycol	Θ_a	61	(20°C)	+0.02 (20-160°C)
		Θ_r	58	(20°C)	-0.02 (20-160°C)

Table 4.5. (Continued)

Solid polymer	Liquid	Contact angles,[a] θ, deg			dθ/dT, deg/°C	
Poly(ethylene terephthalate)	Water	θ_a	82	(20°C)	-0.14	(0-100°C)
		θ_r	54	(20°C)	+0.06	(0-100°C)
		θ_a	55	(120°C)	-0.76	(100-160°C)
		θ_r	25	(120°C)	-1.32	(100-160°C)
	Glycerol	θ_a	75	(20°C)	-0.14	(0-160°C)
		θ_r	41	(20°C)	+0.14	(0-100°C)
	Formamide	θ_a	66	(20°C)	-0.03	(0-80°C)
		θ_r	39	(20°C)	+0.09	(0-100°C)
	Ethylene glycol	θ_a	57	(20°C)	-0.07	(0-100°C)
		θ_r	29	(20°C)	-0.05	(0-100°C)
Polycarbonate	Water	θ_a	84	(20°C)	-0.06	(0-100°C)
		θ_r	68	(20°C)	+0.06	(0-100°C)
		θ_a	63	(120°C)	-0.65	(100-160°C)
		θ_r	51	(120°C)	-1.05	(100-160°C)
	Glycerol	θ_a	72	(20°C)	-0.04	(0-100°C)
		θ_r	56	(20°C)	+0.05	(0-100°C)

Effect of Temperature on Contact Angle

Polymer	Liquid		Angle	Temp	dθ/dT	Range
Polyoxymethylene	Formamide	θ_a	60	(20°C)	+0.07	(0–100°C)
		θ_r	44	(20°C)	+0.05	(0–80°C)
	Ethylene glycol	θ_a	57	(20°C)	−0.03	(0–100°C)
		θ_r	38	(20°C)	+0.12	(0–80°C)
	Water	θ_a	79	(20°C)	−0.14	(0–100°C)
		θ_r	54	(20°C)	+0.04	(0–100°C)
		θ_a	68	(100°C)	−0.335	(100–160°C)
		θ_r	53	(100°C)	−0.237	(100–160°C)
	Glycerol	θ_a	67	(20°C)	−0.06	(0–100°C)
		θ_r	42	(20°C)	+0.08	(0–100°C)
Polyethylene	Formamide	θ_a	61	(20°C)	−0.02	(0–80°C)
		θ_r	46	(20°C)	+0.04	(0–80°C)
	Ethylene glycol	θ_a	56	(20°C)	−0.11	(0–100°C)
		θ_r	34	(20°C)	+0.08	(0–100°C)
	Water	θ_a	96	(20°C)	−0.11	(0–100°C)
		θ_r	62	(20°C)	+0.02	(0–100°C)
		θ_a	72	(120°C)	−0.538	(100–160°C)
		θ_r	46	(120°C)	−1.67	(100–160°C)
	Glycerol	θ_a	80	(20°C)	−0.03	(0–140°C)
		θ_r	62	(20°C)	+0.04	(0–100°C)

Table 4.5. (Continued)

Solid polymer	Liquid	Contact angle,[a] θ, deg		dθ/dT, deg/°C	
	Formamide	θ_a	75 (20°C)	−0.01	(0–140°C)
		θ_r	48 (20°C)	+0.06	(0–100°C)
	Ethylene glycol	θ_a	64 (20°C)	+0.03	(0–80°C)
		θ_r	48 (20°C)	+0.04	(0–60°C)

[a] θ_a, advancing contact angle; θ_r, receding contact angle; θ_Y, Young's (equilibrium) contact angle.
Source: Collected from Refs. 15, 16, and 18.

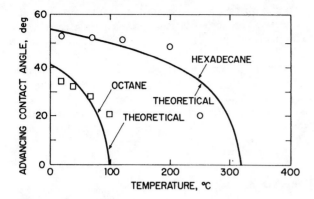

Figure 4.4. Advancing contact angles versus temperature for n-hexadecane and n-octane on FEP. Experimental data are from Figure 4.3. The solid lines are theoretical. (After Ref. 17.)

Differentiation of Young's equation with respect to temperature gives

$$\frac{d \cos \theta}{dT} = \frac{1}{\gamma_{LV}} \left(\frac{d\gamma_{SV}}{dT} - \cos \theta \frac{d\gamma_{LV}}{dT} - \frac{d\gamma_{SL}}{dT} \right) \quad (4.4)$$

which can be used to obtain the temperature coefficient of the contact angle; see also Section 3.2.7.

4.4. EFFECT OF PRIMARY AND SECONDARY TRANSITIONS

The surface tension is discontinuous at primary transition. The temperature coefficient of surface tension is discontinuous at the secondary transition (Section 3.1.4). Contact angles should behave similarly.

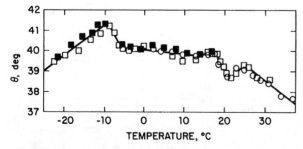

Figure 4.5. Contact angle of n-decane on polytetrafluoroethylene versus temperature. (After Ref. 20.)

As the θ variation is usually quite small, a sensitive method must be used to detect it, such as the tensiometric method with an electrobalance or the capillary rise method on a flat plate, described in Chapter 9.

Contact angles of liquids on polymers have been found to undergo transitions at glass and crystalline transitions [19,20]. The θ versus T plot for n-decane on polytetrafluoroethylene is shown in Figure 4.5. The θ undergoes transitions at three well-known polymorphic transitions of PTFE: the glass transition at −10°C and two crystalline transitions at 17°C and 25°C.

4.5. CONTACT ANGLE AND HEAT OF WETTING

The heat of wetting (immersion) $h_{w(SV)}$ of a solid (initially in equilibrium with the vapor of the wetting liquid) is given by

$$h_{w(SV)} = h_{SL} - h_{SV}$$

$$= -\gamma_{LV} \cos \theta + T \frac{d(\gamma_{LV} \cos \theta)}{dT} \qquad (4.5)$$

which can be obtained by recognizing that $h = \gamma - T(d\gamma/dT)$. The heats of wetting of polytetrafluoroethylene by n-alkanes are calculated from contact angles by the Eq. (4.5). The results agree rather well with calorimetrically measured values [20,21], shown in Table 4.6.

On the other hand, the heat of wetting $h_{w(S)}$ of a clean solid (no initial vapor adsorption) is given by

$$h_w(S) = h_{SL} - h_S = h_{w(SV)} - \pi_e + T\left(\frac{d\pi_e}{dT}\right) \qquad (4.6)$$

which relates $h_{w(S)}$ and $h_{w(SV)}$.

4.6. POLARITY OF LIQUIDS

The polarity of a liquid can be calculated from the interfacial tension or from the contact angle by using the harmonic-mean equation or the geometric-mean equation, whichever is appropriate.

The calculations of γ^d for water from the interfacial tensions between water and alkanes are given in Table 4.7. The harmonic-mean equation gives $\gamma^d = 22.1 \pm 0.3$ dyne/cm (20°C); the geometric-mean equation gives $\gamma^d = 21.8 \pm 0.7$ dyne/cm for water. These compare quite well with the theoretical values of Good and Elbing [22]: $\gamma^d = 19.8$ dyne/cm, $\gamma^i = 2.0$ dyne/cm, and $\gamma^p = 51.1$ dyne/cm, calculated from molecular constants. The γ for water is 72.8 dyne/cm.

Table 4.6. Heat of Wetting of Polytetrafluoroethylene by n-Alkanes

				Heat of wetting, $h_W(S)$, erg/cm^2	
Liquid	θ at 20°C, deg	$d\theta/dT$, deg/°C	$d\gamma_S/dT$, dyne/cm-°C	Calculated by Eq. (4.5)	Experimental calorimetry
n-Hexane	12	—	—	—	-46.9 ± 6.3
n-Octane	26	—	—	—	-39.9 ± 4.0
n-Decane	40.1	-0.106	-0.055	-33.6	-32.8 ± 1.6
n-Undecane	42.1	-0.057	-0.063	-36.0	—
n-Dodecane	42.7	-0.012	-0.068	-38.5	-32.8 ± 1.6
n-Tridecane	43.4	0	-0.068	-38.1	—
n-Tetradecane	44.8	0	-0.068	-38.0	—
n-Pentadecane	46.0	0	-0.065	-36.9	—
n-Hexadecane	46.0	0	-0.064	-36.8	-34.0 ± 1.6

Source: From Refs. 20 and 21.

Table 4.7. Calculation of γ^d for Water (γ = 72.8 dyne/cm) from Interfacial Tension of Water Against Hydrocarbons at 20°C

Hydrocarbon	γ_2,[a] dyne/cm	γ_{12},[b] dyne/cm	γ^d, dyne/cm	
			By harmonic-mean equation	By geometric-mean equation
n-Hexane	18.4	51.1	22.0	21.8
n-Heptane	20.4	50.2	22.7	22.6
n-Octane	21.8	50.8	22.0	22.0
n-Decane	23.9	51.2	21.7	21.6
n-Tetradecane	25.6	52.2	21.0	20.8
Cyclohexane	25.5	50.2	22.8	22.7
Decalin	29.9	51.4	22.4	22.1
White oil	28.9	51.3	22.3	21.3
Average	-	-	22.1 ± 0.3	21.8 ± 0.7

[a] γ_2, surface tension of hydrocarbon.
[b] γ_{12}, interfacial tension between water and hydrocarbon.

The γ^d for mercury (γ = 484 dyne/cm) calculated by the geometric-mean equation from interfacial tensions between mercury and hydrocarbons [25] is found to be 200 ± 7 dyne/cm.

The γ^d of a liquid may also be calculated from its contact angle on a nonpolar solid. In this case, $\gamma_S = \gamma_S^d$ and the polar interaction term drops out. The harmonic-mean equation gives

$$\cos \theta = \frac{4}{\gamma_{LV}} \left(\frac{\gamma_S^d \gamma_{LV}^d}{\gamma_S^d + \gamma_{LV}^d} \right) - 1 \tag{4.7}$$

whereas the geometric-mean equation gives [25]

$$\cos \theta = 2(\gamma_S^d)^{1/2} \frac{(\gamma_{LV}^d)^{1/2}}{\gamma_{LV}} - 1 \tag{4.8}$$

These equations are used to calculate the γ^d for some liquids from contact angles on nonpolar solids such as branched polyethylene (γ_S^d = 35.3 dyne/cm) and hexatriacontane (γ_S^d = 24.9 dyne/cm) at 20°C (Table 4.8).

Table 4.8. Summary and Comparison of γ_{LV}^d of Some Liquids Determined by Various Methods at 20°C

Liquid	γ_{LV}, dyne/cm	γ_{LV}^d, from contact angle data,[a] dyne/cm		γ_{LV}^d from interfacial tension data,[b] dyne/cm	
		Harmonic-mean method	Geometric-mean method	Harmonic-mean method	Geometric-mean method
Tricresyl phosphate	40.9	39.8	39.2	–	–
α-Bromonaphthalene	44.6	47.7	47	–	–
Trichlorobiphenyl	45.3	35.5	44	–	–
Methylene iodide	50.8	49.0	48.5	44.1	49.5
Glycerol	63.4	40.6	37.0	–	–
Formamide	58.2	36.0	39.5	–	–
Water	72.8	22.6	22.5	22.1	21.8

[a] Low-density polyethylene ($\gamma_S = 35.3$ dyne/cm) and triacontane ($\gamma_S = 24.9$ dyne/cm) are used as the nonpolar solids.

[b] From interfacial tension between methylene iodide and water (41.6 dyne/cm [23, 24]) or between water and n-alkenes.

4.7. PREDICTION OF CONTACT ANGLES

The theory of fractional polarity can be used to predict contact angles. Combining the harmonic-mean equation with Young's equation gives

$$(1 + \cos \theta)\gamma_{LV} = 4\left(\frac{\gamma_S^d \gamma_{LV}^d}{\gamma_S^d + \gamma_{LV}^d} + \frac{\gamma_S^p \gamma_{LV}^p}{\gamma_S^p + \gamma_{LV}^p}\right) \quad (4.9)$$

whereas combining the geometric-mean equation with Young's equation gives

$$(1 + \cos \theta)\gamma_{LV} = 2[(\gamma_S^d \gamma_{LV}^d)^{1/2} + (\gamma_S^p \gamma_{LV}^p)^{1/2}] \quad (4.10)$$

where π_e has been neglected. These equations are used to predict the contact angles of some liquids on some polymers in air (Table 4.9). Reasonable agreement with measured values is obtained when both polar and nonpolar interactions are considered. On the other hand, the Fowkes equation, which considers only the nonpolar interaction, gives erratic results.

Contact angles of liquids on solids immersed in other liquids have also been predicted by the fractional polarity equations [26,27]. The harmonic-mean equation is found to give good results, whereas the Fowkes equation is again quite erratic (Table 4.10). However, if the solid is swollen by the immersion liquid, the predicted values will not agree with the measured values with either equation.

4.8. EFFECT OF SOLUTE ADSORPTION ON CONTACT ANGLE

When the wetting liquid is a solution, the solute may adsorb at the liquid-vapor (LV), solid-liquid (SL), and solid-vapor (SV) interfaces and thus affect the contact angle. Two behaviors deserve our attention. First, if there is no specific interaction between the solute and the solid, the adsorptions at liquid-vapor and solid-liquid interfaces will be equal, since in this case the adsorptions arise from repulsion between the solute and the solvent molecules. Second, the solute may also adsorb onto the solid-vapor interface from the vapor phase or by diffusion from the liquid front. Thermodynamic analyses are given below.

Gibbs adsorption isotherm relates the change of interfacial tension to the change of solute activity,

$$\Gamma_j = -\frac{1}{RT} \frac{d\gamma_j}{d \ln a_j} \quad (4.11)$$

Table 4.9. Comparison of Measured and Predicted Contact Angles at 20°C

	γ_{LV}, dyne/cm	γ_{LV}^p, dyne/cm	θ_{exp} deg	Harmonic-mean eq.	Predicted θ, deg Geometric-mean Eq.	Fowkes eq.
On poly(methyl methacrylate), $\gamma = 41.1$ and $\gamma^p = 11.5$ dyne/cm						
Tricresyl phosphate	40.9	1.2	19	40	31	47
Formamide	58.2	22.2	64	50	48	83
Water	72.8	50.7	80	78	68	107
On poly(hexamethylene adipamide), $\gamma = 47.9$ and $\gamma^p = 12.9$ dyne/cm						
Methylene iodide	50.8	6.7	28	28	24	57
Formamide	58.2	22.2	50	39	37	77
Water	72.8	50.7	72	72	62	104
On poly(ethylene terephthalate), $\gamma = 42.1$ and $\gamma^p = 9.3$ dyne/cm						
Trichlorobiphenyl	45.3	10.3	17	22	22	60
Formamide	58.2	22.2	61	51	48	80
Glycerol	63.4	22.9	65	54	52	81
Water	72.8	50.7	81	81	70	105

Table 4.10. Comparison of Measured and Predicted Contact Angles for Some Oils on Solids Immersed in Water at 20°C

	Measured, deg		Predicted θ, deg	
	θ_a	θ_r	Harmonic-mean eq.	Fowkes eq.
Solids not swollen by water				
Oil/water/polytetrafluoroethylene				
n-Octane	40	31	30	0
n-Hexadecane	38	34	37	19
Cyclohexane	42	33	32	9
Mineral oil	37	33	38	20
Oil/water/poly(methyl methacrylate)				
n-Octane	92	75	77	9
n-Hexadecane	92	75	76	0
Mineral oil	93	78	75	9
Solids swollen by water due to hydrophilicity				
Oil/water/nylon 11				
n-Hexadecane	170	-	76	12
n-Hexanol	115	-	180	0
Oil/water/bovine hoof keratin				
n-Hexadecane	170	-	53	12
n-Hexanol	170	-	136	0

Source: From Ref. 26.

where Γ is the surface excess concentration of the solute per unit area, a the solute activity, and j = SV, LV, and SL. At equilibrium, the solute activities in all phases are equal, $a_{SV} = a_{SL} = a_{LV}$. For dilute solutions, the activity can be replaced with the concentration c. Equation (4.11) follows from Eq. (1.11). Differentiation of Young's equation with respect to γ_{LV} and then combining with the Gibbs adsorption isotherm gives

$$\frac{d(\gamma_{LV} \cos \theta)}{d\gamma_{LV}} = \frac{d(\gamma_{SV} - \gamma_{SL})}{d\gamma_{LV}} = \frac{\Gamma_{SV} - \Gamma_{SL}}{\Gamma_{LV}} \qquad (4.12)$$

On the other hand, differentiation of Young's equation with respect to ln c and combining with the Gibbs adsorption isotherm gives

Effect of Solute Adsorption on Contact Angle

$$\frac{d(\gamma_{LV} \cos \theta)}{d \ln c} = \frac{d(\gamma_{SV} - \gamma_{SL})}{d \ln c} = -RT(\Gamma_{SV} - \Gamma_{SL}) \qquad (4.13)$$

In the absence of specific interactions between the solute and the solid, the adsorptions at the liquid-vapor and solid-liquid interfaces will be equal, because the adsorptions originate from the repulsion between the solute and the solvent, as mentioned before. This has been observed in many cases [28-37]. Γ_{LV} can be calculated by the Gibbs adsorption isotherm and the relation

$$\Gamma_{LV} = \gamma_{LV}^\circ - \gamma_{LV} \qquad (4.14)$$

where the superscript ° refers to the pure solvent, and Γ_{SL} can be obtained by

$$\Gamma_{SL} = \gamma_{SL}^\circ - \gamma_{SL} = \gamma_{LV} \cos \theta - \gamma_{LV}^\circ \cos \theta^\circ \qquad (4.15)$$

In the absence of specific interactions, surface excess concentrations are found to be equal, that is, $\Gamma_{LV} = \Gamma_{SL}$.

When physical adsorptions only are involved, a useful parameter k is defined as

$$\frac{\Gamma_{SV} - \Gamma_{SL}}{\Gamma_{LV}} = -k \qquad (4.16)$$

Experimental results show that k is a constant (usually positive). Hence,

$$\gamma_{LV} \cos \theta = k\gamma_{LV} + (k + \cos \theta^\circ)\gamma_{LV}^\circ \qquad (4.17)$$

Since the critical surface tension γ_c can be defined as $\gamma_c = (\gamma_{LV})_{\theta=0}$, Eq. (4.17) becomes

$$\gamma_{LV} \cos \theta = -k\gamma_{LV} + (1 + k)\gamma_c \qquad (4.18)$$

The k usually takes on very simple values. When the adsorptions at liquid-vapor and solid-liquid interfaces are equal and the adsorption at the solid-vapor interface is zero, k = 1. On the other hand, when the adsorptions at all three interfaces (liquid-vapor, solid-liquid, and solid-vapor) are all equal, k = 0. Generally, when there is finite solute adsorption at the solid-vapor interface, $0 \leq k < 1$. This is used to analyze several experimental results below.

4.8.1. Hydrocarbon Surfactant Solutions

Aqueous solutions of hydrocarbon (nonfluorinated) surfactants are used to determine the critical surface tensions of polyethylene and polytetrafluoroethylene [36]. The γ_c values obtained are identical to those obtained with pure liquids, indicating that solute adsorption at the solid-vapor interface is negligible.

Assuming equal adsorptions at the liquid-vapor and solid-liquid interfaces, $k = 1$ and

$$\cos \theta = 2 \frac{\gamma_c}{\gamma_{LV}} - 1 \qquad (4.19)$$

which adequately represents the experimental data (Figure 4.6), where the solid line is drawn according to Eq. (4.19).

4.8.2. Fluorocarbon Surfactant Solutions

On the other hand, significant solute adsorption occurs when aqueous solutions of fluorocarbon surfactants are used to determine the critical

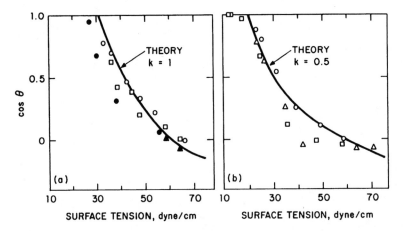

Figure 4.6. Cos θ versus γ_{LV} plots for polyethylene using aqueous solutions of (a) hydrocarbon surfactants (●, sodium di-N-octylsulfosuccinate; ○, sodium di-N-butylsulfosuccinate; ▲ sodium dinonylnaphthalene sulfonate; □, sodium lauryl sulfate); (b) fluorocarbon surfactants (□, φ-octanoic acid; △, φ-decanoic acid; ○, ψ-NH₄ nonanoate). The solid lines are drawn to Eq. (4.18) with k = 1 and 0.5, respectively. (Experimental data are from Refs. 36 and 37.)

surface tension of polyethylene [37]. The γ_c obtained with fluorocarbon surfactant solutions is 20 dyne/cm, whereas that obtained with pure liquids is 31 dyne/cm, indicating that solute adsorption occurs at the solid-vapor interface. Thus, the k value should be between 0 and 1. The experimental data can be represented quite well by letting k = 0.5, that is,

$$\cos \theta = -0.5 + 1.5 \frac{\gamma_c}{\gamma_{LV}} \tag{4.20}$$

shown in Figure 4.6. This indicates that in this case the solute adsorption at the solid-vapor interface is half of those at the liquid-vapor and solid-liquid interfaces, that is, $\Gamma_{LV} = \Gamma_{SL} = 2\Gamma_{SV}$. Consistent with this finding is the fact that the γ_c of a clean polyethylene, after immersion in a fluorocarbon solution, becomes 20 dyne/cm, determined with pure liquids.

In contrast, no solute adsorption at the solid-vapor interface is observed when the fluorocarbon surfactant solutions are used to determine the γ_c of polytetrafluoroethylene [37]. On the other hand, several hydrocarbon surfactants (sodium dodecyl benzene sulfonate and sodium dodecyl sulfate) have also been found to adsorb onto the solid-vapor interfaces on poly(ethylene terephthalate) and polycaprolactam [32].

4.8.3. Aqueous Alcohol Solutions

When determined with aqueous alcohol solutions, the critical surface tensions of many polymers are always found to be about 26 dyne/cm (Table 4.11). This was suggested to arise from adsorption of the alcohols onto the solid-vapor interface [27,38,39]. If equal adsorptions occur at the three interfaces, k = 0 and Eq. (4.18) becomes

$$\gamma_c = \gamma_{LV} \cos \theta \tag{4.21}$$

which is, indeed, found to be true (Table 4.12).

4.9. TABULATION OF EQUILIBRIUM CONTACT ANGLES

Equilibrium contact angles of some pure liquids on smooth and pure low-energy surfaces (20°C) are listed in Table 4.13. They are collected from Refs. 11 and 45-52. Most were reported by Zisman and coworkers [11, 45-51].

Table 4.11. Critical Surface Tension of Some Polymers Determined with n-Alkanol/water Mixtures

Polymer	n-Alkanol/water mixtures	γ_c at 20°C, dyne/cm	References
Nylon 11	Ethanol/water	27	27
	Ethanol/water	26	41
Poly(methyl methacrylate)	Ethanol/water	27	27
	Ethanol/water	26.5	41
	Ethanol/water	26.5	38, 39
	Methanol/water	23	38, 39
	Butanol/water	26	38, 39
	TFE-OH/water [a]	24.7	38, 39
Bovine hoof keratin	Ethanol/water	26.5	27
Human skin	Ethanol/water	27	27
	Ethanol/water	26	43

Tabulation of Equilibrium Contact Angles

Nylon 66	Ethanol/water	28	41
Poly(vinyl chloride)	Ethanol/water	26	41
Polyethylene	Ethanol/water	25.5	41
	Ethanol/water	27.5	36, 37
	Butanol/water	27.5	36, 37
Polystyrene	Ethanol/water	27	41
	Ethanol/water	27	38, 39
	Ethanol/water	27	44
	Methanol/water	21	38, 39
	Methanol/water	27	44
	Propanol/water	27	44
	TFE-OH/water [a]	26.7	38, 39
Poly(ethylene terephthalate)	Ethanol/water	27	41

[a] TFE-OH, 2,2,2-trifluoroethanol.

Table 4.12. Effect of Ethanol Adsorption on the Wettability and Critical Surface Tension of Some Solids Determined with Ethanol/Water Mixtures [a]

Ethanol:water volume ratio	γ_{LV}, dyne/cm	PMMA		PET		Nylon 11		Human skin	
		θ, deg	$\gamma_{LV} \cos \theta$	θ, deg	$\gamma_{LV} \cos \theta$	θ, deg	$\gamma_{LV} \cos \theta$	θ, deg	$\gamma_{LV} \cos \theta$
10:90	51.0	66	20.7	59	26.3	58	27.0	55	29.2
30:70	35.5	44	25.5	55	20.4	42	26.4	43	26.0
50:50	29.7	30	25.7	31	25.5	28	26.2	29	26.0
60:40	27.8	12	27.2	16	26.7	14	27.0	14	27.0
70:30	26.9	0	26.9	0	26.9	0	26.9	0	26.9
γ_c, dyne/cm			27.0		27.0		27.0		27.0

[a] PMMA, poly(methyl methacrylate); PET, poly(ethylene terephthalate). All contact angle data from Ref. 40, except those for PET, which are from Ref. 41.

4.10. DISTORTION OF LIQUID SURFACES: MARANGONI EFFECT

Surface tension gradient (due to composition or temperature variation) can cause local distortion of the liquid surface, known as the *Marangoni effect* [53]. The phenomenon was first reported and correctly explained by Thomson [54]. An excellent review of history and analysis has been given recently [55]. Local variations of surface tension will produce unbalanced tension, causing the liquid to flow from a lower surface tension region to a higher surface tension region. A depression is thus formed on the liquid surface. There are two types of depressions: Benard cells and craters.

4.10.1. Benard Cells

Benard cells usually appear as hexagonal cells with raised edges and depressed centers [56-68]. The driving force for the fluid motion is a slightly lower surface tension at the center of the cells, compared with that at the edges, caused by solvent evaporation.

Theoretical analyses [61-65] establish two characteristic numbers: the Rayleigh number Ra and the Marangoni number Ma, given by

$$Ra = \frac{\rho g \alpha \tau h^4}{K \eta} \quad (4.22)$$

$$Ma = \frac{\tau h^2 (-d\gamma/dT)}{K \eta} \quad (4.23)$$

where ρ is the liquid density, g the gravitational acceleration, α the thermal expansion coefficient, τ the temperature gradient on liquid surface, h the film thickness, K the thermal diffusivity, and η the viscosity. When the critical Rayleigh number is exceeded, the cell

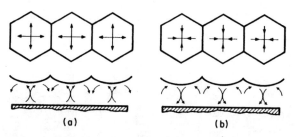

Figure 4.7. Benard cell formation: (a) surface tension gradient-driven; (b) density gradient-driven.

Table 4.13. Equilibrium Contact Angle (Degrees) of Some Liquids on Some Low-Energy Solids at 20°C

Wetting liquid: γ_{LV}, dyne/cm:	Water 72.8	Glycerol 63.4	Formamide 58.2	Methylene iodide 50.8	α-Bromo-naphthalene 44.6	Tricresyl phosphate 40.9	n-Hexadecane 27.6
Organic polymers							
Polyethylene	102	79	77	53	35	34	0
Polytetrafluoroethylene	108	100	92	77	73	75	46
Polytrifluoroethylene	92	82	76	71	61	49	37
Poly(vinylidene fluoride)	82	75	59	63	42	28	24
Poly(vinyl fluoride)	80	66	54	49	33	28	0
Poly(vinylidene chloride)	80	61	61	29	9	10	–
Poly(vinyl chloride)	87	67	66	36	11	14	–
Polystyrene	91	80	74	35	15	–	–

Poly(methyl methacrylate)	80	69	64	41	16	19	0
Polychlorotrifluoroethylene	90	82	82	64	48	44	0
Poly(hexamethylene adipamide)	70	60	50	41	16	–	–
Poly(ethylene terephthalate)	81	65	61	38	15	–	–
Organic compounds							
Paraffin wax	108	96	91	66	47	62	27
Hexatriacontane	111	97	92	77	67	72	46
Monolayers adsorbed on platinum							
Octadecylamine	102	90	81	65	58	65	38
Perfluorodecanoic acid	102	89	–	90	84	82	72
=CCl$_2$ surface	66	56	32	34	5	5	–

Source: Collected from Refs. 11, 45–52.

is formed by a density gradient. The cellular convective flow is downward beneath the central depression and upward beneath the raised edges. On the other hand, when the critical Marangoni number is exceeded, the cell is formed by a surface tension gradient. The cellular convective flow is upward beneath the central depression and downward beneath the raised edges (Figure 4.7). Benard cell formation has also been studied under weightless conditions in space [69]. In polymer coatings, the Benard cell formation is reinforced by two processes: local cooling and local concentration variation. Both give rise to a surface tension gradient.

4.10.2. Craters

Craters are circular depressions on liquid surfaces, arising from a local surface tension gradient caused by low-energy contaminants or surface compositional heterogeneity. Liquids flow from a lower-energy region (depressed crater center) to a higher-energy region (crater rim). The low-energy centers are found sporadically on the liquid surface, forming isolated craters. The theory of Benard cell formation is equally applicable to crater formation.

The flux of fluid q during crater formation is given by [70]

$$q = \frac{h^2(d\gamma/dx)}{2\eta} \qquad (4.24)$$

where $d\gamma/dx$ is the surface tension gradient on the liquid surface. The crater depth d is given by [71]

$$d = \frac{3(d\gamma/dx)R}{\rho g h} = \frac{3(\Delta\gamma)}{\rho g h} \qquad (4.25)$$

where R is the crater radius, and $\Delta\gamma$ is the surface tension difference between high- and low-energy regions. This shows that the gradient $d\gamma/dx$ is the driving force for crater formation. Cratering in polymer coatings is an important industrial problem, deserving serious study. Several discussions of cratering in polymer coatings are available [57,72-74].

4.11. SPREADING OF PARTIALLY SUBMERGED DROPS

A liquid drop resting on a solid and partially submerged in another liquid has wetting behavior quite different from that of a completely submerged liquid drop [75-80]. The wetting of partially submerged drop is important, for instance, in detergency, lubrication, printing, flushing of pigment, and flotation.

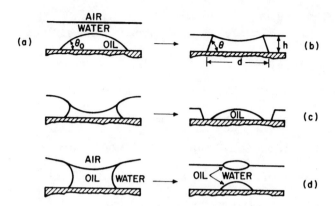

Figure 4.8. Spreading behavior of oil drop completely or partially submerged in water. (After Refs. 75-77.)

The wetting behavior of partially submerged drops is depicted in Figure 4.8. An oil drop resting on a solid and completely submerged in water has a contact angle θ_0, drop height, h_0, and drop base diameter d_0 (Figure 4.8a). As the water level is gradually lowered, the drop shape, θ_0, h_0, and d_0 remain unchanged as long as the water level is higher than the drop height h_0. However, at the instant when water is lowered to h_0, the water film breaks and retracts from the drop apex (the Marangoni effect). The drop elongates and thins so that the drop now has drop base diameter d, drop height h, and contact angle θ (Figure 4.8b). On further lowering of the water level, the oil drop continues to elongate and thin. At the instant when the water level is lowered to a critical disjoining height h_{dc}, the water film retracts rapidly, pulling the oil drop behind. The oil drop is thus forced to spread. Thereafter, the oil film retracts to a drop, exposing bare solid and forming solid-oil-air and solid-water-air interfaces (the disjoining phenomenon) (Figure 4.8c).

On the other hand, if the oil drop is sufficiently large so that the drop height h is greater than a critical splitting thickness h_{sc}, the oil drop will break up into a lens and a submerged drop (splitting phenomenon) (Figure 4.8d). The volumes of the lens and the submerged drop are determined by the works of adhesion at the various interfaces [75-78].

REFERENCES

1. D. P. Graham, J. Phys. Chem., *68*, 2788 (1964).
2. D. P. Graham, J. Phys. Chem., *69*, 4387 (1965).

3. J. W. Whalen, J. Colloid Sci., 28, 443 (1968).
4. E. H. Loeser, W. D. Harkins, and S. B. Twiss, J. Phys. Chem., 57, 251 (1953).
5. G. E. Boyd and H. K. Livingston, J. Am. Chem. Soc., 64, 2383 (1942).
6. P. R. Basford, W. D. Harkins, and S. B. Twiss, J. Phys. Chem., 58, 307 (1954).
7. W. D. Harkins and E. H. Loeser, J. Chem. Phys., 18, 556 (1950).
8. M. E. Tadro, P. Hu, and A. W. Adamson, J. Colloid Interface Sci., 49, 184 (1974).
9. R. E. Johnson, Jr., and R. H. Dettre, J. Colloid Interface Sci., 21, 610 (1966).
10. D. C. Jones and R. H. Ottewill, J. Chem. Soc. Lond., 4076 (1955).
11. H. W. Fox and W. A. Zisman, J. Colloid Sci., 5, 514 (1950).
12. R. J. Good, J. Colloid Interface Sci., 52, 308 (1975).
13. W. D. Harkins, *The Physical Chemistry of Surface Films*, Reinhold, New York, 1952.
14. D. J. Donahue and F. E. Bartell, J. Phys. Chem., 56, 480 (1952).
15. F. D. Petke and B. R. Ray, J. Colloid Interface Sci., 31, 216 (1969). Erratum in J. Colloid Interface Sci., 33, 195 (1970).
16. R. E. Johnson, Jr. and R. H. Dettre, in *Surface and Colloid Science*, Vol. 2, E. Matijevic, ed., Wiley-Interscience, New York, 1969, pp. 85-154.
17. H. Schonhorn, J. Adhes., 1, 38 (1969).
18. A. W. Neumann, G. Haage, and D. Renzow, J. Colloid Interface Sci., 35, 379 (1971).
19. A. W. Neumann, Adv. Colloid Interface Sci., 4, 105 (1974).
20. A. W. Neumann and W. Tanner, J. Colloid Interface Sci., 34, 1 (1970).
21. J. W. Whalen and W. H. Whalen, J. Colloid Interface Sci., 24, 372 (1967).
22. R. J. Good and E. Elbing, Ind. Eng. Chem., 62(3), 54 (1970).
23. E. G. Carter and D. C. Jones, Trans. Faraday Soc., 30, 1027 (1934).
24. W. Fox, J. Am. Chem. Soc., 67, 700 (1945).
25. F. M. Fowkes, Adv. Chem. Ser., 43, 99 (1964).
26. A. F. El-Shimi and E. D. Goddard, J. Colloid Interface Sci., 48, 249 (1974).
27. A. F. El-Shimi and E. D. Goddard, J. Colloid Interface Sci., 48, 242 (1974).
28. F. M. Fowkes and W. D. Harkins, J. Am. Chem. Soc., 62, 3377 (1940).
29. F. M. Fowkes, J. Phys. Chem., 57, 98 (1953).
30. E. H. Lucassen-Reynders, J. Phys. Chem., 67, 969 (1963).

References

31. C. A. Smolders, in *Wetting*, Society of Chemical Industry, London, Monograph No. 25, Gordon and Breach, London, 1967. pp. 318-327.
32. C. A. Smolders, Proc. 4th Int. Congr. Surface Active Subst., 2, 2, 343 (1967).
33. E. Wolfram, Proc. 4th Int. Congr. Surface Active Subst., 2, 351 (1967).
34. E. Wolfram, Kolloid Z. Z. Polym., *211*, 84 (1966).
35. D. Bargeman and F. Van Voorst Vader, J. Colloid Interface Sci., *42*, 467 (1973).
36. M. K. Bernett and W. A. Zisman, J. Phys. Chem., *63*, 1241 (1959).
37. M. K. Bernett and W. A. Zisman, J. Phys. Chem., *63*, 1911 (1959).
38. W. J. Murphy, M. W. Roberts, and J. R. H. Ross, J. Chem. Soc., Faraday Trans., I, *68*, 1190 (1972).
39. S. Frost, W. J. Murphy, M. W. Roberts, and J. R. H. Ross, Faraday Spec. Discuss., *2*, 198 (1972).
40. A. F. El-Shimi, private communication, 1974.
41. J. R. Dann, J. Colloid Interface Sci., *32*, 302 (1970).
42. J. R. Dann, J. Colloid Interface Sci., *32*, 321 (1970).
43. A. W. Adamson, K. Kunichika, F. Shirley, and M. J. Orem, J. Chem. Educ., *45*, 702 (1968).
44. R. H. Ottewill and B. Vincent, J. Chem. Soc., Faraday Trans. I, *68*, 1533 (1972).
45. H. W. Fox and W. A. Zisman, J. Colloid Sci., *7*, 109 (1952).
46. H. W. Fox and W. A. Zisman, J. Colloid Sci., *7*, 428 (1952).
47. A. H. Ellison and W. Z. Zisman, J. Phys. Chem., *58*, 260 (1954).
48. A. H. Ellison and W. A. Zisman, J. Phys. Chem., *58*, 503 (1954).
49. N. L. Jarvis, R. B. Fox, and W. A. Zisman, Adv. Chem. Ser., *43*, 317 (1964).
50. E. G. Shafrin and W. A. Zisman, J. Colloid Sci., *7*, 166 (1952).
51. F. Shulman and W. A. Zisman, J. Colloid Sci., *7*, 465 (1952).
52. S. Wu, Colloid Interface Sci., *71*, 605 (1979).
53. C. G. M. Marangoni, Nuovo Cimento, *2*(5-6), 239 (1971); *3*(3), 97, 193 (1878); Ann. Phys. (Poggendorff), *143*, 337 (1871).
54. J. Thomson, Phil. Mag. Ser., 4, *10*, 330 (1855).
55. L. E. Scriven and C. V. Sterling, Nature, *187*, 186 (1960).
56. C. M. Hansen and P. E. Pierce, Ind. Eng. Chem., Prod. Res. Dev., *12*, 67 (1973).
57. C. M. Hansen and P. E. Pierce, Ind. Eng. Chem., *13*, 218 (1974).
58. H. Jebsen-Marwedel and G. Marwedel, Farbe Lack, *66*, 314 (1960).
59. G. Marwedel and H. Jebsen-Marwedel, Farbe Lack, *71*, 189 (1965).

60. M. J. Block, Nature, *178*, 650 (1956).
61. J. R. A. Pearson, J. Fluid Mech., *4*, 489 (1958).
62. J. C. Berg, in *Advances in Chemical Engineering*, Vol. 6, T. B. Drew and J. W. Hoopes, eds., Academic Press, New York, 1966.
63. L. E. Scriven and C. V. Sterling, J. Fluid Mech., *19*, 324 (1964).
64. D. A. Nield, J. Fluid Mech., *19*, 341 (1964).
65. K. A. Smith, J. Fluid Mech., *24*, 401 (1966).
66. J. N. Anand and R. Z. Balwinski, J. Colloid Interface Sci., *31*, 196 (1969).
67. J. N. Anand, J. Colloid Interface Sci., *31*, 203 (1969).
68. J. N. Anand and H. J. Karam, J. Colloid Interface Sci., *31*, 208 (1969).
69. G. Grodzka and T. C. Bannister, Science, *176*, 506 (1972).
70. P. Fink-Jensen, Farbe Lack, *68*, 155 (1962).
71. A. V. Hersey, Phys. Rev. Ser. 2, *56*, 204 (1939).
72. S. Gusman, Off. Dig., *24*, 296 (1952).
73. F. J. Hahn, J. Paint Technol., *43*(562), 58 (1971).
74. F. J. Hahn and S. Steinhauer, J. Paint Technol., *47*(606), 54 (1975).
75. A. C. Zettlemoyer, M. P. Aronson, and J. A. Lavelle, J. Colloid Interface Sci., *34*, 545 (1970).
76. M. C. Wilkinson, M. P. Aronson, and A. C. Zettlemoyer, J. Colloid Interface Sci., *37*, 498 (1971).
77. M. P. Aronson, A. C. Zettlemoyer, and M. C. Wilkinson, J. Phys. Chem., *77*, 318 (1973).
78. M. P. Aronson, A. C. Zettlemoyer, R. Codell, and M. C. Wilkinson, J. Colloid Interface Sci., *52*, 1 (1975).
79. E. R. Washburn and A. E. Anderson, J. Phys. Chem., *50*, 401 (1946).
80. E. J. Clayfield, D. J. Dean, J. B. Mathews, and T. V. Witham, Proc. 2nd Int. Congr. Surf. Act., *3*, 165 (1957).

5
Surface Tension and Polarity of Solid Polymers

The surface tension of a solid polymer cannot be measured directly, as reversible formation of its surface is difficult. Many indirect methods have been proposed, including the liquid homolog (molecular weight dependence) method, polymer melt (temperature dependence) method, equation of state method, harmonic-mean method, geometric-mean method, critical surface tension, and others. The first four methods give remarkably consistent and reliable results, and are preferred [1]. The last two methods often give variable and low values. In this chapter, these methods are discussed. The constitutive and morphological effects on solid surface tensions are also analyzed.

5.1. DETERMINATION OF SURFACE TENSION AND POLARITY

5.1.1. Liquid Homolog (Molecular Weight Dependence) Method

The molecular-weight dependence of liquid homologs can be accurately fitted into an appropriate equation and extrapolated to a high-molecular-weight range to obtain the surface tension of a solid polymer. Either the $\gamma^{1/4}$ versus M^{-1} or the γ versus $M^{-2/3}$ relation may be used (Sections 3.1.3 and 3.1.5). The standard deviation of the fit is as low as 0.05 dyne/cm when highly accurate data for the liquid homologs are used. The $\gamma^{1/4}$ versus M^{-1} plot gives extrapolated values in remarkable agreement with those obtained by the polymer melt method (Sections 3.1.2 and 3.1.5).

Table 5.1. Comparison of Solid Surface Tensions Obtained by Various Methods

Material	Surface tension, γ_S, at 20°C, dyne/cm					Zisman's critical surface tension $\gamma_{c,Z}$, dyne/cm
	Equation of state method	Polymer melt method	Liquid homolog method	Harmonic-mean method	Geometric-mean method	
Polymers						
Polyethylene	35.9	35.7	34.7	36.1	33.2	31
Polytetrafluoroethylene	22.6	–	23.9	22.5	19.1	18
Polytrifluoroethylene	29.5	–	–	27.3	23.9	22
Poly(vinylidene fluoride)	36.5	–	–	33.2	30.3	25
Poly(vinyl fluoride)	37.5	–	–	38.4	36.7	28
Polyhexafluoropropylene	18.0	–	–	17.0	12.8	17
Poly(vinylidene chloride)	45.2	–	–	45.4	45.0	40
Poly(vinyl chloride)	43.8	–	–	41.9	41.5	39
Polystyrene	43.0	40.7	–	42.6	42.0	33
Poly(methyl methacrylate)	42.5	41.1	–	41.2	40.2	39
Polychlorotrifluoroethylene	32.1	30.9	–	30.1	27.5	31
TFC/CTFE 60/40 [a]	26.1	–	–	25.2	20.9	24
TFE/CTFE 80/20 [b]	22.1	–	–	21.3	16.5	20
E/TFE 50/50 [c]	27.1	–	–	27.6	24.1	26.5

Determination of Surface Tension and Polarity 171

Poly(hexamethylene adipamide)	43.8	46.5	–	44.7	47.0	46
Poly(ethylene terephthalate)	44.0	44.6	–	42.1	41.3	43
Organic solids						
Paraffin wax	32.0	35.0	–	31.0	25.4	23
Hexatriacontane	23.0	–	31.4	23.6	19.1	21
Monolayers (adsorbed on platinum)						
Octadecylamine	30.5	–	–	28.1	25.8	22
Perfluorolauric acid	14.5	–	–	–	–	5.6
Perfluorodecanoic acid	26.2	–	–	19.4	14.1	11
—CF_2H surface (monohydroperfluoroundecanoic acid)	26.5	–	–	–	–	15
=CCl_2 surface (perchloro-2,4-pentadienoic acid)	49.5	–	–	48.4	46.2	43

[a] TFE/CTFE 60/40 = copolymer of tetrafluoroethylene and chlorotrifluoroethylene at 60:40 weight ratio.
[b] TFE/CTFE 80/20 = the previous copolymer at 80:20 weight ratio.
[c] E/TFE 50/50 = copolymer of ethylene and trifluoroethylene at 50:50 weight ratio.
Source: From Ref. 1.

Examples of the liquid homolog method are discussed in section 3.1.5. The values obtained are for amorphous solid surfaces. Macleod's equation may be used to obtain the value for a crystalline surface. The liquid homolog method gives solid surface tensions in excellent agreement with those obtained by the polymer melt method, the equation of state method, and the harmonic-mean method [1] (Table 5.1). The method is regarded as accurate and preferred.

5.1.2. Polymer Melt (Temperature Dependence) Method

The surface tension of polymer melts can be extrapolated to lower temperatures, where the polymer is a solid. The surface tension versus temperature can be accurately fitted with the Guggenheim equation [Eq. (3.1)] and extrapolated to the solid range. However, as the critical temperature T_c of a polymer is usually quite high (about 700°C), a linear relation is sufficiently accurate over ordinary temperature ranges (such as 0-200°C) (Section 3.1.1). In this case, the effect of secondary transition is small and may be neglected. The values obtained are for amorphous solid surfaces. Macleod's equation can be used to obtain those for crystalline surfaces (Section 3.1.4).

Examples of the polymer melt method are discussed in Chapter 3. The results obtained agree remarkably well with those obtained by the liquid homolog method, the equation of state method, and the harmonic-mean method [1] (Table 5.1). The polarity of a polymer can be determined from the interfacial tension between the polymer melt and a nonpolar polymer melt (such as polyethylene) (Section 3.2.4). The polarity has been shown to be independent of temperature (Section 3.2.4). The method is regarded as accurate and preferred.

5.1.3. Equation of State Method

This method uses the contact angles of a series of testing liquids to obtain a spectrum of critical surface tension $\gamma_{c,\phi}$ by using a proposed equation of state. The $\gamma_{c,\phi}$ is then plotted against γ_{LV} to obtain a curve (the equation of state plot), whose maximum value $\gamma_{c,\phi,max}$ equals the surface tension of the solid γ_S. The results obtained agree remarkably well with those obtained by the liquid homolog method, the polymer melt method, and the harmonic-mean method [1]. The method is regarded as accurate and preferred.

Combining the Young equation with the equation of Good and Girifalco gives [1]

$$\cos \theta = 2\phi \left(\frac{\gamma_S}{\gamma_{LV}}\right)^{1/2} - 1 - \frac{\pi_e}{\gamma_{LV}} \qquad (5.1)$$

Determination of Surface Tension and Polarity 173

where θ is the equilibrium contact angle, ϕ the interaction parameter, γ_S the surface tension of the solid in vacuum, γ_{LV} the surface tension of the liquid in equilibrium with its saturated vapor, and π_e the equilibrium spreading pressure. The ϕ has been explicitly expressed in terms of molecular parameters or macroscopic polarities of the two phases (Sections 3.2.2 and 3.2.3). The critical surface tension may be defined as

$$\gamma_c = \lim_{\theta \to 0} \gamma_{LV} \qquad (5.2)$$

which follows from Section 5.1.6. Combining with Eq. (5.1) and expanding in a power series gives [1]

$$\gamma_c = \phi^2 \gamma_S - \pi_e + \frac{\pi_e^2}{4\phi^2 \gamma_S} + \cdots \qquad (5.3)$$

which rapidly converges, and therefore may be truncated to

$$\gamma_{c,\phi} = \phi^2 \gamma_S - \pi_e \qquad (5.4)$$

where $\gamma_{c,\phi}$ is used to replace γ_c, that is, $\gamma_{c,\phi} \equiv \gamma_c$, to emphasize the fact that γ_c is a function of ϕ. Using Eq. (5.4) in (5.1) then gives [1]

$$\cos \theta = 2 \left(\frac{\gamma_{c,\phi}}{\gamma_{LV}} \right)^{1/2} - 1 \qquad (5.5)$$

which is strikingly simple, and is termed the *equation of state* [1].

The equation of state defines a spectrum of critical surface tensions for a given solid when a series of testing liquids are used. Equation (5.5) can be rearranged to

$$\gamma_{c,\phi} = \frac{1}{4}(1 + \cos \theta)^2 \gamma_{LV} \qquad (5.6)$$

which can be used to calculate $\gamma_{c,\phi}$ using the θ of any one testing liquid. A spectrum of $\gamma_{c,\phi}$ values will thus be obtained for a series of testing liquids. Since $\phi_{max} = 1$ when the polarities of the two phases are equal, Eq. (5.6) gives

$$\gamma_{c,\phi,max} = \gamma_S - \pi_e \simeq \gamma_S \qquad (5.7)$$

where the last equality is valid when $\theta \gg 0$.

Thus, if the $\gamma_{c,\phi}$ spectrum is plotted versus the polarity of the testing liquids, a smooth curve should be obtained, having the maximum $\gamma_{c,\phi,max} = \gamma_S$. However, since the polarities of many testing liquids are unknown, $\gamma_{c,\phi}$ may instead be plotted versus γ_{LV}. In this case, the data will be scattered, since $\gamma_{c,\phi}$ depends not only on γ_{LV}, but also on ϕ. However, a smooth curve can be drawn just to encompass all the data points below the curve, such as illustrated in Figures 5.1 to 5.4 for some polymers and organic solids. The maximum of such a curve $\gamma_{c,\phi,max}$ is the surface tension of the solid γ_S. To be accurate, the testing liquids used should cover the polarity range of the solid in question. This can readily be done by selecting a series of liquids ranging from nonpolar hydrocarbons to very polar liquids such as water. Some useful testing liquids are listed in Table 5.2. Zisman's critical surface tensions, designated here as $\gamma_{c,Z}$, are also shown in Figures 5.1 to 5.4. It can be seen that the Zisman's critical surface tension can in fact be determined with a single testing liquid having a low contact angle by using Eq. (5.6). Thus Zisman plot is unnecessary.

As examples, Table 5.2 gives the $\gamma_{c,\phi}$ spectra for polyethylene and polytetrafluoroethylene. The $\gamma_{c,\phi,max}$ occurs at large θ, so that $\pi_e \sim 0$ should be valid. Further, note for instance that $\gamma_{c,\phi}$ of polyethylene is 15.8 dyne/cm if water is used as the testing liquid. This means that the critical surface tension for polyethylene will be 15.8 dyne/cm if a series of testing liquids of varying surface tension but of the same polarity as water are used as the testing liquids with a Zisman plot.

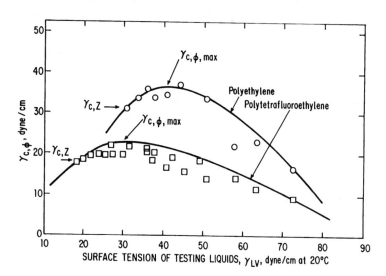

Figure 5.1. Equation of state plots for polyethylene and polytetrafluoroethylene. (After Ref. 1.)

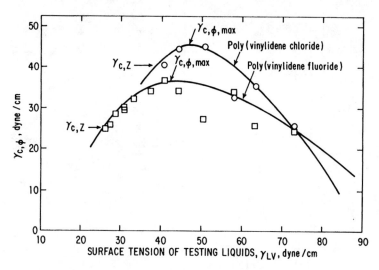

Figure 5.2. Equation of state plots for poly(vinylidene chloride) and poly(vinylidene fluoride). (After Ref. 1.)

The γ_S values obtained by the equation of state method for some polymers and organic solids are listed in Table 5.1. Excellent agreement of the equation of state method with the liquid homolog method, the polymer melt method, and the harmonic-mean method can be seen [1].

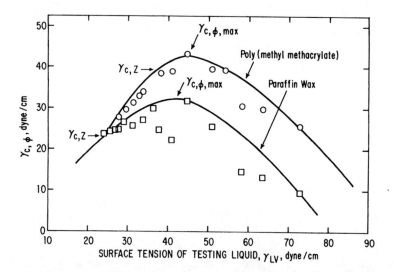

FIGURE 5.3. Equation of state plots for poly(methyl methacrylate) and paraffin wax. (After Ref. 1.)

Table 5.2. Calculation of Critical Surface Tension Spectra for Polyethylene and Polytetrafluoroethylene at 20°C

Liquids	γ_{LV}, dyne/cm	Polyethylene		Polytetrafluoroethylene	
		θ, deg	$\gamma_{c,\phi}$ dyne/cm	θ, deg	$\gamma_{c,\phi}$ dyne/cm
n-Alkanes					
Hexane	18.4	—	—	12	18.0
Heptane	20.3	—	—	21	19.0
Octane	21.8	—	—	26	19.7
Decane	23.9	—	—	35	19.8
Dodecane	25.4	—	—	42	19.3
Hexadecane	27.6	—	—	46	19.8
n-Alkybenzenes					
Hexylbenzene	30.0	—	—	52	19.6
Benzene	28.9	—	—	46	20.7
Halocarbons					
Methylene iodide	50.8	52	33.2	88	13.6
sym-Tetrabromomethane	49.7	—	—	79	17.6

Tetrachlorobiphenyl	45.3	–	–	81	15.1
α-Bromonaphthalene	44.6	35	36.9	73	18.6
Hexachloropropylene	38.1	–	–	65	19.3
Perchlorocyclopentadiene	37.5	–	–	67	18.1
sym-Tetrachloroethane	36.3	10	35.7	56	22.1
Hexachlorobutadiene	36.0	–	–	60	20.2
Tetrachloroethylene	31.7	–	–	49	21.7
Carbon tetrachloride	26.8	–	–	36	21.9
Esters					
Tricresyl phosphate	40.9	34	34.2	75	16.2
Benzylphenyl undecanoate	37.7	28	33.4	67	18.2
Bis(2-ethylhexyl)-phthalate	31.2	5	31.1	63	16.5
Bis(2-ethylhexyl)-sebacate	31.1	–	–	62	16.8
Miscellaneous					
t-Butyl naphthalene	33.7	7	33.4	65	17.0
Formamide	58.2	77	21.8	92	13.5
Glycerol	63.4	79	22.5	100	10.8
Water	72.8	94	15.8	108	8.7
Mercury	485	–	–	150	2.2

Source: From Ref. 1.

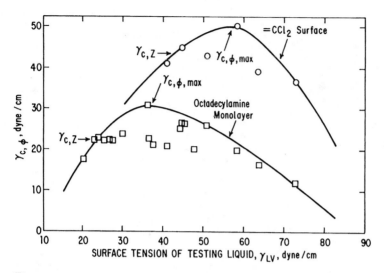

Figure 5.4. Equation of state plots for =CCl$_2$ surface (adsorbed monolayer of perchloro-2,4-pentadienoic acid) and —CH$_3$ surface (adsorbed monolayer of octadecylamine). (After Ref. 1.)

5.1.4. Harmonic-Mean Method

This method uses the contact angles of two testing liquids and the harmonic-mean equation [2]. The results agree remarkably well with the liquid homolog method, the polymer melt method, and the equation of state method. The method is regarded accurate and preferred. The harmonic-mean equation has already been shown to predict the interfacial tensions between polymer melts and liquids reliably (Section 3.2.5).

Using the harmonic-mean equation in the Young's equation gives [2]

$$(1 + \cos \theta_1)\gamma_1 = 4\left(\frac{\gamma_1^d \gamma_S^d}{\gamma_1^d + \gamma_S^d} + \frac{\gamma_1^p \gamma_S^p}{\gamma_1^p + \gamma_S^p}\right) \quad (5.8)$$

$$(1 + \cos \theta_2)\gamma_2 = 4\left(\frac{\gamma_2^d \gamma_S^d}{\gamma_2^d + \gamma_S^d} + \frac{\gamma_2^p \gamma_S^p}{\gamma_2^p + \gamma_S^p}\right) \quad (5.9)$$

where $\gamma = \gamma^d + \gamma^p$ and the subscripts 1 and 2 refer to the testing liquids 1 and 2, respectively. If γ_j^d and γ_j^p of the testing liquids (j = 1 and 2) are known, the dispersion and polar components of solid surface tension ($\gamma_S{}^d$ and $\gamma_S{}^p$) can be obtained from the contact angles θ_1 and θ_2 by solving the two simultaneous equations. Water and methylene iodide are two convenient testing liquids, whose preferred γ^d and γ^p values are listed in Table 5.3. These values are determined from interfacial tensions between water and n-alkanes, or between water and methylene iodide, using the harmonic-mean equation.

Equations (5.8) and (5.9) are a set of simultaneous quadratic equations, which can be solved readily by hand. The algebraic manipulations are, however, rather tedious, although straightforward. Therefore, it is convenient to do the calculations with a computer. An interactive computer program for this purpose is given in Appendix I. Readers would find this program quite useful.

The surface tension and polarity of some materials obtained by the harmonic-mean method are compared with those by the polymer melt method (Table 5.4). Both the surface tensions and polarities obtained by the two methods agree remarkably well. In contrast, the geometric-mean method gives too low values. The surface tensions obtained by various methods are compared in Table 5.1. The harmonic-mean

Table 5.3. Preferred Values of Surface Tension and Its Components for Water and Methylene Iodide Used for the Calculation of Surface Tension of Solid Polymers from Contact Angles

Liquid	Surface tension at 20°C, dyne/cm			Remark
	γ	γ^d	γ^p	
Harmonic-mean equation				
Water	72.8	22.1	50.7	a
Methylene iodide	50.8	44.1	6.7	b
Geometric-mean equation				
Water	72.8	21.8	51.0	a
Methylene iodide	50.8	49.5	1.3	b
		48.5	2.3	c

^a From interfacial tension between water and hydrocarbon; see Table 4.7.

^b From interfacial tension between water and methylene iodide, γ_{12} = 41.6 dyne/cm [3,4].

^c From contact angles on nonpolar solids; see Table 4.8.

Table 5.4 Comparison of Surface Tension and Polarity as Determined by Various Methods at 20°C [a]

Polymer	Contact angle, deg		Harmonic-mean method		Geometric-mean method		Molten polymer data	
	θ_w	θ_m	γ	x^p	γ	x^p	γ	x^p
Polyethylene	102	53	36.1	0.022	33.2	0	35.7	0
Polytetrafluoroethylene	108	77	22.5	0.089	19.1	0.026	26.5	-
Polyhexafluoropropylene	112	90	17.0	0.174	12.8	0.066	-	-
Polystyrene	91	35	42.6	0.099	42.0	0.014	40.7	0.168
Poly(methyl methacrylate)	80	41	41.2	0.245	40.2	0.107	41.1	0.281
Poly(n-butyl methacrylate)	91	52	34.6	0.180	33.3	0.060	31.2	0.160
Polychlorotrifluoroethylene	90	64	30.1	0.282	27.0	0.141	31.1	-
Poly(vinylidene fluoride)	82	63	33.2	0.376	30.3	0.234	36.5	-
Poly(vinyl fluoride)	80	49	38.4	0.292	36.7	0.147	37.5	-
Poly(vinylidene chloride)	80	29	45.4	0.196	45.0	0.067	45.2	-
Poly(vinyl chloride)	87	36	41.9	0.146	41.5	0.036	43.8	0.11
Poly(ethylene terephthalate)	81	38	42.1	0.221	41.3	0.085	44.6	-
Poly(hexamethylene adipamide)	70	41	44.7	0.344	43.2	0.211	46.5	-
Paraffin wax	108	66	31.0	0	25.4	0	35.0	0

[a] θ_w, contact angle of water; θ_m, contact angle of methylene iodide. Polarity $x^p = \gamma^p/\gamma$.

Source: From Ref. 2.

Determination of Surface Tension and Polarity

method is remarkably consistent with the liquid homolog method, the polymer melt method, and the equation of state method.

5.1.5. Geometric-Mean Method

This is based on the contact angles of two testing liquids and the geometric-mean equation, which is unsatisfactory for low-energy surfaces but is preferred for high-energy surfaces (Section 3.2.3). Therefore, the method is inadequate for low energy surfaces, but should be preferred for high-energy surfaces.

Combining the geometric-mean equation with Young's equation gives

$$(1 + \cos \theta_1)\gamma_1 = 2[(\gamma_1^d \gamma_S^d)^{1/2} + (\gamma_1^p \gamma_S^p)^{1/2}] \qquad (5.10)$$

$$(1 + \cos \theta_2)\gamma_2 = 2[(\gamma_2^d \gamma_S^d)^{1/2} + (\gamma_2^p \gamma_S^p)^{1/2}] \qquad (5.11)$$

where the subscripts 1 and 2 refer to the testing liquids 1 and 2, respectively. When two testing liquids of known surface tension and its components are used to measure the contact angle, Eqs. (5.10) and (5.11) can be solved simultaneously to give γ_S^d and γ_S^p of the solid. Water and methylene iodide are a convenient choice for the testing liquids [5]. The preferred surface tension and its components for water and methylene iodide are listed in Table 5.3. Other liquid pairs have also been used [6]. However, different liquid pairs tend to give different results. For instance, for poly(hexamethylene adipamide) at 20°C, the γ varies from 35.7 to 45.8 dyne/cm, the γ^d from 20.7 to 40.2 dyne/cm, and the γ^p from 3.0 to 14.9 dyne/cm with different pairs of testing liquids [6].

The surface tensions and polarities for some low-energy solids obtained by the geometric-mean method are compared in Tables 5.1 and 5.4 with those obtained by other methods. Both the surface tension and the polarity values obtained by the geometric-mean method are often much too low.

5.1.6. Critical Surface Tension

The concept of critical surface tension was first proposed by Fox and Zisman [7-9]. An empirical rectilinear relation was found between $\cos \theta$ and γ_{LV} for a series of testing liquids on a given solid. When homologous liquids are used as the testing liquids, a straight line is often obtained. When nonhomologous liquids are used, however, the data are often scattered within a rectilinear band or give a curved line.

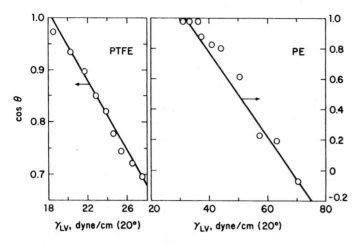

Figure 5.5. Zisman plots for polytetrafluoroethylene (PTFE) using n-alkanes as the testing liquids, and for polyethylene (PE) using various testing liquids. (After Refs. 7 and 9.)

The intercept of the line at cos θ = 1 is the critical surface tension γ_c. When a band is obtained, the intercept of the lower line of the band is defined as the *critical surface tension*. The cos θ versus γ_{LV} plot is known as the *Zisman plot*. Some examples are given in Figs. 5.5 and 5.6.

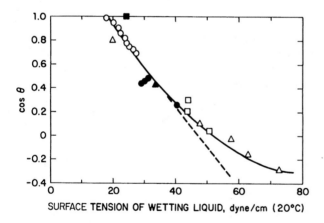

Figure 5.6. Zisman plot for polytetrafluoroethylene using various testing liquids: ○, n-alkanes; ▲, miscellaneous hydrocarbons; ●, esters; □, nonfluoro halocarbons; ■, fluorocarbons; △, miscellaneous liquids (After Ref. 8.)

Determination of Surface Tension and Polarity

The critical surface tension equals the surface tension of the liquid which just exhibits a zero contact angle on the solid. Therefore,

$$\gamma_c = \lim_{\theta \to 0} \gamma_{LV}$$

$$= \gamma_S - \gamma_{SL} - \pi_e \tag{5.12}$$

which follows from the Young equation and indicates that γ_c is smaller than γ_S by the amount $(\gamma_{SL} + \pi_e)$. The interfacial tension γ_{SL} varies with the nature of the testing liquid, and will be very large when the polarities of the testing liquid and the solid are very different (see Chapter 3). Furthermore, the π_e also varies with the nature of the testing liquid, and can be very large when the contact angle is small (particularly near the zero contact angle) (see Chapter 4). Therefore, γ_c will vary with the testing liquids used, and tends to be much smaller than γ_S. For instance, depending on the testing liquids used, the γ_c of poly(ethylene terephthalate) varies from 27 to 46 dyne/cm at 20°C [10], and the γ_c of perfluorodecanoic acid monolayer (adsorbed on platinum) varies from 11 to 27.5 dyne/cm at 20°C [11]. Therefore, γ_c values must be used with due caution. Critical surface tensions and surface tensions of many materials at 20°C are collected and compared in Tables 5.1 and 5.5.

The empirical linear relation between $\cos \theta$ and γ_{LV} may be expressed by

$$\cos \theta = 1 + k(\gamma_c - \gamma_{LV}) \tag{5.13}$$

where the constant k is the slope of the Zisman plot. The k varies with the testing liquids used on a given solid, but is often about 0.03 to 0.04 cm/dyne [27]. Several attempts were made to justify the linear relation theoretically, but none were successful [28-37]. Thus, the relation above is purely empirical. A correct theoretical relation is the equation of state given in Eq. (5.5). Using $k = -d \cos \theta / d\gamma_{LV}$ in Eq. (5.5) gives

$$k = \phi \left(\frac{\gamma_S}{\gamma_{LV}^3} \right)^{1/2} = \left(\frac{\gamma_c}{\gamma_{LV}^3} \right)^{1/2} \tag{5.14}$$

which shows that the slope k cannot be constant in the Zisman plot. The curvatuve in Zisman plot was originally thought to arise from specific interactions (such as hydrogen bonding) between the liquids and the solid [8]. A curved Zisman plot for polytetrafluoroethylene is shown in Figure 5.6. However, any curvature in a Zisman plot can, in fact, be quantitatively explained by the variability of k as given by Eq. (5.14) (Figure 5.7).

Table 5.5. Comparison of Surface Tension and Critical Surface Tension at 20°C

Material	Surface tension		Critical surface tension	
	γ, dyne/cm	Reference	γ_c, dyne/cm	Reference
Hydrocarbon polymers				
Polyethylene	35.7	12	31	13
Polypropylene	30.1	12	32	14
Polyisobutylene	33.6	12	27	14
Polystyrene	40.7	12	33	13
Halogenated hydrocarbon polymers				
Polychloroprene	43.6	12	38	14
Poly(vinyl chloride)	42.9	1	39	13
Poly(vinylidene chloride)	45.2	1	40	13
Poly(vinyl fluoride)	37.5	1	28	13
Poly(vinylidene fluoride)	36.5	1	25	13
Polytetrafluoroethylene	23.9	1	19	13
Polyhexafluoropropylene	17.0	1	16	13
Polychlorotrifluoroethylene	32.1	1	31	13
Polytrifluoroethylene	29.5	1	22	13
TFE/CTFE 60/40 [a]	26.2	1	24	8
TFE/CTFE 80/20 [b]	22.1	1	20	8
E/TFE 50/50 [c]	27.1	1	26.5	8

Vinyl polymers				
Poly(vinyl acetate)	36.5	12	33	15
Poly(vinyl alcohol)	-	-	37	13
Poly(vinyl formal)	-	-	39	14
Poly(vinyl butyral)	-	-	38	15
Acrylic polymers				
Poly(methyl acrylate)	40.1	15	35	15
Poly(ethyl acrylate)	37.0	15	33	15
Poly(butyl acrylate)	33.7	15	31	15
Poly(ethylhexyl acrylate)	30.2	15	31	15
Methacrylic polymers				
Poly(methyl methacrylate)	41.1	12	39	13
Poly(ethyl methacrylate)	35.9	15	31.5	16
Poly(n-propyl methacrylate)	33.2	15	32	15
Poly(n-butyl methacrylate)	31.2	12	32	15
Poly(n-hexyl methacrylate)	30.0	16	27.5	16
Poly(octyl methacrylate)	28.8	16	23.5	16
Poly(lauryl methacrylate)	32.8	-	21.3	16
Poly(stearyl methacrylate)	36.3	-	20.8	16
Poly(phenyl methacrylate)	-	-	35	17
Poly(benzyl methacrylate)	-	-	36	17
Poly(hydroxyethyl methacrylate)	-	-	37	15
Poly(dimethylaminoethyl-methacrylate)	-	-	36	15
Poly(t-butylaminoethyl methacrylate)	-	-	34	15

Table 5.5 (Continued)

Material	Surface tension		Critical surface tension	
	γ, dyne/cm	Reference	γc, dyne/cm	Reference
Polyethers				
Polyoxymethylene	-	-	38	14
Poly(ethylene oxide)-diol	42.9	12	43	14
Poly(ethylene oxide)-dimethylether	37.1	12	-	-
Poly(propylene oxide)-diol	30.9	12	32	14
Poly(propylene oxide)-dimethylether	30.2	12	-	-
Poly(tetramethylene oxide)-diol	31.9	12	-	-
Polyepichlorohydrin	43.2	12	35	14
Polyester				
Poly(ethylene terephthalate)	42.1	12	43	13
Polyamides				
Polycaproamide, nylon 6	-	-	42	14
Poly(hexamethylene adipamide), nylon 66	44.7	1	46	13
Poly(heptamethylene pimelamide), nylon 77	-	-	43	14

Determination of Surface Tension and Polarity

Poly(octamethylene suberamide), nylon 88	–	34	14
Poly(nonamethylene azelamide), nylon 99	–	36	14
Poly(decamethylene sebacamide), nylon 1010	–	32	14
Polyundecylamide, nylon 11	–	33	14
Silicones			
Polydimethylsiloxane	19.9	24	14
Polydiethylsiloxane	25.7	–	–
Polymethylphenylsiloxane	26.1	–	–
Miscellaneous polymers			
Polyacrylonitrile	–	50	15
Polymethacrylonitrile	–	39	15
Nitrocellulose	–	38	15
Cellulose acetate butyrate	–	34	15
Ethyl cellulose	–	32	15
Amine-cured epoxy resin	–	44	15
Urea-formaldehyde resin	–	61	14
Wool	–	45	14
Organic solids			
Paraffin wax	32.0	23	9
Hexatriacontane	23.3	21	9

Table 5.5. (Continued)

Material	Surface tension		Critical surface tension	
	γ, dyne/cm	Reference	γ_c, dyne/cm	Reference
Monolayers (adsorbed on platinum)				
Octadecylamine	30.5	1	22	18
Perfluoroaluric acid	14.5	1	5.6	19
Perfluorodecanoic acid	26.2	1	11	11
—CF_2H surface (monohydro-perfluoroundecanoic acid)	26.5	1	15	20
=CCl_2 surface (perchloro-2,4-pentadienoic acid)	49.0	1	43	21
Organic pigments				
Metal-free phthalocyanine	52.8	22	35.6	23
Copper phthalocyanine	46.9	22	31.3	23

Determination of Surface Tension and Polarity

Chlorinated copper phthalocyanine (green)	42.0	22	27.5	23
Toluidine red	53.0	22	27.5	23
Liquids				
$F(CF_2)_2CH_2OH$	17.4	24	13.2	24
$H(CF_2)_2CH_2OH$	26.0	24	17.8	24
$H(CF_2)_4CH_2OH$	23.8	24	17.5	24
n-Hexadecane	27.3	25	19.5	25
Mineral oil	29.9	25	18.5	25
Squalene	31.4	25	26.7	25
Water	72.8	24	19.1 ~ 21.7	24, 26

[a] TFE/CTFE 60/40, copolymer of tetrafluoroethylene and chlorotrifluoroethylene at 60:40 weight ratio.
[b] TFE/CTFE 80/20, the previous copolymer at 80:20 weight ratio.
[c] E/TFE 50/50, copolymer of ethylene and tetrafluoroethylene at 50:50 weight ratio.

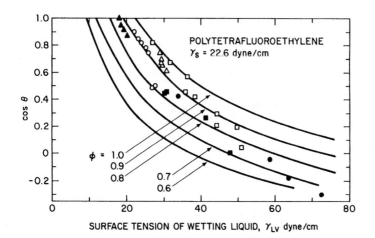

Figure 5.7. Theoretical cos θ versus γ_{LV} plots. Solid lines are drawn to the equation of state at several different ϕ values. Experimental data are for polytetrafluoroethylene. The wetting liquids used are: ○, n-alkanes; △, n-alkyl benzenes; ▲, siloxanes; □, halocarbons; ■, esters; ●, miscellaneous liquids.

The relation between γ_c and γ_S can also be given as

$$\gamma_S = \frac{(2\gamma_c + \pi_e)^2}{4\phi^2 \gamma_c} \tag{5.15}$$

which is obtained by combining Eqs. (5.1) and (5.2) and rearranging. If π_e is small, Eq. (5.15) simplifies to

$$\gamma_c \simeq \phi^2 \gamma_S \tag{5.16}$$

which can also be obtained from Eq. (5.4).

The concept of critical surface tension may also be applied to liquid substrates [24-26]. In this case, either the cos θ versus γ_{LV} plot or the λ_{ab} versus γ_{LV} plot may be used to determine the γ_c for a liquid, where λ_{ab} is the initial spreading coefficient. An example for water as the substrate is shown in Figure 5.8. The γ_c for water thus obtained with n-alkanes as the testing liquids is 19.1 to 21.7 dyne/cm at 20°C [24,26]. This turns out to be the dispersion component of the surface tension of water, as expected, since nonpolar liquids are used as the testing liquids.

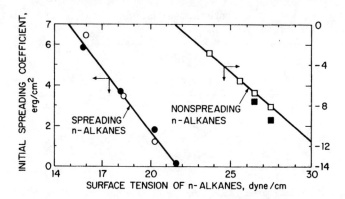

Figure 5.8. Initial spreading coefficient versus surface tension of n-alkanes on water at 20°C. (After Ref. 26.)

The critical surface tension decreases slowly with increasing temperature [38,39] (Figure 5.9). The $-(d\gamma_c/dT)$ values are similar to the $-(d\gamma_s/dT)$ values (Table 5.6 and Figure 5.10). Here, the γ_c for molten polyethylene is obtained by using other molten polymers as the testing liquids on the molten polyethylene in the $\cos \theta$ versus γ_{LV} plot (Figure 5.11). Although $-(d\gamma_c/dT)$ and $-(d\gamma_s/dT)$ are nearly identical, the γ_c is consistently much lower than the γ_s (see also Table 5.5).

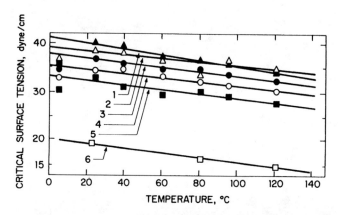

Figure 5.9. Temperature dependence of critical surface tension: (1) poly(ethylene terephthalate); (2) polyoxymethylene; (3) polyethylene; (4) polycarbonate; (5) polystyrene; (6) polytetrafluoroethylene. (After Ref. 38.)

Table 5.6. Comparison of Temperature Coefficients for Critical Surface Tension and Surface Tension

Polymer	$-(d\gamma_c/dT)$,[a] dyne/cm-°C	$-(d\gamma_s/dT)$,[b] dyne/cm-°C
Polyethylene	0.05	0.057
Polystyrene	0.06	0.072
Polytetrafluoroethylene	0.05	0.058
Polyoxymethylene	0.04	-
Poly(ethylene terephthalate)	0.06	0.065
Polycarbonate	0.04	-

[a] $-(d\gamma_c/dT)$ values from Ref. 38.
[b] $-(d\gamma_s/dT)$ values from Ref. 12 and Table 3.1.

Originally, contact angles at constant temperatures were used in the Zisman plot. This may be termed the *isothermal method*. Recently, Rhee [40-42] proposed a nonisothermal method. The cos θ and γ_{LV} of a given liquid metal on a given high-energy solid at various temperatures are used to construct the Zisman plot. The temperature at which the liquid metal just spreads (exhibits a zero contact angle) on the solid is defined as the *critical temperature for spreading*, T_{cs}. The γ_{LV} of the liquid metal at T_{cs} is the *critical surface tension* of the solid γ_{cs} at T_{cs}. Shafrin and Zisman [39] showed that the nonisothermal method applies equally well to low-energy solids.

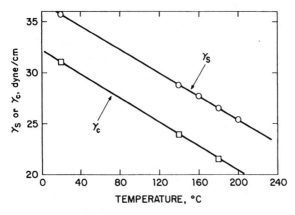

Figure 5.10. Comparison of temperature dependence of surface tension and critical surface tension for linear polyethylene.

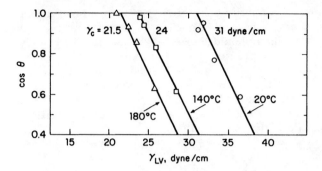

Figure 5.11. Cos θ versus γ_{LV} plots for polyethylene at 20°C, 140°C, and 180°C using molten poly(vinyl acetate), polyisobutylene, poly(n-butyl methacrylate), and poly(tetramethylene oxide) as the testing liquids. The data at 20°C are extrapolated from the melt data. (Data from Ref. 12.)

As pointed out in Section 5.1.3, a Zisman plot is, however, unnecessary. The γ_c can be obtained from the contact angle of a single liquid [1]. If the liquid used has a low contact angle, Eq. (5.6) gives the Zisman's critical surface tension directly.

5.1.7. Remarks on Some Materials

The γ_s values for some materials obtained by various methods are compared here. Those obtained by the equation of state method, the harmonic-mean method, the liquid homolog method, and the polymer melt method are shown to be consistent and preferred.

Polyethylene

The γ_s at 20°C is 35.9 dyne/cm by the equation of state method, 35.7 dyne/cm by the melt method, 34.7 dyne/cm by the liquid homolog method, and 36.1 dyne/cm by the harmonic-mean method. These values are quite consistent with one another. In contrast, the geometric-mean method gives 32.4 dyne/cm, and Zisman's critical surface tension is 31 dyne/cm, which are apparently too low.

Polytetrafluoroethylene

The γ_s at 20°C is 22.6 dyne/cm by the equation of state method, 23.9 dyne/cm by the liquid homolog method, and 22.5 dyne/cm by the harmonic-mean method. In contrast, the geometric-mean method gives 19.1 dyne/cm, and Zisman's critical surface tension is 18 dyne/cm, which are too low; this is discussed below.

The surface tension of a fluorocarbon oil, $C_{21}F_{44}$, has been directly measured to be 21.5 dyne/cm at 20°C [43]. As the molecular weight increases, the surface tension should increase. Thus, the γ_S of polytetrafluoroethylene should be greater than 21.5 dyne/cm. It is thus noteworthy that the γ_S values for polytetrafluoroethylene obtained by the equation of state method, the liquid homolog method, and the harmonic-mean method are all in the range 22-24 dyne/cm, consistent with expectation. In contrast, the geometric-mean method and the critical surface tension give values that are too low.

Paraffin Wax

The γ_S at 20°C is 32.0 dyne/cm by the equation of state method, 35.0 dyne/cm by the melt method, and 31.0 dyne/cm by the harmonic-mean method. In contrast, the geometric-mean method gives 25.4 dyne/cm, and Zisman's critical surface tension is 23 dyne/cm.

The surface tension of molten paraffin wax has been directly measured to be 32 dyne/cm at its melting point of 65°C [44]. As the molten paraffin wax cools and solidifies, the surface tension should increase. Thus, it is noteworthy that the γ_S values for solid paraffin wax at 20°C obtained by the equation of state method, the melt method, and the harmonic-mean method are all in the range 31-35 dyne/cm, consistent with expectation. On the contrary, Fowkes's value of 25.5 dyne/cm and Zisman's critical surface tension of 15-23 dyne/cm [9] are much too low.

5.1.8. Miscellaneous Methods

Several other methods are discussed below.

Interaction Parameter Method

This method was proposed by Good and coworkers [28,29]. Neglecting π_e and rearranging Eq. (5.1) gives

$$\gamma_S = \frac{\gamma_{LV}(1 + \cos \theta)^2}{4\phi^2} \tag{5.17}$$

where the interaction parameter ϕ may be estimated from liquid homologs of the solid or from molecular constants by Eq. (3.49). Reasonable results can be obtained, as shown in Table 5.7.

Vapor Adsorption Method

When a liquid has a zero contact angle on a solid, Young's equation becomes

Table 5.7. Surface Tensions Calculated by the Equation of Good and Girifalco

Material	φ Value used	Calculated γ_S, at 20°C, dyne/cm
Polytetrafluoroethylene	φ = 0.818 Estimated from low-molecular-weight homologs	27.8
	Estimated from molecular constants using μ = 1.2 debyes	24.0
	Estimated from molecular constants using μ = 0 debye	21.0
Polyhexfluoropropylene	Estimated from molecular constants using μ = 1.2 debyes	19.0
Polychlorotrifluoroethylene	Estimated from molecular constants using μ = 1.2 debyes	38.0
Perfluorolauric acid monolayer	Estimated from molecular constants using μ = 1.0 debye	13.0
Octadecylamine monolayer	φ = 0.92 Estimated from low-molecular-weight homologs	27.8

Source: From Refs. 28 and 29.

$$\pi_e = \gamma_S - \gamma_{LV} - \gamma_{SL} \qquad (5.18)$$

Using Fowkes's equation in Eq. (5.18) gives [36]

$$\gamma_S^d = \frac{(\pi_e + 2\gamma_{LV})^2}{4\gamma_{LV}^d} \qquad (5.19)$$

which assumes that polar interactions are absent.

The π_e can be obtained by integration of the vapor adsorption isotherm (Sections 1.5 and 4.1), and has thus been measured on a number of polymers and high-energy solids (Table 4.1). Fowkes [36] has used these values to calculate the γ_S^d values for some solids (Table 5.8). Reasonable results can be obtained, provided that θ is zero and polar interactions are absent. The γ_S^d of 123 dyne/cm for graphite agrees well with the value of 109 dyne/cm obtained from contact angles. However, the γ_S^d 28.5 dyne/cm at 90 K obtained for polyporpylene is much too low as compared with the value of 42 dyne/cm by the four preferred methods (the equation of state method, the melt method, the liquid homolog method, and the harmonic-mean method).

Immersed Contact Angle Method

Instead of using the contact angles in air, several authors [45-47] used the contact angles of oils on the solid immersed in water. These immersed contact angles are then used in the harmonic-mean equation or the geometric-mean equation to calculate the surface tension of the solid. The reported γ_S values thus obtained for several polymers are, however, invariably much too high (often about twice as high as the presently accepted values). These unusually high values appear to result from adsorption of the oils at water/polymer interfaces. Such effects were unaccounted for in the treatments reported.

Parabolic Plot Method and Method of Sell and Neumann

In this method, $\gamma_{LV} \cos \theta$ is plotted against γ_{LV}. A parabolic curve is obtained whose intersection with the 45° line, that is, the $\gamma_{LV} \cos \theta = \gamma_{LV}$ line, is suggested as the γ_S [48,49]. Contrary to the claim, this method in fact gives Zisman's critical surface tension γ_c, and offers no advantage over the much simpler Zisman plot, as already pointed out by others [50-52].

Algebraic analysis of this method readily gives the relation

$$\gamma_{LV} \cos \theta = (1 + k\gamma_c)\gamma_{LV} - k\gamma_{LV}^2 \qquad (5.20)$$

Table 5.8. Values of γ_S^d Determined from π_e of Adsorbed Vapors

Solid	Vapor	Temperature, °C	π_e, dyne/cm	γ_S^d, dyne/cm
Polypropylene	Nitrogen	−195	12	26
	Argon	−183	13	28.5
Graphite	Nitrogen	−195	51	123
	n-Heptane	25	59	123
Copper	n-Heptane	25	29	60
Iron	n-Heptane	25	53	108
	Argon	−183	47	106
	Nitrogen	−195	40	89
Ferric oxide	n-Heptane	25	54	107
TiO$_2$ (anatase)	n-Heptane	25	46	92
	n-Butane	0	43	89
	Nitrogen	−195	56	141
Silica	n-Heptane	25	39	78
BaSO$_4$	n-Heptane	25	38	76

Source: From Ref. 36.

which follows from Eq. (5.13) and has been shown to conform to the experimental parabolic curve. At the intersection of the parabolic curve with the 45° line, we have $\gamma_{LV} \cos \theta = \gamma_{LV}$, which is equivalent to a zero contact angle, that is, $\cos \theta = 1$. Therefore, this method gives $\gamma_c = \gamma_{LV}$ at $\theta = 0$, which is *exactly* Zisman's critical surface tension.

Sell and Neumann [48,49] suggested empirically that the interaction parameter ϕ is a function of the interfacial tension only. This is immediately contradictory to the theoretical basis of the parameter, which requires that the ϕ depends on the polarities of the two phases (Section 3.2.2). Therefore, the equation of Sell and Neumann is purely empirical, and in fact, is often inapplicable. For instance, the γ values of polyethylene and poly(vinyl acetate) at 140°C are 28.8 and 28.6 dyne/cm, respectively [1]. The equation of Sell and Neumann predicts that the interfacial tension between the two polymers would be about 0 dyne/cm. In fact, however, the measured interfacial tension is as high as 11.3 dyne/cm at 140°C [1].

Polymer Solution Method

If the surface tension of polymer solution is extrapolated to 100% polymer solid, the γ_S of the solid polymer may presumably be obtained [53,54]. The γ_S values obtained depend greatly on the extrapolation method. The variation of surface tension is greatest at the first few percentages of solvent addition. Therefore, an accurate theoretical function is required to make a reliable extrapolation.

Several empirical extrapolation methods have been proposed [53,54]. All give values that are much too low, apparently because of inadequate extrapolation methods. Perhaps the equation of Prigogine and Gaines and the equation of Belton and Evans would give more reliable results (Section 3.1.6).

Chan's Geometric-Mean Method

Fowkes equation is extended to include a polar term by certain modifications and assumptions [55,56]. Contact angles of a series of liquids are used to obtain the surface tension and polarity. The results obtained are similar to those obtained by the geometric-mean method of Section 5.1.5.

5.2. SURFACE TENSION AND POLARITY OF ORGANIC PIGMENTS

The surface tension and polarity of some organic pigments have been determined from the contact angles of water and methylene iodide by the harmonic-mean method [22]. The pigments are pressed into optically smooth pellets, on which the contact angles are measured. The results are listed in Table 5.9. The chemical structures of the pigments are shown in Figure 5.12.

Table 5.9. Surface Tension and Its Components for Some Organic Pigments at 20°C

Pigment	Contact angle, deg		Surface tension,[a] dyne/cm			Critical surface tension,[b] dyne/cm
	Water	Methylene iodide	γ^d	γ^p	γ	
Indanthrone	40.0	22.5	33.2	30.0	63.2	–
Manganese β-oxynaphthoic acid derivative	55.0	41.0	27.8	24.1	51.9	–
Thioindigoid red	65.0	24.0	35.1	16.3	51.4	–
Isoindolinone	69.0	34.0	32.2	15.0	47.2	–
γ-Quinacridone	70.5	25.0	35.7	13.4	49.1	–
Toluidine red	73.0	5.0	39.7	13.3	53.0	27.5
Metal-free phthalocyanine	70.0	0	40.1	12.7	52.8	35.6
Chlorinated copper phthalocyanine	87.0	36.0	35.8	6.2	42.0	24.7
Copper phthalocyanine	83.0	23.0	40.0	6.9	46.9	31.3

[a] Surface tension and its components calculated from contact angles by the harmonic-mean method, and taken from Ref. 22.
[b] Critical surface tension taken from Ref. 23.

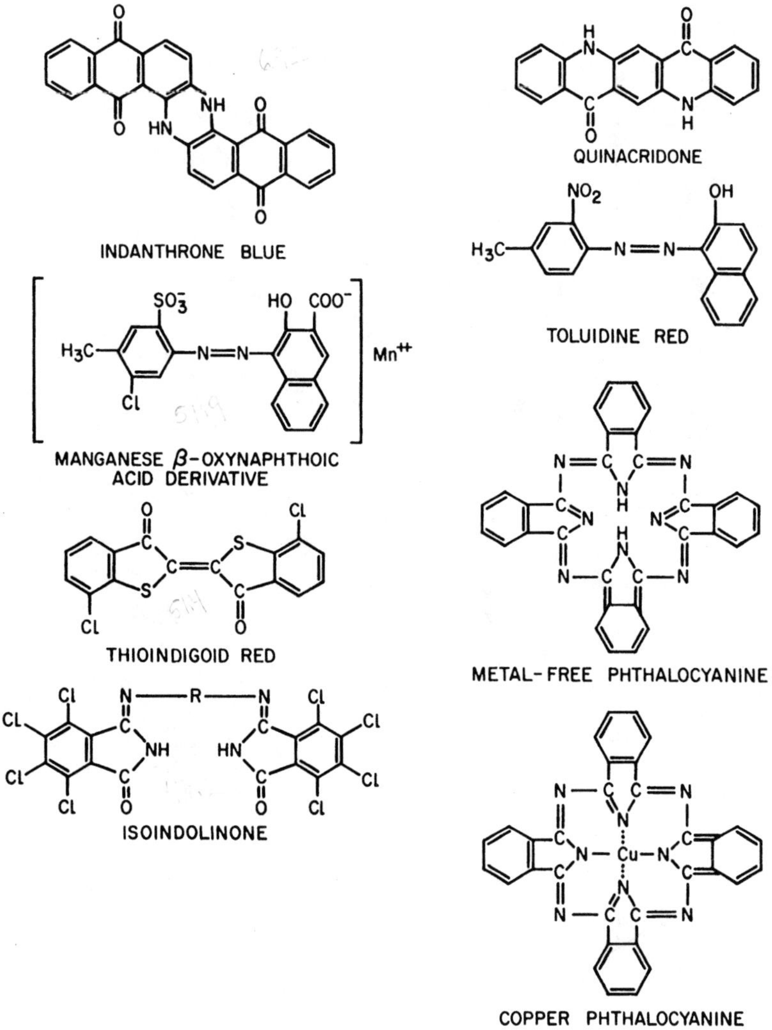

200   5 / Surface Tension and Polarity of Solid Polymers

INDANTHRONE BLUE

QUINACRIDONE

MANGANESE β-OXYNAPHTHOIC ACID DERIVATIVE

TOLUIDINE RED

THIOINDIGOID RED

METAL-FREE PHTHALOCYANINE

ISOINDOLINONE

COPPER PHTHALOCYANINE

Figure 5.12. Chemical structures of some organic pigments.

The surface tensions are found to be in the range 40-65 dyne/cm, and the polarities in the range 0.15-0.50 at 20°C. These are consistent with their chemical structures, and are higher than those of most organic liquids as expected. Organic pigments contain chromophore groups (such as nitro, azo, and aromatic rings, double bonds, and sterically strained groups), auxochrome groups (electron donors, such as amino and hydroxyl groups), and covalently bonded metal atoms.

Morphological Effect on Surface Tension

These are moderately to highly polar groups. Furthermore, organic pigments have tightly packed aromatic rings, having densities of 1.3-1.6 g/ml, which are higher than those of most common organic liquids. Therefore, colored pigments should have high surface tension and high polarity. Their critical surface tensions are much lower than their surface tensions [22,23].

5.3. CONSTITUTIVE EFFECT ON SURFACE TENSION

Shafrin and Zisman [57] noted that surfaces having similar chemical constitutions have similar γ_c values. A constitutive law is thus stated as: "Surfaces having similar chemical compositions have similar surface tensions." This is, however, only an approximation, since the Macleod-Sugden equation already states that the surface tension depends not only on surface constitution (parachor), but also on molecular packing (density) (Section 3.1.2). However, most organic polymers have similar molecular packing except with certain halocarbon polymers. Thus, to a first approximation, the proposed constitutive law is valid (Table 5.10).

To obtain an adequate picture, however, both the effects of chemical constitution and molecular packing should be considered as illustrated by the effect of chlorination and fluorination on the surface tension of polyethylene (Table 5.11). Successive chlorination increases both the surface tension and the polarity of polyethylene, whereas the nonpolar component is little changed. On the other hand, introduction of the first fluorine into the ethylene unit increases both the surface tension and the polarity, whereas the nonpolar component is lowered. Further fluorination decreases both the nonpolar and the polar components of the surface tension. All of these results can be explained in terms of the effects of chlorine and fluorine atoms on the parachor and the density.

5.4. MORPHOLOGICAL EFFECT ON SURFACE TENSION

The surface tension changes discontinuously at the crystal-melt transition, and is proportional to the fourth power of the density (Section 3.1.4). Thus, the surface tension of a semicrystalline polymer should vary with its surface crystallinity. The crystalline density is usually higher than the amorphous density. Therefore, the crystalline surface tension is usually much higher than the amorphous surface tension. For instance, the amorphous surface tension of polyethylene is 35.7 dyne/cm at 20°C, and its amorphous density is 0.855 g/ml. The crystalline density is 1.000 g/ml. Therefore, its crystalline surface tension is calculated to be 66.8 dyne/cm at 20°C, which agrees very well with experimental values (Table 5.12).

Table 5.10. Surface Constitution and Surface Tension at 20°C

Surface constitution	Surface tension, γ_S, dyne/cm	Zisman's critical surface tension, γ_c, dyne/cm
Hydrocarbon surfaces		
—CH$_3$ (crystal)	23.0	21
—CH$_3$ (monolayer)	30.5	22
—CH$_2$—	35.9	31
—CH$_2$— and ∺CH∺	43.0	33
∺CH∺ (phenyl ring edge)	45.1	35
Fluorocarbon surfaces		
—CF$_3$	14.5	6
—CF$_2$H	26.5	15
—CF$_3$ and —CF$_2$—	17.0	16

—CF$_2$—	22.6	18
—CH$_2$CF$_3$	22.5	20
—CF$_2$—CFH—	29.5	22
—CF$_2$—CH$_2$—	36.5	25
—CFH—CH$_2$—	37.5	28
Chlorocarbon surfaces		
—CHCl—CH$_2$—	43.8	39
—CCl$_2$—CH$_2$—	45.2	40
=CCl$_2$	49.5	43
Silicon surfaces		
—O—Si(CH$_3$)$_2$—O—	19.8	—
—O—Si(CH$_3$)(C$_6$H$_5$)—O—	26.1	—

Table 5.11. Effect of Progressive Halogenation on the Surface Tension of Polyethylene

Surface tension at 20°C, dyne/cm

Polymer	Equation of state method γ_S	Harmonic-mean method			Zisman's critical surface tension, γ_c
		γ_S	γ_S^d	γ_S^p	
Polyethylene	35.9	36.1	36.1	0	31
Progressive chlorination					
Poly(vinyl chloride)	43.8	41.9	35.8	6.1	39
Poly(vinylidene chloride)	45.2	45.4	36.5	8.9	40
Polytrichloroethylene	53	–	–	–	–
Polytetrachloroethylene	55	–	–	–	–
Progressive fluorination					
Poly(vinyl fluoride)	37.5	38.4	27.2	11.2	28
Poly(vinylidene fluoride)	36.5	33.2	20.7	12.5	25
Polytrifluoroethylene	29.5	27.3	18.5	8.8	22
Polytetrafluoroethylene	22.6	22.5	20.5	2.0	18

Table 5.12. Surface Tension and Surface Crystallinity of Polyethylene Molded Against Various Surfaces

Mold (nucleating) surface	Surface crystallinity, %	Surface tension at 20°C, dyne/cm
Nitrogen	0	36.2
Polytetrafluoroethylene	0	36.2
Poly(ethylene terephthlate)	0	36.2
Copper	5.1	37.4
Nickel	53.3	51.3
Tin	60.1	53.8
Aluminum	63.2	54.9
Glass	63.2	54.9
Chromium	66.2	56.1
Mercury	86.4	64.8
Gold	96.3	69.6
Tantalum	78.5	61.4
NaCl	86.4	64.8
KBr	86.4	64.8
KCl	86.4	64.8
CaF_2	91.4	67.3

Source: From Refs. 58-60.

When a polymer melt cools and solidifies, an amorphous surface is usually formed, although its bulk phase may be semicrystalline. Fractions not accommodated in the crystalline structure are rejected to the surface. However, if the polymer is cooled and solidified against a nucleating surface, surfaces having various degrees of crystallinity can be obtained. Some high-energy mold surfaces were found to nucleate polyethylene, giving transcrystalline surfaces, whereas low-energy mold surfaces gave amorphous surfaces [58-60]. The surface tensions and surface crystallinites of some polymers are listed in Tables 5.12 and 5.13. The various crystalline polymer surfaces were obtained by molding against certain high-energy mold surfaces which were later removed by dissolution in chemicals (aqueous sodium cyanide or mercury for gold, aqueous sodium hydroxide or hydrochloric acid for aluminum, etc.).

Crystalline polymer surfaces may be transcrystalline, spherulitic, or lamellar, depending on the cooling rate and nucleating activity of the mold surface [61-64] (see also Chapter 9). A transcrystalline layer is formed when the mold surface nucleates massive numbers of nuclei, which are so crowded that they are forced to grow normal to the surface. A transcrystalline surface layer formed on a polyethylene molded against aluminum foil is shown in Figure 5.13. The formation of transcrystalline surface layers are well known in polyethylene and polypropylene [59-70], in polyamides [70,71], and in polyurethanes [70]. These transcrystalline surface layers are usually 10 to 100 μm thick and are effectively macroscopic. They have mechanical properties very different from those of the bulk phase [66,67] (Chapter 9). On the other hand, if nuclei are formed only sporadically on the surface, a spherulitic morphology will be obtained.

Only high-energy mold surfaces were originally thought to be effective in nucleation; low-energy mold surfaces were thought to be ineffective [59,60]. This has been shown not to be true [61-64]. The nucleating activity has nothing to do with the surface energy, but depends on unknown nuances of surface chemical composition. The crystalline morphology obtained depends on both the cooling rate and the nucleating activity. For instance, polypropylene develops a spherulitic surface layer when molded against aluminum oxide or copper oxide, but develops a transcrystalline surface layer when molded against poly(ethylene terephthlate) or polytetrafluoroethylene [61-64]. When molded against FEP (fluorinated ethylene-propylene copolymer), polypropylene develops a spherulitic surface layer, whereas when molded against polytetrafluoroethylene, a transcrystalline surface layer is formed [61-64]. Further, fast cooling gives a spherulitic surface layer, whereas slow cooling gives a transcrystalline surface layer when polypropylene is molded against poly(ethylene terephthalate) [61-64]. These are discussed further in Chapter 9.

There is, however, a controversy as to the exact composition of transcrystallized surfaces nucleated against high-energy mold surfaces. Although internal reflection IR spectroscopy (relatively insensitive for surface analysis) showed that the transcrystallized surfaces are not contaminated [59,60], recent sensitive ESCA analyses revealed the ex-

Table 5.13. Surface Tension and Surface Crystallinity of Some Polymers Molded (Nucleated) Against Gold or Nitrogen

Polymer	Mold (nucleating) Surface	Surface crystallinity, %	Surface tension at 20°C, dyne/cm
Polyethylene	Gold	96.3	69.6 [a]
	Nitrogen	0	36.2 [a]
Poly(hexamethylene adipamide)	Gold	94.1	74.4 [a]
	Nitrogen	0	46.5 [b]
Polychlorotrifluoro-ethylene	Gold	100	58.9 [a]
	Nitrogen	0	31.1 [b]
Isotactic polypropylene	Gold	100	39.5 [a]
	Nitrogen	0	30.1 [b]
Atactic polypropylene	Gold	0	28.0 [a]
	Nitrogen	0	28.0 [a]
FEP, fluorinated-ethylene propylene copolymer	Gold	—	40.4 [a]
	Nitrogen	0	18.8 [a]

[a] From Refs. 58-60.
[b] From Ref. 12.

Figure 5.13. Polarized optical photomicrograph of a thin cross section of polyethylene molded against aluminum foil, showing transcrystalline surface layer of about 25 μm thickness and spherulitic interior phase. (Reprinted with permission from H. Schonhorn, Macromolecules, *1*, 145 (1968). Copyright © 1968 by the American Chemical Society.)

istence of a thin layer (20-60 Å) of polar oxygenated hydrocarbons on FEP transcrystallized against gold foil [72] and on polyethylene transcrystallized against aluminum foil [73]. These polar contaminants may come from thermal oxidation of the polymer surfaces during melting (catalyzed by the high-energy surfaces) or from residual oxide of the high-energy mold surfaces. Thus, the increased wettability and bondability of polymer surfaces transcrystallized against high-energy surfaces may be due to the presence of these polar contaminants, rather than to increased surface density and cohesive strength by transcrystallization. Additional work on this subject would be welcome.

On the other hand, injection-molded polymers tend to have lamellar surfaces [74]. Polymer molecules tend to orient in the direction of flow during injection molding, forming lamellar surface morphology. This has been observed on polyoxymethylene [74], polyethylene [75], and polypropylene [64]. Injection-molded polyoxymethylene shows three morphological zones [74]. In the surface zone, about 5 mils thick, the lamellae are oriented perpendicular to the surface and the direction of flow. Below this is an intermediate zone where the lamellae are oriented perpendicular to the surface only. In the bulk interior, spherulites are formed and oriented randomly. Injection-molded polyethylene exhibits many morphological zones, which have been investigated in detail [75]. The lamellar surface structure formed on an injection-molded polypropylene is shown in Chapter 9.

5.5. EFFECTS OF ADDITIVES, POLYMER BLENDS, COPOLYMERS, CONFORMATION, AND TACTICITY ON SURFACE TENSION

These effects for solid polymers are similar to those for polymer melts and liquids (Chapter 3). Several examples are cited below.

Low-energy additives can migrate to a solid surface and drastically lower its surface tension [76,77]. For instance, 0.25% by weight of some fluorocarbon additives drastically reduces the surface tensions of polystyrene, poly(methyl methacrylate), and poly(vinylidene chloride) to about 15-20 dyne/cm, resembling those for pure fluorocarbon surfaces [77] (Table 5.14).

The surfaces of mixtures of two poly(fluoroalkyl methacrylates), differing in fluoroalkyl side chain length, have been investigated by contact angle measurement and ESCA [78]. The lower-energy component (having a longer fluoroalkyl side chain) is found to concentrate on the surface. In other examples, fluorocarbon polymers are shown to exhibit pronounced surface activity when blended with hydrocarbon polymers [79]. Surface activity of the lower-energy component in a copolymer has also been reported [17].

Surface molecular conformations are found to affect the surface tension. Poly(γ-methyl L-glutamate) can exist in two conformations. In

Table 5.14. Critical Surface Tension of Some Polymer Films Containing Fluorocarbon Additives

Fluorocarbon additive [a] at 0.25% by weight	γ_c at 20°C, dyne/cm		
	Polystyrene	Poly(methyl methacrylate)	Poly(vinylidene chloride)
None	32.5	39.5	40.5
A: $(F_{59}H_{31}C_{45}O_{18})$	21.8	14.8	17.3
B: $(F_{59}H_{10}C_{34}O_{12}N)$	22.4	14.7	17.3
C: $(F_{59}H_{16}C_{36}O_{10}N_2I)$	14.1	14.3	15.7

[a] Additive A, $F[C(CF_3)F-CF_2-O]_9-C(CF_3)F-CO_2CH_2-CH_2(OCH_2CH_2)_6OCH_3$; additive B, $F[C(CF_3)F-CF_2-O]_9-C(CF_3)F-CON(CH_2CH_2OH)_2$; additive C, $F[C(CF_3)FCF_2O]_9-C(CF_3)F-$ $CONH(CH_2)_3N^+(CH_3)_3I^-$.

Source: From Ref. 77.

the coiled conformation, the γ_c is 48 dyne/cm. In the extended conformation, the γ_c is 36 dyne/cm [80].

The effect of tacticity on surface tension has not been investigated adequately. Any effect is expected to arise from the density effect. Very small differences in the γ_c of some polymethacrylates of different tacticity have been reported [81].

REFERENCES

1. S. Wu, J. Colloid Interface Sci., 71, 605 (1979). Erratum in J. Colloid Interface Sci., 73, 590 (1980).
2. S. Wu, J. Polym. Sci., C34, 19 (1971).
3. E. G. Carter and D. C. Jones, Trans. Faraday Soc., 30, 1027 (1934).
4. W. Fox, J. Am. Chem. Soc., 67, 700 (1945).
5. D. K. Owens and R. C. Wendt, J. Appl. Polym. Sci., 13, 1741 (1969).
6. D. H. Kaelble, *Physical Chemistry of Adhesion*, Wiley-Interscience, New York, 1971.
7. H. W. Fox and W. A. Zisman, J. Colloid Sci., 5, 514 (1950).
8. H. W. Fox and W. A. Zisman, J. Colloid Sci., 7, 109 (1952).
9. H. W. Fox and W. A. Zisman, J. Colloid Sci., 7, 428 (1952).
10. J. R. Dann, J. Colloid Interface Sci., 32, 302, 321 (1970).
11. F. Schulman and W. A. Zisman, J. Colloid Sci., 7, 465 (1952).
12. S. Wu, in *Polymer Blends*, Vol. 1, D. R. Paul and S. Newman, eds., Academic Press, New York, 1978, pp. 243-293.
13. W. A. Zisman, Adv. Chem. Ser., 43, 1 (1964).
14. E. G. Shafrin, in *Polymer Handbook*, 2nd ed., J. Brandrup and E. H. Immergut, eds., Wiley-Interscience, New York, 1975, pp. III. 221-228.
15. S. Wu, Org. Coat. Plast. Chem., 31(2), 27 (1971).
16. K. Kamagata and M. Toyama, J. Appl. Polym. Sci., 18, 167 (1974).
17. M. Toyama, A. Watanabe, and T. Ito, J. Colloid Interface Sci., 47, 802 (1974).
18. E. G. Shafrin and W. A. Zisman, J. Colloid Sci., 7, 166 (1952).
19. E. F. Hare, E. G. Shafrin, and W. A. Zisman, J. Phys. Chem., 58, 236 (1954).
20. A. H. Ellison, H. W. Fox, and W. A. Zisman, J. Phys. Chem., 57, 622 (1953).
21. A. H. Ellison and W. A. Zisman, J. Phys. Chem., 58, 260 (1954).
22. S. Wu and K. J. Brzozowski, J. Colloid Sci., 37, 686 (1971).
23. P. R. Buehler, G. L. Brown, V. P. Parikh, and H. J. Salmon, J. Paint Technol., 45(577), 60 (1973).
24. R. E. Johnson, Jr., and R. H. Dettre, J. Colloid Interface Sci., 21, 610 (1966).

25. N. J. Jarvis and W. A. Zisman, J. Phys. Chem., 63, 727 (1959).
26. E. G. Shafrin and W. A. Zisman, J. Phys. Chem., 71, 1309 (1967).
27. A. W. Adamson, *Physical Chemistry of Surfaces*, 2nd ed., Wiley-Interscience, New York, 1967.
28. R. J. Good and L. A. Girifalco, J. Phys. Chem., 64, 561 (1960).
29. R. J. Good, Adv. Chem. Ser., 43, 74 (1964).
30. R. J. Good, L. A. Girifalco, and G. Krauss, J. Phys. Chem., 62, 1418 (1958).
31. V. R. Gray, Chem. Ind., 969 (June 5, 1965); in *Aspects of Adhesion*, Vol. 2, D. J. Alner, ed., CRC Press, Cleveland, Ohio, 1966, pp. 42-48; in *Wetting*, Society of Chemical Industry, London, Monograph No. 25, Gordon and Breach, London, 1967, pp. 99-119.
32. T. C. Patton, J. Paint Technol., 42(551), 666 (1970).
33. C. C. Harris, as quoted by M. Rosoff in "Physical Methods in Macromolecular Chemistry," B. Carroll, ed., Vol. I, Marcel Dekker, New York, 1969, pp. 1-108.
34. H. E. Garrett, in *Aspects of Adhesion*, Vol. 2, D. J. Alner, ed., CRC Press, Cleveland, Ohio, 1966, pp. 18-41.
35. F. M. Fowkes, *Chemistry and Physics of Interfaces*, Vol. 1, American Chemical Society, Washington, D.C., 1965, pp. 1-12.
36. F. M. Fowkes, in *Chemistry and Physics of Interfaces*, Vol. 2, American Chemical Society, Washington, D.C., 1971, pp. 153-168.
37. E. Wolfram, Kolloid Z. Z. Polym., 182, 75 (1962).
38. F. D. Petke and B. R. Ray, J. Colloid Interface Sci., 31, 216 (1969). Erratum in J. Colloid Interface Sci., 33, 195 (1970).
39. E. G. Shafrin and W. A. Zisman, J. Phys. Chem., 76, 3259 (1972).
40. K. S. Rhee, J. Am. Ceram. Soc., 53, 386 (1970).
41. K. S. Rhee, J. Am. Ceram. Soc., 54, 332 (1971).
42. K. S. Rhee, J. Am. Ceram. Soc., 55, 157 (1972).
43. R. H. Dettre and R. E. Johnson, Jr., J. Phys. Chem., 71, 1529 (1967); J. Colloid Interface Sci., 31, 568 (1969).
44. J. F. Padday, Proc. 2nd Int. Congr. Surf. Act., 3, 136 (1957).
45. G. J. Crocker, Rubber Chem. Technol., 42, 30 (1969).
46. Y. Tamai, T. Matsunaga, and K. Horiuchi, J. Colloid Interface Sci., 60, 112 (1977); T. Matsunaga, J. Appl. Polym. Sci., 21, 2847 (1977).
47. J. D. Andrade, S. M. Ma, R. N. King, and D. E. Gregonis, J. Colloid Interface Sci., 72, 488 (1979).
48. P. J. Sell and A. W. Neumann, Z. Phys. Chem., 41, 191 (1964); Angew. Chem. (Int. Ed.), 5(3), 299 (1966).
49. A. W. Neumann, R. J. Good, C. J. Hope, and M. Sejpal, J. Colloid Interface Sci., 49, 291 (1974).
50. E. Wolfram, Z. Phys. Chem., 44, 367 (1965).
51. M. C. Phillips and A. C. Riddiford, Z. Phys. Chem., 47, 17 (1965).
52. J. R. Huntsberger, J. Adhes., 7, 289 (1976).

References

53. J. E. Marian, ASTM STP *340*, 122 (1963).
54. Yu. S. Lipatov and A. E. Feinerman, J. Adhes., *3*, 3 (1971).
55. R. K. S. Chan, J. Colloid Interface Sci., *32*, 492 (1970).
56. R. K. S. Chan, J. Colloid Interface Sci., *32*, 499 (1970).
57. E. G. Shafrin and W. A. Zisman, J. Phys. Chem., *64*, 519 (1960).
58. H. Schonhorn and F. W. Ryan, J. Phys. Chem., *70*, 3811 (1966).
59. H. Schonhorn, Polym. Lett., *5*, 919 (1967).
60. H. Schonhorn, Macromolecules, *1*, 145 (1968).
61. D. R. Fitchmun and S. Newman, J. Polym. Sci., Polym. Lett. Ed., *7*, 301 (1969).
62. D. R. Fitchmun and S. Newman, J. Polym. Sci., A-2, *8*, 1545 (1970).
63. D. R. Fitchmun, S. Newman, and R. Wiggle, J. Appl. Polym. Sci., *14*, 2441 (1970).
64. D. R. Fitchmun, S. Newman, and R. Wiggle, J. Appl. Polym. Sci., *14*, 2457 (1970).
65. J. R. Shaner and R. D. Corneliussen, J. Polym. Sci., A-2, *10*, 1611 (1972).
66. T. K. Kwei, H. Schonhorn, and H. L. Frisch, J. Appl. Phys., *38*, 2512 (1967).
67. H. L. Frisch, H. Schonhorn, and H. L. Frisch, J. Elastoplast., *3*, 214 (1971).
68. R. K. Eby, J. Appl. Phys., *35*, 2720 (1964).
69. V. A. Kargin, T. I. Sogolova, and T. K. Shaposhuikova, Dokl. Akad. Nauk SSSR, *180*, 901 (1968). English translation: Trans. Acad. Sci. USSR, *180*, 406 (1968).
70. E. Jenckel, E. Teege, and W. Hinricks, Kolloid Z., *129*, 19 (1952).
71. R. J. Barriaut and L. F. Gronholz, J. Polym. Sci., *18*, 393 (1955).
72. D. W. Dwight and W. M. Riggs, J. Colloid Interface Sci., *47*, 650 (1974).
73. D. Briggs, D. M. Brewis, and M. B. Konieczko, J. Mater. Sci., *12*, 429 (1977).
74. E. S. Clark, SPE J., *23*, 46 (July, 1967).
75. V. Tan and M. R. Kamal, J. Appl. Polym. Sci., *22*, 2341 (1978).
76. N. L. Jarvis, R. B. Fox, and W. A. Zisman, Adv. Chem. Ser., *43*, 317 (1964).
77. M. K. Bernett, Polym. Eng. Sci., *17*, 450 (1977).
78. R. W. Phillips and R. H. Dettre, J. Colloid Interface Sci., *56*, 251 (1976).
79. M. Langsam and G. J. Mantel, J. Appl. Polym. Sci., *19*, 2235 (1975).
80. R. E. Baier and W. A. Zisman, Macromolecules, *3*, 70 (1970).
81. M. Toyama and T. Ito, J. Colloid Interface Sci., *49*, 139 (1974).

6
Wetting of High-Energy Surfaces

6.1. INTRODUCTION

Surfaces are classified into two types: high-energy surfaces and low-energy surfaces [1]. High-energy materials include metals, metal oxides, and inorganic compounds (oxides, nitrides, silica, sapphire, and diamond). They are usually hard, refractory, and dense, having surface tensions in the range 200–5000 dyne/cm. On the other hand, low-energy materials include organic compounds, organic polymers, and water. They are usually soft, low-melting, and light, having surface tensions below 100 dyne/cm. Table 6.1 lists the surface tensions for some high-energy materials.

Low-energy materials tend to adsorb strongly onto high-energy surfaces, as this will greatly decrease the surface energy of the system. Thus, a high-energy surface exposed to an ordinary environment tends to be covered with a film of adsorbed water or organic contaminants, and is changed into a low-energy surface. The strong affinity of water and organic materials for a high-energy surface is manifested in the large free energy of adsorption ($\pi_e = \gamma_S - \gamma_{SV}$) (Table 4.1). The affinity of organic compounds for high-energy surfaces is roughly four times greater than that for low-energy surfaces. Therefore, high-energy surfaces can be preserved only in an ultraclean environment. On the other hand, high-energy surfaces are

Table 6.1. Surface Tension of Some Molten Elements, Molten Salts, Glasses, and Solid Refractory Materials

Material	Temperature, °C	Surface tension, dyne/cm	Experimental method	Reference
Molten elements				
Aluminum (mp 660°C)	660	873	Sessile drop	2
	1600	725		2
Gold (mp 1064°C)	1064	1130	Sessile drop	2
	1300	1020		2
Copper (mp 1083°C)	1083	1300	Pendent drop	3
Iron (mp 1535°C)	1535	1760	Sessile drop	2
	2000	1597		
Mercury	20	484	Sessile drop	4
	25	484.2		5
	50	479.0		5
	75	474.5		5
Sodium (mp 97.8°C)	123	198	Drop volume	2
	617	144	Bubble pressure	2
Lead (mp 327.5°C)	327.5	470	Bubble pressure	2
Sulfur (mp 112.8°C)	112.8	60.9	Pendent drop	2
	250	51.1		
Silicon (mp 1410°C)	1450	725	Pendent drop	2
Titanium (mp 1660°C)	1660	1650	Pendent drop	2
	1680	1582	Drop weight	2

Introduction

Molten salts			
$BiCl_3$	271	Capillary pressure	3
$NaCl$	801	-	6
$NaNO_3$	308	Bubble pressure	3
$Ba(NO_3)_2$	595	Bubble pressure	3
Molten glasses			
$Na_2O, CaO, 6SiO_2$	1050	Bubble pressure	6
$Na_2O, CaO, 5SiO_2$	1050	Bubble pressure	6
$1.4Na_2O, 0.9CaO, 6SiO_2$	1200	Bubble pressure	6
$1.4Na_2O, 0.9CaO, 0.03$			
$BaO, 6SiO_2$	1200	Bubble pressure	6
$1.4Na_2O, 0.9CaO, 0.03$			
$PbO, 6SiO_2$	1200	Bubble pressure	6
$Na_2O, 2CaO, 7SiO_2$	1206	Wilhelmy plate	6
$2Na_2O, CaO, 7SiO_2$	1206	Wilhelmy plate	6
$32.7Na_2O, 67.3SiO_2$	1100	Wilhelmy plate	7
$15.1K_2O, 7.5CaO, 77.8SiO_2$	1300	Fiber tension	8
Solid glasses and refractory materials			
Soda-lime glass	900	Pendent drop	9
	600	Fiber tension	10
TiC	1100	-	11
	3150	Critical surface tension	12
TiB_2	2850	Critical surface tension	13
AlN	2450	Critical surface tension	12

Table 6.1. (Continued)

Material	Temperature, °C	Surface tension, dyne/cm	Experimental method	Reference
TaC	1605	1098	Critical surface tension	14
Graphite	720	230	Critical surface tension	15
Metal oxides				
Al_2O_3	1850	950	Thermal etching	16
	2080	700	–	6
	1500	1050	Critical surface tension	17
	1230	440	Critical surface tension	15
PbO	900	79	–	6
FeO	1420	585	–	6

Spreading on High-Energy Surfaces

also quite hydrophilic. The affinity of water for high-energy surfaces is often more than an order of magnitude greater than that for low-energy surfaces (Table 4.3). Therefore, water tends to invade the interface between a polymer adhesive and a high-energy adherend (Section 15.3).

6.2. SPREADING ON HIGH-ENERGY SURFACES

All pure, low-energy liquids always spread spontaneously (exhibit zero contact angle) on a high-energy surface unless the liquid is autophobic or decomposes upon contact with the high-energy surface to release a more strongly adsorbing low-energy product which converts the high-energy surface into a low-energy surface [18].

6.2.1. Autophobic Liquids

High-energy surfaces strongly adsorb low-energy materials, as this greatly lowers the surface energy of the system. Thus, adsorbed monolayers will inevitably form at the solid-vapor interface. Autophobic liquids have surface tensions greater than those of their own adsorbed monolayers, and therefore cannot spread on them. Autophobic liquids usually have amphipathic molecules, whose polar ends adsorb onto the high-energy surface, leaving the nonpolar low-energy portions exposed on the monolayer surface. Such nonpolar surfaces tend to have surface tensions lower than those of the bulk liquids, which therefore form finite contact angles on them [19-21]. Some autophobic liquids and their contact angles on some high-energy surfaces are listed in Table 6.2.

6.2.2. Hydrolysis of Esters upon Adsorption

All organic liquids, except some esters, exhibit a zero contact angle on metals, silica, and alumina [18]. The nonspreading esters are hydrolyzed instantly upon adsorption onto glass, silica, and metal oxides. The hydrolyzates (acids and bases) rapidly adsorb to form low-energy monolayers covering the high-energy surfaces on which the liquid esters cannot spontaneously spread.

Hydrated glass and silica have greater hydrolytic activity than do hydrated metal oxides [18]. For instance, bis(2-ethylhexyl)sebacate spreads spontaneously on metals, but form drops of finite contact angles on silica, glass, and alumina. The diester readily hydrolyzes on these hydrated high-energy surfaces to form 2-ethylhexanoic acid. The critical surface tension of the adsorbed 2-ethylhexanoic acid is 28 dyne/cm, which is lower than the surface tension of the diester,

Table 6.2. Some Autophobic Liquids and Their Contact Angles on High-Energy Surfaces at 20°C

Liquid	γ_{LV}, dyne/cm	θ, deg			
		18.8 Stainless steel	Platinum	Fused silica	α-Al_2O_3
1-Octanol	27.8	35	42	42	43
2-Octanol	26.7	14	29	30	26
n-Octanoic acid	29.2	34	42	32	43
2-Ethyl-1-hexanol	26.7	5	20	26	19
Tricresyl phosphate	40.9	-	7	14	18

Source: From Ref. 21.

Spreading on High-Energy Surfaces

31.1 dyne/cm (20°C). However, on metal surfaces, the diester adsorbs without hydrolysis. The molecules lie flat, giving a monolayer of higher critical surface tension; hence, the diester can spread on it [18].

6.2.3. Wetting of Gold by Water

All pure low-energy liquids exhibit a zero contact angle on high-energy surfaces, except autophobic liquids and hydrolytic esters. Thus, water should spread spontaneously on gold. However, White [22] reported that water exhibits a finite contact angle on pure gold. Erb [23] reported contact angles of water on pure gold to be 55–85°, measured with a closed distillation system. Fowkes [24] asserted that water should exhibit a finite contact angle on gold, since only dispersion force can interact between water and a noble metal. These cases of nonzero contact angles have, however, been shown to be due to contamination of the high-energy gold surface by low-energy materials, or inadequacy of the theoretical model. Water should exhibit a zero contact angle on pure gold (discussed below).

Bewig and Zisman [25] reported zero contact angles of water on pure gold and platinum. The metals were polished to a mirrorlike gloss with carborundum. Organic impurities were scrupulously removed. Any possible oxide films were eliminated by heating in hydrogen. White and Drobek [26] suggested that an alumina polishing agent could have been embedded in Bewig and Zisman's metal samples, causing the complete spreading of water. They thus used diamond paste as the polishing agent, and heated the samples in pure oxygen at 1000°C to eliminate any residual diamond particles. They obtained a contact angle of 61° for water on gold. Erb [27] asserted that the distillation method should be the most reliable, giving 61–65° as the correct contact angle of water on pure gold. Thelen [28] adopted this value in some theoretical calculations. However, Zisman [29] pointed out that the 61–65° values are practically identical to those for water on gold and platinum whose surfaces have been precoated with adsorbed monolayers of hexane and benzene, reported earlier by Bewig and Zisman [30]. The gold samples of Erb and White appear to have been contaminated with organic impurities.

Subsequently, Bernett and Zisman [31] showed that pure water always exhibits a zero contact angle on pure gold from which organic contaminants and oxide polishing particles have been scrupulously removed. Nuclear activation analysis and X-ray microprobe were used to confirm the purity of the gold surfaces. Schrader [32] found that pure water exhibits a zero contact angle on pure gold by the ultrahigh-vacuum technique. Pure water was also found to exhibit a zero contact angle on pure copper and pure silver, wherein the purity of the metal surfaces is ensured by the ultrahigh-vacuum technique [33].

Theoretical calculation based on Lifshitz theory of van der Waals forces shows that the presence of one monolayer of hydrocarbon contaminant will render the gold surface hydrophobic, with a large water contact angle [34].

The controversy discussed above illustrates the experimental difficulty in obtaining pure and uncontaminated high-energy surfaces.

6.3. EFFECT OF WATER AND ORGANIC CONTAMINATIONS

6.3.1. Adsorptivity of High-Energy Surfaces

High-energy surfaces tend to adsorb water and organic materials rapidly from an ambient environment, and become low-energy surfaces, as already pointed out. The detection and control of organic contamination of high-energy surfaces have been reviewed recently elsewhere [35].

The contact angle of a liquid on a freshly cleaned high-energy surface will increase rapidly with time until reaching a plateau value, as the high-energy surface is exposed to laboratory air and adsorbs low-energy contaminants. Such behavior is shown in Figure 6.1 for the contact angles of water on several freshly cleaned high-energy surfaces. The metal surfaces were polished to a mirror finish with alumina powder, degreased with trichloroethylene and acetone, and then heated to 400–500°C for 300 min in air to remove any organic contaminants and form oxide layers on the surfaces. The glass surface was cleaned with sulfochromate solution and then fired in air as for the metals. Clean mica surface was obtained by cleaving a piece of mica immediately before use. The initial contact angles of water on these freshly prepared

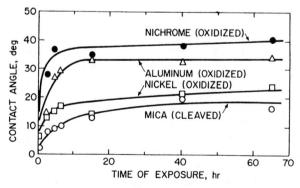

Figure 6.1. Contact angles of water on high-energy surfaces exposed to laboratory air versus exposure time. (After Ref. 35.)

surfaces are quite low, between 3 and 8°. Nichrome and aluminum have higher initial rates of contact angle change, and attain higher final contact angles. Different high-energy surfaces have different adsorptivity and different final states of adsorbed contaminant films.

The changes in the contact angles of water on several high-energy surfaces exposed to Nujol mineral oil vapor for 2 hr are listed in Table 6.3. The contact angle changes are least for glass and mica, greatest for transition metal oxides. The trends are consistent with the adsorptivity of fatty acids from solutions onto mica, platinum, gold, chromium, nickel, iron, and copper [36-38].

Different adsorptivity of high-energy surfaces leads to a useful technique for storing clean high-energy surfaces [35]. As oxidized metal surfaces are more adsorptive than glass and mica, the former can be used as contaminant scavengers when storing the latter. A piece of mica that had been contaminated by exposure to laboratory air was placed in an aluminum container which had been heated in air to 500°C to remove any organic contaminants and to form oxide layers on the walls. The contact angle of water on the mica was checked periodically and found to decrease with storage time (Table 6.4). This suggests that organic contaminants were gradually desorbed from the mica surface and adsorbed onto the more adsorptive oxidized aluminum walls. The contact angle was not changed when a freshly prepared aluminum specimen was placed in the oxidized aluminum container for more than 110 hr [35]. The effectiveness of various con-

Table 6.3. Contact Angles of Water on Some High-Energy Surfaces

Metal	θ, deg	
	Initial	After exposure to Nujol vapor for 2 hr
Mica (cleaved)	3.7	7.4
Glass (chromic acid-cleaned)	5.7	6.3
Aluminum (oxidized)	4.2	10.1
Nickel (oxidized)	7.0	17.9
Nichrome	4.9	32
Zinc (oxidized)	7.6	20.2
Iron (oxidized)	4.9	19.7

Source: From Ref. 35.

Table 6.4. Desorption of Contaminants from Mica Stored in Oxidized Aluminum Container

Time of storage, hr	Contact angle of water on mica, deg
0	15
16	13
24	11
40	10.5
104	7.5

Source: From Ref. 35.

tainers in maintaining a clean and high-adsorptive Nichrome surface is given in Table 6.5.

6.3.2. Effect of Water Adsorption

Water vapor adsorbs rapidly onto any high-energy surface. Significant adsorption occurs even at as low as 0.6% relative humidity. The adsorbed water forms partial or multiple monolayers and converts a high-

Table 6.5. Effect of Storage Condition on the Variation of Contact Angle of Water on Nichrome

Storage environment	Contact angle of water on Nichrome after 16 hr of storage, deg
Initial (after cleaning)	6
Laboratory air	35
Stoppered glass bottle	30
Empty aluminum desiccator	18
Desiccator with activated charcoal	16
Desiccator with activated alumina	12
Desiccator with aluminum shots	6

Source: From Ref. 35.

energy surface into a low-energy surface [29,38–41]. The contact angle increases, and the surface tension decreases, with increasing relative humidity (RH). The changes are reversible, indicating that the variation in thickness of the physically adsorbed water layer is responsible. The wettability and surface tension of high-energy surfaces are therefore determined mainly by the amount of adsorbed water. All high-energy surfaces (including glass, quartz, sapphire, and various metals) have practically identical surface tensions in ordinary air. The critical surface tensions are always about 36 dyne/cm at 95% RH and about 45 dyne/cm at 0.6% RH (20°C), regardless of the nature of the high-energy materials.

Soda-Lime Glass

The contact angles of some liquids on a soda-lime glass at several relative humidities are given in Table 6.6. The critical surface tension of the soda-lime glass was found to be 46 dyne/cm at 0.6–1% RH [39]. Olsen and Osteras [42] reported a value of 75 dyne/cm at about 0.001% RH [39]. The critical surface tension varies linearly with relative humidity (water vapor activity) (Figure 6.2).

At 98% RH, the critical surface tension is 30 dyne/cm, which is 8 dyne/cm higher than that of bulk water (22 dyne/cm). Thus, it appears that the surface field of force of dry glass is not completely suppressed by the existence of a few monolayers of water [39]. Only after many additional monolayers of water have been adsorbed will the surface field of force approach that of bulk water.

The surface tension of glass at the room temperature is about 500 dyne/cm (Table 6.1). The conversion of glass from a high-energy to a low-energy surface occurs at the chemisorption of the first water layer. Additional water molecules are adsorbed physically. The variation of the thickness of physically adsorbed water layer accounts for the variation of surface tension with the relative humidity [29].

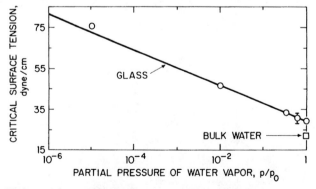

Figure 6.2. Effect of water vapor activity on critical surface tension of soda-lime glass. (After Ref. 39.)

Table 6.6. Wettability of Soda-Lime Glass by Some Pure Liquids Having Negative Spreading Coefficients on Bulk Water at 20°C

Liquid	γ_{LV}, dyne/cm	On duplex water film	Contact angle, θ deg			
				On glass equilibrated at:		
			95% RH	63–53% RH	1% RH	
Methylene iodide	50.8	37	36	31	13	
Tetrabromoethane	47.5	31	36	9	9	
1-Methylnaphthalene	38.7	22	7	–	5	
Dicyclohexyl	32.8	31	21	21	0	
Isopropyl bicyclohexyl	30.9	33	13	–	5	
n-Hexadecane	27.6	23	0	0	0	

Source: From Ref. 29.

Table 6.7. Wettability of Some Glasses at 0.6% and 95% RH at 20°C (Specimens Predried at 120°C for 10 Min)

Liquid	γ_{LV}, dyne/cm	Contact angle, θ deg					
		0.6% RH			95% RH		
		Pyrex	Quartz	Sapphire	Pyrex	Quartz	Sapphire
α-Methyl naphthalene	38.7	5	5	5	14	6	16
p-Octadecyl toluene	31.5	5	9	8	28	16	25
n-Hexadecane	27.6	0	0	0	28	28	29
n-Dodecane	25.4	0	0	0	0	5	5
n-Decane	23.9	0	0	0	0	0	0
Methylene iodide	50.8	19	22	24	36	37	38
sym-Tetrabromoethane	47.5	8	17	18	33	35	33
α-Bromonaphthalene	44.6	9	5	9	36	33	35
1,2,3-Tribromobutane	39.9	–	–	–	27	18	24
1,3-Dibromopropane	38.9	0	0	0	10	8	6
Bromotrichloromethane	30.3	0	0	0	0	0	0

Source: From Ref. 29.

Borosilicate Glass, Quartz, and Sapphire

These high-energy surfaces have similar wettability and surface tension as soda-lime glass in air. Contact angles of some liquids on them are listed in Table 6.7. The critical surface tensions are about 36 to 37 dyne/cm at 95% RH, and 46 to 47 dyne/cm at 0.6% RH for all of them [29,41], much like those for soda-lime glass. The surface properties are not affected by drying the specimens between 65 and 215°C, indicating that the chemisorbed first water layer is not removed at these temperatures.

However, heating above 400°C will dehydrate the surfaces. If a polished silica (cleaned by degreasing and boiling in 30% hydrogen peroxide) is heated in oxygen at various temperatures, the contact angles of water (measured immediately after cooling to room temperature in a clean and dry environment) will increase with the temperature of heating (Figure 6.3). The change arises from dehydration of hydrophilic silanol (Si-OH) groups to form hydrophobic siloxane (Si-O-Si) bonds [35]. If the silica is precleaned by etching with HF-NH$_4$F, the increased hydrophobicity arises from formation of hydrophobic Si-F bonds, owing to reaction with residual fluoride [35].

Metals and Metal Oxides

These also have surface properties very similar to soda-lime glass, quartz, and sapphire [29,40]. Again, the critical surface tensions are about 46–47 dyne/cm at 0.6% RH and 36–37 dyne/cm at 95% RH. However, there are differences in the slope of the Zisman plot, reflecting differences in the nature of the surfaces.

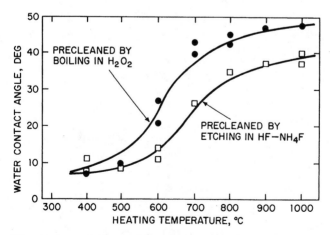

Figure 6.3. Effect of heating in dry oxygen on the contact angle of water on silica plate. (After Ref. 35.)

6.4. KINETICS OF SURFACE-ENERGY VARIATION

The rate of surface-energy reduction as a high-energy surface is exposed to humidity follows first-order kinetics [44]. The reduction in surface energy results mainly from reduction of polar component, while the nonpolar (dispersion force) component is little changed. Such behavior for a sulfochromate-treated aluminum exposed to humid air at 23°C is shown in Figure 6.4. The rates of change at very high (100% RH) and very low (0.016% RH) humidities are slower than at intermediate humidities.

The kinetics can be analyzed in terms of the stochastic model for cumulative damage [44]. A special form of the Halpin-Polley [45] is given by [44]

$$\phi = \frac{M_0 - M_t}{M_0 - M_\infty} = 1 - \exp\left[\frac{-K(t - t_0)^b}{a_T a_M a_C}\right] \tag{6.1}$$

where ϕ is the fractional degradation, M_0 the measured property at $t = t_0$, M_∞ the measured property at infinite time, M_t the measured property at time t, K a system constant, a_T the time shift factor for thermal stress, a_M the time shift factor for mechanical stress, a_C the time shift factor for chemical stress, and b the time exponent for the degradation kinetics. For the humidity degradation of surface energy, first-order kinetics applies, and the effects of thermal and mechanical

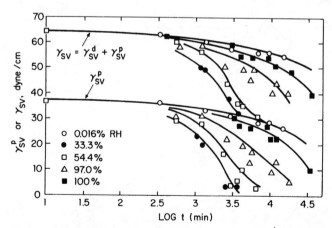

Figure 6.4. Effect of exposure time t on the surface tension γ_{SV} and its polar component $\gamma_{SV}{}^p$ for a sulfochromate-treated 2024-T3 aluminum at 23°C. (After Ref. 44.)

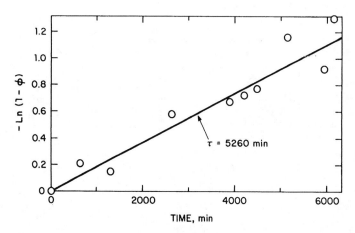

Figure 6.5. First-order kinetics for the surface aging of sulfochromate-treated 2024-T3 aluminum at 97% RH and 23°C. (After Ref. 44.)

stresses may be neglected. In this case, $b = 1$ and $a_T = a_M = 1$. The relaxation time is then $\tau = a_C/K$, and $M_t = \gamma_S$ or $M_t = \gamma_S^p$. For this special case, Eq. (6.1) becomes

$$-\ln(1 - \phi) = \frac{t}{\tau} \tag{6.2}$$

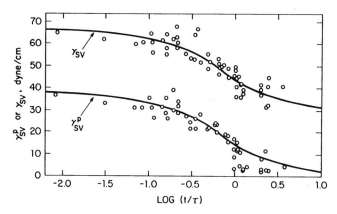

Figure 6.6. Master curves for exposure time-relative humidity superposition of γ_{SV} and γ_{SV}^p for a sulfochromate-treated 2024-T3 aluminum. (After Ref. 44.)

Table 6.8. Relaxation Time for Surface Energy Reduction of Chromic Acid-Etched 2024-T3 Aluminum at 23°C in Ambient Air of Various Relative Humidities

RH, %	P_{H_2O}, torr	Relaxation time τ, min
0.016	0.0034	43,700
12.2	2.6	1,830
33	7.0	1,890
54	11.5	3,000
97	20.4	5,260
100	21.1	22,000

Source: From Ref. 44.

which is the standard relation for first-order kinetics. A linear plot of $-\ln(1 - \phi)$ versus t for the data at 97% RH at 23°C is given in Figure 6.5. The relaxation time τ is found to be 5260 min. The relaxation times at various relative humidities thus found are given in Table 6.8.

An important consequence of Eq. (6.1) is that the values of γ_S^p and γ_S may be shifted along the log t axis to superimpose on a single master curve as a function of the reduced time, log (t/τ). This master curve is given by

$$M_t = M_\infty + (M_0 - M_\infty) \exp\left(\frac{-t}{\tau}\right) \tag{6.3}$$

Time-humidity superposition master curves for γ_S and γ_S versus log (t/τ) for the sulfochromated aluminum are given in Figure 6.6.

REFERENCES

1. H. W. Fox and W. A. Zisman, J. Colloid Interface Sci., 5, 514 (1950).
2. G. Lang, in *Handbook of Chemistry and Physics*, 58th ed., R. C. Weast, ed., CRC Press, Cleveland, Ohio, 1977 pp. F-25—42.
3. As quoted by A. W. Adamson, *Physical Chemistry of Surfaces*, 2nd ed., Interscience, New York, 1967, pp. 42—45.
4. M. E. Nicholas, P. A. Joyner, B. M. Tessem, and B. M. Olson, J. Phys. Chem., 65, 1373 (1961).

5. C. Kemble, Trans. Faraday Soc., 42, 526 (1946).
6. G. W. Morey, The Properties of Glass, 2nd ed., Reinhold, New York, 1954, pp. 191–209.
7. A. E. J. Vickers, J. Soc. Chem. Ind., 57, 14 (1938).
8. G. Keppeler and A. Albrecht, Glastech. Ber., 18, 275 (1940).
9. J. K. Davis and F. E. Bartell, Anal. Chem., 20, 1182 (1948).
10. N. M. Parikh, J. Am. Ceram. Soc., 41, 18 (1958).
11. D. T. Livey and P. Murray, Plansee Proc., 2nd Seminar, Reutte-Tyrol, 1956, p. 375.
12. S. K. Rhee, J. Am. Ceram. Soc., 53, 639 (1970).
13. S. K. Rhee, J. Am. Ceram. Soc., 53, 386 (1970).
14. S. K. Rhee, J. Am. Ceram. Soc., 55, 157 (1972).
15. S. K. Rhee, J. Am. Ceram. Soc., 54, 376 (1971).
16. W. D. Kingery, J. Am. Ceram. Soc., 37, 42 (1954).
17. J. C. Eberhart, J. Phys. Chem., 71, 4125 (1967).
18. H. W. Fox, E. F. Hare, and W. A. Zisman, J. Phys. Chem., 59, 1097 (1955).
19. E. F. Hare and W. A. Zisman, J. Phys. Chem., 59, 335 (1955).
20. E. G. Shafrin and W. A. Zisman, J. Phys. Chem., 64, 519 (1960).
21. W. A. Zisman, Adv. Chem. Ser., 43, 1 (1964).
22. M. L. White, J. Phys. Chem., 68, 3083 (1964).
23. R. A. Erb, J. Phys. Chem., 69, 1036 (1965).
24. F. M. Fowkes, in Chemistry and Physics of Interfaces, American Chemical Society, Washington, D.C., 1965; pp. 1–12; also, Ind. Eng. Chem., 56, 40 (1964).
25. K. W. Bewig and W. A. Zisman, J. Phys. Chem., 69, 4238 (1965).
26. M. L. White and J. Drobek, J. Phys. Chem., 70, 3432 (1966).
27. R. A. Erb, J. Phys. Chem., 72, 2412 (1968).
28. E. Thelen, J. Phys. Chem., 71, 1946 (1967).
29. W. A. Zisman, in Adhesion Science and Technology, Vol. 9A, L. H. Lee, ed., Plenum Press, New York, 1975, pp. 55–91.
30. K. W. Bewig and W. A. Zisman, J. Phys. Chem., 68, 1804 (1964).
31. M. K. Bernett and W. A. Zisman, J. Phys. Chem., 74, 2309 (1970).
32. M. E. Schrader, J. Phys. Chem., 74, 2313 (1970).
33. M. E. Schrader, J. Phys. Chem., 78, 87 (1974).
34. V. A. Parsegian, G. H. Weiss, and M. E. Schrader, J. Colloid Interface Sci., 61, 356 (1977).
35. M. L. White, in Clean Surfaces, G. Goldfinger, ed., Marcel Dekker, New York, 1970, pp. 361–373.
36. J. R. Miller and J. E. Berger, J. Phys. Chem., 70, 3070 (1966).
37. G. L. Gaines, Jr., J. Colloid Interface Sci., 15, 321 (1960).

References

38. H. D. Cooke and H. E. Ries, Jr., J. Phys. Chem., *63*, 226 (1959).
39. E. G. Shafrin and W. A. Zisman, J. Am. Ceram. Soc., *50*, 478 (1967).
40. M. K. Bernett and W. A. Zisman, J. Colloid Interface Sci., *28*, 243 (1968).
41. M. K. Bernett and W. A. Zisman, J. Colloid Interface Sci., *29*, 413 (1969).
42. D. A. Olsen and A. J. Osteras, J. Phys. Chem., *68*, 2730 (1964).
43. E. G. Shafrin and W. A. Zisman, J. Phys. Chem., *71*, 1309 (1967); R. E. Johnson, Jr., and R. H. Dettre, J. Colloid Interface Sci., *21*, 610 (1966).
44. D. H. Kaelble and P. J. Dynes, J. Colloid Interface Sci., *52*, 562 (1975).
45. J. C. Halpin and H. W. Polley, J. Compos. Mater., *1*, 64 (1967); J. C. Halpin, ibid., *6*, 208 (1972).

7
Dynamic Contact Angles and Wetting Kinetics

7.1. INTRODUCTION

The angle formed at a stationary liquid front is termed the *static contact angle*. The angle formed at a moving liquid front is termed the *dynamic contact angle*. Static contact angles are determined by the equilibrium of interfacial energies. Dynamic contact angles are determined by the balance of interfacial driving force and viscous retarding force. There are two types of dynamic contact angles: those in spontaneous motion (where the liquid front moves spontaneously, driven by interfacial forces) and those in forced motion (where the motion is imposed externally).

Examples of spontaneous motion include (1) penetration of a liquid into a capillary, and (2) spreading of a liquid drop toward an equilibrium shape. Spontaneous spreading of a liquid drop placed on a rigid surface is shown in Figure 7.1. Initially, the instantaneous contact angle θ is almost 180°, when the drop has just been placed on the surface. Sometime later, the angle becomes smaller, as the drop spontaneously spreads. Finally, the drop comes to an equilibrium shape, and the liquid front ceases to move. The contact angle now reaches its equilibrium value θ_e.

Examples of forced motion include (1) running a rod or a plate into a liquid, and (2) forcing a liquid to run up a capillary by apply-

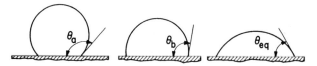

Figure 7.1. Spontaneous spreading of a liquid drop on a rigid surface.

ing hydrostatic pressure. The forced spreading of a liquid by running a rod into it is shown in Figure 7.2. Initially (stage a), the system is at rest. The contact angle at the three-phase boundary is at its equilibrium value θ_e. In stage b, the rod has just been moved into the liquid. A transitory situation is created in which the liquid meniscus is distorted, producing a higher contact angle and a high-pressure region at P. The three-phase boundary will flow upward to prevent the contact angle from increasing indefinitely. In state c, a steady state is reached in which the liquid pushed down by the moving rod is replenished by the liquid pushed up by interfacial forces. The liquid meniscus remains somewhat distorted, and the contact angle comes to a steady dynamic value θ_d, which is greater than θ_e.

7.2. KINETICS OF SPONTANEOUS MOTION

7.2.1. Precursor Films at Liquid Front

There are two types of spreading liquid front: (1) front with precursor films, and (2) front without precursor films (Figure 7.3).

Precursor films may consist of a thin primary film (less than 1000–10,000 Å thick), followed by a somewhat thicker secondary film (less than 1–2 μm thick). The bulk liquid follows the secondary film. When the system approaches the equilibrium, the secondary film will merge into the bulk. However, the primary film may persist, or merge with the bulk liquid. These precursor films are in-

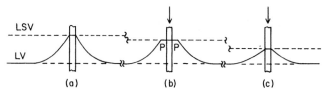

Figure 7.2. Forced motion of a liquid on a rigid surface by running a rod into the bulk liquid.

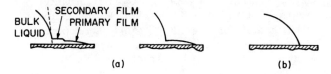

Figure 7.3. Spontaneously spreading liquid front.

visible by conventional techniques, but can be detected by interference microscopy or ellipsometry. In many cases, only a single projecting film is observed, primary and secondary films being indistinct. Precursor films have been observed in the spontaneous spreading of water on clean glass, squalene on steel [1-3], polydimethylsiloxane on silicon (precursor film about 7000 Å thick by interference microscopy) [4], polyethylene melt on glass, and molten glass on Fernico metal (precursor film about 1 μm thick by SEM) [5]. The rate-determining step may be the spreading of the precursor film (in light fluids) or the flow of the bulk liquid (in viscous fluids). The macroscopic contact angle is that formed between the solid surface and the projection of the bulk liquid front.

On the other hand, no precursor films are observed in the spreading of polyethylene melts and poly(ethylene-vinyl acetate) melts on aluminum and FEP (fluorinated ethylene-propylene copolymer) [6].

7.2.2. Shape of Spreading Drop

Two shapes have been observed: (1) spherical cap, and (2) cap with foot (Figure 7.4). Examples of spherical cap are polyethylene melts and poly(ethylene-vinyl acetate) melts on aluminum, steel, and FEP [6-9], and glycerol on aluminum [7]. Examples of cap with foot are molten fluoroalkyl methacrylate polymers on glass [10].

A spherical cap will be formed when the rate of contact angle change is rate-determining. A cap with a foot will be formed when the flow rate of bulk liquid is rate-determining. This is consistent

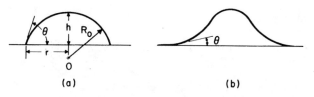

Figure 7.4. Shapes of spontaneously spreading liquid drops: (a) spherical cap; (b) cap with foot.

with two experimental observations. A drop of molten fluoroalkyl methacrylate polymer spreads on glass with a spherical-cap shape at 160°C, but with a cap-with-foot shape at 110°C [10]. A drop of molten polyethylene spreads on steel with a spherical cap shape when the equilibrium contact angle is large, and with a cap-with-foot shape when the equilibrium contact angle is small [11].

The geometry of a spherical cap is characterized by the relations

$$V_0 = \frac{\pi}{3}(3R_0 - h)h^2 \tag{7.1}$$

$$\frac{h}{r} = \tan\frac{\theta}{2} = \frac{1 - \cos\theta}{\sin\theta} \tag{7.2}$$

$$\frac{r^3}{V_0} = \frac{3}{\pi}\left[\frac{\sin\theta(1+\cos\theta)}{(1-\cos\theta)(2+\cos\theta)}\right] \tag{7.3}$$

where V_0 is the drop volume, h the drop height, R_0 the radius of curvature, r the base radius, and θ the instantaneous contact angle. These relations can be used to check whether the drop is spherical.

7.2.3. Kinetics of Spontaneous Spreading

The driving force in spontaneous spreading is the unbalanced interfacial forces, given by

$$f_d = \gamma_{SV} - \gamma_{SL} - \gamma_{LV}\cos\theta \tag{7.4}$$

where f_d is the driving force acting on a unit length of the liquid front, γ_{SV} the surface tension of the solid in equilibrium with the saturated liquid vapor, γ_{SL} the interfacial tension between the solid and the liquid, γ_{LV} the liquid surface tension, and θ the instantaneous contact angle. Applying Young's equation gives

$$f_d = \gamma_{LV}(\cos\theta_e - \cos\theta) \tag{7.5}$$

where θ_e is the equilibrium contact angle. By definition, $\theta_e = \theta_\infty$, the contact angle at infinite time.

The retarding force is the viscous resistance of the liquid, given by

$$f_r = \eta\left(\frac{dv}{dz}\right)L \tag{7.6}$$

Kinetics of Spontaneous Motion

where f_r is the viscous retarding force per unit length of liquid front, η the liquid viscosity, v the interfacial velocity, z the direction perpendicular to the liquid-front flow direction, and L the jump distance (the length of a flow unit in the flow direction). Assume Couett flow (constant shear rate); then $dv/dz = v/b$, where b is a characteristic height. Equating f_d and f_r gives

$$v = \frac{b\gamma_{LV}}{\eta L}(\cos \theta_e - \cos \theta) \tag{7.7}$$

It can be shown that $v/b = d\cos\theta/dt$. Therefore, Eq. (7.7) can be rewritten as

$$\frac{d\cos\theta}{dt} = \frac{\gamma_{LV}}{\eta L}(\cos\theta_e - \cos\theta) \tag{7.8}$$

which was originally derived by Cherry and Holmes [8] by assuming that the liquid front moves between successive equilibrium positions with intervening energy barriers.

Newman [12] empirically suggested the relation

$$\cos\theta = \cos\theta_e(1 - ae^{-ct}) \tag{7.9}$$

where the constant a is given by

$$a = 1 - \frac{\cos\theta_0}{\cos\theta_e} \tag{7.10}$$

where θ_0 is the contact angle at zero time. Equation (7.9) is in fact the integral form of Eq. (7.8), wherein the constant c is given by

$$c = \frac{\gamma_{LV}}{\eta L} \tag{7.11}$$

These relations represent quite well the spreading kinetics of drops of molten polyethylene and poly(ethylene-vinyl acetate) on aluminum, mica, and FEP [8,12]. The rate of spreading is predicted to be dependent on viscosity and equilibrium contact angle, but independent of drop mass, consistent with some experiments. In other cases, however, the spreading rate is found to depend on drop mass (discussed later).

Schonhorn and coworkers [6,7] found empirically that the spreading rate is independent of drop mass, and can be superimposed to give a single master curve, that is,

$$\frac{r}{r_0} = F(a_T t) \tag{7.12}$$

$$\frac{\cos \theta}{\cos \theta_e} = G\left(\frac{r}{r_0}\right) = H(a_T t) \tag{7.13}$$

where r is the base radius, r_0 that at $\theta = 90°$, t the time, and a_T the time-temperature shift factor given by

$$a_T = \frac{\gamma_{LV}}{\eta L_W} \tag{7.14}$$

where L_W is the scaling length, which is independent of drop mass and temperature but is dependent on the nature of the liquid-solid pair. F, G, and H are universal functions of the reduced time $a_T t$ or the reduced base radius r/r_0. Comparison with Eq. (7.11) shows that

$$a_T = c = \frac{\gamma_{LV}}{\eta L} \tag{7.15}$$

and thus $L_W = L$. Figure 7.5 shows that the spreading rate of molten drops of poly(ethylene-vinyl acetate) on aluminum at 170°C is independent of drop mass. Figures 7.6 and 7.7 show the master curves for r/r_0 and $\cos \theta/\cos \theta_e$ versus $a_T t$.

Ogarev and coworkers [13] analyzed the rate of drop spreading by assuming that the drop takes a conical shape to simplify the mathematics. They derived

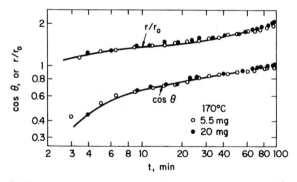

Figure 7.5. Kinetics of spontaneous spreading of molten poly(ethylene-vinyl acetate) on sulfochromated aluminum at 170°C. (After Ref. 6.)

Kinetics of Spontaneous Motion

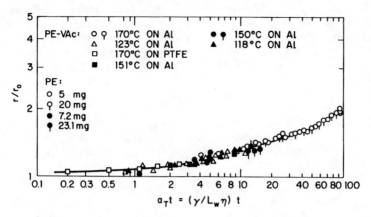

Figure 7.6. Master curve for spontaneous spreading of liquid drops. (After Ref. 6.)

$$t = \frac{(3V_0/\pi)^{1/3}(\eta/8)}{\gamma_{LV}(\cos\theta_e - \cos\theta)(\tan\theta)^{4/3}} \qquad (7.16)$$

$$r^4 = \frac{24V_0}{\pi\eta}\gamma_{LV}(\cos\theta_e - \cos\theta)t \qquad (7.17)$$

where t is the time, V_0 the drop volume, and r the drop base radius. These equations fit the experimental data quite well, for instance, in

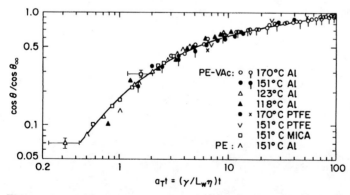

Figure 7.7. Master curve for spontaneous spreading of liquid drops. (After Ref. 6.)

the spreading of polydimethylsiloxane on silicon [4], glass, and fluoropolymer [13]. These equations indicate that spreading rate is dependent on drop volume; the smaller the drop, the faster the rate. This drop-volume effect has been confirmed experimentally in many cases [4,9,13]. An example for molten polyethylene on aluminum is shown in Figure 7.8. Generally, spreading rate depends on drop volume. It is, however, not known why the contrary is sometimes observed (Figure 7.5).

Several other analyses have also been reported. Van Oene and coworkers [9] analyzed the drop spreading by adapting the flow of free films. The results predict a drop-volume dependence of spreading rate. Lau and Burns [14,15] proposed an empirical relation,

$$\frac{dA}{dt} = k\gamma_{LV}(\cos\theta_e - \cos\theta) \tag{7.18}$$

where A is the drop base area and k is a rate constant independent of drop volume but dependent on temperature and viscosity. The relation is confirmed experimentally for the spreading of molten polystyrenes on glass.

Caneva and coworkers [11] reported that the spreading rate of molten polyethylene on steel follows first-order kinetics,

$$\frac{d\theta}{dt} = K(\theta_e - \theta) \tag{7.19}$$

where K is a rate constant. This relation has also been observed in the spreading of water on paraffin wax [16] and carbon tetrachloride

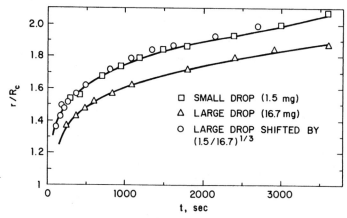

Figure 7.8. Effect of drop volume on spontaneous spreading of molten polyethylene drops on aluminum at 120°C. $R_c = (3V_0/2\pi)^{1/3}$, where V_0 is the drop volume. (After Ref. 9.)

Kinetics of Spontaneous Motion

on mercury immersed in water or aqueous surfactant solutions [17,18]. On the contrary, second-order kinetics was observed with certain surfactant solutions used as the immersion liquid [18]. Apparently, first-order kinetics will be observed if the surfactant tightly absorbs onto the substrate and is not displaced by the spreading liquid. On the other hand, if the surfactant is displaced by the spreading liquid, second-order kinetics will be observed [18–20]. Other analyses have also been reported [21,22].

7.2.4. Kinetics of Capillary Penetration

The rate of spontaneous penetration of a liquid into a capillary is analyzed below. First, the Washburn equation based on a quasi-steady-state approximation (assuming a steady dynamic angle and neglecting the transient initial rate) is discussed. Next, the effect of variable dynamic angle is included in the Washburn analysis. Finally, a general hydrodynamic equation is given, taking into account the transient behavior at short times.

Washburn Equation

Consider the laminar flow of a Newtonian fluid through a capillary. Poiseuille's law gives the average linear velocity of the fluid $\bar{v}$ as

$$\bar{v} = \frac{Q}{\pi R^2} \tag{7.20}$$

$$= \frac{R^2 \Delta P}{8\eta h} \tag{7.21}$$

where Q is the volumetric flow rate, R the radius of the capillary, ΔP the pressure drop along axial distance h, and η the viscosity of the fluid. The pressure drop is composed of two terms, that is,

$$\Delta P = \Delta P_c + \Delta P_g \tag{7.22}$$

where ΔP_c is the capillary pressure and ΔP_g is the hydrostatic pressure. The capillary pressure is given by

$$\Delta P_c = \gamma_{LV}\left[\frac{1}{G_1} + \frac{1}{G_2}\right] \tag{7.23}$$

where G_1 and G_2 are the two principal radii of curvature of the fluid surface. In small cylindrical tubes where the fluid meniscus does not depart from a spherical shape,

$$\Delta P_c = \frac{2\gamma_{LV}\cos\theta}{R} \tag{7.24}$$

For horizontal tubes, the effect of hydrostatic pressure is zero, that is,

$$\Delta P_g = 0 \quad \text{(horizontal tubes)} \tag{7.25}$$

But for vertical tubes, the hydrostatic pressure is given by

$$\Delta P_g = \pm \rho g h \quad \text{(vertical tubes)} \tag{7.26}$$

where ρ is the fluid density, g the gravitational acceleration, the minus sign is for up flow, and the plus sign is for down flow.

Combining Eqs. (7.21), (7.24), and (7.26) gives

$$\frac{dh}{dt} = \frac{R^2}{8\eta h}\left(\frac{2\gamma_{LV}\cos\theta}{R} \pm \rho g h\right) \tag{7.27}$$

which describes the rate of fluid rise (or descent) in a vertical capillary, and is known as the *Washburn equation* [23]. For horizontal capillaries, of course, the term $\rho g h$ is put to zero. If the contact angle θ is assumed to be constant, that is, $\theta = \theta_e$, Eq. (7.27) can readily be integrated to give the relation for a fluid rising in a vertical capillary [12,24],

$$\left[\ln\left(\frac{h_\infty}{h_\infty - h}\right)\right]\frac{h}{h_\infty} = \frac{R^2 \rho g t}{8\eta h_\infty} \tag{7.28}$$

where h_∞ is the equilibrium height (that is, at $t = \infty$) achieved in a vertical capillary, that is,

$$h_\infty = \frac{2\gamma_{LV}\cos\theta_e}{R\rho g} \tag{7.29}$$

obtainable by letting $dh/dt = 0$ in Eq. (7.27). On the other hand, the relation for a fluid flowing in a horizontal capillary is given by

$$h^2 = \frac{R\gamma_{LV}(\cos\theta_e)t}{2\eta} \tag{7.30}$$

Kinetics of Spontaneous Motion

Effect of Dynamic Contact Angle

Dynamic contact angle is not constant, but varies with time during capillary flow, particularly for viscous molten polymers. For horizontal capillaries, combining Eqs. (7.9) and (7.27) gives the relation [12]

$$\frac{dh}{dt} = \frac{R\gamma_{LV}(\cos\theta_e)(1 - ae^{-ct})}{4\eta h} \tag{7.31}$$

which can readily be integrated to give

$$h^2 = \frac{R\gamma_{LV}\cos\theta_e}{2\eta}\left[\left(t - \frac{a}{c}\right) + \frac{ae^{-ct}}{c}\right] \tag{7.32}$$

which reduces to Eq. (7.30) if $\cos\theta = \cos\theta_e$.

For a fluid flowing in a vertical capillary, the relation is given by [12]

$$\frac{dh}{dt} = \frac{R^2}{8\eta h}\left[\frac{2\gamma_{LV}(\cos\theta_e)(1 - ae^{-ct})}{R} \pm \rho g h\right] \tag{7.33}$$

which has no general solution of an elementary form. Special cases for short and long times or fine capillaries can, however, be solved readily. But intermediate times require numerical solution or approximation.

General Hydrodynamic Equation

The Washburn equation is derived based on a quasi-steady-state approximation; that is, the kinetic energy and the frictional work within the fluid are neglected. This renders the Washburn equation inadequate for short times when the liquid momentum changes rapidly with time and height of penetration. Inspection of Eq. (7.27) gives

$$\frac{dh}{dt} \to \infty \quad \text{as } t \to 0 \text{ and } h \to 0 \tag{7.34}$$

which is obviously untrue.

Consideration of energy conservation and frictional work of the fluid gives a more general and rigorous hydrodynamic equation for the penetration of a fluid into a vertical capillary [24,25]:

$$\left(h + \frac{7}{6}R\right)\frac{d^2h}{dt^2} + 1.225\left(\frac{dh}{dt}\right)^2 + \left(\frac{8\eta}{\rho R^2}\right)\frac{dh}{dt}$$

$$= \frac{1}{\rho}\left(\frac{2\gamma_{LV}\cos\theta}{R} - \rho gh\right) \qquad (7.35)$$

Equation (7.35) has been solved numerically by assuming a constant contact angle (Figure 7.9). As can be seen,

$$\frac{dh}{dt} = 0 \quad \text{as } t \to 0 \text{ and } h \to 0 \qquad (7.36)$$

which is the correct boundary condition, as opposed to Eq. (7.34). At short times, the general equation and the Washburn equation are quite different. At long times, the two equations merge together. In other words, the Washburn equation is only an asymptotic solution, not valid for short times. Strictly, the contact angle in Eq. (7.35) should be treated as time-dependent.

7.2.5. Kinetics of Slit Penetration

The equations for penetration into slits differ from those for cylindrical capillaries only in the numerical constants. For penetration

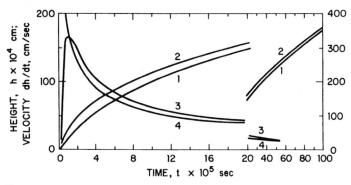

Figure 7.9. Kinetic plots for a liquid rising in a vertical cylincrical capillary; $R_0 = 0.001$ cm; $P_c = 50/R_0$; $\rho = 1$ g/ml; $\eta = 0.01$ g/cm-sec. (1) h versus t by the complete equation; (2) h versus t by the Washburn equation; (3) dh/dt versus t by the complete equation; (4) dh/dt versus t by the Washburn equation. (After Ref. 25.)

into a slit between two parallel plates of width W and distance H between the two plates, the average flow velocity is given by [12]

$$\frac{dh}{dt} = \frac{2}{3}\left(\frac{H^2 \Delta P}{8\eta h}\right) \qquad (7.37)$$

The capillary pressure is given by

$$\Delta P_c = 2\frac{\gamma_{LV}}{H}\cos\theta \qquad (7.38)$$

Other equations follow accordingly.

7.3. KINETICS OF FORCED MOTION

7.3.1. Earlier Works

Ablett [26] was apparently the first to report that contact angles vary with interfacial velocity in forced motion. A cylinder coated with paraffin wax was partially immersed in water. The static angle of water on paraffin wax was 104.57°. When the cylinder is rotated, the contact angle changed with rotation speed. The advancing angle increased to the maximum value of 113.15°, and the receding angle decreased to the minimum value of 96.33°, at interfacial velocities greater than 26 mm/min. Later, Rose and Heins [27] investigated the dynamic contact angles by pushing liquid slugs through glass tubes at various velocities. Their results showed a linear relationship between $\cos\theta$ and velocity within the velocity ranges studied (0–144 mm/min).

7.3.2. Three Velocity Regimes in Forced Motion

Generally, dynamic contact angles in forced motion fall into three characteristic velocity regimes. In the low-velocity range, the contact angle is independent of interfacial velocity, termed the *Hansen-Miotto-Elliott-Riddiford* mode. In the intermediate-velocity range, the contact angle varies with interfacial velocity, wherein the retarding force is the interfacial viscosity, termed the *Blake-Haynes* mode. The advancing angle increases and the receding angle decreases with increasing interfacial velocity. In the high-velocity range, surging of the liquid occurs, wherein the retarding force is the bulk viscosity, termed the *Friz* mode or *surging* mode. These three regimes are discussed below.

7.3.3. Low-Velocity Regime

This is the HMER mode (Hansen-Miotto-Elliott-Riddiford mode) or the molecular relaxation mode. Hansen and Miotto [28] suggested that the velocity effect on dynamic angle arises from lagging of molecular relaxation behind the impressed motion. If the molecular relaxation time is τ and the displacement length is L, the natural recovery velocity is $v_n = L/\tau$. If the interfacial velocity v is lower than v_n, the molecules will have sufficient time to recover from the imposed displacement. In this case, the dynamic angle will be independent of interfacial velocity and identical to the static angle. On the other hand, if v is greater than v_n, the molecules will be displaced and the dynamic angle will vary with interfacial velocity.

Riddiford and coworkers [29-32] reported experimental observations consistent with the model described above. The critical interfacial velocity occurs at about 1 mm/min. Below the critical rate, the dynamic angle is independent of velocity and is identical to the static angle. Above the critical rate, the advancing dynamic angle increases and the receding dynamic angle decreases with increasing velocity. However, these dynamic angles reach plateau values at about 10 mm/min. Such behaviors for water on siliconed glass and polyethylene are shown in Figure 7.10. Other examples include water on fluoropolymer [29], and glycerol, formamide, and methylene iodide on siliconed glass [32].

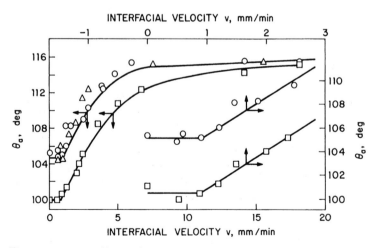

Figure 7.10. Effect of interfacial velocity on advancing dynamic contact angle of water on siliconed glass (O at 22°C and △ at 42°C) and on polyethylene (□ at 22°C). (After Ref. 31.)

Kinetics of Forced Motion

Lowe and Riddiford [29] suggested that the critical rate coincides with the relaxation times for nonpolar and polar molecular forces. Below the critical rate, both nonpolar and polar forces operate. But above the critical rate, dipole relaxation lags behind and only the dispersion force can operate. Thus, dynamic angles could be used to resolve polar and nonpolar interactions. This has been applied to the interactions between water and polytetrafluoroethylene [29]. Interestingly, the relaxation time of dynamic contact angle (as the impressed motion is stopped) correlates well with the NMR spin-lattice relaxation time of the liquids [32].

However, molecular relaxation times in simple liquids should be much faster than the interfacial velocities covered above. Therefore, the observed velocity effects on dynamic contact angles cannot be accounted for in terms of the molecular relaxation effects [33]. Instead, the velocity effects cited could be due to the hysteresis of contact angle arising from roughness or heterogeneity, at least at velocity time scales much longer than the molecular relaxation times. In this case, the maximum advancing dynamic angle should not exceed the maximum advancing static angle. The minimum receding dynamic angle should not be lower than the receding static angle. Further, if there is no hysteresis, the dynamic angle should be independent of the interfacial velocity in the ranges covered.

This hysteresis theory appears to be supported by experimental observations. Dynamic contact angles of water and hexadecane on some polymers and monolayers are measured by the Wilhelmy plate method [33]. The substrates include a poly(fluoroalkyl methacrylate), and FEP copolymer, a siliconed glass, a monolayer of trimethyloctadecylammonium chloride, and a perfluorodecanoic acid monolayer. Advancing dynamic angles of water on these substrates over the velocity range of 0.1−250 mm/min are shown in Figure 7.11. Also given for comparison are the data of Elliott and Riddiford [31] cited earlier for water on siliconed glass and polyethylene. The present data show that a significant velocity effect on dynamic angle is observed only when the static angle exhibits large hysteresis, such as on trimethyloctadecylammonium chloride monolayer (Table 7.1). The constant dynamic angle at low velocity and increasing dynamic angle at higher velocity as reported by Elliott and Riddiford are not observed.

7.3.4. Intermediate-Velocity Regime

This is the Blake-Haynes mode. The advancing dynamic angle increases and the receding dynamic angle decreases with increasing interfacial velocity. The magnitudes of these dynamic angles depend on the interfacial viscosity that gives rise to the retarding force. A linear relation between $\cos \theta$ and v is usually observed, consistent with Eq. (7.7) obtained by equating the interfacial driving force to the inter-

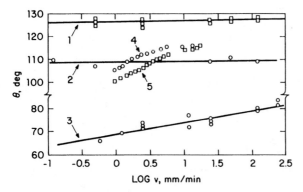

Figure 7.11. Effect of interfacial velocity on advancing dynamic contact angle of water on (1) poly(fluoroalkyl methacrylate); (2) siliconed glass; (3) monolayer of trimethyloctadecylammonium chloride on glass; (4) siliconed glass; (5) polyethylene. (Curves 1 to 3 are from Ref. 33; curves 4 and 5 are from Ref. 31.)

facial retardation force. The data of Rose and Heins [27] up to a rate of 144 mm/min, and of Riddiford and coworkers [29–32] between 1 and 10 mm/min, conform quite well with the linear relation.

Blake and Haynes [34] pictured the liquid motion as the sliding of liquid molecules over energy barriers on the solid surface at the three-phase boundary. They obtained

$$v = 2K\lambda \sinh \left[\frac{\gamma_{LV}}{nkT} (\cos \theta_e - \cos \theta) \right] \quad (7.39)$$

Table 7.1. Static Advancing and Receding Contact Angles of Water on Some Solids

	Contact angles of water, deg		
Solid	θ_a	θ_r	$\Delta\theta$
Poly(fluoroalkyl methacrylate)	125.4	113.1	12.3
Siliconed glass	107.5	93.8	13.7
Trimethyloctadecyl-ammonium chloride monolayer	72.1	32.9	39.2

Source: From Ref. 33.

Kinetics of Forced Motion

where K is the number of molecular displacement per unit time per unit length of liquid front, λ the average distance between energy barriers (surface sites), n the concentration of surface sites, k the Boltzmann constant, and T the temperature. When the bracketed quantity is small, Eq. (7.39) simplifies to

$$v = \frac{2K\lambda \gamma_{LV}}{nkT} (\cos \theta_e - \cos \theta) \tag{7.40}$$

which is similar to Eq. (7.7). Thus, the equation of Cherry and Holmes is a special case of the equation of Blake and Haynes. On the other hand, when the bracketed quantity is large, Eq. (7.39) simplifies to

$$v = K\lambda \exp \left[\frac{\gamma_{LV}}{nkT} (\cos \theta_e - \cos \theta) \right] \tag{7.41}$$

which predicts a linear plot for $\cos \theta$ versus $\log v$. The dynamic angles in the system benzene-water-siliconed glass give a linear plot of $\cos \theta$ versus $\log v$ in the range 0–600 mm/min (Figure 7.12) [34]. The dynamic angles in the system octane-polytetrafluoroethylene fit Eq. (9.39) quite well in the range 0–18,000 mm/min [35]. At higher velocities, deviations occur as surging flow sets in.

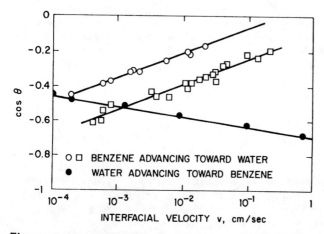

Figure 7.12. Blake-Haynes plot for benzene-water-siliconed glass in two slightly different tubes. (After Ref. 34.)

7.3.5. High-Velocity Regime

This is the Friz mode or the surging mode. At high velocities, surging flow occurs. Hydrodynamic analysis by Friz [36] gives

$$\tan \theta = 3.4 \left(\frac{v\eta}{\gamma_{LV}}\right)^{1/3} \quad (\theta < 90°) \tag{7.42}$$

where the main retarding force is the bulk viscosity. The relation has been confirmed experimentally, for instance, for various liquids in prewetted rectangular glass tubes in a weightless environment at interfacial velocities 1.4–28 cm/sec [37].

Schwartz and Tejada [35] suggested that a general empirical relation is given by

$$\tan \theta = \alpha \left(\frac{v\eta}{\gamma_{LV}}\right)^{\beta} \tag{7.43}$$

where α and β are adjustable constants. Dynamic angles of many systems at high velocities have been found to conform to Eq. (7.43), indicating a surging mode [35].

7.3.6. Experimental Correlations

Forced dynamic angles can be conveniently analyzed by plotting ($\cos \theta_e - \cos \theta$) versus log v [35]. This is termed the *Blake-Haynes plot*. In such a plot, the HMER mode is a straight horizontal line. The Blake-Haynes mode is a sloped straight line. The Friz mode is also a straight line but with a different slope. Strictly, the Friz mode is curved in such a plot. However, the mathematical nature of the Friz equation is such that it is nearly identical to the Blake-Haynes equation for θ between 38 and 73°.

Forced dynamic angles versus velocity (0.01–100 cm/sec) have been investigated in many systems [35]. The liquids include di(2-ethylhexyl) sebacate, diethyl phthalate, α-bromonaphthalene, benzyl alcohol, isopropanol, methylene iodide, Freon TF, hexadecane, octane, hexane, and water. The solids include stainless steel, titanium, aluminum, nylon, poly(methyl methacrylate), and polytetrafluoroethylene. The results are analyzed by the Blake-Haynes plot. The HMER mode (molecular relaxation mode) is observed only with low-viscosity nonpolar liquids such as hexane. For all other liquids, the HMER is absent. The Blake-Haynes mode extends to very low velocity ranges. There is a transition to the Friz mode (surging mode) at high-velocity ranges. The Blake-Haynes plot for the system hexane-nylon is given

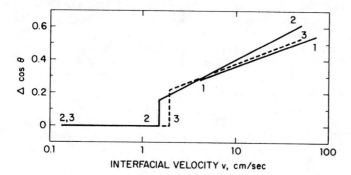

Figure 7.13. Blake-Haynes plot for hexane-nylon, showing the molecular relaxation HMER mode at low velocities and the Blake-Haynes mode at higher velocities. (1) Smooth nylon surface; (2) roughened with No. 240 grits; (3) roughened with No. 120 grits. (After Ref. 35.)

in Figure 7.13, showing the HMER mode at low velocities and the Blake-Haynes mode at other velocities. The Blake-Haynes plots for hexane on stainless steel and poly(methyl methacrylate), α-bromonaphthalene on nylon, octane on PTFE, and di(2-ethylhexyl) sebacate on PTFE are given in Figure 7.14. In these systems, the HMER mode is absent. The Blake-Haynes mode extends down to very low velocity

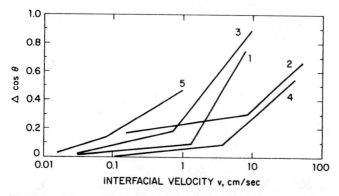

Figure 7.14. Balke-Haynes plots for (1) hexane-stainless steel, $\theta_e = 0°$; (2) hexane-poly(methyl methacrylate), $\theta_e = 0°$; (3) α-bromonaphthalene-nylon, $\theta_e = 16°$; (4) hexane-polytetrafluoroethylene, $\theta_e = 26°$; (5) Di(2-ethylhexyl) sebacate-polytetrafluoroethylene, $\theta_e = 61°$. The lower legs are the Blake-Haynes mode. The upper legs are the Friz surging mode. (After Ref. 35.)

ranges. The upper legs of the straight lines are the Friz mode. Interestingly, surface roughness has little effect on the forced dynamic behavior.

Hoffman [38] correlated the dynamic angles with the capillary number ($\eta v/\gamma$), which is the ratio of viscous force to interfacial force. A shift factor can be used to obtain a master curve. Jiang and coworkers [39] suggested an empirical equation

$$\frac{\cos \theta_e - \cos \theta}{1 + \cos \theta_e} = \tanh \left(4.96 \frac{\eta v}{\gamma}\right) \qquad (7.44)$$

Other analyses and correlations have also been given [40,41].

REFERENCES

1. W. B. Hardy, *Collected Works*, Cambridge University Press, London, 1936, p. 38, 667.
2. D. H. Bangham and Z. Saweris, Trans. Faraday Soc., *34*, 554 (1938).
3. W. D. Bascom, R. L. Cottington, and C. R. Singleterry, Adv. Chem. Ser., *43*, 355 (1964).
4. G. C. Sawicki, in *Wetting, Spreading and Adhesion*, J. F. Padday, ed., Academic Press, London, 1978, pp. 361–375.
5. W. Radigan, H. Ghiradella, H. L. Frisch, H. Schonhorn, and T. K. Kwei, J. Colloid Interface Sci., *49*, 241 (1974).
6. H. Schonhorn, H. L. Frisch, and T. K. Kwei, J. Appl. Phys., *37*, 4967 (1966).
7. T. K. Kwei, H. Schonhorn, and H. L. Frisch, J. Colloid Interface Sci., *28*, 543 (1968).
8. B. W. Cherry and C. M. Holmes, J. Colloid Interface Sci., *29*, 174 (1969).
9. H. van Oene, Y. F. Chang, and S. Newman, J. Adhes., *1*, 54 (1969).
10. R. H. Dettre and R. E. Johnson, Jr., J. Adhes., *2*, 61 (1970).
11. C. Caneva, G. Gusmano, G. Maura, and G. Rinaldi, Ann. Chim., *60*(8–9), 570 (1970).
12. S. Newman, J. Colloid Interface Sci., *26*, 209 (1968).
13. V. A. Ogarev, T. N. Timonina, V. V. Arslanov, and A. A. Trapezikov, J. Adhes., *6*, 337 (1974).
14. W. W. Y. Lau and C. M. Burns, J. Colloid Interface Sci., *45*, 295 (1973).
15. W. W. Y. Lau and C. M. Burns, J. Polym. Sci.—Phys., *12*, 431 (1974).
16. T. A. Elliott and O. M. Ford, J. Chem. Soc., Faraday Trans. I, 1814 (1972).

17. G. A. Wolstenholme and J. H. Schulman, Trans. Faraday Soc., 46, 488 (1950).
18. M. C. Wilkinson and T. A. Elliott, J. Colloid Interface Sci., 48, 225 (1974).
19. T. A. Elliott and M. Morgan, J. Chem. Soc., 5A, 558 (1966).
20. T. A. Elliott and M. Morgan, J. Chem. Soc., 5A, 570 (1966).
21. T. P. Yin, J. Phys. Chem., 73, 2413 (1969).
22. R. V. Dyba, J. Phys. Chem., 74, 2040 (1970).
23. E. W. Washburn, Phys. Rev., 17, 273 (1921).
24. J. R. Ligenza and R. B. Bernstein, J. Am. Chem. Soc., 73, 4636 (1951).
25. J. Szekely, A. W. Newmann, and Y. K. Chuang, J. Colloid Interface Sci., 35, 273 (1971).
26. R. Ablett, Phil. Mag., 46, 244 (1923).
27. W. Rose and R. W. Heins, J. Colloid Sci., 17, 39 (1962).
28. R. S. Hansen and M. Miotto, J. Am. Chem. Soc., 79, 1765 (1957).
29. A. C. Lowe and A. C. Riddiford, J. Chem. Soc. D, 387 (1970).
30. G. E. P. Elliott and A. C. Riddiford, Nature, 195, 795 (1962).
31. G. E. P. Elliott and A. C. Riddiford, J. Colloid Interface Sci., 23, 389 (1967).
32. M. C. Phillips and A. C. Riddiford, J. Colloid Interface Sci., 41, 77 (1972).
33. R. E. Johnson, Jr., R. H. Dettre, and D. A. Brandreth, J. Colloid Interface Sci., 62, 205 (1977).
34. T. D. Blake and J. M. Haynes, J. Colloid Interface Sci., 30, 421 (1969).
35. A. M. Schwartz and S. B. Tejada, J. Colloid Interface Sci., 38, 359 (1972).
36. G. Friz, Z. Angew. Phys., 19(4), 374 (1965).
37. T. A. Coney and W. J. Masica, NASA Tech. Note, TND-5115 (1969).
38. R. L. Hoffman, J. Colloid Interface Sci., 50, 228 (1975).
39. T. S. Jiang, S. G. Oh, and J. C. Slattery, J. Colloid Interface Sci., 69, 74 (1979).
40. N. van Quy, Int. J. Eng. Sci., 9, 101 (1971).
41. V. Ludviksson and E. N. Lightfoot, AIChE J., 14, 674 (1968).

8
Experimental Methods for Contact Angles and Interfacial Tensions

8.1. MEASUREMENTS OF CONTACT ANGLES

Generally, the accuracy of contact angle measurement is not limited by the experimental technique, but by the reproducibility of the surfaces investigated. Instruments that measure to an accuracy of 1° are usually more than adequate [1]. Usually, there is good agreement among the results obtained by various methods if appropriate precautions are taken. The methods described here are for liquids on solids in air. Most of these can also be used, with slight modifications, for liquids on solids immersed in other liquids.

8.1.1. Drop-Bubble Methods

The methods use liquid drops or air bubbles resting on plane solid surfaces. They are also known as *sessile drop methods* when liquid drops are used, or *captive bubble methods* when air bubbles are used. Knowledge of liquid surface tension and density are not required.

Direct Observation—Tangent Method

The most commonly used method for measuring the contact angle involves direct observation of the profile of a liquid drop or an air bubble

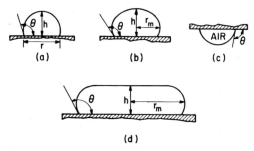

Figure 8.1. Profiles of sessile drops and captive bubbles.

resting on a plane solid surface (Figure 8.1). The contact angle is obtained by measuring the angle made between the tangent to the profile at the point of contact with the solid surface. This can be done on a projected image [2], or a photograph [3] of the drop profile, or directly measured using a telescope fitted with a goniometer eyepiece [4,5]. Commercial apparatus is available [6-8]. Contact angles accurate to ±1 or 2° can readily be obtained. The uncertainties are higher for small angles (less than 10°) and large angles (larger than 160°), because of the difficulty in locating the point of contact for constructing a tangent. An error analysis has been reported [9].

Reflected Light Method

Alternatively, the contact angle may be measured by reflected light [10,11]. A light beam emitted from a microscope will be reflected back into the microscope when the light beam is at right angles to the drop surface. Thus, if the microscope is focused on a three-phase contact point, the angle which the microscope makes with the vertical direction when the liquid drop appears bright is equal to the contact angle. The method is, however, restricted only to angles less than 90°; all unwanted reflections from the drop surface must be eliminated. However, error analysis shows that this method is no more accurate than the direct goniometric measurement [9].

Interference Microscopy Method

Contact angles less than about 10° are generally difficult to measure accurately by the tangent method or the reflected light method. Recently, Sawicki [12] reported an interference microscopy method capable of measuring contact angles less than 10° to an accuracy of ±0.1°. The profile of the leading edge of a drop can also be observed. A monochromatic light (5461-Å green mercury line) is reflected with a beam splitter onto the reflecting substrate to form interference bands parallel to the drop edge. The contact angle is calculated from the

Determination from Drop Dimensions

The contact angles can also be determined by measuring the dimensions of a liquid drop. For very small drops, of the order of 10^{-4} ml [13,14], the distorting effect of gravity is negligible, and the drop takes the shape of a spherical segment. In this case, the contact angle θ can be calculated by

$$\tan \frac{\theta}{2} = \frac{h}{r} \tag{8.1}$$

or

$$\sin \theta = \frac{2hr}{h^2 + r^2} \tag{8.2}$$

where h is the drop height and r is the radius of the drop base. The drop height is usually smaller than the base radius, and more difficult to measure. In such a case, the contact angle may instead be calculated, for a spherical drop, from the drop volume and drop base radius by

$$\frac{r^3}{V_0} = \frac{3 \sin^3 \theta}{\pi(2 - 3 \cos \theta + \cos^3 \theta)} \tag{8.3}$$

where V_0 is the drop volume. Bikerman [15] suggested measuring the base radii for drops of different volumes, and using the value of r^3/V_0 (extrapolated to $V_0 = 0$) to calculate the contact angle.

For larger drops, the drop shapes are distorted by gravity. The drop profiles are described by the equation of Bashforth and Adams [16]. The equation has been integrated numerically. Tables and graphs are available for calculating the contact angles from drop geometry [17–25]. On the other hand, polynomial functions have been obtained and used to calculate the contact angle from drop volume and drop-base radius [26]. A method in which the drop profile is assumed to be elliptical has also been proposed [27].

For large sessile drops (Figure 8.1d), Poisson [28] derived the relation

$$\cos \theta = 1 - \sqrt{2} \left[\frac{h}{a} - \left(\frac{a}{3r_m} \right) \frac{1 - \cos^3(\theta/2)}{\sin(\theta/2)} + \frac{a}{G_0} \right] \tag{8.4}$$

where h is the height of the drop, r_m its maximum radius, G_0 its radius of curvature at the apex, and a the capillary constant defined by

$$a = \left[\frac{2\gamma}{g(\rho_L - \rho_g)}\right]^{1/2} \tag{8.5}$$

where γ is the liquid surface tension, g the gravitational acceleration, ρ_L the liquid density, and ρ_g the surrounding gas density. When r_m is larger than 2 cm, the term containing G_0 can be neglected. The resulting equation is similar to that derived by Ferguson [28] for a drop having a flat apex. For very large drops, the term containing r_m may also be neglected, but the drop size required to do so depends on the magnitude of the contact angle. For example, for water, if θ is 20°, the term can be neglected with an error of about 1° when r_m is about 3 cm, but if θ is 140°, r_m must be greater than 25 cm.

Formation of Drops

Liquid drops suitable for contact angle measurement can readily be placed onto solid surfaces by using a microsyringe. Advancing and receding contact angles can be obtained by increasing or decreasing the drop volume until the three-phase boundary moves over the solid surface. To obtain reproducible results, care must be taken to avoid vibration and distortion of the drop during the volume change [29]. The deleterious effects can be reduced by keeping the capillary pipette of the microsyringe immersed in the drop during the entire measurement [30] (Figure 8.2a–c). The presence of the pipette in the liquid drop does not affect the contact angle [30,31].

The drop volume may also be changed by adding or withdrawing the liquid through a capillary in the solid (Figure 8.2.d and e). Alternatively, a liquid drop may be placed on a solid whose angle of tilt can be changed. The angles formed as the drop begins to slide over the surface are the advancing and receding contact angles [32] (Figure 8.2f).

Similar techniques can also be used to place captive bubbles or drops in contact with solids. These techniques have been discussed elsewhere [1,33–35].

8.1.2. Tensiometric (Wilhelmy Plate) Method

This is perhaps the most versatile and is suitable for both static (advancing and receding) and dynamic contact angle measurements on flat plates or filaments. The method was first used by Wilhelmy [36] and is often known as the *Wilhelmy plate method*.

Measurement of Contact Angles

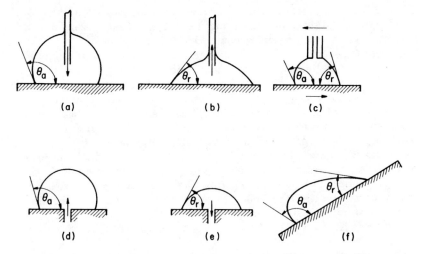

Figure 8.2. Techniques for forming sessile drops. (After Ref. 1.)

The force acting on a thin plate, vertically and partially immersed in a liquid (Figure 8.3a), is

$$F = p\gamma \cos \theta - \rho g A d \tag{8.6}$$

where p is the plate perimeter, γ the liquid surface tension, θ the contact angle, ρ the liquid density, g the gravitational acceleration, A the cross-sectional area of the plate, and d the immersion depth. The term $p\gamma \cos \theta$ is the weight of the liquid supported by the plate; the term $\rho g A d$ is the buoyancy correction. When the end of the plate is exactly on the level of the general liquid surface (Figure 8.3b), the force F_0 is

$$F_0 = p\gamma \cos \theta \quad (\text{at } d = 0) \tag{8.7}$$

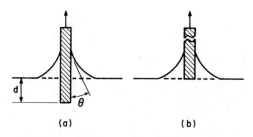

Figure 8.3. Profile of a verticle plate.

where the buoyancy correction is zero. Thus, a plot of F versus d should give a straight line, which can be extrapolated to zero depth so that the buoyancy correction is not necessary.

This method has been used extensively [1,37-43]. Static angles are obtained when the plate is at rest. The advancing angle is obtained by immersing the plate into the liquid. The receding angle is obtained by withdrawing the plate. Dynamic angles are obtained when the plate is in motion. Relaxation of dynamic angles with time can be followed as the plate motion is stopped [44]. The force-depth curve forms a loop, whenever there is contact angle hysteresis [1]. The sides of the loops are given by Eq. (8.6), where θ is either static (advancing or receding) or dynamic angle. The contact angles can readily be determined by extrapolating each side of the loop to zero depth, thus eliminating the need for buoyancy correction. The contact angles obtained usually agree with those measured by the sessile drop method within 1° [1]. The method has also been used for molten polymers [40].

The tensiometric method can also be used for fibers and filaments. The governing relations are the same as for plates. As the fiber perimeter is quite small, the force to be measured is also quite small. Therefore, several fibers may be mounted on a specimen rack and measured at the same time. An electrobalance is used to measure the force accurately. The fiber perimeter can be measured by microscopy, or calculated from the cross-sectional area (for round cross sections), which can be obtained from the slope of the F versus d plot. This method has been successfully used on fibers as thin as 10 μm in diameter [45].

8.1.3. Level-Surface Methods

Titled Plate Method

In this method, a solid plate is partially immersed in the liquid. The tilt angle of the solid is adjusted to obtain a level liquid surface (Figure 8.4a). The contact angle is then obtained from the tilt angle.

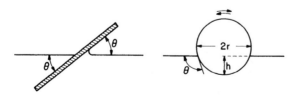

TILTED PLATE METHOD   CYLINDER METHOD

Figure 8.4. (a) Tilted plate method; (b) cylinder method.

Measurement of Contact Angles

Adam and Jessop [46] and Harkins [47] advocated this method as being the most accurate and reproducible. Advancing and receding angles can be obtained by lowering or raising the plate. Means for sweeping the liquid surface to remove impurities can be provided in the apparatus [48,49].

Cylinder Method

The level of liquid around a partially immersed, horizontal cylinder is adjusted until the liquid touches the cylinder without any curvatuve [50] (Figure 8.4b). The contact angle is calculated by

$$\cos \theta = \frac{h}{r} - 1 \tag{8.8}$$

where r is the cylinder radius and h is the immersion depth. Dynamic angles can be measured by rotating the cylinder. Ablett [50] used this method to measure the dynamic advancing and receding contact angles of water on paraffin (Section 7.3.1). A modification of the above method uses a sphere instead of a cylinder [51,52].

8.1.4. Capillary Rise Method

Capillary Rise on Flat Plate

The contact angle of a liquid meniscus on a vertical flat plate can be given by [53]

$$\sin \theta = 1 - \frac{\rho g h^2}{2\gamma} \tag{8.9}$$

where h is the height of the meniscus above the general liquid level (Figure 8.5). Thus, if the meniscus height h, the liquid density ρ, and the liquid surface tension γ are known, the contact angle can be determined. The advancing and receding angles can be obtained by lowering or raising the plate. The cell can be thermostated for measurement at elevated temperatures [54].

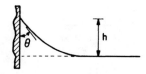

Figure 8.5. Capillary rise on a flat plate.

Cylindrical and Slit Capillary Methods

For a cylindrical capillary, the capillary rise relation is given by (Section 7.2.4)

$$\cos \theta = \frac{\rho g h r}{2\gamma} \qquad (8.10)$$

where h is the capillary height and r is the radius of the capillary. For a slit capillary, the capillary rise relation is given by (Section 7.2.5)

$$\cos \theta = \frac{\rho g h w}{2\gamma} \qquad (8.11)$$

where w is the distance between the two parallel flat plates. Applications of these methods have been discussed elsewhere [55,56].

8.1.5. Contact Angles on Fibers

Measurement of the contact angle on a thin fiber presents a special problem. Some liquids tend to surround a single fiber, forming a symmetrical unduloid, whereas others may remain on one side of a fiber, forming a clamshell profile [57,58]. In some cases, contact angles were found to vary with drop size and fiber daimeter [59]. Even in those cases where "true" contact angles can be formed, the measurement method must be appropriately modified for thin fibers.

Contact angles on thin fibers (as small as 10 μm diameter) are best measured by tensiometric method. As the fiber is very thin, the wetting force to be measured is quite small. Such small forces can, however, be measured accurately with an electrobalance [45] (Section 8.1.2).

Other methods are often not suitable. The direct tangent method is difficult, as the radius of curvature of the liquid meniscus at the three-phase boundary is quite small (of the order of fiber radius) and a tangent line is nearly impossible to be drawn. The difficulty is somewhat alleviated by using the tilted-plate method [60,61] or by the reflected light method [62]. Contact angles reproducible within ±1−2° have been reported on cylindrical [60−62] and trilobal [60] fibers of about 10 μm diamteter.

Contact angles on filaments have also been measured by the height of meniscus on a vertical filament partially immersed in a liquid by using an approximate equation [63], and by the drop profile method [59,64].

Measurement of Contact Angles

8.1.6. Contact Angles on Powders

Two methods have been used: the capillary height method and the displacement pressure method. In both, the powders must be packed into a column. Uniform and reproducible packing of a column is difficult, however, and is usually the limiting factor. Moreover, as the wetting liquid enters the column, the powder packing is altered.

In the capillary height method, Eq. (8.10) is recast as

$$\cos\theta = \frac{\rho k g h r}{2\gamma} \qquad (8.12)$$

where k is the column packing factor. The column height can be measured optically or, preferably, gravimetrically. If a liquid having a zero contact angle on the powder is used, then

$$k = \frac{2\gamma_0}{\rho_0 g h_0 r_0} \qquad (8.13)$$

where the subscript 0 refers to the liquid having a zero contact angle. If the capillary height for any other liquid is now measured, its contact angle is given by

$$\cos\theta = \frac{\gamma_0 \rho h}{\gamma \rho_0 h_0} \qquad (8.14)$$

In the displacement pressure method, the pressure ΔP needed to prevent a liquid from moving through a packed column is measured [65–67]. The effective radius r_e of the packed column is given by

$$r_e = \frac{2\gamma_0}{\Delta P_0} \qquad (8.15)$$

where the subscript 0 refers to a liquid having a zero contact angle on the powder. For any other liquid, the contact angle is given by

$$\cos\theta = \frac{\gamma_0 \Delta P}{\gamma \Delta P_0} \qquad (8.16)$$

which is strictly valid only for cylindrical capillaries. Different equations should be used for other geometries [1]. The exact capillary geometry in a packed column is unknown. Therefore, this method gives only relative values.

8.2. MEASUREMENTS OF INTERFACIAL AND SURFACE TENSIONS OF POLYMER LIQUIDS AND MELTS

All the methods developed for light fluids are, in principle, applicable to polymer liquids and melts. However, the rate of equilibration is a limiting factor, as polymers are usually highly viscous. Prolonged heating at high temperature tends to cause thermal degradation, and should be avoided. Therefore, only a few methods are practical.

Those methods based on drop profiles are usually preferred for both interfacial and surface tension measurements, especially the pendent drop method and the sessile drop method [68]. These methods do not require contact angle data, but require density data. The tensiometric method has the advantage of not requiring the density data, but requires a zero contact angle, which can readily be achieved with clean glass or roughened platinum plate in surface tension measurement, but is generally difficult in interfacial tension measurement. The pendent drop method [69–72] and the tensiometric method [73] have yielded practically identical surface tension values for polyethylenes. Other methods, such as the capillary height method, the ring method, and the maximum bubble pressure method, are rather sluggish in attaining equilibrium and are less useful for viscous polymers.

8.2.1. Pendent Drop-Bubble Method

This is the most simple, reliable, and versatile: applicable to the measurements of both surface and interfacial tensions for polymer liquids and melts. The rate of equilibration is the fastest among the various methods, because the solid-liquid contact area is quite small. The equilibrium process involves mainly the movement of liquid-liquid interface, which is much faster than the movement of solid-liquid interface. The method has been used extensively by Wu [69–71] and Roe [72,74] to measure the surface and interfacial tensions of many polymer liquids and melts. These results constitute the bulk of the available data. Experimental techniques and apparatus have been discussed [69].

The profile of a pendent drop (or bubble) at hydrostatic and interfacial equilibrium (Figure 8.6) is governed by the equation of Bashforth and Adams [17,75] (Section 1.3). The interfacial (or surface) tension is given by

$$\gamma = \frac{\Delta \rho g (d_e)^2}{H} \qquad (8.17)$$

where γ is the interfacial (or surface) tension, $\Delta \rho$ the density difference of the two phases (the drop and the surrounding medium), g

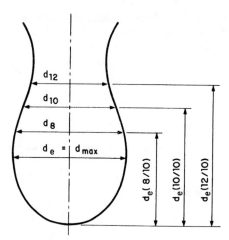

Figure 8.6. Profile of a pendent drop.

the gravitational acceleration, d_e the equatorial diameter of the drop, and H is the correction factor determined by the shape parameter λ_s, defined by

$$\lambda_s = \frac{d_s}{d_e} \tag{8.18}$$

where d_s is the horizontal diameter at a selected plane measured at a distance of d_e up from the apex (bottom) of the drop. Values of 1/H as a function of λ_s have been obtained by numerical solution of the equation of Bashforth and Adams, and tabulated [76–80]. An extensive tabulation has been collected by Padday [24,75]. Thus, from the two diameters and the densities, the interfacial (or surface) tension can be calculated. The method is independent of contact angle.

It is important to assure that the drop attains its equilibrium configuration. This is usually attained instantaneously in light fluids, but may take minutes to hours in highly viscous polymers, particularly in interfacial tension measurement.

The equilibrium may be regarded as attained when the drop profile (characterized by d_e and 1/H) ceases to change with time. Alternatively, a series of shape parameters λ_n may first be determined, as defined by

$$\lambda_n = \frac{d_n}{d_e} \tag{8.19}$$

where d_n is the diameter of the drop on the plane measured at a distance of $d_e(n/10)$ up from the apex of the drop, where n = 8, 9, 10, 11, or 12 (Figure 8.6). Next, 1/H values corresponding to λ_n values are found. Tables of 1/H values as a function of λ_n values are available [79,80]. At equilibrium, the 1/H values are independent of λ_n values. Thus, if the series of 1/H values is constant, equilibrium is attained. Examples of the application of this technique for checking the equilibrium of dependent drops have been discussed [69,72]. The density data required may be obtained using a dilatometer, for instance, as described by Bekkedahl [81]. A disadvantage of the pendent drop method is that an initially stable drop may detach if the interfacial (or surface) tension decreases during the course of measurement. This problem is alleviated in the sessile drop method, but the equilibration rate is slower.

8.2.2. Sessile Drop-Bubble Method

This is also based on drop profile measurement. The rate of equilibration is slower than in the pendent drop method, because the equilibration process involves mainly the movement of solid-liquid interface, which is much more sluggish than the movement of liquid-liquid interface. However, an advantage is that the possibility of drop detachment is eliminated. Sessile drops on solid planes are commonly used for liquids of high surface tension such as molten inorganic glasses and metals. Sessile bubbles under flat plates wetted by liquids are often used for liquids of low surface tension, because these liquids tend to exhibit low contact angles, making sessile drop profiles inadequate for measurement. In the case of interfacial tension measurement, the bubbles are replaced with drops of the second liquids. Experimental techniques have been discussed [82–84].

Figure 8.7 shows the profile of a sessile drop (or bubble) on a flat plate. The dimensions measured are usually the equatorial radius r and the distance between the equator and the apex h. The interfacial (or surface) tension is then given by

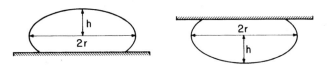

SESSILE DROP OR BUBBLE   CAPTIVE DROP OR BUBBLE

Figure 8.7. Profile of a sessile drop.

$$\gamma = \frac{\Delta\rho g r^2}{BF^2} = \frac{\Delta\rho g h^2}{BG^2} \tag{8.20}$$

where B, F, and G are functions of r/h and have been computed and tabulated [19]. The method is independent of contact angle, but requires density data. The constancy of r and r/h may be used as the criterion for the attainment of equilibrium. Alternatively, the interfacial (or surface) tension may be calculated by the profile-fitting method [75,85].

8.2.3. Rotating Drop-Bubble Method

This is also based on drop profile, but under centrifugal force. The apparatus and operation are complicated in that the cell containing the drop (or bubble) must be rotated at a constant high speed (about 5000 rpm). The method was first devised by Vonnegut [86], and refined by Princen and coworkers [53,87]. Patterson, Hu, and Grindstaff [88] have used this method to measure the surface and interfacial tensions of several polymers.

When a cylindrical tube containing a liquid and a bubble (or a drop of lighter, immiscible second liquid) is rotated horizontally about its long axis, the bubble (or the drop) will elongate under the centrifugal force to take the shape of a long cylinder with rounded ends. The equilibrium shape of the bubble is determined by the surface (or interfacial) tension, density and the centrifugal force. The surface (or interfacial) tension is given by

$$\gamma = \frac{\Delta\rho\omega^2}{4C} \tag{8.21}$$

where ω is the angular velocity in rad/sec and C is a constant to be calculated from

$$L_0 = \frac{(4/3)(Cr^3 + 1)}{(Cr^3)^{1/3}} \tag{8.22}$$

where L_0 is the equilibrium length of the rotating bubble (or drop) and r is the initial radius.

Patterson, Hu, and Grindstaff [88] have described an apparatus and operating procedure for polymer liquids and melts. The results obtained compare reasonably well with those obtained by other methods. The apparatus is, however, complicated, and the rate of equilibration is slow. For instance, for liquids of "medium" viscosity (3000–5000 poises), equilibrium was not reached after more than 3 hr at 6100 rpm

[88]. A plot of the logarithm of the drop length versus reciprocal time was found to give a straight line after an induction period of about 10 min. The equilibrium drop length was then obtained by extrapolation to infinite time.

8.2.4. Tensiometric (Wilhelmy Plate) Method

This method is suitable for surface tension measurement, but not for interfacial tension measurement, because of the requirement for a zero contact angle. The equilibration process primarily involves movement of the solid-liquid interface, which would be extremely slow for interfacial tension measurements. Dettre and Johnson [40] have developed a technique for measuring the surface tension of polymer melts. The results obtained are nearly identical to those obtained by the pendent drop method [68]. An advantage of this method is that density data are not required.

The force acting on a vertical plate partially immersed in a liquid has been discussed in Section 8.1.2. The contact angle should be zero. The condition of zero contact angle can be achieved readily using an appropriate plate. Many liquids exhibit zero contact angles on clean glass plate, particularly for receding contact angles. Thus, to ensure zero contact angle, the plate is usually immersed to a certain depth, and then withdrawn to the position such that the bottom of the plate is exactly on the same level as the bulk liquid surface (Figure 8.3b). The liquid film above the meniscus drains slowly. At equilibrium, a thin and "tenacious" liquid film will remain attached to the plate above the meniscus, because of the high viscosity of the liquid. The force acting on the plate is then given by [40]

$$F = p\gamma + mg \qquad (8.23)$$

where F is the force on the vertical plate at equilibrium, p the perimeter of the plate, m the mass of the thin, adhering liquid film above the meniscus, and g the gravitational acceleration. If measurements are made for several different immersion depths, then

$$mg = xA \qquad (8.24)$$

where x is the immersion depth and A is the weight of the thin, adhering film per unit immersion depth. Thus, a plot of F/p versus x will give a straight line with the slope A/p. Extrapolation to x = 0 corrects for the weight of the retained film.

8.2.5. Capillary Height Method

Although this method is usually regarded as the most accurate for low-viscosity liquids [75], its utility for viscous liquids is limited because of the slow rate of equilibration. Edwards [89] reported that 4 weeks is required for equilibration of a polyisobutylene of 10^5 poises. Harford and White [90] reported that 4 days is required for equilibration of a polystyrene melt at 200°C. Such prolonged exposure to high temperatures would cause severe thermal degradation of the polymer melt. In contrast, only about 10−30 min is required for equilibration in the pendent drop method.

The method requires a zero contact angle and density data. The requirement for a zero contact angle limits its utility to the measurement of surface tension. The method has been used to measure the surface tension of polychlorotrifluoroethylene melts [91], liquid polyisobutylenes [89], and polystyrene melts [90]. The latter results deviate significantly from those reported by others using other methods [68].

Let the height of the bottom of the meniscus above the level of free surface of the liquid be h and the contact angle of the liquid on the capillary wall be θ (Figure 8.8). Equating the liquid column weight to the surface tension force on the meniscus periphery gives

$$\gamma \cos \theta = \frac{\rho g r h}{2} \quad \text{(approximate)} \tag{8.25}$$

where r is the radius of the capillary. This is only approximate, since the weight of the liquid above the bottom of the meniscus (that is, the volume AOA' is neglected. As the contact angle is difficult to measure, the method is used only when the contact angle is zero. The condition of zero contact angle can be achieved readily using glass tubes.

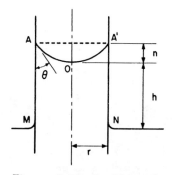

Figure 8.8. Capillary rise in a cylindrical tube.

Various methods for end-volume correction have been proposed. Jurin [92] assumed that the meniscus is spherical, and obtained

$$a^2 = r\left(h + \frac{r}{3}\right) \tag{8.26}$$

where a is the capillary constant defined in Eq. (8.5). This is applicable to small-diameter tubes (less than 0.01 cm). For intermediate tubes (diameters 0.01–0.1 cm), Rayleigh [93] proposed an equation that assumes an elliptical meniscus,

$$a^2 = rh\left[1 + \frac{r}{3h} - 0.1288\left(\frac{r}{h}\right)^2 + 0.1312\left(\frac{r}{h}\right)^3\right] \tag{8.27}$$

For larger-diameter tubes, Rayleigh [93] obtained

$$\frac{r\sqrt{2}}{a} - \ln\frac{a}{h\sqrt{2}} = 0.8381 + 0.2798\left(\frac{2}{r\sqrt{2}}\right)$$

$$+ \frac{1}{2}\ln\frac{r\sqrt{2}}{a} \tag{8.28}$$

applicable for $r/a \geq 6$, that is, for a tube of diameter 1.6 cm when water is used. Sugden [94] pointed out that the correct capillary rise equation is

$$a^2 = bh \quad \text{(exact)} \tag{8.29}$$

where b is the radius of curvature at the bottom of the meniscus. This equation is exact for all tubes, and is obtained from the Adams and Bashforth equation [17]. Tables have been developed with which Eq. (8.29) can be solved by successive approximation [75,94]. This method is the most satisfactory for obtaining the capillary constant from experimental data.

8.2.6. DuNouy Ring Method

Sondhauss [95] and Timberg [96] appear to have been the first to measure the surface tension of liquids from the maximum force required to pull a ring from the surface of a liquid. Du Nouy [97] described a convenient form of apparatus, and his name has become associated with the method. The method has been used to measure the surface tension of liquid epoxy resins (up to 200 poises) [98], polyethylene and polypropylene melts [99,100], and liquid polyorganosiloxanes [101].

The maximum pull on the ring is given approximately by

$$\gamma = \frac{mg}{4\pi R} \quad \text{(approximate)} \tag{8.30}$$

where R is the radius of the ring. This relation is only approximate, since the shape of the meniscus at rupture is different from that formed between a liquid and a flat plate, and the liquid ruptures at the plane below the plane of the ring so that a small but significant volume of the liquid always attaches to the ring. A correct equation is then

$$\gamma = \frac{mg}{4\pi R} f \tag{8.31}$$

where f is the correction factor that has been tabulated [75,102,103].

Films of viscous liquids tend to stretch considerably before rupture. The force-displacement curve thus depends not only on surface tension, but also on the viscoelastic properties of the viscous liquid. Automatic recording of the force at various constant rates of displacement is therefore necessary for polymer liquids and melts. Such an apparatus has been described [98].

8.2.7. Breaking Thread Method

This is based on breakup of a fluid thread, and has been used by Chappelear [104] in an attempt to measure the interfacial tension between some polymer melts. Zero-shear viscosities of the two phases are needed, which are often difficult to measure. The reported interfacial tensions have probable errors of ±30%. However, the method does not require density data, and would be useful in cases where the densities of the two phases are nearly identical. Further development of the method would be welcome.

A cylindrical fluid thread embedded in another fluid is unstable to perturbations of wavelength greater than the circumference of the thread. The growth of the disturbance (Figure 8.9) is given by [105]

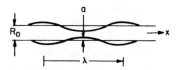

Figure 8.9. Breakup of a fluid thread.

$$\alpha = \alpha_0 \exp(qt) \tag{8.32}$$

$$q = \frac{\gamma \Omega(\chi, p)}{\eta_m R_0} \tag{8.33}$$

where α is the amplitude at time t, α_0 that at zero time, q the rate constant, γ the interfacial tension, R_0 the initial thread diameter, η_m the matrix viscosity, η_t the thread viscosity, and Ω a function of χ and p, where

$$\chi = \frac{\pi R_0}{\lambda} \tag{8.34}$$

$$p = \frac{\eta_t}{\eta_m} \tag{8.35}$$

and λ the wavelength. The function $\Omega(\chi, p)$ has been tabulated [105,106].

A plot of log $(2\alpha/R_0)$ versus t should be linear, with a slope q/2.303, from which γ can be obtained. This eliminates the need to know α_0, which is difficult to measure.

8.2.8. Miscellaneous Methods

Other methods include separation of spheres [107], healing scratches [108], the blister method [109], and the heat of fusion method [110]. These need further refinement [68].

REFERENCES

1. R. E. Johnson, Jr., and R. H. Dettre, in *Surface and Colloid Science*, Vol. 2, E. Matijevic, ed., Academic Press, New York, 1969, pp. 85–153.
2. E. Kneen and W. W. Benton, J. Phys. Chem., *41*, 1195 (1937).
3. P. A. Thiessen and E. Schoon, Z. Elektrochem., *46*, 170 (1940).
4. P. C. Hewlett and C. A. Pollard, in *Adhesion*, Vol. 1, K. W. Allen, ed., Applied Science Publishers, London, 1977, pp. 33–52.
5. W. C. Bigelow, D. L. Pickett, and W. A. Zisman, J. Colloid Sci., *1*, 513 (1946); H. W. Fox and W. A. Zisman, J. Colloid Sci., *5*, 514 (1950).
6. Rame-Hart, Inc., 43 Bloomfield Ave., Mountain Lakes, N.J. 07046.

References

7. Kernco Instruments Co., 19 Walt Whitman Rd., Huntington Station, N.Y. 11746.
8. Imass, Inc., Box 134, Accord (Hingham), Mass. 02018.
9. J. J. Bikerman, *Physical Surfaces*, Academic Press, New York, 1970, p. 250.
10. I. Langmuir and V. J. Schaefer, J. Am. Chem. Soc., *59*, 2400 (1937).
11. T. Fort, Jr., and H. T. Patterson, J. Colloid Sci., *18*, 217 (1963).
12. G. C. Sawicki, in *Wetting, Spreading and Adhesion*, J. F. Padday, ed., Academic Press, London, 1978, pp. 361–375.
13. G. L. Mack, J. Phys. Chem., *40*, 159 (1936).
14. F. E. Bartell and H. H. Zuidema, J. Am. Chem. Soc., *58*, 1449 (1936).
15. J. J. Bikerman, Ind. Eng. Chem., Anal. Ed., *13*, 443 (1941).
16. F. Bashforth and J. C. Adams, *An Attempt to Test the Theories of Capillary Action*, Cambridge University Press, London, 1883.
17. B. S. Ellefson and N. W. Taylor, J. Am. Ceram. Soc., *21*, 193 (1938).
18. G. L. Mack and D. A. Lee, J. Phys. Chem., *40*, 169 (1936).
19. D. N. Staicopolus, J. Colloid Interface Sci., *17*, 439 (1962); *18*, 793 (1963); *23*, 453 (1967).
20. R. Ehrlich, J. Colloid Interface Sci., *28*, 5 (1968).
21. C. Maze and G. Burnet, Sur. Sci., *13*, 451 (1969).
22. J. F. Padday, Nature, *198*, 378 (1963).
23. J. F. Padday, in *Surface and Colloid Science*, Vol. 1, E. Matijevic, ed., Academic Press, New York, 1969, pp. 151–251.
24. K. J. Baumeister and T. D. Hamill, *Liquid Drops—Numerical and Asymptotic Solutions of Their Shapes*, NASA Tech. Note, TND-4779, September 1968.
25. K. G. Parvatikar, J. Colloid Interface Sci., *23*, 274 (1967).
26. L. R. Fisher, J. Colloid Interface Sci., *72*, 200 (1979).
27. D. J. Ryley and B. H. Khoshaim, J. Colloid Interface Sci., *59*, 243 (1977).
28. S. D. Poisson, *Nouvelle théorie de l'action capillaire*, Paris, 1831.
29. F. E. Bartell and G. B. Hatch, J. Phys. Chem., *39*, 11 (1935).
30. F. E. Bartell and C. W. Bjorklund, J. Phys. Chem., *56*, 453 (1952).
31. A. M. Gaudin, A. F. Witt, and T. G. Decker, Trans. AIME, *226*, 107 (1963).
32. G. Macdougall and C. Ockrent, Proc. R. Soc. Lond., *A180*, 151 (1942).
33. G. R. M. del Guidice, Eng. Mining J., *137*, 291 (1936).
34. A. F. Taggart, T. C. Taylor, and C. R. Ince, Trans. AIME, *87*, 285 (1930).

35. I. W. Wark and A. B. Cox, Trans. AIME, *112*, 189 (1934).
36. L. Wilhelmy, Ann. Phys., *119*, 177 (1863).
37. J. Gaustalla, J. Chim. Phys., *51*, 83 (1954).
38. J. Gaustalla, Proc. 2nd Int. Congr., Surf. Act. Lond., *3*, 143 (1957).
39. R. E. Johnson, Jr., and R. H. Dettre, J. Colloid Sci., *20*, 173 (1965).
40. R. H. Dettre and R. E. Johnson, Jr., J. Colloid Interface Sci., *21*, 367 (1966).
41. R. H. Dettre and R. E. Johnson, Jr., Adv. Chem. Soc., *43*, 136 (1964).
42. R. H. Dettre and R. E. Johnson, Jr., J. Phys. Chem., *69*, 1507 (1965).
43. R. E. Johnson, Jr., R. H. Dettre, and D. A. Brandreth, J. Colloid Interface Sci., *62*, 205 (1977).
44. M. C. Phillips and A. C. Riddiford, J. Colloid Interface Sci., *41*, 77 (1972).
45. B. Miller and R. A. Young, Text. Res. J., *45*, 359 (1975).
46. N. K. Adam and G. Jessop, J. Chem. Soc., 1865 (1925).
47. W. D. Harkins, *The Physical Chemistry of Surface Films*, Reinhold, New York, 1952.
48. F. M. Fowkes and W. D. Harkins, J. Am. Chem. Soc., *62*, 3377 (1940).
49. N. K. Adam and R. S. Morrell, J. Soc. Chem. Ind. Lond., *53*, 255 T (1934).
50. R. Ablett, Phil. Mag., *46*, 244 (1923).
51. G. D. Yarnold, Proc. Phys. Soc. Lond., *58*, 120 (1946).
52. G. D. Yarnold and B. J. Mason, Proc. Phys. Soc. Lond., *62B*, 125 (1949).
53. H. M. Princen, in *Surface and Colloid Science*, Vol. 2, E. Matijevic, ed., Wiley-Interscience, New York, 1969, pp. 1–84.
54. A. W. Neumann, Adv. Colloid Interface Sci., *4*, 105 (1974).
55. C. H. Bosanquet and Hartley, Phil. Mag., *42*, 456 (1921).
56. F. E. Bartell and A. D. Wooley, J. Am. Chem. Soc., *55*, 3518 (1933).
57. F. W. Minor, A. M. Schwartz, E. A. Wulkow, and L. C. Buckles, Text. Res. J., *29*, 940 (1959).
58. A. M. Schwartz and C. A. Rader, Proc. 4th Int. Congr. Surf. Act., *2*, 383 (1964).
59. J. I. Yamaki and Y. Katayama, J. Appl. Polym. Sci., *19*, 2897 (1975).
60. T. H. Grindstaff, Text. Res. J., *39*, 958 (1969).
61. T. H. Grindstaff and H. T. Patterson, Text. Res. J., *45*, 760 (1975).
62. W. C. Jones and M. C. Porter, J. Colloid Interface Sci., *24*, 1 (1967).
63. B. Deryagin, (C.R. Dokl.) Acad. Sci., USSR, *51*, 519 (1946).

64. B. J. Carroll, J. Colloid Interface Sci., *57*, 488 (1976).
65. F. E. Bartell and H. J. Osterhof, Colloid Symp. Monogr., *5*, 113 (1928).
66. F. E. Bartell and H. J. Osterhof, Ind. Eng. Chem., *19*, 1277 (1927).
67. F. E. Bartell and H. Y. Jennings, J. Phys. Chem., *38*, 495 (1934).
68. S. Wu, J. Macromol. Sci., *C10*, 1 (1974).
69. S. Wu, J. Colloid Interface Sci., *31*, 153 (1969).
70. S. Wu, J. Phys. Chem., *74*, 632 (1970).
71. S. Wu, J. Polym. Sci., *C34*, 19 (1971).
72. R. J. Roe, J. Phys. Chem., *72*, 2013 (1968).
73. R. H. Dettre and R. E. Johnson, Jr., J. Colloid Interface Sci., *21*, 367 (1966).
74. R. J. Roe, J. Colloid Interface Sci., *31*, 228 (1969).
75. J. F. Padday, in *Surface and Colloid Science*, Vol. 1, E. Matijevic, ed., Wiley, New York, 1969, pp. 101–149.
76. D. O. Niederhauser and F. E. Bartell, *Report of Progress: Fundamental Research on the Occurrence and Recovery of Petroleum*, American Petroleum Institute, Lord Baltimore Press, Baltimore, 1950, p. 114.
77. S. Fordham, Proc. R. Soc., Lond., *A194*, 1 (1948).
78. C. E. Stauffer, J. Phys. Chem., *69*, 1933 (1965).
79. R. J. Roe, V. L. Bacchetta, and P. M. G. Wong, J. Phys. Chem., *71*, 4190 (1967).
80. The tables are available from the Library of Congress, ADI Auxiliary Publications Project, Document ADI 9668.
81. N. Bekkedahl, J. Res. Natl. Bur. Stand., *42*, 145 (1949).
82. T. Sakai, Polymer, *6*, 659 (1965).
83. Y. Oda and T. Hata, Prepr., Meet. High Polymer Soc. Japan, 17th Ann. Meet., May, 1968; p. 267; 6th Symp. Adhes. Adhes., Osaka, Japan, June 1968, pp. 69–70.
84. T. Hata, Hyomen [Surface, Japan], *6*, 281 (1968).
85. N. E. Dorsey, J. Wash. Acad. Sci., *18*, 505 (1928).
86. B. Vonnegut, Rev. Sci. Instrum., *13*, 6 (1942).
87. H. M. Princen, I. Y. Z. Zia, and S. G. Mason, J. Colloid Interface Sci., *23*, 99 (1967).
88. H. T. Patterson, K. H. Hu, and T. H. Grindstaff, J. Polym. Sci., *C34*, 31 (1971).
89. H. Edwards, J. Appl. Polym. Sci., *12*, 2213 (1968).
90. J. R. J. Harford and E. F. T. White, Plast. Polym., *37* (127) 53 (1969).
91. H. Schonhorn, F. W. Ryan, and L. H. Sharpe, J. Polym. Sci., A-2, *4*, 538 (1966).
92. J. Jurin, Phil. Trans., *29–30*, 739 (1718).
93. Lord Rayleigh (J. W. Strutt), Proc. R. Soc. Lond., *A92*, 184 (1915).

94. S. Sugden, J. Chem. Soc., 1483 (1921).
95. Sondhauss, Ann. Phys., Erg. Bd., 8, 266 (1878).
96. Timberg, Ann. Phys., Erg. Bd., 30, 545 (1887).
97. L. Du Nouy, J. Gen. Physiol., 1, 521 (1919).
98. S. B. Newman and W. L. Lee, Rev. Sci. Instrum., 29, 785 (1958).
99. H. Schonhorn and L. H. Sharpe, J. Polym. Sci., A3, 569 (1965).
100. H. Schonhorn and L. H. Sharpe, J. Polym. Sci., B3, 235 (1965).
101. H. W. Fox, P. W. Taylor, and W. A. Zisman, Ind. Eng. Chem., 39, 1401 (1947).
102. W. D. Harkins and H. F. Jordan, J. Am. Chem. Soc., 52, 1756 (1930).
103. B. B. Freud and H. Z. Freud, J. Am. Chem. Soc., 52, 1772 (1930).
104. D. C. Chappelear, Polym. Prepr., 5, 363 (1964).
105. S. Tomotika, Proc. R. Soc. Lond., A150, 322 (1935); A153, 302 (1936).
106. F. D. Rumscheidt and S. G. Mason, J. Colloid Sci., 17, 260 (1962).
107. H. Tarkow, J. Polym. Sci., 28, 35 (1958).
108. L. B. Harris and J. P. Vernon, J. Polym. Sci. Polym. Phys. Ed., 10, 499 (1972).
109. M. T. Lilburne, J. Mater. Sci., 5, 351 (1960).
110. T. G. Mahr, J. Phys. Chem., 74, 2160 (1970).

9
Modifications of Polymer Surfaces: Mechanisms of Wettability and Bondability Improvements

Polymer surfaces are often difficult to wet and bond, because of low surface energy, incompatibility, chemical inertness, or the presence of contaminants and weak boundary layers. Surface treatments are used to change the chemical composition, increase the surface energy, modify the crystalline morphology and surface topography, or remove the contaminants and weak boundary layers. Many processes have been developed to modify polymer surfaces, including chemical treatments, photochemical treatments, plasma treatments, heterogeneous nucleation, and surface grafting. These processes generally cause physical or chemical changes in a thin surface layer (100 Å to 100 μm thick) without affecting the bulk properties.

The chemical compositions of these modified surface layers are often difficult to analyze. Internal reflection IR spectroscopy is usually inadequate, because of its deep penetration (~ 10 μm); chemical changes in a thin surface layer may be missed. On the other hand, recent advances in ESCA with shallow penetration (~ 10 Å) greatly increase the sensitivity of surface analysis. Recent ESCA results show that surface compositions are changed in most surface treatments, where previous results by internal reflection IR spectroscopy failed to detect any chemical changes.

Many treatment processes are used commercially, such as acid etching of plastics prior to metal plating [1-3], UV irradiation of EPDM rubber prior to painting [4], and plasma treatments of polymer films and boards prior to adhesive bonding and printing [5]. Guides to the selection of various treatments for polymers have been discussed elsewhere [6–10]. Here, the physical and chemical changes induced in the surface layers and the mechanisms of wettability and bondability improvements by various surface treatments are discussed.

9.1. CHEMICAL TREATMENTS

Chemical treatments can cause physical and chemical changes on the surfaces. Sodium etching of fluoropolymers causes mainly defluorination, forming surface unsaturations. Chromic acid etching of polyolefins causes dissolution of amorphous regions and thus increases surface roughness and crystallinity. Some surface oxidation also occurs, introducing surface carbonyls and hydroxyls. Iodine treatment of nylon causes primarily polymorphic crystalline transformations.

9.1.1. Sodium Etching of Fluoropolymers

Fluoropolymers such as polytetrafluoroethylene (PTFE), fluorinated ethylene-propylene copolymer (FEP), and polychlorotrifluoroethylene (PCTFE) have low wettability and bondability. Sodium-ammonia complex in liquid ammonia [11,12] and sodium-naphthalene complex in tetrahydrofuran [13,14] were the first commercially successful surface treatments for fluoropolymers. Other alkali metals have also been used in liquid ammonia, with naphthalene in tetrahydrofuran, or as mercury amalgams [14–16]. These solutions are intensely colored and highly conductive, containing sodium cations and electrons [17–19].

Both sodium-ammonia and sodium-naphthalene complexes are equally effective. The sodium-naphthalene complex is more widely used, because of better stability and ease of handling. It is an equimolar complex of sodium and naphthalene dissolved in tetrahydrofuran [7,10,14, 20,21], and is commercially available [7,10,21]. Fluoropolymers are treated by immersion in the complex solution for 1–5 min. The treatment darkens the polymer surface to a depth of about 1 μm. The surface tension, polarity, wettability, and bondability of fluoropolymers are dramatically increased by the treatment [22–24] (Figure 9.1 and Tables 9.1 and 9.2). Extended treatment will produce a spongelike surface structure [22].

Surface chemical analysis by ESCA strikingly shows complete disappearance of a fluorine peak, the appearance of an intense oxygen peak, and broadening and shifting of C_{1s} peak to a lower binding energy (Figure 9.1). These indicate complete defluorination and carbonization of the treated surfaces to a depth of 50–100 Å. Extensive amounts of C=C unsaturation, carbonyl, and carboxyl groups are introduced [22,25]. Bromine uptake shows the presence of 2×10^{-6} moles C=C double bonds per cm^2 of geometric area [15]. Various vinyl monomers

Table 9.1. Surface Tension, Work of Adhesion, and Intrinsic (Equilibrium) Fracture Energy for Sodium Naphthalene-Treated Fluorinated Ethylene-Propylene Copolymer Bonded with SB Rubber Adhesive (90°-Peel Test) [a]

	Surface tension, dyne/cm			Work of adhesion, W_a, erg/cm^2	Intrinsic fracture energy, G_0, erg/cm^2
	γ^d	γ^p	γ		
FEP, untreated	19.6	0.4	20.0	48.4	21.9
Sodium naphthalene-treated					
10 sec	35.4	5.3	40.7	68.0	851
20 sec	36.6	7.7	44.3	70.2	1170
60 sec	33.8	13.4	47.2	69.8	1290
90 sec	34.9	14.5	49.4	71.1	1620
120 sec	34.4	15.9	50.3	71.1	1780
500 sec	34.2	15.7	49.9	72.2	2420
1000 sec	35.4	15.1	50.5	71.8	1990
Helium plasma-treated (30 min)	24.2	4.6	28.8	58.8	68.5
SB rubber adhesive	27.8	1.3	29.1	—	—

[a] SB rubber adhesive is a copolymer of styrene and butadiene (40:60) weight ratio, $T_g = -40°C$) lightly cross-linked in situ on the FEP adherends. Intrinsic fracture energy is the fracture energy at zero rate obtained by extrapolation of FEP/SB peel energy. W_a is the work of adhesion between FEP and SB rubber. See Section 14.4.3 for more details.

Source: From Ref. 24.

Table 9.2. Wettability, Surface Elemental Composition, and Peel Strength of Untreated and Sodium-Etched Fluorinated Ethylene-Propylene Copolymer (FEP) [a]

FEP film	Peel strength lb/in.	Water contact angles, deg		ESCA binding energy (eV) and intensity (counts/sec)			
		θ_a	θ_r	295 eV, C(F)	629 eV, F	286 eV, C(H, O)	532 eV, O
Untreated control	0	109	93	17,000	140,000	0	0
Etched in Na/NH$_3$	8.5	52	16	0	3,000	17,500	17,500
Etched and subjected to abrasion	4.5	66	25	4,500	48,000	13,500	14,500
96 hr at 200°C aging	2.5	101	74	10,000	120,000	10,000	6,500
100 hr UV exposure	0	91	36	12,500	102,000	5,000	5,500
16 days in boiling water	5.0	54	0	0	2,000	28,000	13,000

[a] Peel strengths were measured on joints formed by sandwiching FEP films between aluminum coupons using an acrylic adhesive.
Source: From Ref. 22.

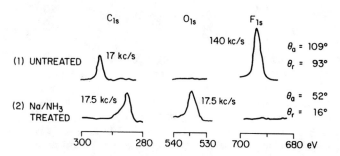

Figure 9.1. ESCA spectra and water contact angles of fluorinated ethylene-propylene copolymer (FEP); (1) untreated; (2) treated with sodium-ammonia solution. kc/s = kilocounts per second. (After Ref. 22.)

can be grafted onto these double bonds [15]. Some amino groups are also formed when treated with sodium-ammonia complex.

Similar results are obtained by IR spectroscopy [26] (Figure 9.2). The strong bands at 1200–1350 cm^{-1} are due to CF vibrations. The bands at 1600 and 1700 cm^{-1} are due to conjugated C=C bonds, at 2150 cm^{-1} due to carbonyls, at 2853 and 2925 cm^{-1} due to CH$_2$ groups, at 2950 cm^{-1} due to CH$_3$, 3300 cm^{-1} due to hydroxyl, and at 3400 cm^{-1} due to NH groups.

Certain correlations among wettability, bondability, and surface chemical composition can be seen in Tables 9.1 and 9.2. These correlations are, however, incomplete and at times contradictory. The fracture strength of an adhesive joint cannot be simply related to wettability or surface chemical composition alone.

On the other hand, heating and UV exposure tend to cause a treated surface to revert to its original untreated conditions (Table 9.2).

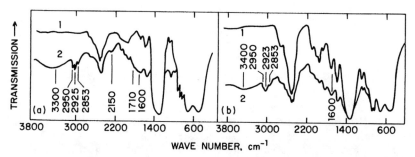

Figure 9.2. IR spectra of polytetrafluoroethylene films treated with (a) sodium-naphthalene complex in tetrahydrofuran; (b) sodium-ammonia solution: (1) untreated; (2) treated. (From Ref. 26.)

9.1.2. Chromic Acid Etching of Polyolefins, ABS, and Other Polymers

Chromic acid is used commercially to etch polypropylene and ABS prior to metal plating [1–3]. It has also been used to treat polyethylene [7–10,27–31], poly(phenylene oxide) [7], polyoxymethylene [7,8,42], polyether [42], and polystyrene [8]. A typical chromic acid bath contains potassium dichromate, water, and concentrated sulfuric acid (sp. gr. 1.84) at a weight ratio of 4.4:7.1:88.5 [27–33,41,42]. Various combinations of chromic acid (CrO_3), sulfuric acid, and phosphoric acid have also been used [2,7,9,10,34–37,43].

Chromic acid etching preferentially removes amorphous or rubbery regions. Depending on surface crystalline morphology, highly complex rootlike cavities may be formed on the etched surface. Some surface oxidation also occurs. Dramatic improvements of wettability and bondability arise, however, mainly from the complex topography rather than from polar groups introduced by surface oxidation.

The dissolution of hydrocarbon polymer by hexavalent chromium in acid medium proceeds through the formation of a tetravalent chromium intermediate which hydrolyzes to an alcohol. Further oxidation causes chain scission to give an olefin, and aldehyde, a ketone, or a carboxylic acid [44–46], that is,

$$\sim CH_2-\underset{\underset{H}{|}}{\overset{\overset{R}{|}}{C}}-CH_2\sim \xrightarrow{\text{CHROMIC ACID}} \sim CH_2-\underset{\underset{\underset{(OH)_3}{|}}{\underset{Cr(IV)}{|}}}{\overset{\overset{R}{|}}{C}}-CH_2\sim \xrightarrow{H_2O}$$

$$\sim CH_2-\underset{\underset{OH}{|}}{\overset{\overset{R}{|}}{C}}-CH_2\sim \xrightarrow{Cr(VI)} \begin{array}{l} \sim CH_2-\overset{\overset{H}{|}}{C}=O + O=\overset{\overset{R}{|}}{C}-CH_2\sim \\ \sim CH_2-\overset{\overset{O}{\|}}{C}-CH_2\sim + \text{etc.} \end{array}$$

The relative rates of reaction for C——H, ——CH$_2$——, and ——CH$_3$ groups are about 4600:75:1 for alkanes [47]. Accordingly, the etching rate is the fastest with polypropylene (about 140 µg/cm^2-hr), intermediate with branched polyethylene (about 98 µg/cm^2-hr), and slowest with linear polyethylene (about 75 µg/cm^2-hr) [27].

On ABS surfaces, chromic acid preferentially attacks butadiene rubber particles, producing numerous undercutting cavities on the surfaces (Figure 9.3). Such surfaces have numerous mechanical anchoring sites, and are used for metal plating [2,38].

Chemical Treatments

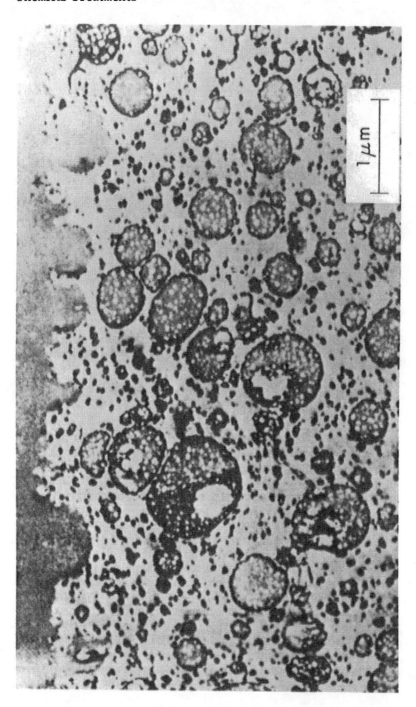

Figure 9.3. TEM photomicrograph of a cross section of chromic acid-etched ABS. (From Ref. 38.)

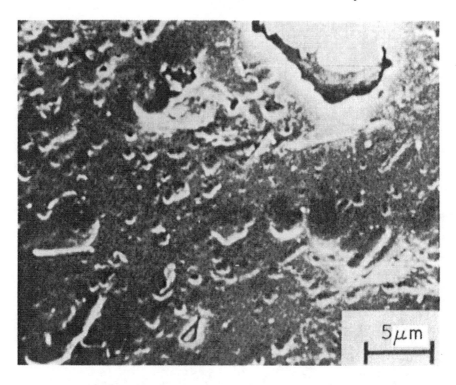

Figure 9.4. SEM photomicrograph of a chromic acid-etched pure polypropylene. (From Ref. 2.)

However, on polypropylene surfaces, chromic acid etches both the amorphous and the crystalline regions at similar rates [2,40]. Therefore, a chromic acid-etched polypropylene surface has few, shallow features (Figure 9.4). Such a surface has few mechanical anchoring sites and is not suitable for metal plating [2]. To achieve preferential etching of the amorphous region, polypropylene is pretreated by immersion in a warm hydrocarbon [48] or an aqueous emulsion of terpentine [36,37] before etching. The hydrocarbons swell the amorphous region and increase its susceptibility.

The surface topography of etched polypropylene depends on its surface crystalline morphology, which is affected by mold surface and cooling rate (see Sections 5.4 and 9.5). At ordinary cooling rates, polypropylene develops spherulitic surface morphology (Figure 9.5) when compression-molded against aluminum oxide, copper oxide, and FEP; transcrystalline surface morphology (Figure 9.6) when compression-molded against PTFE and poly(ethylene terephthalate) films; and lamellar surface morphology (Figure 9.7) when injection-molded [36,

Chemical Treatments

37,48]. Different cooling rates may give different surface crystalline morphologies (Sections 5.4 and 9.5).

On spherulitic surfaces, chromic acid preferentially etches the less ordered (amorphous) regions between the arms of spherulites, producing deep cavities 10−15 μm deep (Figure 9.5). On transcrystalline surfaces, chromic acid etching produces shallow pebblelike textures (Figure 9.6), since transcrystalline layers are formed by massive nucleation and are therefore highly crystalline. On lamellar surfaces, chromic acid preferentially attacks interlamellar regions, producing deep cavities 5−20 μm deep and transverse to the flow direction (Figure 9.7). The deeply etched spherulitic and lamellar surfaces have numerous mechanical anchoring sites for metal platings to adhere strongly (25−40 lb/in. peel strength), whereas the shallowly etched transcrystalline surface has few anchoring sites, giving low adhesion of metal plating (0−10 lb/in. peel strength) (Figure 9.8).

On the other hand, deeply etched and complex surface topography can also be obtained on platable-grade polypropylene without pretreatment before chromic acid etching (Figure 9.9). Platable-grade polypropylene contains additives such as unsaturated hydrocarbons, petroleum resins, polyterpenes, and the like [2]. The additives probably plasticize the amorphous regions, making them more easily attacked by chromic acid.

In the case of polyethylenes, fine roughness and lamellar packets are revealed on both branched (low-density) and linear (high-density) polyethylenes after etching with 70°C chromic acid for 1 min (Figure 9.10). Prolonged etching (1−5 hr) causes extensive pitting [34].

Hydroxyl, carbonyl, and sulfonic acid groups are detected on chromic acid-etched polyolefin surfaces by ESCA [49], UV spectroscopy [34,39], and IR spectroscopy [2,27,28]. These surface polar groups result from oxidation, and are most abundant on branched polyethylene, intermediate on linear polyethylene, and least abundant on polypropylene. Apparently, oxidative products are constantly removed from polypropylene, which should be the most susceptible, whereas linear polyethylene is the least attacked by chromic acid.

Wettabilities of etched polyolefins are greatly increased [27−29,33, 34,39] (Figure 9.11 and Table 9.3). For instance, the surface tension of a branched polyethylene is increased from 34.2 dyne/cm to 52.3 dyne/cm after chromic acid treatment (Table 9.3).

Bondabilities of etched polyolefins are also greatly increased [2,7, 10,27−32,35−37]. Some examples are shown in Figure 9.12, where the treated polymers are bonded to a nylon fabric with an acrylic adhesive [27]. The apparent correlation between peel strength (Figure 9.12) and wettability (Figure 9.11), however, should not be construed as typical. The fracture strength should depend not only on wettability, but also (among other characteristics) on surface topography, which is affected by specific etching condition and polymer morphology.

Chemical Treatments

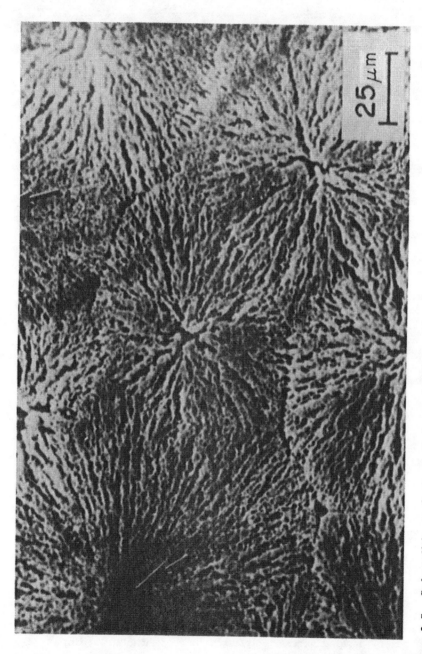

Figure 9.5. Spherulitic surface morphology of polypropylene obtained by compression molding against aluminum foil: top, polarized light photomicrograph of a thin cross section normal to the surface; bottom, SEM photomicrograph of a molded surface etched with chromic acid. (From Ref. 48.)

Figure 9.6. Transcrystalline surface morphology of polypropylene compression molded against poly(ethylene terephthalate). Upper photo: polarized light photomicrograph of a thin cross section normal to the surface; lower photo: SEM photomicrograph of a molded surface etched with chromic acid. (From Ref. 48.)

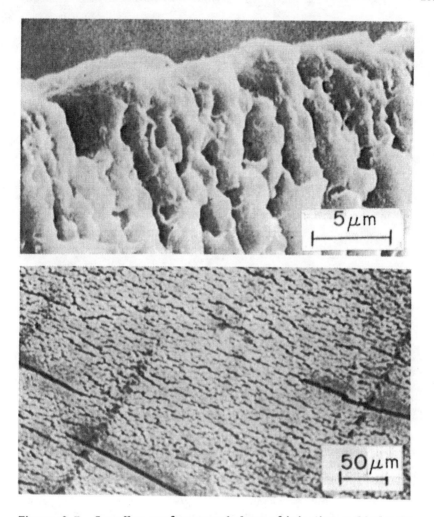

Figure 9.7. Lamellar surface morphology of injection-molded polypropylene. Upper photo: SEM photomicrograph of chromic acid-etched cross section normal to the surface and parallel to the extrusion direction; lower photo: SEM photomicrograph of a chromic acid-etched surface. (From Ref. 37.)

9.1.3. Iodine Treatment for Nylon

Nylon 6 and nylon 66 are made metal-platable by treating with aqueous iodine-potassium iodide solutions (0.25—0.5 M in iodine) between 20 and 80°C [50]. Treated surfaces are relatively smooth, so mechanical anchoring is not important [2,50]. In the case of nylon 6, x-ray diffraction shows that the treatment changes the surface crystallinity

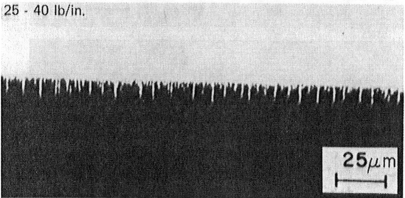

Figure 9.8. Relationship between interfacial topography and peel strength of copper plating (lighter area) on chromic acid-etched polypropylene (darker area). (From Ref. 36.)

from α form (where the N-H groups lie flat on the surface) to γ form (where the N-H groups stand vertical to the surface) after the treatment [51–56]. This crystalline transformation occurs only below 40°C. Above 80°C, the α form persists, and metal platings adhere poorly to such a surface. The increased adhesion of metal plating to the treated nylon 6 surface thus arises from increased chemical interaction with the γ-form amide groups.

9.1.4. Other Chemical Treatments

Numerous other chemicals have been used to treat polymer surfaces, primarily for polyolefins and vulcanized rubbers. In some cases, both the wettability and the bondability are improved. In other cases,

Chemical Treatments

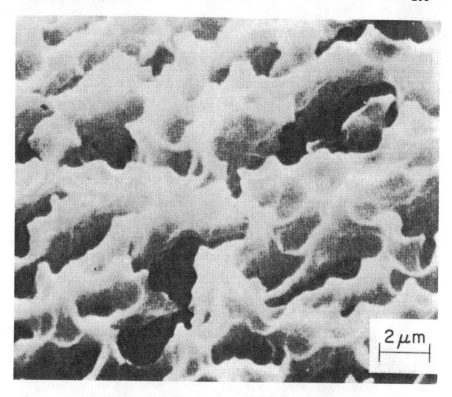

Figure 9.9. SEM photomicrograph of a chromic acid-etched platable-grade polypropylene. (From Ref. 2.)

only the former or the latter is improved, while the other is little affected. The improved wettability and bondability will be lost if the treated surfaces are heated to near the softening point of the polymers. Apparently, the bulk material, having lower surface energy than the treated surface, tends to migrate to the surface when rendered mobile on heating [57–59]. This occurs at about 85°C for polyethylene [59].

Polyolefins have been treated with potassium chlorate-sulfuric acid mixtures [57,59], potassium permanganate-sulfuric acid mixtures [28], nitric acid [60–64], p-toluene sulfonic acid [7,31], sulfuric acid [27, 33,65], fuming sulfuric acid [66,67], sulfur, carbenes, and nitrines [68–73], alkyl peroxides [35], peroxysulfate [74], fluorine gas [75, 76], and ozone [77–81]. Sulfuric acid often produces only moderate changes in wettability and bondability [27,33,65]. Fuming sulfuric acid introduces sulfonic acid groups onto polyolefin surfaces, which can be further modified by reacting with other reagents [66,67]. Nitric acid preferentially attacks the amorphous regions and can reveal crystalline morphology [60–64]. Sulfur, carbenes, and nitrines can undergo C—C insertion reactions [68–73]. Fluorine gas causes

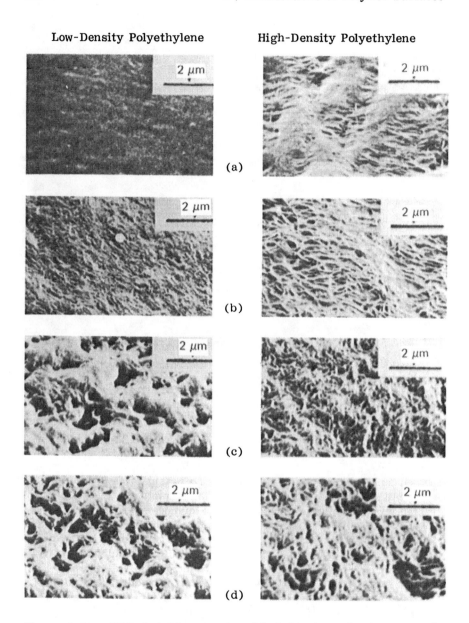

Figure 9.10. SEM photomicrographs of low-density polyethylene (left column) and high-density polyethylene (right column) etched with crhomic acid at 70°C. (a) unetched; (b) etched for 1 min; (c) etched for 1 hr; (d) etched for 5 hr. (From Ref. 34.)

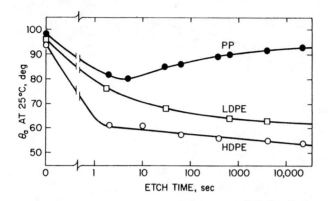

Figure 9.11. Advancing contact angle of water on sulfochromic acid-etched polyolefins. PP, polypropylene; LDPE, low-density polyethylene; HDPE, high-density polyethylene. (After Ref. 27.)

hydrogen abstraction, fluorination, formation of C=C double bonds, and cross-linking on polyolefin surfaces [75,76]. Ozone oxidizes polyolefin surfaces to produce hydroxyl, ketone, aldehyde, and carboxylic acid groups [77–81].

Chlorination has been used to treat polyolefins, nitrile rubber, and butyl rubber [82]. Amines and caustics have been used to treat

Table 9.3. Contact Angle and Surface Tension of Chromic Acid-Treated and Untreated Branched Polyethylene [a]

		Contact angle, degree	
Testing liquid	γ_{LV}, dyne/cm	Untreated	Chromic acid-treated, 71°C, 15 min
Water	72.8	94	46
Glycerol	63.4	79	52
Formamide	58.2	77	33
Liquid epoxy	48.3	61	29
Poly(ethylene glycol)	44.5	49	20
Tricresyl phosphate	40.9	34	8
γ_S, dyne/cm	—	34.2	52.3

[a] Contact angles from Ref. 29. Surface tension γ_S by the equation of state method.

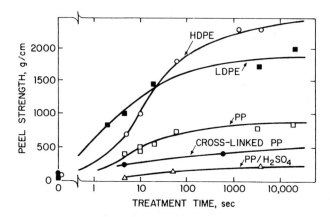

Figure 9.12. Peel strength versus chromic-acid treatment time. HDPE, high-density polyethylene; LDPE, low-density polyethylene; PP, polypropylene; PP/H_2SO_4, polypropylene treated with sulfuric acid. (After Ref. 27.)

poly(ethylene terephthalate) and polycarbonate [83–85]. Imidoalkylene substitution reaction was used to treat polycarbonate [86]. Zirconium hydrocarbyl catalysts were used to treat poly(ethylene terephthalate) [87]. p-Toluene sulfonic acid is recommended for treating polyacetals [7,31].

9.2. FLAME AND THERMAL TREATMENTS

In flame and thermal treatments, polymer surfaces are oxidized and polar oxygenated groups are introduced. The oxidation proceeds by a free radical mechanism, accompanied by chain scission. The improved wettability results from formation of polar groups, whereas the improved bondability results from increased wettability (due to polar groups) and interdiffusion (due to scission products).

9.2.1. Flame Treatment

Flame treatments are used commercially to make polyolefins, polyacetals, and poly(ethylene terephthalate) printable and bondable [7,81,88–90]. Oxidizing flames having oxygen in slight excess of oxygen/fuel stoichiometry usually give the optimum results [88]. Typical flame contact time is 0.01–0.1 sec [89]. Polymer surfaces are oxidized to a depth of about 40–90 Å [91]. ESCA analysis shows that hydroxyl, carbonyl, carboxyl, and amide groups are introduced on flame-treated polyethylene [91].

Flames contain excited species of O, NO, OH, and NH [91,92]. These free radicals can abstract hydrogen from polymer surfaces. Subsenquent surface oxidation propagates by a free radical mechanism common to thermal and plasma oxidations. Antioxidants in polymers do not affect flame and plasma oxidations, but do retard thermal oxidation. This is because flame and plasma oxidations are initiated by the abundant free radicals present in the flame or the plasma, but thermal oxidation is initiated by slow generation of polymer radicals by thermal activation (Sections 9.2.2 and 9.3). Antioxidants can quench a thermal oxidation by scavenging polymer radicals whose generation is the rate-determining step. On the other hand, although antioxidants can still scavenge polymer radicals in flame and plasma oxidations, the generation of polymer radicals is not the rate-determining step in these cases. Chain scission occurs during the free radical propagation, giving polar and mobile scission products.

The improved wettability results from the introduction of surface polar groups. The improved bondability results from the increased wettability (due to polar groups) and increased interfacial diffusivity (due to scission products). The effect of interfacial diffusivity on bondability is discussed in Section 9.3.

9.2.2. Thermal Treatment

Polyethylene is made wettable and bondable by exposing to a blast of hot ($\sim$500°C) air [93]. IR spectroscopy shows the introduction of carbonyl groups onto the treated surfaces [93]. Melt-extruded polyethylene (melt temperatures 280–300°C) is oxidized with the introduction of carbonyl, carboxyl, and some amide groups, shown by ESCA [94] and IR [95]. Some hydroperoxides are also formed. The amounts of terminal vinyl ($RCH{=}CH_2$) and internal vinyl ($RCH{=}CHR$) groups decrease, while the side-chain vinylidene [R—C($=CH_2$)—R] groups increase. The total C$=$C unsaturation, however, remains unchanged [96–99].

Thermal oxidation proceeds by a free radical machanism [96–99]. The initiation step is the slow generation of polymer radicals by thermal activation,

$$RH \xrightarrow{\text{thermal activation}} R\cdot + H\cdot \quad \text{or} \quad R_1\cdot + R_2\cdot$$

which is the rate-determining step. Antioxidants can scavenge the polymer radicals and thus quench the reaction. The propagation steps include reaction of the polymer radicals with O_2 in the air to form hydroperoxide intermediates, and subsequent radical transfer, dissociation, disproportionation, and recombination, accompanied by chain scission and some cross-linking [100].

The improved wettability arises from the introduction of surface polar groups [35]. The improved bondability arises from increased wettability (due to polar groups) and interfacial diffusivity (due to scission products).

9.3. PLASMA TREATMENTS

Plasma treatments are widely used to improve the wettability and bondability of polyolefins, polyesters, and many other polymers. The treatments cause chain scission, ablation, cross-linking, and oxidation to a depth of typically 50–500 Å. The polymer radicals formed during plasma treatments are long-lived, and can react with oxygen and nitrogen upon exposure to the air after treatments. Therefore, even when carried out in inert gases (helium, argon, nitrogen, and the like), plasma treatments always introduce polar oxygen and nitrogen groups onto polymer surfaces, shown by ESCA analyses [101–103]. The wettability is always improved to various degrees.

The bondability improvement has been attributed variously to introduction of polar groups (improved wettability), surface cross-linking (strengthening of weak boundary layer), electret formation, and interfacial diffusion (lowered surface viscosity and increased molecular mobility by chain scission). Many objections have been raised to the strengthening of the weak boundary layer by surface cross-linking and the electrostatic effect by electret formation as the mechanisms of bondability improvement. The preferred mechanism is that both the increased wettability and increased molecular mobility (lowered surface viscosity) improve the bondability. In fact, the most striking feature of bondability improvement is that the bonding temperature required to produce a strong adhesive bond is greatly lowered by plasma treatment. In all cases investigated, untreated polymers can form equally strong bonds when bonded above their melting points. Plasma treatments mainly make the polymers bondable below their melting points. This suggests that increased interfacial flow and diffusion are the paramount factor. In addition, polar groups can also increase interfacial attraction, and thus improve the bond strength. On the other hand, strengthening of the supposed weak boundary layer should play a minor role, and the electret effect has been shown to be irrelevant (this is discussed later).

Two types of plasma are used mainly: cold plasma (glow discharge in reduced pressure) and hybrid plasma (corona discharge at atmospheric pressure) in various gases (helium, argon, hydrogen, oxygen, nitrogen, air, nitrous oxide, ammonia, carbon dioxide, and others). Both types of plasma modify the surface composition, wettability, and bondability similarly, differing only in details.

Plasma treatments on many polymers have been reported, including polyethylene [76,81,101,103–137], polypropylene [107–110,118,119,

122,123,132,138−142], poly(4-methyl-1-pentene) [108,109], polyisobutylene [122], polybutadiene [143], polytetrafluoroethylene [101, 104,105,122,123,139], fluorinated ethylene-propylene copolymer [108, 109,144,145], ethylene-tetrafluoroethylene copolymer [75,102], poly(vinyl fluoride) [107,108,122,123], poly(vinylidene fluoride) [105, 108], polychlorotrifluoroethylene [123], poly(vinyl chloride) [118, 126,139], poly(vinylidene chloride) [115,126], chlorinated polyethylene [122], polydimethylsiloxane [107,112,113,131,146,147], polyoxymethylene [101,107,108,110,122,148], polyacrylonitrile [101], nylon 6 [101, 107,108,110,122,144], nylon 66 [108], poly(ethylene terephthalate) [101,107,108,115,122,123,149−151], cellulose [118,126,142,152,153], cellophane [109], cellulose acetate [101], cellulose acetate butyrate [107,109], sulfite pulp [152], birch wood [152], polystyrene [101, 107−109,122,126,142], poly(methyl methacrylate) [118,139], polycarbonate [107,109,122,139], polyurethane [139], poly(acrylic acid) [110], polysulfide [122], polyimide [122], natural rubber [122], poly(vinyl alcohol) [110], and wool [154,155].

9.3.1. Generation and Reaction of Plasma

Plasmas are ionized gases that contain ions, electrons, radicals, excited molecules, and atoms [156]. These ionized gases are luminous, electrically neutral, and are generated by electrical discharges, high-frequency electromagnetic oscillations, shock waves, high-energy radiations (such as α and γ rays), and so on. There are three types of plasmas: thermal plasma, cold plasma, and hybrid plasma [115].

Thermal plasmas are produced by atmospheric arcs, sparks, and flames. The constituent gas molecules, ions, and electrons are in thermal equilibrium, having temperatures of many thousands of kelvin. If a power supply of low impedance is used, large current flows and an arc is obtained. On the other hand, if a power supply of high impedance is used, small current flows and a corona (hybrid plasma) is obtained.

Cold plasmas are generated by glow discharges at reduced pressures (0.01−10 torr). The gaseous ions and molecules are at the ambient temperature, whereas the electrons are at tens of thousands of kelvin. The electron temperature T_e is defined as $T_e = (2/3k)e_k$, where e_k is the kinetic energy of the electron. Glow discharges can be generated between point-plate or wire-plate electrodes, using dc or ac (low frequency, $\sim$60 Hz) voltages of 500 to several thousand volts. On the other hand, electrodeless glow discharges are generated by high-frequency oscillations introduced into the gases by means of coils wound around vessels, or electrodes attached outside the vessels. High-frequency oscillations are supplied by a spark-gap generator (10−50 kHz), radio-frequency generator (1.5−50 MHz), or microwave generator (150−10,000 MHz) [156].

Hybrid plasmas are produced by corona discharges, ozonizer, and so on, at atmospheric pressure or moderately reduced pressures. Corona discharges consist of numerous tiny thermal sparks uniformly distributed. The average temperature of the gas in a corona is slightly above the ambient temperature. High-impedance power supplies are used, having dc or ac voltages of 5–50 kV.

Many different types of plasma reactors have been used to treat polymer surfaces (Figure 9.13). There are three ways in which a plasma and a solid are arranged to interact. The first is that the plasma and the solid are in physical contact but are isolated electrically. An example is a polymer specimen placed inside an RF glow discharge. The second is that the plasma and the solid are in both physical and electrical contact. An example is a polymer specimen placed between the electrodes in a corona or a glow discharge. The third is that the plasma and the solid are in nonsteady physical contact but are not in electrical contact. An example is dropping a polymer specimen through the core of a plasma.

Plasmas contain UV radiation, radicals, excited molecules, and atoms, ions, and electrons [156–159]. Typical examples are

$$He \longrightarrow h\nu + He + He^+ + e$$

$$O_2 \longrightarrow h\nu + O_2^* + O + O_2^+ + O^+ + e$$

where an asterisk denotes a metastable neutral species. Chemical reactions are induced mainly by UV radiation and the neutrals (metastables and radicals). Ions and electrons are relatively unimportant. In some cases, plasma reactions are caused mainly by the action of UV radiation, such as in the glow-discharge treatment of polyethylene [116,117].

Plasma-polymer reactions proceed by free radical mechanisms. In the initiation step, polymer radicals are formed by several different processes:

By UV radiation:

$$\left.\begin{array}{l}RH\\RF\end{array}\right] \xrightarrow{h\nu} \left[\begin{array}{l}R\cdot + H\cdot\\R\cdot + F\cdot\end{array}\right.$$

By excited noble gas atoms:

$$RH \xrightarrow{He^*} \left[\begin{array}{l}RH^* + He\\R\cdot + H\cdot + He\\R_1 + R_2\cdot + He\end{array}\right.$$

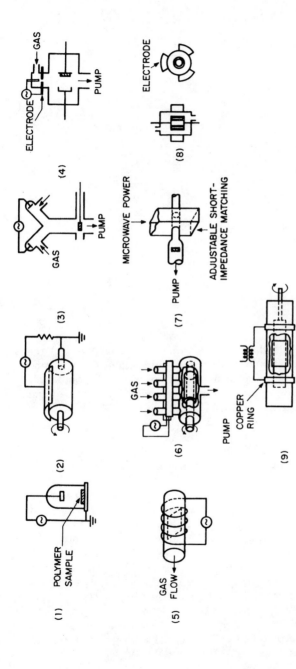

Figure 9.13. Typical plasma reactors for surface treatment of polymers: (1) O_2 RF corona at 1 atm, 5–10 W, or glow discharge in O_2 at reduced pressure [124]; (2) air corona at 1 atm, 2.6 kHz, 5 kV [135]; air corona at 1 atm, 20 kHz, 16 kV [151]; (3) N_2 ac glow discharge at 1 torr, 300 W [120]; (4) O_2 dc plasma jet at 1 ∿ 38 torr, 400 W [123]; (5) inert-gas RF electrodeless glow discharge at 1 torr, 15 MHz, 100 W [104]; (6) inert gas RF electrodeless glow discharge at 1 torr, 15 MHz, 500 W [112]; (7) microwave plasmas at 1 torr, 100 W [107]; (8) N_2 RF corona at 1 atm, 0.5 MHz, 10 W [140]; (9) electrodeless glow discharge in O_2, H_2, N_2, or Ar at 3 torr, 60 Hz, 15 kV [146].

$$RF \xrightarrow{He^*} \begin{bmatrix} RF^* + He \\ R\cdot + F\cdot + He \\ R_1\cdot + R_2\cdot + He \end{bmatrix}$$

By hydrogen radicals:

$$\begin{matrix} RH \\ RF \end{matrix} \xrightarrow{H\cdot} \begin{bmatrix} R\cdot + H_2 \\ R\cdot + HF \end{bmatrix}$$

By oxygen radicals:

$$RH \xrightarrow{O\cdot} \begin{bmatrix} R\cdot + H\cdot + O_2 \\ R_1\cdot + R_2O\cdot \\ R\cdot + OH\cdot \end{bmatrix}$$

$$RF \xrightarrow{O\cdot} \begin{bmatrix} R\cdot + F\cdot + O_2 \\ R_1\cdot + R_2O\cdot \end{bmatrix}$$

These polymer radicals then undergo chain scission, radical transfer, oxidation, disproportionation, and recombination. These reactions give rise to a combination of degradation, ablation, oxidation, and cross-linking in the surface layer. The modified surface layers are typically 50–500 Å thick [101,105,117,141].

9.3.2. Chain Scission and Ablation

Chain scission occurs simultaneously with cross-linking and oxidation. Small-molecule degradation products are constantly removed by evaporation and sputtering, while polymeric scission products remain intermeshed with undegraded polymers and cross-linked networks. The nature and the amount of these polymeric scission products depend on polymer properties, plasma gas, discharge power, and treatment time. These polymeric scission products have lower molecular weight, lower glass transition temperature, and lower viscosity (but no too low to form weak boundary layers) than do the undegraded polymers, and therefore can improve bondability by promoting interfacial flow and interdiffusion. In addition, they may contain polar groups, which can increase interfacial attraction through specific interactions. The amounts and properties of these polymeric scission products are the most important factor in improving the bondability.

Plasma Treatments

The ratio of polymeric to small-molecule scission product depends on polymer properties and plasma-gas properties, among other factors. Polyporpylene produces mostly small-molecule degradation products and few polymeric scission products in "inert"-gas (such as helium and nitrogen) plasmas, but produces appreciable amounts of polymeric scission products in oxygen and nitrous oxide plasmas. Therefore, the bondability of polypropylene is little improved with an inert-gas plasma, but is greatly improved with oxygen or nitrous oxide plasma.

As small-molecule degradation products are constantly removed, the polymer specimen loses weight continually. The amount of weight loss (ablation) increases linearly with time [101,110,122,123] (Figures 9.14 and 9.15). In some cases, polymer surfaces are roughened, or form circular mounds, which however, have little effect on bondability [126-128,140,151].

9.3.3. Surface Cross-Linking

Surface cross-linking, produced by recombination of polymer radicals, occurs simultaneously with chain scission. Cross-linking increases the molecular weight, whereas scission decreases it. After removing gel fractions, the molecular weights of plasma-treated surface layers are indeed found to be decreased, such as in polyethylene, polycarbonate, and poly(4-methyl-1-pentene) exposed to helium or oxygen plasma [109].

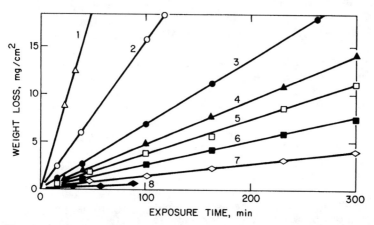

Figure 9.14. Ablation of various polymers exposed to O_2 RF plasma at 1 torr: (1) polysulfide; (2) polyoxymethylene; (3) polypropylene; (4) low-density polyethylene; (5) poly(ethylene terephthalate); (6) polystyrene; (7) polytetrafluoroethylene; (8) sulfur-vulcanized natural rubber. (After Ref. 122.)

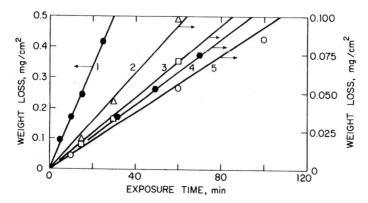

Figure 9.15. Ablation of various polymers exposed to helium plasma (30 W at 0.15 torr): (1) polyoxymethylene; (2) poly(ethylene terephthalate); (3) polyethylene; (4) nylon 6; (5) polypropylene. (After Ref. 110.)

The existence of a cross-linked skin is vividly demonstrated by heating a block of plasma-treated polyethylene and a block of untreated polyethylene above its melting point. The untreated specimen melts and flows, but the plasma-treated specimen retains its rectangular shape [104,105]. The thickness of the cross-linked skin is determined from the gel content after solvent extraction. Typical thicknesses are 50–500 Å, but can be as thick as 6 μm after prolonged treatment (Table 9.4). The skin thickness grows with square root of treatment time, consistent with the light attenuation model [116,117, 157] (Figure 9.16).

ESCA analysis shows that the surface layer of polytetrafluoroethylene is defluorinated and cross-linked on exposure to argon or nitrogen plasma (Figure 9.17). The F_{1s} (689.5 eV) peak decays rapidly with exposure time. The C_{1s} (higher binding energy, 292.2 eV, CF_2 structure) peak decays at about the same rate. The C_{1s} (lower binding energy, 287.5 eV, CF structure) increases at about the same rate as the decay of F or CF_2 [101]. Similar behavior is also observed on an ethylene-tetrafluoroethylene copolymer [75,102,160].

The tendency to surface cross-linking depends on polymer properties and plasma-gas types. High-density polyethylene is one of the most readily cross-linked, whereas low-density polyethylene is only moderately cross-linked [104,105,109,116,117,119,128,133,135] (Table 9.4). Polycarbonate [109], polytetrafluoroethylene [101,104,105], fluorinated ethylene-propylene copolymer [144,145], and ethylene-tetrafluoroethylene copolymer [75,102] are also readily cross-linked by plasmas.

Table 9.4. Thickness of Cross-linked Surface Layer Resulting from Plasma Treatment of Some Polymers[a]

Polymers	He glow discharge 1 torr, 100 W		Glow discharge, 0.6 torr, 160 W, 60 min	He or O_2 glow discharge, 0.4 torr, 100 W, 60 min	O_2 or N_2O glow discharge, 1 torr, 1–4000 sec	N_2 corona, 1 bar, 100 sec
	1 sec	15 min				
High-density polyethylene	300 Å	0.8 μm	6.2 μm	4.4 μm	—	—
Low-density polyethylene	—	—	—	0.34 μm	—	—
Polypropylene	0	0	—	—	300 Å	400 Å
Poly(4-methyl-1-pentene)	—	—	—	0.76 μm	—	—
Polycarbonate	—	—	—	3.56 μm	—	—

[a]The thickness of cross-linked surface skin is determined by the gel content; Refs. 105, 109, 116, 138, and 141.

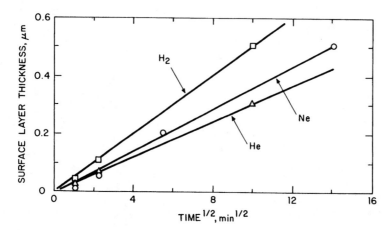

Figure 9.16. Thickness of cross-linked surface layer on polyethylene versus square root of exposure time for polyethylene exposed to various inert-gas plasmas. (After Ref. 157.)

On the other hand, polypropylene mainly undergoes ablation with little cross-linking in He, Ar, H_2, N_2, CO_2, and NH_3 plasmas [104, 105,108,119,138,141], but is cross-linked to a depth of 300 Å with simultaneous extensive oxidation in O_2 and N_2O plasmas [138]. Polyoxymethylene also mainly undergoes chain scission and ablation with little cross-linking [101,110,122,148].

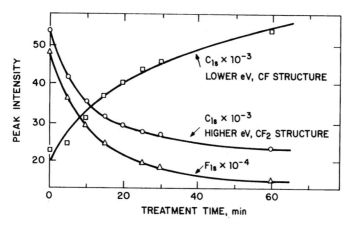

Figure 9.17. Changes of ESCA peaks of polytetrafluoroethylene as a function of plasma treatment time (RF argon plasma at 0.04 torr and 13.56 MHz). (After Ref. 101.)

9.3.4. Incorporation of Polar Groups

Polar groups are always incorporated onto plasma-treated polymer surfaces regardless of gas types (He, Ar, H_2, N_2, O_2, air, etc.). These polar groups are readily detected by ESCA, but are often missed by internal reflection IR spectroscopy. Argon plasma introduces oxygen groups (carbonyls, carboxyls, and hydroxyls), and nitrogen plasma introduces oxygen groups and nitrogen groups (amides, imides, and nitriles), onto polyethylene, polystyrene, poly(ethylene terephthalate), polyoxymethylene, cellulose acetate, polyacrylonitrile, nylon 6, polytetrafluoroethylene, and ethylene tetrafluoroethylene copolymer [101–103]. ESCA spectra for polyethylene and polytetrafluoroethylene treated with argon and nitrogen plasma are shown in Figures 9.18 and 9.19, respectively.

In the case of inert-gas-plasma treatment, the surface oxygen groups are formed most likely by the reaction of long-lived polymer radicals with oxygen when the specimen is exposed to air after the plasma treatments. ESR studies show that polymer radicals formed by plasma treatments do not decay appreciably in 24 hr [113,118]. Other possible but less likely origins of oxygen include oxygen contamination in the gas, decomposition of oxygenated additives, sputtering of reactor walls, and oxygen desorbed from the reactor walls [102,132].

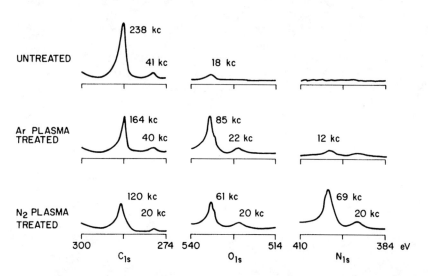

Figure 9.18. ESCA spectra of polyethylene treated with argon or nitrogen plasma (13.56 MHz, 30 W, 0.04 torr, 1 min); kc = kilocounts. (After Ref. 101.)

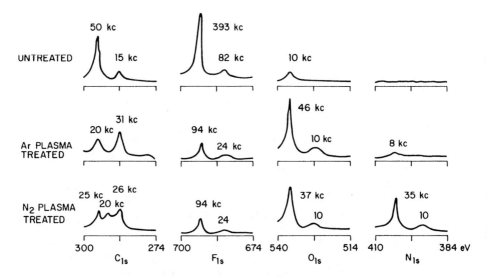

Figure 9.19. ESCA spectra of polytetrafluoroethylene treated with argon or nitrogen plasma (13.56 MHz, 30 W, 0.04 torr, 1 min); kc = kilocounts. (After Ref. 101.)

On the other hand, air and oxygen plasmas can cause extensive surface oxidation. The polar groups introduced are so abundant that they are readily detected by IR spectroscopy. Various carbonyl and carboxyl groups are introduced to oxygen-plasma-treated polyethylene, similar to those introduced by thermal oxidation [76,77,124,125,127] (Figure 9.20). The mechanisms of plasma oxidation in some polymers are given below. Air corona causes chain scission, decarboxylation, and formation of terminal m- and p-phenol groups on poly(ethylene terephthalate) [161–163], that is,

$$-\underset{\underset{O}{\|}}{C}-\underset{}{\bigcirc}-\underset{\underset{O}{\|}}{C}-O-\underset{\underset{H}{|}}{\overset{\overset{H}{|}}{C}}-\underset{\underset{H}{|}}{\overset{\overset{H}{|}}{C}}-O- \xrightarrow{h\nu} -\underset{\underset{O}{\|}}{C}-\underset{}{\bigcirc}-\underset{\underset{O}{\|}}{C}\bullet \;+\; \bullet O-\underset{\underset{H}{|}}{\overset{\overset{H}{|}}{C}}-\underset{\underset{H}{|}}{\overset{\overset{H}{|}}{C}}-$$

$$-\underset{\underset{O}{\|}}{C}-\underset{}{\bigcirc}-\underset{\underset{O}{\|}}{C}\bullet \longrightarrow -\underset{\underset{O}{\|}}{C}-\underset{}{\bigcirc}\bullet \;+\; CO$$

$$-\underset{\underset{O}{\|}}{C}-\underset{}{\bigcirc}\bullet \xrightarrow[OH]{H_2O} -\underset{\underset{O}{\|}}{C}-\underset{}{\bigcirc}-OH$$

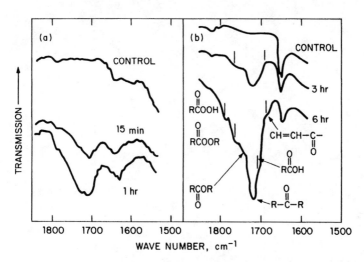

Figure 9.20. Carbonyl absorption bands for polyethylene: (a) oxygen plasma-treated; (b) thermally oxidized in air. (After Ref. 99 and 127.)

Oxygen plasma introduces hydroxyl groups (in the form of Si—CH_2OH) onto polydimethylsiloxane [131]. Air plasma introduces carbonyl groups onto polyoxymethylene [148]. NH_3 plasma introduces —NH_2 groups onto polypropylene, polytetrafluoroethylene, poly(vinyl chloride), polycarbonate, and others [139]. All are detectable by IR spectroscopy.

9.3.5. Effect on Wettability

Since polar groups are always incorporated onto polymer surfaces, even with the inert-gas plasmas, wettabilities are always improved. Several examples are given in Tables 9.5 and 9.6. For instance, the surface tension of polyethylene increases from 34.2 dyne/cm to 68.9 dyne/cm (20°C) after argon plasma treatment; the surface tension of polydimethylsiloxane increases from 20.9 dyne/cm to 71.6 dyne/cm (20°C) after argon plasma treatment (Table 9.5). Many other examples of improved wettability with inert gases (helium, argon, hydrogen, nitrogen), oxygen, air, nitrous oxide, carbon dioxide, and ammonia plasmas have been reported [76,104,105,109,113,115,121–128, 134,139,140,144,149,150,153]. The extent of wettability improvement depends on gas type, gas pressure, treatment duration, specimen temperature and position, discharge power and frequency, and other factors [108,127]. The effect of specimen temperature is quite pronounced, for instance, with polypropylene [108,127].

Table 9.5. Wettability of Polyethylene and Polydimethylsiloxane Before and After Argon Plasma Treatment[a]

Liquids	Liquid surface tension γ_{LV}, dyne/cm	Contact angle, deg			
		HD polyethylene		Polydimethylsiloxane	
		Untreated	Treated	Untreated	Treated
Water	72.8	97.5	19.0	99.5	10.6
Glycerol	63.4	85.2	10.5	91.5	6.8
Formamide	58.2	77.3	0	85.5	0
α-Bromonaphthalene	44.6	—	0	—	0
Tricresyl phosphate	40.9	42.0	0	64.5	0
n-Hexadecane	27.6	—	—	38.0	0
Surface tension at 20°C, dyne/cm	—	34.2	68.9	20.9	71.6

[a] RF argon plasma, 150 W at 0.6 torr for 10 min; Ref. 113.

Table 9.6. Effect of Various Gas Plasmas on the Contact Angles (Degrees) of Water on Some Polymers[a]

Polymer	Untreated	Treated by gas plasma				
		O_2	Ar	He	N_2	H_2
Polyethylene, low-density	97	50	—	46	—	—
Polyethylene, high-density	96	41	—	38	—	—
Poly(4-methyl-1-pentene)	104	60	—	92	—	—
Polystyrene	88	0	—	10	—	—
FEP	103	103	—	83	—	—
Polydimethylsiloxane	117	46	83	—	20	44
	99.5	—	10.6	—	—	—

[a] RF plasma, 50 ∿ 150 W, 0.4 ∿ 0.6 torr, 10 ∿ 30 min; Ref. 109.

Some authors reported that the wettability was not changed when treated with inert-gas plasmas, for instance, polyethylene treated with helium, argon, nitrogen, or hydrogen plasma [76,104,105,125–128, 144]. Under similar conditions, however, a great majority of authors found improved wettability [108,109,113–115]. Such discrepancy could be due to lack of careful wettability measurements in those cases where the wettability was reported to be unchanged. Most contact angle data were apparently not critically evaluated. Hysteresis effect, expected to be present, was not reported or analyzed. As polar groups are always introduced, the wettability should always be changed. This should be detected best by hysteresis measurements. Even if the advancing angle is unchanged, the receding angle should always be lowered.

9.3.6. Effect on Bondability

Both autohesive and adhesive bondabilities can be greatly increased with inert-gas or oxygen-containing gas plasmas. Some examples are given in Table 9.7. Treatment times required to achieve the maximum improvement depend on polymer types, among other factors. For polyethylene, a few seconds is sufficient (Figure 9.21). For polytetrafluoroethylene, a few thousand seconds is required (Figure 9.22). In many cases, the bondability versus treatment time follows the empirical relation [108].

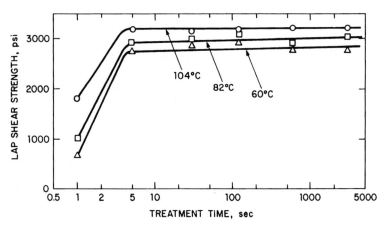

Figure 9.21. Lap shear strength of aluminum/epoxy adhesive/helium-plasma-treated polyethylene/epoxy adhesive/aluminum joint versus plasma treatment time. RF helium plasma at 100 W and 1 torr. The temperatures are the bonding temperatures. (After Ref. 105.)

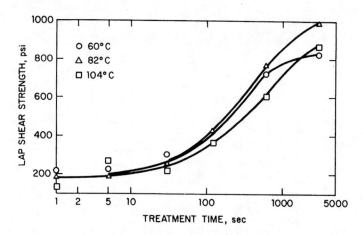

Figure 9.22. Lap shear strength of aluminum/epoxy adhesive/polytetrafluoroethylene/epoxy adhesive/aluminum joint versus treatment time of polytetrafluoroethylene in neon plasma (15 MHz, 100 W, 1 torr). (After Ref. 105.)

$$\sigma_t = \sigma_0 + \frac{t}{a + bt} \tag{9.1}$$

where σ_t is the joint strength with the polymer treated for a time period t, σ_0 that with untreated polymer, and a and b are empirical constants (Figure 9.23).

Diverse views have been expressed as to the mechanisms of bondability improvement. The proposed mechanisms include improved wettability due to incorporation of polar groups, strengthening of the weak boundary layer, increased interfacial attraction and interdiffusion,

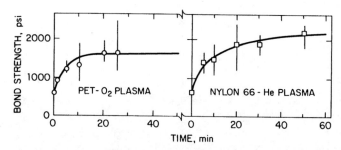

Figure 9.23. Effect of plasma treatment time on bond strength. The solid lines are drawn to Eq. (9.1). (After Ref. 108.)

Table 9.7. Effect of Plasma Treatments on the Joint Strengths of Some Polymers

Polymer	Type of joint	Untreated control	Joint strength Polymer pretreated in the plasma of:					
			He	Ar	N_2	O_2	CO_2	NH_3
Polyethylene[a]	Peel strength, lb/in, PE/butyl rubber adhesive/PE bonds	1.58	5.88	—	4.30	—	4.14	4.72
Polyethylene[b,c]	Lap shear strength, psi, aluminum/epoxy adhesive/polymer/aluminum	372	1382	—	—	1446	—	—
Polypropylene[b,d]	Same as above	370	450	—	—	1870	—	—
Poly(vinyl fluoride)[b,d]	Same as above	278	1290	—	—	1370	—	—
Nylon 6[b,c]	Same as above	846	1238	—	—	1520	—	—
Polycarbonate[b,d]	Same as above	410	660	—	—	800	—	—
Polystyrene[b,e]	Same as above	566	4015	—	—	3118	—	—

Plasma Treatments

Material	Description						
Poly(ethylene terephthalate)[b,e]	Same as above	530	1660	—	—	1215	—	—
Poly(ethylene terephthalate)[f]	Peel strength, lb/in, PET/RFL adhesive/rubber	2.5	11.5	14.5	16.9	7.5	—	21.0
Poly(vinyl chloride)[g]	Shear strength, kg/cm^2, PVC/corona-treated cellulose	0	—	—	22.3	5.3	—	—
Polyethylene[g]	Shear strength, kg/cm^2, PE/corona-treated cellulose	1.3	—	—	19.1	6.5	—	—
Polyoxymethylene[h]	Lap shear strength, psi, POM/epoxy adhesive steel	50	1700	1800	1700	1700 (air)	—	—

[a] 14-MHz RF glow discharge, 100 W, 1 torr for 1000 sec, from Ref. 111.
[b] 13.56-MHz RF glow discharge, 50 W, 0.4 torr, from Ref. 108.
[c] 1-min treatment time.
[d] 30-sec treatment time.
[e] 30-min treatment time.
[f] 13.56-MHz RF glow discharge, 100 w, 0.5 torr for 9.4 sec, from Ref. 149.
[g] 60-Hz 15-kV corona at 1 atm, 50 W, 15-min treatment time, from Ref. 126.
[h] 450-kHz glow discharge, 0.5–2.0 torr, 10 min, from Ref. 148.

hydrogen bonding, and electret effect. Our preferred view is that the improved bondability arises from the increased interfacial attraction and interdiffusion due to surface oxidation and formation of polymeric scission products, mentioned in Section 9.3.2 and discussed further later.

The improved bondability of polyethylene treated with an oxygen plasma was attributed to the introduction of surface carbonyl groups (detected by IR spectroscopy) as early as 1956 [124]. Surface polar groups can increase the wettability and introduce specific interactions across the interface. These effects should tend to improve the bondability. This view is certainly quite reasonable, and is widely advocated [109,113,114,121,123,135,147,151]. Typical correlations between bond strength and wettability are shown in Figures 9.24 and 9.25. Rapid increase of bond strength and wettability occurs with the introduction of surface carbonyl groups (absorbance at 1720 cm^{-1}, Figure 9.24). The correlations are, however, sometimes inconsistent (Figure 9.25).

Schonhorn and coworkers [104,105,144,145] reported that the bondabilities of polyethylene, polytetrafluoroethylene, and some other polymers are greatly increased by treatment with inert-gas (helium, argon, and hydrogen) plasmas without surface oxidation (examined by internal reflection IR spectroscopy), and the wettability is unchanged. The only significant change is the cross-linking of surface layer to a depth of about 50–500 Å (Section 9.3.3). They thus concluded that the improved bondability arises from strengthening of the supposed weak boundary layer on the surfaces by cross-linking. The forma-

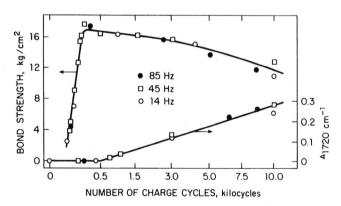

Figure 9.24. Autohesive bond strength (heat-sealed at 55°C for 3 min and 7 kg/cm^2 pressure) and number of surface carbonyl groups (absorbance at 1720 cm^{-1}) versus number of oxygen corona discharge cycles (19 kV and 1 atm) for plasma treatment of polyethylene. The abscissa is a square-root scale. (After Ref. 125.)

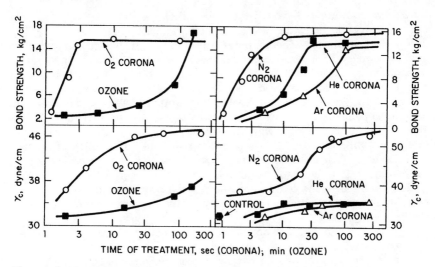

Figure 9.25. Time of corona or ozone treatment versus bond strength and surface tension for low-density polyethylene (bonded with a polyamide adhesive). (After Ref. 77.)

tion of cross-linked skin has been confirmed by others. However, under similar conditions, many other authors found that the wettability is also increased. ESCA analysis definitively established that surface oxidation occurs in all cases, discussed before. Thus, the proposed weak-boundary-layer concept is not the only possible explanation.

On the other hand, Schonhorn and coworkers [105,138] reported that the bondability of polypropylene was not improved by treatment with inert-gas plasmas, but was improved by treatment with oxygen or nitrous oxide plasma. They attributed this to the effect that inert-gas plasmas mainly cause ablation without cross-linking on polypropylene. They reported that a cross-linked skin of about 300 Å was formed on polypropylene when treated with oxygen and nitrous oxide plasmas, although surface oxidation also occurred. On the contrary, other authors found that polypropylene is indeed cross-linked to a depth of about 400 Å by inert-gas plasmas, but that the bondability is not improved [108,119,141]. On the other hand, no cross-linked skins are found on polyoxymethylene after treatments with helium, argon, nitrogen, and air plasmas, but the bondability is greatly increased [148]. Furthermore, excessive surface cross-linking during plasma treatment was found to degrade the bondability of polyethylene [133,135]. Thus, surface cross-linking can be only an incidental factor at best in bondability improvement.

Interfacial hydrogen bonding was also advocated as the mechanism of bondability improvement [103,134,150,151,164]. The autohesion of plasma-treated polyethylene was suggested to arise from interfacial hydrogen bonding between the enolic and the ketonic forms of the carbonyl groups introduced by plasma treatment [103,134]. The autohesion of plasma-treated poly(ethylene terephthalate) was suggested to arise from interfacial hydrogen bonding between the terminal phenol groups produced by plasma treatment [150,151]. When washed with acetyl chloride, treated polyethylene and treated poly(ethylene terephthalate) lose their autohesive bondability. This is assumed to arise from acetylation of the enolic carbonyls and the terminal phenols, respectively, thus eliminating the active hydrogen. On the other hand, washing with bromine water destroys the bondability of treated polyethylene, but does not affect that of treated poly(ethylene terephthalate). This is assumed to arise from displacement of the enolic hydrogen by bromine in the case of treated polyethylene, but bromination occurs on the phenyl ring without affecting the phenolic OH group in the case of treated poly(ethylene terephthalate). These explanations are plausible, but are not consistent with basic features of bonding behavior (discussed further later).

Electret formation was detected on some corona-treated polyethylenes, leading to the suggestion that the increased bondability arises from an electrostatic effect [127,129,130]. A correlation between the zeta potential of aqueous suspension of treated polymer powder and joint strength was reported [111]. In many cases, however, bondabilities are improved without detectable electrostatic effects [76, 125,146]. The proposed electret mechanism is not confirmed.

Our preferred and most likely mechanism attributes the improved bondability to both surface oxidation and formation of polymeric scission product. The polar groups introduced can increase the interfacial attraction and wettability through specific interactions across the interface. The polymeric scission products have lower molecular weight, lower glass transition temperature, and lower viscosity. They can therefore promote interfacial flow and interdiffusion. Small-molecule scission products would tend to form a weak boundary layer, and are undesirable. They are, however, constantly removed by evaporation, sputtering, and scouring during plasma treatment.

As discussed before, chain scission, ablation, cross-linking, and oxidation occur simultaneously during plasma treatment. Low-molecular-weight scission products are constantly ablated. After separation of the cross-linked gels, the molecular weight of the soluble surface layer is found to be lower than the untreated bulk material, for instance, in polyethylene, poly(4-methyl-1-pentene), and polycarbonate [109].

Furthermore, the most striking feature of the bondability improvement (for autohesion and adhesion) is that the bonding temperatures required to produce strong bonds are greatly reduced. In all the polymers investigated, untreated polymers can form equally strong bonds if bonded above their melting points or softening points. For instance, strong adhesive bonds can be formed between untreated polyethylene

Plasma Treatments

and epoxy resin if the bonding temperature is above the melting point (130°C) of the polyethylene. But when plasma-treated polyethylene is used, strong adhesive bonds can be formed at as low as 25°C (Figure 9.26). Untreated polyethylene can adhere strongly to butyl rubber only if bonded above the melting point of polyethylene. But helium or ethylene plasma-treated polyethylene adheres strongly to butyl rubber when bonded at as low as 50°C [165] (Figure 9.27). Air corona-treated polyethylene can be heat-sealed as low as 75°C, whereas an untreated specimen requires a temperature near melting point (130°C) for heat sealing [103,134]. Air corona-treated poly(ethylene terephthalate) can be heat-sealed as low as 140°C, whereas untreated specimen requires a temperature near the melting point (265°C) [150]. These are consistent with the diffusion mechanism.

To form hydrogen bonds, the two phases must be brought into atomic contact within, say, 5 Å. At temperatures far below the melting point, polymer viscosities are so high that any appreciable interfacial flow cannot occur. However, interfacial flow must occur before any hydrogen bond can form across the interface. Therefore, softening of the polymer surface layer by the formation of polymeric scission product is the primary mechanism of bondability improvement. These polymeric scission products are likely to be polar, and can thus also increase the wettability and the interfacial attraction through specific interactions (such as dipole-dipole and hydrogen-bonding interactions) across the interface.

Washing of poly(ethylene terephthalate) film with aqueous KOH followed by a dilute HCl dip is known to hydrolyze the polyester and

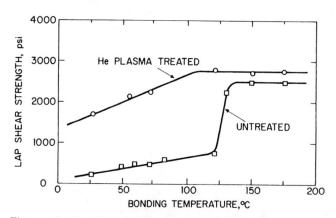

Figure 9.26. Lap shear strength of aluminum/epoxy adhesive/polyethylene/epoxy adhesive/aluminum joint versus bonding temperature. The untreated polyethylene has a melting point at 130°C. RF helium plasma, 15 MHz, 1 torr, 100 W for 1 sec. (After Ref. 104.)

Table 9.8. Effects of Oxygen Corona Treatments on the Joint Strengths of Some Polymer/Cellulose Bonds[a]

	\multicolumn{4}{c}{Shear strength, kg/cm^2}			
	Polyethylene	Polystyrene	Poly(vinyl chloride)	Poly(vinylidene chloride)
Temperature of pressing	90°C	110°C	110°C	110°C
Untreated control	1.3	0	0	0.5
Cellulose treated only	1.3	1.3	0.5	2.0
Polymer treated only	4.4	6.7	3.3	2.2
Both treated	6.5	15.2	5.3	3.8

[a]Corona, 60 Hz, 15 kV, 1 atm, 50 W, from Ref. 126. The joints are prepared by pressing the polymer against the cellulose at a pressure of 5.7 kg/cm^2 for 2 min at the temperature specified.

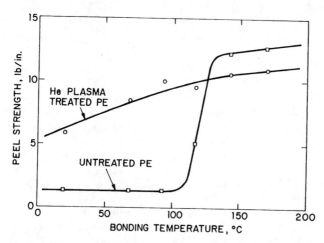

Figure 9.27. Effect of bonding temperature on peel strengths (23°C) of helium plasma-treated polyethylene to butyl rubber adhesive bonds and untreated polyethylene to butyl rubber adhesive bonds. (After Refs. 111 and 165.)

produce terminal p-carboxyl groups on the phenyl ring. These carboxyl groups should have strong hydrogen-bonding capability. However, when two pieces of such hydrolyzed PET films are pressed at 140°C against each other, only a marginal bond strength is obtained [150]. Similar observations are also reported in bonding various plasma-treated and untreated polymers to cellulose (Table 9.8). Thus, hydrogen bonding cannot be the primary mechanism of bondability improvement by plasma treatment.

The effects of post treatments with acetyl chloride, bromine water, and several other chemicals, mentioned earlier, can be explained in terms of increased glass transition temperature and viscosity of the surface layers caused by reaction with these chemicals. This interpretation is consistent with the known effects of chemical substituents on glass transition temperature and viscosity of polymers.

The improved bondability will rapidly diminish if a treated specimen is aged above the softening point of the polymeric scission product. Such a temperature is about 80°C for plasma-treated polyethylene [58,113,134,164]. The polymeric scission product is likely to be polar and have higher surface tension than the bulk polymer, and therefore tends to migrate away from the surface when heat-aged. On the other hand, if the aging temperature is far below the softening point, the improved bondability could be preserved for months.

9.4. PHOTOCHEMICAL TREATMENTS

UV irradiation of polymer surface produces chemical, wettability, and bondability modifications similar to those induced by plasma treatment. In some cases, the UV radiation in a plasma is the main mode of plasma-polymer reaction. For instance, essentially the same chemical, wettability, and bondability changes are obtained on polyethylene and poly-(ethylene terephthalate) when treated with UV radiation or plasma [116,117,150,166]. To be effective, however, correct UV wavelengths should be used. For instance, UV radiation of a wavelength of 1840 Å cross-links the polyethylene surface, whereas 2537 Å does not [116, 117]. UV lamps emit relatively narrow wavelengths, whereas a plasma emits a wide range of UV wavelengths. Therefore, the plasma is more widely effective and less selective than is the UV lamp.

UV irradiation causes chain scission, cross-linking, and oxidation on polymer surfaces even in an inert gas, just as in plasma treatment [121,150]. Carbonyls are introduced onto polyethylene; terminal p- and m-phenol groups are formed by chain scission in poly(ethylene terephthalate). Wettabilities and bondabilities are greatly increased [121,150,166-168]. UV irradiation in either helium or air gives similar bondability improvements to polyethylene (Figure 9.28).

Many different mechanisms of bondability improvement have been proposed, including improved wettability [121], strengthening of the weak boundary layer by surface cross-linking [166], and hydrogen bonding [150]. The preferred view is that the bondability is improved by the formation of polar polymeric scission products, which promote interfacial flow, interdiffusion, and polar interactions, just as in the case of plasma modification (Section 9.3.6).

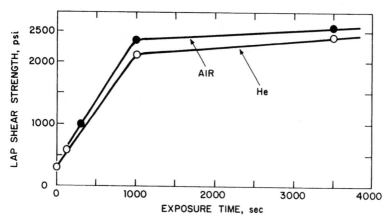

Figure 9.28. Lap shear strength of polyethylene bonded with an epoxy adhesive versus exposure time of the polyethylene in 1849-Å UV radiation chamber in air or helium at 1 torr. (After Ref. 166.)

Photochemical Treatments

The effectiveness of UV treatment can be promoted by using a photosensitizer. Benzophenone serves this purpose very well. It can sublime off after the UV treatment. Benzophenone-sensitized UV treatments of polyolefins, EPDM rubber, and fluoropolymers are used commercially to promote wettability and bondability [167,168]. A commercial setup has been described [4].

Benzophenone catalyzes the initiation of photolysis by forming resonance-stabilized diphenylhydroxymethyl radicals, that is,

$$RH \xrightarrow[h\nu]{\phi-CO-\phi} R\cdot \;+\; \phi-\underset{\cdot}{\overset{OH}{\underset{|}{C}}}-\phi \leftrightarrow H\langle\underline{\cdot}\rangle = \overset{OH}{\underset{|}{C}}-\phi \text{ etc.}$$

The alkyl radical R· can then undergo propagation and terminal reactions, as in the case of thermal and plasma oxidations.

9.5. CRYSTALLINE MODIFICATIONS OF POLYMER SURFACES

Surfaces of semicrystalline polymers can be crystallized to form transcrystalline, spherulitic, or lamellar structure by molding against certain mold surfaces (Sections 5.4 and 9.1.2). Photomicrographs of various morphologies are shown in Figures 5.13 and 9.5 to 9.10. Certain characteristics of these crystalline surface layers are discussed in Sections 3.1.4, 5.4, and 9.1.2.

The crystalline structures obtained depend on cooling rate, mechanical stress, mold surface, and other factors. For instance, either spherulitic or transcrystalline surface morphology is obtained on polypropylene, molded against poly(ethylene terephthalate), depending on cooling rate (Table 9.9). The particular surface morphology obtained has nothing to do with the surface energy of the mold surface, but rather is affected by the subtle nuances of the chemical compositions of the mold surface. For instance, polypropylene develops transcrystalline surface morphology when molded against polytetarfluoroethylene, poly(ethylene terephthalate), or polyamide, but develops spherulitic surface morphology when molded against fluorinated ethylene-propylene copolymer, aluminum oxide, or copper oxide [36,37,48,169,170].

Transcrystallized polymer surfaces are highly crystalline, as columnar spherulites are tightly packed together. These transcrystalline surface layers on polyethylene [170-177], polypropylene [36,37,48, 169,170,172-182], polyamide [183,184], and polyurethane [183] are often 10-100 μm thick, and have mechanical properties, wettability, and bondability very different from those of the bulk phase.

Dynamic mechanical properties of the transcrystalline surface layers on polyethylene and polypropylene have been measured [175-177].

Table 9.9. Effect of Cooling Rate on Polypropylene Surface Morphology

Mold surface	Cooling rate	Surface morphology	Radius of spherulite, μm		Thickness of transcrystalline surface layer, μm
			Surface	Interior	
Aluminum oxide	Fast	Spherulitic	1–2	5–6	—
	Normal	Spherulitic	35–40	35–40	—
	Slow	Spherulitic	80–85	80–85	—
Poly(ethylene terephthalate)	Fast	Spherulitic	1–2	5–6	—
	Normal	Transcrystalline	—	35–40	45–50
	Slow	Transcrystalline	—	80–85	95–100

Source: From Ref. 48.

Crystalline Modifications of Polymer Surfaces

Thin films (about 1 mil thick) which comprise almost entirely of transcrystalline materials are used to measure the dynamic moduli. Alternatively, the dynamic mechanical properties of films containing surface layers of constant thickness are measured as a function of the total film thickness. Both the storage modulus E' and the loss modulus E" are found to decrease with increasing film thickness, approaching asymptotic values at large thicknesses. Such data can be analyzed in terms of a parallel model in which the bulk phase is sandwiched between two surface layers. The complex modulus E* of the composite is given by

$$E^* = E^*_b + (E^*_s - E^*_b) \frac{t_s}{t} \qquad (9.2)$$

where E^*_b is the complex modulus for the bulk phase, E^*_s that for the surface phase, t_s the thickness of the two surface layers combined, and t the total film thickness. Since $E^* = E' + iE''$, the same relation also applies to E' and E", respectively. Thus,

$$E' = E'_b + (E'_b - E'_s) \frac{t_s}{t} \qquad (9.3)$$

$$E'' = E''_b + (E''_b - E''_s) \frac{t_s}{t} \qquad (9.4)$$

Thus, a plot of E versus t^{-1} will give a straight line with a slope of $(E_b - E_s)t_s$ and an intercept E_b on the ordinate. From such plots,

Table 9.10. Dynamic Mechanical Properties of Transcrystalline Surface Layer and Spherulitic Bulk Phase for Polyethylene and Polypropylene[a]

	Polyethylene	Polypropylene
Surface layer (transcrystalline)		
E', dyne/cm^2	19.7×10^9	15.3×10^9
E", dyne/cm^2	6.7×10^8	12.0×10^8
Bulk phase (spherulitic)		
E', dyne/cm^2	9×10^9	9.7×10^9
E", dyne/cm^2	2×10^8	3×10^8

[a]E', storage modulus; E", loss modulus.
Source: From Ref. 175.

Table 9.11. Adhesive Joint Strength, Surface Tension, and Surface Crystallinity of Some Polymers Having Transcrystalline Surface Layers Obtained by Molding Against Nucleating Mold Surfaces[a]

Nucleating mold surface	Surface tension at 20°C, dyne/cm	Surface crystallinity %	Tensile shear strength of adhesive joint, psi
Polyethylene			
Nitrogen	36.2	0	330
PTFE	36.2	0	500
Aluminum	54.9	63.2	2300
Gold	69.6	96.3	2640
Helium-plasma-treated	68.9	0	2720
Chromic acid-etched	52.3	—	2740
Perfluorinated ethylene-propylene copolymer, FEP			
Nitrogen	18.8	0	125
Aluminum	—	—	1900
Gold	40.4	—	2874
Helium-plasma-treated	28.8	—	1240
Sodium-etched	50	—	2968
Nylon 6			
Nitrogen	47.9	0	745
Gold	74.4	94.1	1968
Helium-plasma-treated	—	—	3281

[a] Adhesive joints are aluminum/epoxy adhesive/polymer/epoxy adhesive/aluminum lap shear joints; Refs. 144, 185, and 186. Other data from various sources.

E_b, E_s, and t_s can be found. The dynamic moduli of the transcrystalline surface phase and of the spherulitic bulk phase of polyethylene and polypropylene are listed in Table 9.10.

Transcrystallized polymer surfaces have very high surface tension and bondability. The greatly increased surface tension can be quantitatively explained by the increase of surface density (Sections 3.1.4 and 5.4). The surface tensions of some transcrystalline polymer surfaces are given in Tables 5.11 and 5.12. There is, however, a controversy here. Since ESCA analysis of transcrystallized polyethylene and FEP surfaces revealed the presence of polar oxygenated groups (probably formed by surface oxidation catalyzed by the mold surface during melt processing), the increased surface tension was also attributed to these polar contaminants (Section 5.4).

The bondability is dramatically increased by surface transcrystallization, such as on polyethylene, polypropylene, perfluorinated ethylene-propylene copolymer, and nylon 6 [138,144,185,186]. The improved bondability was attributed to the strengthening of the weak boundary layer by surface transcrystallization [138,144,185,186]. However, since the wettability is also greatly increased, it is not imperative to invoke the concept of the weak boundary layer, whose existence has not been proved and is inferred only. Surface tension, surface crystallinity, and adhesive joint strength for some polymers are compared in Table 9.11. Good correlation between surface tension and joint strength can be seen. See also Chapter 12.

9.6. MISCELLANEOUS MODIFICATIONS

Various vinyl monomers have been grafted onto polymer surfaces [15, 187–191]. Grafting sites on polymers are produced by exposure to plasma, UV radiation, or chemical agents. Polymer radicals (formed in the absence of oxygen) or polymer peroxides (formed in the presence of oxygen) by plasma or UV exposure serve as the initiation and grafting sites. Grafting can be done by preirradiation of the polymer surface followed by exposure to monomers, or by simultaneous irradiation of the polymer and the monomer. The kinetics of grafting reaction has been analyzed [188]. Alternatively, grafting sites may be generated by chemical means. For instance, polytetrafluoroethylene is treated with potassium-NH_3 solution to produce $C{=}C$ double bonds (Section 9.1.1), to which various vinyl monomers can be polymerized to render the fluoropolymer surface nonthrombogenic [190].

Abrasion of polymer surface breaks chemical bonds and produces polymer radicals. When abraded in air, the radicals formed react instantly with oxygen, nitrogen, or water to form polar groups. On the other hand, when abraded in a liquid, radicals formed will instantly react with the liquid molecules. For instance, liquid epoxy resins are grafted onto polyethylene, polypropylene, and polytetrafluoroethylene when they are abraded in liquid epoxy resins [192–194]. These grafted surfaces show dramatically increased wettability and bondability, whereas surfaces similarly abraded in air show only marginal improvements.

Various colloidal metal oxides, silica, and carbon black can be deposited onto polymer surfaces to increase their wettability [195–201]. Deposits of organic titanates are reported to improve both wettability and bondability [200,201].

RFL (resorcinol-formaldehyde-latex) has been widely used commercially to prime the surfaces of rayon, nylon, polyester, glass fiber, and aramid fiber tire cords to achieve strong fiber-rubber adhesion [202–206]. A typical RFL is a blend of resorcinol-formaldehyde resin and a rubber latex (such as a terpolymer of styrene, butadiene, and vinylpyrrolidone or natural rubber). RFL adheres strongly to rayon and nylon directly, but weakly to polyester, glass fiber, and aramid fiber, probably because rayon and nylon have many reactive surface groups, whereas polyester, aramid, and glass fiber lack reactive surface groups. Therefore, rayon and nylon cords are dipped in RFL directly, whereas polyester cord requires a polyisocyanate or polyepoxide pretreatment before RFL dip; aramid cord requires a polyepoxide pretreatment; glass fiber requires a silane pretreatment.

Many different opinions have been expressed as to the mechanism of RFL-tire cord adhesion, including chemical bonding across the interface [207,208], hydrogen bonding [209], dipole-dipole interaction [210], and interdiffusion [208,211]. Many different opinions have also been expressed as to the mechanism of rubber-RFL adhesion, including interdiffusion [212], convulcanization [213], and ionic interaction [214]. A preferred view is that chemical bonding is predominant at the RFL-tire cord interface, and interdiffusion and covulcanization is predominant at the rubber-RFL interface.

REFERENCES

1. W. Goldie, *Metallic Coating of Plastics*, Vols. 1 and 2, Electrochemical Publications Ltd., Middlesex, England, 1968.
2. I. A. Abu-Isa, Polym.-Plast. Technol. Eng., *2*(1), 29 (1973).
3. N. Feldstein, Plating, 57, 803 (1970).
4. R. E. Knox, Mod. Plast., 56 (February 1972); Rubber World, 31 (February 1972).
5. J. R. Hollahan and A. T. Bell, eds., *Technique and Applications of Plasma Chemistry*, Wiley-Interscience, New York, 1974.
6. J. S. Mijovic and J. A. Koutsky, Polym.-Plast. Technol. Eng., *9*(2), 139 (1977).
7. R. C. Snogren, *Handbook of Surface Preparation*, Palmerton, New York, 1974.
8. C. V. Cagle, *Adhesive Bonding: Techniques and Applications*, McGraw-Hill, New York, 1968.
9. R. C. Snogren, Adhes. Age, *12*(7), 26 (1969); *12*(8), 36 (1969).
10. N. J. DeLollis and O. Montoya, Adhes. Age, *6*(1), 32 (1963).
11. Belgian Patent 548,516 (1957); British Patent 793,731 (1958).
12. U.S. Patent 2,789,063 (1957).
13. U.S. Patent 2,809,130 (1957).

References

14. E. R. Nelson, T. J. Kilduff, and A. A. Benderly, Ind. Eng. Chem., *50*, 329 (1958).
15. M. L. Miller, R. H. Postal, P. N. Sawyer, J. G. Martin, and M. J. Kaplit, J. Appl. Polym. Sci., *14*, 257 (1970).
16. J. Jansta, F. P. Dousek, and J. Riha, J. Appl. Polym. Sci., *19*, 3201 (1975).
17. Collected Papers, Colloque Weyl IV, *Electrons in Fluids—The Nature of Metal-Ammonia Solutions*, J. Phys. Chem., *79*(26) (December 18, 1975).
18. G. Lepoutre and M. J. Sienko, eds., *Metal-Ammonia Solutions: Physicochemical Properties*, W. A. Benjamin, New York, 1964.
19. J. J. Lagowski and M. J. Sienko, eds., *Metal-Ammonia Solutions*, Butterworths, London, 1970.
20. A. A. Benderly, J. Appl. Polym. Sci., *6*, 221 (1962).
21. M. C. St. Cyr, Adhes. Age, *5*(8), 31 (1962).
22. D. W. Dwight and W. M. Riggs, J. Colloid Interface Sci., *47*, 650 (1974).
23. E. H. Cirlin and D. H. Kaelble, J. Polym. Sci., Polym. Phys. Ed., *11*, 785 (1973).
24. E. H. Andrews and A. J. Kinlock, Proc. R. Soc. Lond., *A332*, 385 (1973).
25. H. Brecht, F. Mayer, and H. Binder, Agnew. Mackromol. Chem., *33*, 89 (1973).
26. F. K. Borisova, G. A. Galkin, A. V. Kiselev, A. Ya. Korolev, and V. I. Lygin, Kolloid Zh., *27*, 320 (1965); Colloid J. USSR, *27*, 265 (1965).
27. P. Blais, D. J. Carlsson, G. W. Csullog, and D. M. Wiles, J. Colloid Interface Sci., *47*, 636 (1974).
28. F. K. Borisova, A. V. Kiselev, A. Ya. Korolev, V. I. Lygin, and I. N. Solomonova, Kolloid Zh., *28*, 792 (1966); Colloid J. USSR, *28*, 639 (1966).
29. H. W. Rauhut, Adhes. Age, *12*(12), 28 (1969); *13*(1), 34 (1970).
30. A. T. Devine and M. J. Bodnar, Adhes. Age, *12*(5), 35 (1969).
31. M. D. Anderson and M. J. Bodnar, Adhes. Age, *17*(11), 26 (1964).
32. M. J. Barbarisi, Tech. Rep. 3456, Picatinny Arsenal, Dover, N.J., 1967; U.S. Government Unclassified Document AD 651093.
33. T. Tsunoda, T. Seimiya, and T. Sasaki, Bull. Chem. Soc. Jpn., *35*, 1570 (1962).
34. K. Kato, J. Appl. Polym. Sci., *18*, 3087 (1974); *20*, 2451 (1976); *21*, 2735 (1977).
35. D. M. Brewis, J. Mater. Sci., *3*, 262 (1968).
36. D. R. Fitchmun, S. Newman, and R. Wiggle, J. Appl. Polym. Sci., *14*, 2441 (1970).
37. D. R. Fitchmun, S. Newman, and R. Wiggle, J. Appl. Polym. Sci., *14*, 2457 (1970).

38. K. Kato, Polymer, 9, 419 (1968).
39. K. Kato, J. Appl. Polym. Sci., 19, 1593 (1975).
40. V. J. Armond and J. R. Atkinson, J. Mater. Sci., 3, 332 (1968).
41. E. Groshart, Met. Finish., 70(2), 85 (1972).
42. ASTM Annual Standards, 22, 625 (1977).
43. A. Rantell, Electroplat. Met. Finish., 24(11), 5 (1971).
44. K. B. Wiberg and R. Eisenthal, Tetrahedron, 20, 1151 (1964).
45. W. F. Sager, J. Am. Chem. Soc., 78, 4970 (1956).
46. F. Holloway, M. Cohen, and F. H. Westheimer, J. Am. Chem. Soc., 73, 65 (1951).
47. F. Mares and J. Rocek, Coll. Czech. Chem. Commun., 26, 2370 (1961).
48. D. R. Fitchmun and S. Newman, J. Polym. Sci., A-2, 8, 1545 (1970).
49. D. Briggs, D. M. Brewis, and M. B. Konieczko, J. Mater. Sci., 11, 1270 (1976).
50. I. A. Abu-Isa, J. Appl. Polym. Sci., 15, 2865 (1971).
51. S. Ueda and T. Kimura, Kobunshi Kagaku, 15, 243 (1958).
52. M. Tsuruta, H. Armito, and M. Ishibashi, Kobunshi Kagaku, 15, 619 (1958).
53. Y. Kinoshita, Makromol. Chem., 33, 1 (1959).
54. D. Vogelsang, J. Polym. Sci., A1, 1055 (1963).
55. E. M. Bradburry, L. Brown, A. Elliot, and D. A. D. Parry, Polymer, 6, 465 (1965).
56. I. A. Abu-Isa, J. Polym. Sci., A-1, 9, 199 (1971).
57. A. Baszkin and L. Ter Minassian-Saraga, J. Polym. Sci., C34, 243 (1971).
58. A. Baszkin and L. Ter Minassian-Saraga, J. Colloid Interface Sci., 43, 473, 478 (1973).
59. A. Baszkin, M. Nishino, and L. Ter Minassian-Sarage, J. Colloid Interface Sci., 54, 317 (1976).
60. R. P. Plamer and A. J. Cobbold, Makromol. Chem., 74, 174 (1964).
61. A. Peterlin and G. Meinel, J. Polym. Sci., B3, 1059 (1965).
62. T. Hinton and A. Keller, J. Appl. Polym. Sci., 13, 745 (1969).
63. V. J. Armond and J. R. Atkinson, J. Appl. Polym. Sci., 14, 1833 (1970).
64. R. C. Ferguson, H. J. Stoklosa, W. W. Yau, and H. H. Hoehn, J. Appl. Polym. Sci., C34, 119 (1978).
65. T. R. Bott, A. P. Harvey, and D. A. Palmer, J. Appl. Polym. Sci., 14, 1833 (1970).
66. D. A. Olsen and A. J. Osteras, J. Polym. Sci., A-1, 7, 1921 (1969).
67. D. A. Olsen and A. J. Osteras, J. Polym. Sci., A-1, 7, 1927 (1969).

References

68. D. A. Olsen and A. J. Osteras, J. Polym. Sci. A-1, 7, 1913 (1969).
69. D. A. Olsen and A. J. Osteras, J. Appl. Polym. Sci., 13, 1523 (1969).
70. A. J. Osteras and D. A. Olsen, J. Appl. Polym. Sci., 13, 1537 (1969).
71. A. J. Osteras and D. A. Olsen, Nature, 221, 1140 (1969).
72. D. A. Olsen and A. J. Osteras, J. Colloid Interface Sci., 32, 19 (1970).
73. D. A. Olsen and A. J. Osteras, J. Colloid Interface Sci., 32, 12 (1970).
74. C. E. M. Morris, J. Appl. Polym. Sci., 14, 2171 (1970).
75. H. Schonhorn and R. H. Hansen, J. Appl. Polym. Sci., 12, 1231 (1968).
76. D. T. Clark and W. J. Weast, J. Macromol. Sci., $C12(2)$, 191 (1975).
77. M. Stradal and D. A. I. Goring, Polym. Eng. Sci., 17, 38 (1977).
78. G. D. Cooper and M. Prober, J. Polym. Sci., 44, 397 (1960).
79. H. C. Beachell and S. P. Nemphos, J. Polym. Sci., 21, 113 (1956).
80. F. H. Kendall and J. Mann, J. Polym. Sci., 19, 503 (1956).
81. D. J. Buckley and S. B. Robison, J. Polym. Sci., 19, 145 (1956).
82. S. F. Bloyer, Mod. Plast., 32(11), 105 (1955).
83. C. M. Chu and G. L. Wilkes, J. Macromol. Sci., Phys., B10, 551 (1974).
84. J. R. Caldwell and W. J. Jackson, Jr., J. Polym. Sci., C24, 15 (1968).
85. P. R. Lewis and R. J. Ward, J. Colloid Interface Sci., 47, 661 (1974).
86. G. F. Trott, J. Appl. Polym. Sci., 18, 1411 (1974).
87. B. Dobbs, R. Jackson, J. N. Gaitskell, R. F. Gray, G. Lynch, D. T. Thompson, and R. Whyman, J. Polym. Sci., Chem., 14, 1429 (1976).
88. R. L. Ayres and D. L. Shofuer, SPE J., 28(12), 51 (1972).
89. R. E. Greene, Tappi, 48(9), 80A (1965).
90. A. J. G. Allan, J. Polym. Sci., 38, 297 (1959).
91. D. Briggs, D. M. Brewis, and M. B. Konieczko, J. Mater. Sci., 14, 1344 (1979).
92. A. G. Gaydon, *The Spectroscopy of Flames*, 2nd ed., Chapman & Hall, London, 1974.
93. W. H. Kreidl and F. Hartmann, Plast. Technol., 1, 31 (1955).
94. D. Briggs, D. M. Brewis, and M. B. Kouieczko, Eur. Polym. J., 14, 1 (1978).
95. H. A. Willis and V. J. I. Zichy, Proc. Polym. Surf. Symp., Durham, England, March 1977.

96. F. M. Rugg, J. J. Smith, and R. C. Bacon, J. Polym. Sci., *13*, 535 (1954).
97. H. C. Beachell and G. W. Tarbet, J. Polym. Sci., *45*, 451 (1960).
98. B. Baum, J. Appl. Polym. Sci., *2*, 281 (1959).
99. J. P. Luongo, J. Polym. Sci., *42*, 139 (1960).
100. E. M. Bevilacqua, J. Polym. Sci., *C24*, 285 (1968).
101. H. Yasuda, H. C. Marsh, S. Brandt, and C. N. Reilley, J. Polym. Sci., Polym. Chem. Ed., *15*, 991 (1977).
102. D. T. Clark and A. Dilks, in *Characterization of Metal and Polymer Surfaces*, Vol. 2, L. H. Lee, ed., Academic Press, New York, 1977, pp. 101–132.
103. A. R. Blythe, D. Briggs, C. R. Kendall, D. G. Rance, and V. J. I. Zichy, Polymer, *19*, 1273 (1978).
104. R. H. Hansen and H. Schonhorn, J. Polym. Sci., *B4*, 203 (1966).
105. H. Schonhorn and R. H. Hansen, J. Appl. Polym. Sci., *11*, 1461 (1967).
106. R. R. Sowell, as quoted in Rubber Chem. Technol., *46*, 549 (1973).
107. J. R. Hall, C. A. L. Westerdahl, A. T. Devine, and M. J. Bodnar, J. Appl. Polym. Sci., *13*, 2085 (1969).
108. J. R. Hall, C. A. L. Westerdahl, M. J. Bodnar, and D. W. Levi, J. Appl. Polym. Sci., *16*, 1465 (1972).
109. C. A. L. Westerdahl, J. R. Hall, E. C. Schramn, and D. W. Levi, J. Colloid Interface Sci., *47*, 610 (1974).
110. H. Yasuda, C. E. Lamaze, and K. Sakaoku, J. Appl. Polym. Sci., *17*, 137 (1973).
111. N. Saka, G. Y. Yee, and N. P. Suh, 35th ANTEC, SPE, 1977, pp. 337–339.
112. J. R. Hollahan, J. Phys. E., J. Sci. Instrum., *2*, 203 (1969).
113. R. R. Sowell, N. J. DeLollis, H. J. Gregory, and O. Montoya, J. Adhes., *4*, 15 (1972).
114. B. W. Malpass and K. Bright, in *Aspects of Adhesion*, Vol. 5, D. J. Alner, ed., CRC Press, Cleveland, Ohio, 1969, pp. 214–225.
115. A. Bradley and J. D. Fales, Chem. Tech., 232 (April 1971).
116. M. Hudis and L. E. Prescott, J. Polym. Sci., *B10*, 179 (1972).
117. M. Hudis, J. Appl. Polym. Sci., *16*, 2397 (1972).
118. C. H. Bamford and J. C. Ward, Polymer, *2*, 277 (1961).
119. J. L. Weininger, J. Phys. Chem., *65*, 941 (1961).
120. J. L. Weininger, Nature, *186*, 546 (1960).
121. N. J. DeLollis, Rubber Chem. Technol., *46*, 549 (1973).
122. R. H. Hansen, J. V. Pascale, T. De Benedicts, and P. M. Rentzepis, J. Polym. Sci., *3*, 2205 (1965).
123. R. M. Mantell and W. L. Ormand, Ind. Eng. Chem., Prod. Res. Dev., *3*, 300 (1964).

124. K. Rossmann, J. Polym. Sci., *19*, 141 (1956).
125. M. Stradal and D. A. I. Goring, Can. J. Chem. Eng., *53*, 427 (1975).
126. C. Y. Kim, G. Suranyi, and D. A. I. Goring, J. Polym. Sci., *C30*, 533 (1970).
127. C. Y. Kim, J. M. Evans, and D. A. I. Goring, J. Appl. Polym. Sci., *15*, 1365 (1971).
128. C. Y. Kin and D. A. I. Goring, J. Appl. Polym. Sci., *15*, 1357 (1971).
129. J. M. Evans, J. Adhes., *5*, 1 (1973).
130. J. M. Evans, J. Adhes., *5*, 29 (1973).
131. J. R. Hollahan and G. L. Carlson, J. Appl. Polym. Sci., *14*, 2499 (1970).
132. J. M. Evans, J. Adhes., *5*, 9 (1973).
133. M. Stradal and D. A. I. Goring, J. Adhes., *8*, 57 (1976).
134. D. K. Owens, J. Appl. Polym. Sci., *19*, 265 (1975).
135. H. E. Wechsberg and J. B. Webber, Mod. Plast., *36*(11), 101 (1959).
136. J. C. von der Heide and H. L. Wilson, Mod. Plast., *38*(5), 199 (1961).
137. L. R. Hougen, Nature, *188*, 577 (1960).
138. H. Schonhorn, F. W. Ryan, and R. H. Hansen, J. Adhes., *2*, 93 (1970).
139. J. R. Hollahan, B. B. Stafford, R. D. Falb, and S. T. Payne, J. Appl. Polym. Sci., *13*, 807 (1969).
140. P. Blais, D. J. Carlsson, and D. M. Wiles, J. Appl. Polym. Sci., *15*, 129 (1971).
141. D. J. Carlsson and D. M. Wiles, Can. J. Chem. Eng., *48*, 2397 (1970).
142. P. B. Noll and J. E. McAllister, Paper, Film Foil Converter, *37*(10), 46 (1963).
143. M. L. Kaplan and P. G. Kelleher, Science, *169*, 1206 (1970).
144. H. Schonhorn and F. W. Ryan, J. Polym. Sci., A-2, *7*, 105 (1969).
145. K. Hara and H. Schonhorn, J. Adhes., *2*, 100 (1970).
146. M. Hudis and L. E. Prescott, J. Appl. Polym. Sci., *19*, 451 (1975).
147. N. J. DeLollis and O. Montoya, J. Adhes., *3*, 57 (1971).
148. R. M. Lerner, Adhes. Age, *12*(12), 35 (1969).
149. E. L. Lawton, J. Appl. Polym. Sci., *18*, 1557 (1974).
150. A. Bradley and T. R. Heagney, Anal. Chem., *42*, 894 (1970).
151. B. Leclercq, M. Sotton, A. Baszkin, and L. Ter-Minassian-Saraga, Polymer, *18*, 675 (1977).
152. D. A. I. Goring, Pulp Paper Mag. Can., *68*(8), T372 (1967).

153. P. F. Brown and J. W. Swanson, Tappi, 54, 2012 (1971).
154. M. M. Millard, in *Characterization of Metal and Polymer Surfaces*, Vol. 2, L. H. Lee, ed., Academic Press, New York, 1977, pp. 85–100.
155. A. E. Pavlath, in *Techniques and Applications of Plasma Chemistry*, J. R. Hollahan and A. T. Bell, eds., Wiley-Interscience, New York, 1974, pp. 149–176.
156. F. K. McTaggart, *Plasma Chemistry in Electrical Discharges*, Elsevier, Amsterdam, 1967.
157. M. Hudis, in *Techniques and Applications of Plasma Chemistry*, J. R. Hollahan and A. T. Bell, eds., Wiley-Interscience, New York, 1974, pp 113–147.
158. J. R. Hollahan, J. Chem. Educ., 43, A401 (1966).
159. A. Tsukamoto and N. N. Lichtin, J. Am. Chem. Soc., 84, 1601 (1962).
160. D. T. Clark and A. Dilks, J. Polym. Sci., Polym. Chem. Ed., 16, 911 (1978).
161. F. B. Marcotte, D. Campbell, J. A. Cleaveland, and D. T. Turner, J. Polym. Sci., A-1, 5, 481 (1967).
162. C. V. Stephenson, B. C. Moses, R. E. Burks, Jr., W. C. Coburn, Jr., and W. S. Wilcox, J. Polym. Sci., 55, 465 (1961).
163. C. V. Stephenson, J. C. Lacey, Jr., and W. S. Wilcox, J. Polym. Sci., 55, 477 (1961).
164. A. Baszkin and L. Ter-Minassian-Saraga, Polymer, 19, 1083 (1978); 15, 759 (1974).
165. N. H. Sung, Polym. Eng. Sci., 19, 810 (1979).
166. H. Schonhorn and F. W. Ryan, J. Appl. Polym. Sci., 18, 235 (1974).
167. R. A. Bragole, J. Elastoplast., 4, 226 (1972).
168. R. A. Bragole, Adhes. Age, 17(4), 24 (1974).
169. D. Fitchmun and S. Newman, Polym. Lett., 7, 301 (1969).
170. V. A. Kargin, T. I. Sugolova, and T. K. Shaposhnikova, Dokl. Akad. Nauk SSSR, 180, 901 (1968); Trans. Acad. Sci. USSR, 180, 406 (1968).
171. R. K. Eby, J. Appl. Phys., 35, 2720 (1964).
172. H. Schonhorn, J. Polym. Sci. Polym. Lett. Ed., 5, 919 (1967).
173. H. Schonhorn, Macromolecules, 1, 145 (1968).
174. H. Schonhorn, in *Adhesion Fundamentals and Practice*, The Ministry of Technology, U.K., Gordon and Breach, London, 1969, pp. 12–21.
175. T. K. Kwei, H. Schonhorn, and H. L. Frisch, J. Appl. Phys., 38, 2512 (1967).
176. S. Matsuoka, J. H. Daane, H. E. Bair, and T. K. Kwei, J. Polym. Sci., Polym. Lett. Ed., 6, 87 (1968).
177. H. L. Frisch, H. Schonhorn, and T. K. Kwei, J. Elastoplast., 3, 214 (1971).

178. M. R. Kantz and R. D. Corneliussen, J. Polym. Sci., Polym. Lett. Ed., *11*, 279 (1973).
179. J. R. Shaner and R. D. Corneliussen, J. Polym. Sci., A-2, 10, 1611 (1972).
180. D. G. Gray, J. Polym. Sci., Polym. Lett. Ed., *12*, 509 (1974).
181. S. Y. Hobbs, Nature (Phys. Sci.), *234*, 12 (1971).
182. D. G. Gray, J. Polym. Sci., Polym. Lett. Ed., *12*, 645 (1974).
183. E. Jenckel, E. Teege, and W. Hinrichs, Kolloid Z., *129*, 19 (1952).
184. R. J. Barriault and L. F. Gronholz, J. Polym. Sci., *18*, 393 (1955).
185. H. Schonhorn and F. W. Ryan, J. Polym. Sci., A-2, *6*, 231 (1968).
186. H. Schonhorn and F. W. Ryan, Adv. Chem. Ser., *87*, 140 (1968).
187. D. J. Angier, in *Chemical Reactions of Polymers*, E. M. Fettes, ed., Interscience, New York, 1964, pp. 1009–1053.
188. A. J. Restaino and W. N. Reed, J. Polym. Sci., *36*, 499 (1959).
189. A. Bradley, Chem. Tech., 507 (August 1973).
190. R. I. Leininger, R. D. Falb, and G. A. Grode, Ann. N.Y. Acad. Sci., *146*, 11 (1968).
191. A. F. Lewis and L. J. Forrestal, ASTM STP *360*, 59 (1963).
192. C. H. Lerchenthal, M. Brenman, and N. Yitshaq, J. Polym. Sci., Polym. Chem. Ed., *13*, 737 (1975).
193. M. Brenman and C. H. Lerchenthal, Polym. Eng. Sci., *16*, 747 (1976).
194. C. H. Lerchenthal and M. Brenman, Polym. Eng. Sci., *16*, 760 (1976).
195. J. K. Marshall and J. A. Kitchener, J. Colloid Interface Sci., *22*, 342 (1966).
196. R. K. Iller, J. Colloid Interface Sci., *21*, 569 (1966).
197. R. K. Iller, J. Am. Ceram. Soc., *47*, 194 (1964).
198. J. T. Kenney, W. P. Townsend, and J. A. Emerson, J. Colloid Interface Sci., *42*, 589 (1973).
199. F. Galembeck, J. Polym. Sci., Polym. Lett. Ed., *15*, 107 (1977).
200. G. R. DeHoff and T. F. McLaughlin, Jr., Mod. Plast., *31*, 107 (November 1963).
201. S. J. Monte and G. Sugarman, in *Additives for Plastics*, Vol. 1, R. B. Seymour, ed., Academic Press, New York, 1978, pp. 169–192.
202. T. Takeyama and J. Matsui, Rubber Chem. Technol., *42*, 159 (1969).
203. D. W. Anderson, Rubber Age, 69 (September 1971).
204. F. H. Sexsmith and E. L. Polaski, in *Adhesion Science and Technology*, Vol. 9A, L. H. Lee, ed., Plenum, New York, 1975, pp. 259–279.

205. E. P. Plueddemann, Adhes. Age, *18*(6), 36 (1975).
206. Y. Iyengar, J. Appl. Polym. Sci., *22*, 801 (1978).
207. A. L. Miller and S. B. Robison, Rubber World, *137*, 397 (1957).
208. H. Moult, in *Handbook of Adhesives*, I. Skeist, ed., Reinhold, New York, 1962, pp. 495–504.
209. H. Patterson, Adhes. Age, *6*(9), 38 (1963).
210. M. W. Wilson, Tappi, *43*(2), 129 (1960).
211. J. Mather, Br. Polym. J., *3*(3), 58 (1971).
212. V. E. Basin, A. A. Berlin, and R. V. Uzina, Sov. Rubber Technol., *21*(9), 12 (1962).
213. M. I. Dietrick, Rubber World, *136*, 847 (1957).
214. D. B. Boguslavskii, I. L. Shmurak, K. N. Borodushkino, A. A. Berlin, and R. V. Uzina, Sov. Rubber Technol., *21*(12), 15 (1962).

10
Adhesion: Basic Concept and Locus of Failure

10.1. DEFINITIONS

Various definitions for adhesion have been proposed [1–7]. However, none is completely satisfactory or generally accepted. Any satisfactory definition must account for both the thermodynamic and the mechanical aspects of adhesion. The following definitions are proposed and used in this book.

Adhesion refers to the state in which two dissimilar bodies are held together by intimate interfacial contact such that mechanical force or work can be transferred across the interface. The interfacial forces holding the two phases together may arise from van der Waals forces, chemical bonding, or electrostatic attraction. The mechanical strength of the system is determined not only by the interfacial forces, but also by the mechanical properties of the interfacial zone and the two bulk phases. When an adhesively bonded structure breaks under low applied stress, the structure is often said to have "poor adhesion." Such usage can be misleading, since the fracture may have occurred cohesively near the interface rather than along the interface.

Thermodynamic adhesion refers to equilibrium interfacial forces or energies associated with reversible processes, such as ideal adhesive strength, work of adhesion, and heat of wetting. The term was first proposed by Eley [3]. *Chemical adhesion* refers to adhesion involving chemical bonding at the interface. *Mechanical adhesion* arises from microscopic mechanical interlocking over substantial portions of the interface [1].

Adherend, a term coined by de Bruyne [8], refers to a body that is bonded to another body with an adhesive. The word is used throughout the industry, but has not been recognized by lexicographers. *Substrate* refers to any material on which an adhesive is applied. "Adherend" and "substrate" may be used interchangeably. *Cohesion* refers to adhesion or attraction within a bulk phase. *Autohesion*, a term coined by Voyutskii [9], refers to adhering or bonding of two identical bodies which are separate before bonding.

10.2. BASIC CONCEPT OF ADHESIVE BOND STRENGTH

10.2.1. Ideal Adhesive Strength

The free energy required to separate *reversibly* two phases from their equilibrium position to infinity at constant temperature and pressure is the work of adhesion. The maximum force per unit area required in such a process is the ideal adhesive strength σ^a, which relates to the work of adhesion W_a by [10,11]

$$\sigma^a = \frac{16}{9(3)^{1/2}} \frac{W_a}{z_0} \qquad (10.1)$$

where z_0 is the equilibrium separation between the two phases; see Eq. (2.75).

For polymers, typically $W_a = 100$ erg/cm^2 and $z_0 = 2$ Å for dispersion forces [12]. The ideal adhesive strength is thus calculated to be about 80,000 psi (550 MPa), which is at least two orders of magnitude greater than the real adhesive strength usually obtainable. This has often been quoted as proof that if two phases are in intimate molecular contact, dispersion forces alone are sufficient to give strong adhesive bond. Since dispersion attractions are universally present between any two bodies, it was suggested that the nature and magnitude of interfacial attraction is unimportant and that wetting is the only requirement for a strong adhesive bond [3,8,13–15].

Such an argument is, however, defective. It ignores the fact that the ideal adhesive strength can be realized only when the fracture process is reversible. This is, however, almost never the case in practice. In almost all real systems, the fracture process is irreversible, accompanied by large viscoelastic dissipation. As discussed below, the real mechanical strength of an adhesive bond is determined largely by the nature of existing flaws and viscoelastic properties of the system.

10.2.2. Real Adhesive Strength

The real adhesive strength is usually at least more than an order of magnitude lower than the ideal adhesive strength, because of the ubiquitous presence of flaws in the interfacial zone and the bulk phases. Stress concentration will occur around a flaw, so that local stress is much greater than the average stress. When the local stress exceeds the local strength, fracture will occur. This can be restated as: When the elastic strain energy released by extension of the crack is sufficient to supply the energy required to create new crack surface, the crack will extend, resulting in fracture. That is, fracture will occur, when [16]

$$-\frac{\partial U}{\partial A} \geq G \qquad (10.2)$$

where U is the elastic strain energy of the specimen as a whole, A the interfacial area of the crack, and G the energy required to create one unit interfacial area of the crack and is termed the *fracture energy*. Note that one unit interfacial area consists of two units of surface area. Details are discussed in Sections 14.1 through 14.4.

Fracture energy consists mainly of reversible work of adhesion or cohesion and irreversible plastic work, that is,

$$G = W_c + W_p \qquad \text{for cohesive fracture} \qquad (10.3)$$

or

$$G = W_a + W_p \qquad \text{for adhesive fracture} \qquad (10.4)$$

where W_c is the work of cohesion, W_a the work of adhesion, and W_p the plastic work (viscoelastic dissipation). For a perfectly brittle material, $G = W_c$ or W_a, since plastic yielding does not occur. This is approximated by silicate glass and oligomeric polystyrene (Table 10.1). On the other hand, extensive plastic yielding usually occurs around the crack tip in many materials, such as poly(methyl methacrylate), polystyrene, and steel, whether they are commonly regarded as brittle or ductile. The plastic work will, of course, be larger in a ductile material than in a brittle material. However, even in a brittle material, the plastic work is usually several orders of magnitude greater than the reversible work of adhesion or cohesion, that is, $G \sim W_p$, since $W_p \gg W_a$ or W_c; Tables 10.1 and 10.2.

For a linear-elastic material containing a central elliptical crack (Figure 10.1), Eq. (10.2) becomes [21–23]

Table 10.1. Comparison Between Cohesive Fracture Energy G and Surface Tension γ for Some Materials

Material	$G/2$, erg/cm^2	γ, erg/cm^2
Poly(methyl methacrylate)	2×10^5 [a]	41.1 [b]
Polystyrene	7×10^5 [c]	40.7 [b]
Oligomeric polystyrene (MW ~3000)	40 [d]	40.7 [b]
Steel	1×10^6 [a]	2300
Glass	550 [a]	549

[a] From Ref. 17.
[b] From Ref. 18.
[c] From Ref. 19.
[d] From Ref. 20.

Table 10.2. Comparison of Ideal and Real Cohesive Strengths for Some Materials

	Tensile strength, MPa	
Material	Ideal	Real
Nylon 66		
Molded plastics	240 [a]	83 [b]
Drawn yarn	29,000	1,000 [c]
Polyethylene, molded	180 [a]	38 [b]
Polystyrene, molded	210 [a]	69 [b]
Aramid, yarn	7,900	2,760 [c]
Steel		
Drawn yarn	9,800	1,960 [c]
Block	2,000	600 [d]
E-glass, yarn	10,000	1,518 [c]
Graphite, whisker	100,000 [d]	24,000 [d]
Al_2O_3, whisker	54,000 [d]	15,000 [d]

[a] Calculated by Eq. (2.98).
[b] From Ref. 25.
[c] From Ref. 26.
[d] From Ref. 27.

Basic Concept of Adhesive Bond Strength

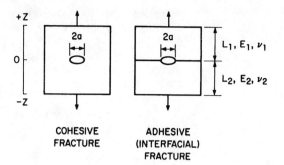

Figure 10.1. Comparison of cohesive and adhesive (interfacial) fractures.

$$f = \left(\frac{EG}{\pi a}\right)^{1/2} \quad \text{plane stress} \tag{10.5}$$

where f is the fracture stress, E the elastic modulus, and 2a the length of the central crack; see also Sections 14.1 through 14.4. This is the Griffith criterion for cohesive fracture. The relation has been verified by showing that f varies with $a^{-1/2}$ [17]. For plane strain, E is replaced with $E/(1 - \nu^2)$, where ν is Poisson's ratio.

The relation is equally applicable to adhesive (interfacial) fracture. In this case, two dissimilar crack surfaces are created. The fracture stress becomes [24,28]

$$f = \left(\frac{E_{12}G_a}{\pi a}\right)^{1/2} \quad \text{adhesive fracture} \tag{10.6}$$

where G_a is the adhesive fracture energy, a the half-length of the central interfacial crack, and E_{12} a composite elastic modulus given by

$$E_{12} = \frac{E_1 E_2}{\phi_1 E_2 + \phi_2 E_1} \tag{10.7}$$

where ϕ_j is the fractional length of phase j.

Thus, the real adhesive (or cohesive) strength is determined by the flaw size, fracture energy, and viscoelastic properties of the two phases, and is therefore dependent on rate and temperature. At ordinary rates, the real fracture process is irreversible, involving largely plastic work. For brittle materials, typically E = 0.5 Mpsi, G = 10^5 erg/cm^2, and the intrinsic flaw size a_0 is about 50 µm [17]. The fracture strength is thus calculated to be about 10,000 psi.

On the other hand, if the work of adhesion or cohesion (typically 100 erg/cm^2) alone contributes to the fracture energy and the plastic work is negligible, the fracture strength would be only about 200 psi, although the ideal strength is as high as 220,000 psi, as calculated in Section 10.2.1. This shows that any correct theory of adhesion cannot be based on consideration of reversible intermolecular energies or ideal adhesive strength alone without regard to the flaw size, plastic dissipation, and the irreversible nature of the fracture process.

10.2.3. Interfacial Structure and Adhesion: Necessary and Sufficient Conditions for Strong Adhesion

Van der Waals attraction between two planar macroscopic bodies diminishes rapidly with distance by z^{-3}, where z is the distance of separation (Section 2.5). The equilibrium interfacial separation is typically 2–5 Å. Therefore, intimate molecular contact at the interface is necessary to obtain strong interfacial attraction. Without intimate molecular contact, interfacial attraction will be very weak, and the applied stress that can be transmitted from one phase to the other through the interface will accordingly be very low. Therefore, intimate molecular contact at the interface is *necessary* to form a strong adhesive bond. However, intimate molecular contact alone is not sufficient to give a strong adhesive bond in many cases, as discussed below.

Liquids and solids of similar chemical compositions have intermolecular forces of similar magnitudes. Yet liquids have poor mechanical strength, and solids have good mechanical strength. Furthermore, the mechanical strength of a polymer drops off sharply below its critical molecular weight for entanglement. This is because the mechanical strength of a polymer derives mainly from molecular entanglement. Of course, highly ordered crystals can have good mechanical strength without molecular entanglement by virtue of tight molecular packing and lack of flaws. However, molecular entanglement is all-important in common polymers.

Viscoelastic dissipation constitutes the major component of fracture energy, as discussed previously. The magnitude of viscoelastic dissipation depends on the extent of molecular entanglement as well as on the magnitude of intermolecular force. The volume of the dissipation zone around the crack tip can be large only if molecules are entangled.

Therefore, it is important to recognize that the interface is a region of finite thickness, wherein the segments of macromolecules may interpenetrate. Its mechanical strength will very much depend on its structure. For the present purpose, we classify the interface into two types; sharp and diffuse. Three types of adhesive behavior can be visualized; as described below.

Basic Concept of Adhesive Bond Strength

Sharp Interface with Weak Molecular Force

If the interface is molecularly sharp and the molecular force is weak (such as the dispersion force between a nonpolar polymer and a polar polymer), viscoelastic dissipation will be small and the interface will be mechanically weak, because of lack of molecular entanglement and the likelihood of interfacial slippage. Strong specific interaction or chemical bonding across a sharp interface will tie the dissimilar molecules across the interface to the entangled bulk phases, and thus improve the mechanical strength.

Sharp Interface with Strong Molecular Force

If the interface is molecularly sharp and the molecular force is strong (such as the dispersion force between a nonpolar polymer and a high-energy material or molecular forces involving specific interactions), interfacial slippage will not occur and the interface will be mechanically strong.

Table 10.3. Comparison of Various Attractive Energies

Type of force	Typical energy range, kcal/mole	Reference
Van der Waals forces		
Dispersion force	5	
Dipole-dipole	Up to 10	
Dipole-induced dipole	Up to 0.5	
Hydrogen bond		
Acetic acid	15.9 (2)[a]	29
Ethanol	4.0 (2)[a]	29
$CHCl_3$-ether	6.0 (1)[a]	29
H_2O	6.8 (2)[a]	29
HF	40.8 (6)[a]	29
Phenylazo-1-naphthylamine	22.1 (1)[a]	29
Chemical bonds		
Covalent bond	15–170	30
Ionic bond	140–250	31
Metallic bond	27–83	32

[a] The number in parentheses is the number of hydrogen bonds per molecule.

Diffuse Interface with Any Molecular Force

If the interface is diffuse with sufficient molecular diffusion and entanglement, interfacial slippage will not occur and the interfacial zone will be mechanically strong regardless of the nature of molecular force. Interdiffusion of molecular segments should be sufficient; interdiffusion of entire macromolecules is not necessary, as discussed further in Section 11.3.

Therefore, both primary (chemical bonds) and secondary (van der Waals attractions) bonds are important in adhesion, depending on interfacial structure. Various primary and secondary bonds are discussed elsewhere [33,34]. Typical bond energies are listed in Table 10.3.

10.3. LOCUS OF FAILURE

When an adhesive bond breaks at a low applied stress, it is commonly said to have a "poor" adhesion. This usage can be quite misleading, since the fracture may have occurred exactly at the interface, in a thin layer very close to the interface but within a bulk phase, well within a bulk phase, or in a mixed interfacial and cohesive failure. Cohesive fracture in a bulk phase far from the interface can easily be identified. However, great care must be exercised to ascertain if an apparent interfacial failure is indeed a true interfacial failure, a cohesive failure in a thin layer close to the interface (but not at the interface), or a mixed interfacial cohesive failure. When a fracture occurs within 100– 1000 Å of the interface, identification of the true locus of failure can be quite difficult. Several techniques for identification are discussed later.

Correct identification of the locus of failure is of great theoretical and practical importance. It is the first step in analyzing an adhesion problem. If the failure occurs cohesively in a thin layer next to the interface, efforts should be directed toward strengthening this weak boundary layer rather than increasing the interfacial attraction. On the other hand, if the failure occurs at the interface, the remedy would be to increase interfacial attraction or interfacial diffusion. Incorrect identification of the failure locus will lead to wasted efforts.

Unfortunately, there is controversy as to whether true interfacial failure can occur in a "proper" adhesive joint. A proper adhesive joint is one whose interface is not contaminated with a cohesively weak layer. Bikerman [7] argued that true interfacial separation is practically impossible in a proper joint. However, recently both experimental [35–44] and theoretical [45,46] results clearly show that true interfacial failure can occur in a proper joint. Bikerman's doctrine is reviewed first, and the possibility of true interfacial failure is discussed next.

Locus of Failure   345

10.3.1. Bikerman's Doctrine

In 1947, Bikerman [7,47] proposed: "In a proper joint, true interfacial failure practically *never* occurs. What is taken for interfacial failure is actually separation in a weak boundary layer, that is, a thin layer greater than atomic dimensions with mechanical strength much weaker than that of either bulk phase. In the absence of weak boundary layer, failure always occurs within the weaker of the two phases." More recently, Bikerman [48] made a rather broad statement: "Rupture so rarely proceeds exactly between the adhesive and the adherend that these events (that is, 'failure in adhesion') need not be treated in any theory of adhesive joints." If a system appears to have failed at a phase boundary, the failure must actually have occurred in an unsuspected layer of material at the interface, which had a low cohesive strength: that is, a weak boundary layer [7]. Such a layer may be a brittle oxide of a metal, a low-molecular-weight fraction that migrates to the surface of a polymer, a minor component of one phase, such as an antioxidant or a lubricant that accumulates at the interface, or a contaminant such as a grease [7,49].

Bikerman [7] offered four proofs. The *first* is based on probability. The probability of an interfacial crack advancing between the next two unlike atoms at the interface versus turning between two like atoms in either phase is 1/3. To advance past n atoms at the interface, the probability is $(1/3)^n$, a vanishingly small number for large n. The *second* states that the attraction between two unlike molecules A_{12}^d is always intermediate in magnitude between those between two like molecules, A_{11}^d and A_{22}^d, that is,

$$A_{11}^d \geq A_{12}^d \geq A_{22}^d \tag{10.8}$$

which follows from the London theory of dispersion force (see Section 2.1). The *third* states that interfacial strength must be greater than that of a bulk phase, since the interfacial volume is quite small and, therefore, the probability of having a critical flaw in the interfacial zone is vanishingly small. The *fourth* contends that the probability of an interfacial crack following a tortuous path (Figure 10.2) is negligible [7,49]. Examples of separation in the weak boundary layer have been cited [7,49] (Section 11.1).

However, Bikerman's doctrine cannot be true. There is no question that many apparent interfacial failures are actually failures in thin weak boundary layers near the interface. However, this alone cannot logically be generalized to prove that *all* apparent interfacial failures are cohesive failures in weak boundary layers. Many cases of true interfacial failure have been reported by others [35–44], and are discussed later.

Figure 10.2. Model for a tortuous interface.

Furthermore, Bikerman's four proofs are theoretically unsound [45]. In the first proof, the probability of crack path should be weighted by bond energies. The ratio of the probability of breaking an A——A bond to that of breaking an A——B bond should correctly be given as [45,46]

$$\frac{P(A\text{——}A)}{P(A\text{——}B)} = \exp\left(-\frac{U_{AA} - U_{AB}}{RT}\right) \tag{10.9}$$

where U is the bond energy. Consider U_{AA} = 80 kcal/mole (chemical bonding), and U_{AB} = 5 kcal/mole (dispersion force). This gives $P(A\text{——}A)/P(A\text{——}B) = 5 \times 10^{-55}$ at room temperature. The number of atoms n which the crack must pass before the crack has 1% probability of departing the interface is $(1 - 5 \times 10^{-55})^n = 0.99$, which gives $n = 2 \times 10^{51}$. This corresponds to a distance of 4×10^{41} m (assuming an atomic diameter of 2 Å), a distance larger than the diameter of the universe! Thus, the odds that the crack will follow a tortuous path at the interface are overwhelming [45]. Bikerman's first and fourth proofs are thus invalid.

Bikerman's second proof is also invalid. Equation (10.8) is valid only for dispersion forces. When specific or polar interactions are present, the attraction between unlike atoms can be smaller than that between like atoms. Generally, the attraction constant between unlike bodies is correctly given as [12]

$$A_{12} = \phi(A_{11}A_{22})^{1/2} \tag{10.10}$$

where ϕ is the interaction parameter (Section 3.3). When the polarities of the two bodies are identical, ϕ has the maximum value of unity. In this case, A_{12} will indeed be intermediate between A_{11} and A_{22}. However, when the polarities of the two bodies are unequal, ϕ is less than unity, and A_{12} will be smaller than both A_{11} and A_{22}. Equation (10.10) can be rewritten as

Locus of Failure

$$W_a = \phi(W_{c1}W_{c2})^{1/2} \qquad (10.11)$$

which follows from Section 3.3. Many examples where W_a is smaller than both W_{c1} and W_{c2} have been given [18,54]. For instance, W_c for poly(methyl methacrylate) is 64.0 erg/cm^2, that for polyethylene is 57.6 erg/cm^2, but W_a between the two polymers is 51.1 erg/cm^2, all at 104°C [18]. See also Chapter 3.

Bikerman's third proof ignores the fact that the probability of having a critical flaw at the interface can be greater than in bulk phase, since complete interfacial wetting is difficult.

The weak materials that allegedly form weak boundary layers have never been isolated and identified. The mechanisms by which they lower the adhesive joint strength have not been investigated. Rather, the existence of a weak boundary layer was inferred indirectly [7,47, 50,51]. Such an approach has an inherent danger of circular reasoning [45].

Recently, Robertson [52] elegantly showed that the mere presence of a weak boundary layer at the interface does not necessarily seriously weaken an adhesive joint. A thin layer of weak oligomeric polystyrene (MW 10,000) is sandwiched between two plates of high-molecular-weight poly(methyl methacrylate). Fracture toughness in cleavage of the joint is measured. When the polystyrene layer is less than 1 µm thick, the fracture toughness is essentially identical to that for bulk poly(methyl methacrylate). Only when the polystyrene layer exceeds 1 µm does the fracture toughness of the joint fall precipitously, by about 85%. Thus, the "weak" boundary layer is not weak until its thickness exceeds 1 µm, which is macroscopic. When the polystyrene layer is thin (less than 1 µm), the fracture path sinuates between the polystyrene and the poly(methyl methacrylate). This clearly shows that a thin layer of cohesively weak material does not necessarily cause serious degradation in adhesive joint strength. The critical thickness should, however, decrease with decreasing molecular weight of the boundary material.

DeLollis and Montoya [53] showed that an epoxy adhesive strongly adheres to a cross-linked silicone rubber which contains a fraction of extractable, cohesively weak, low-molecular-weight silicone rubber on its surface. See also Chapter 12.

10.3.2. True Interfacial Failure Can Occur

The interface may be sharp or diffuse. When interfacial diffusion is sufficient, the interface is diffuse, such as between thermoplastic polymers of similar polarities. In this case, clean interfacial separation is obviously impossible (except perhaps at zero rate), as molecular segments are entwined. On the other hand, when little or no interfacial diffusion occurs, the interface will be sharp, such as between polar and nonpolar polymers, between a thermoplastic polymer and a cross-linked polymer, or between a polymer and an inorganic material. In this case, true interfacial separation may occur, as discussed below. The structure of interface is discussed in Sections 3.3 and 11.3.

Interfacial separation will occur when the interfacial strength is weaker than the bulk strength. In terms of the Griffith criterion, interfacial failure will occur when

$$\left(\frac{E_{12}G_a}{a_{12}}\right)^{1/2} < \left(\frac{E_2 G_2}{a_2}\right)^{1/2} \tag{10.12}$$

where phase 1 is stronger than phase 2. In the absence of interfacial chemical bonding, $E_{12} \sim (E_1 E_2)^{1/2}$ and $G_a \sim (G_1 G_2)^{1/2}$. Therefore,

$$\frac{a_2}{a_{12}} < \left(\frac{E_2 G_2}{E_1 G_1}\right)^{1/2} \tag{10.13}$$

is the condition for interfacial separation.

Fracture can occur in any of the five zones (Figure 10.3) in an adhesive joint: cohesive failure in the bulk of phase 1, cohesive failure in the bulk of phase 2, cohesive failure in a thin layer in phase 1 very near the interface within δ (region 3), cohesive failure in a thin layer in phase 2 very near the interface within δ (region 5), or true interfacial failure (along 4).

Good [45] considered a two-phase adhesive bond with a smooth and sharp interface at $z = 0$ and under tension in the z direction. The two phases are of equal cross section and volume (Figure 10.3). Assume

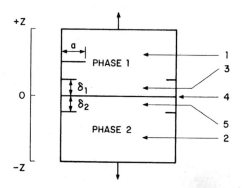

Figure 10.3. Five possible failure loci in an adhesive joint.

Locus of Failure

that the stress concentrations around the specimen edges due to Poisson's contractions are less than those at crack tips. Edge cracks of length a exist along z. As the tensile stress is increased, failure will start by crack extension at the weakest point. Following Good's method [45], consider for simplicity cracks of equal length. Then, failure will proceed in the region where the product $E(z)G(z)$ is the least, where $E(z)$ denotes that E varies with z and $G(z)$ denotes that G varies with z. Three general cases, with eight subcases, are analyzed below.

Case I: Strong or Moderate Interfacial Forces

Both bulk phases are homogeneous. Both $E(z)$ and $G(z)$ are constant in the bulk phases, and vary smoothly from one phase to the other in the interfacial zone between δ_1 and δ_2. There is no weak boundary layer.

(Ia) ΔE and ΔG same sign. This is depicted in Figure 10.4a, where $\Delta E = E_2 - E_1$ and $\Delta G = G_2 - G_1$. The product $E(z)G(z)$ has the lowest value in the bulk of the weaker phase. Cohesive fracture in the bulk of the weaker phase is predicted.

(Ib) ΔE and ΔG opposite sign (situation I). This is depicted in Figure 10.4b. The product $E(z)G(z)$ has a minimum in the thin layer in phase 2 near the interface. Cohesive fracture in this thin layer is predicted. Bikerman's doctrine would have interpreted this as proof of the existence of a weak boundary layer. However, the model specifies that there is no weak boundary layer. Thus, in this case, Bikerman's doctrine would invariably lead to incorrect conclusion.

(Ic) ΔE and ΔG opposite sign (situation II). This is depicted in Figure 10.4c, and is a variant of case Ib. The existence of a minimum in $E(z)G(z)$ depends on the relative magnitudes and slopes of $E(z)$ and $G(z)$ near $z = 0$. In addition to the case sketched in Figure 10.4c,

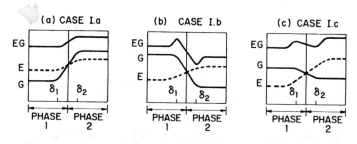

Figure 10.4. Analysis of fracture locus for strong or moderate interfacial forces: (a) case Ia, ΔE and ΔG same sign; (b) case Ib, ΔE and ΔG opposite sign: situation I; (c) case Ic, ΔE and ΔG opposite sign: situation II.

it is possible to have no minimum at all near $z = \delta_1$ or δ_2, or below the level of $E(z)G(z)$ in phase 1. Thus, cohesive fracture in the bulk of either phase is predicted.

Case II: Weak Interfacial Forces

Both $E(z)$ and $G(z)$ are constant in the bulk phases. $E(z)$ varies smoothly from one phase to the other. However, because of weak interfacial forces, $G(z)$ has a minimum at the interface at $z = 0$. There is no weak boundary layer.

(IIa) ΔE and ΔG opposite sign. This is depicted in Figure 10.5a. The product $E(z)G(z)$ has a minimum at the interface at $z = 0$. True interfacial failure is predicted. The interfacial weakness may be remedied by specific interaction, chemical bonding, or interdiffusion.

(IIb) ΔE and ΔG same sign (very weak forces). This is depicted in Figure 10.5b. The product $E(z)G(z)$ has a minimum at the interface at $z = 0$. True interfacial failure is predicted.

(IIc) ΔE and ΔG same sign (moderately weak forces). This is depicted in Figure 10.5c. Although $E(z)G(z)$ has a minimum at the interface, the minimum is above the level of $E(z)G(z)$ in phase 1. Thus, cohesive fracture in the bulk of the weaker phase is predicted, although the interfacial forces are weak. Only when the interfacial forces are sufficiently weak that there is a sharp minimum in $G(z)$ at the interface will true interfacial failure occur.

Case III: Weak or Strong Boundary Layer

Both $E(z)$ and $G(z)$ are constant in the bulk phases, but have minimum or maximum in the interfacial zone because of the existence of a weak or a strong boundary layer, respectively.

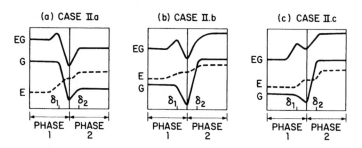

Figure 10.5. Analysis of fracture locus for weak interfacial forces: (a) case IIa, weak interfacial forces with ΔE and ΔG opposite sign; (b) case IIb, very weak interfacial forces with ΔE and ΔG same sign; (c) case IIc, moderately weak interfacial forces with ΔE and ΔG of same sign.

Locus of Failure

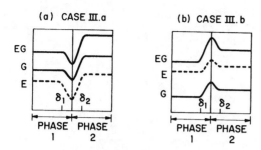

Figure 10.6. Analysis of fracture locus: (a) case IIIa, weak boundary layer; (b) case IIIb, strong boundary layer.

(IIIa) Weak boundary layer. This is depicted in Figure 10.6a. Both $E(z)$ and $G(z)$ have minimum in the interfacial zone between $z = \delta_1$ and δ_2, because of the existence of a third phase (a weak boundary layer). The product $E(z)G(z)$ has a minimum in the interfacial zone. Cohesive fracture in the weak boundary layer is predicted. This is the case where Bikerman's doctrine applies.

(IIIb) Strong boundary layer. This is depicted in Figure 10.6b. Both $E(z)$ and $G(z)$ have maximum in the interfacial zone because of the existence of third phase (a strong boundary layer). Cohesive fracture in the weaker bulk phase is predicted. The strong boundary layer may be induced by interfacial cross-linking, crystallization, or chemical bonding.

10.3.3. Experimental Evidence of True Interfacial Failure

Transitions from cohesive to interfacial failures have been observed to accompany changes in the rate and temperature of testing in peeling of adhesives from rigid adherends [35-39]. Interfacial failures are favored at low temperatures [35-38] and high rates [35,38]. Cohesive failures are favored at high temperatures and low rates [35-38]. Mixed interfacial and cohesive failures can occur at intermediate temperatures and rates [37]. A master curve can be obtained for the strength of an adhesive bond by superimposing the data at various temperatures and rates by using shift factors [35,38]. Gent and Petrich [35], for instance, obtained a single master curve by plotting σ/T versus $\log Ra_T$, where σ is the adhesive bond strength, T the temperature, R the testing rate, and a_T the WLF shift factor [35].

Interfacial failure has been confirmed by wettability measurements [38-40,43], electron microscopy [40-42], interferometry [44], and combined scanning electron microscopy and ESCA on adhesives and

Table 10.4. Comparison of Some Surface Analytical Techniques

	Ion scattering spectrometry (ISS)	Secondary ion mass spectrometry (SIMS)
Principle	Binary elastic collision of ions with surface atoms	Sputtering of surface atoms by ion beam
Incident beam		
Beam	Ions	Ions
Energy	1–3 keV	1–3 keV
Emitted beam	Rebound ions	Sputtered surface ions
Energy	—	0–20 eV
Sampling depth	First atomic layer on surface	Successive surface layers can be sputtered
Sampling diameter	—	10–300 μm
Elemental analysis	Yes	Yes
Quantitative?	Yes	No
Information on chemical combination?	No (yes, only in fine features)	No (yes, in some cases)
Incident beam induced surface damage	Sputtering damage	Removal of surface atoms by sputtering
Background pressure, torr	—	10^{-10}

adherends before bonding and after separation. For instance, the surface tensions of polytetrafluoroethylene adherend and of polycaprolactam adhesive are found to be unchanged before bonding and after separation [38]. Both the surface topography (by scanning electron microscopy) and surface chemical compositions (by ESCA) are unchanged for a poly(vinyl fluoride) on a cross-linked alkyd before bonding and after separation.

Auger electron spectroscopy (AES)	Electron spectroscopy for chemical analysis (ESCA)	Electron microprobe analyzer (EMA)
Ejection of Auger electron	Ejection of photoelectrons by X-ray	Emission of X-ray by electron beam
Electron 0.1–5 keV	X-ray 1–10 keV	Electron 0.5–40 keV
Auger electrons	Photoelectrons	X-ray
20–2000 eV	200–10,000 eV	200–10,000 eV
20 Å	10–20 Å	0.1–1 µm
0.1–1 mm	1–3 mm	10^{-3}–0.25 mm
Yes	Yes	Yes
Yes	Yes	Yes
Yes	Yes	Yes
Useful for metals, not useful for polymers due to extensive surface damage by incident electron beam	No	—
10^{-9}–10^{-10}	10^{-6}–10^{-10}	10^{-5}–10^{-10}

10.3.4. Determination of Failure Locus

Determination of the exact locus of failure is often a difficult task. Combinations of various methods are advisable. Each method must be adapted to any specific problem. Useful techniques are contact angle measurement [38–43], optical microscopy [55], scanning and transmission electron microscopy [41,42,56–60], internal reflection IR spectroscopy [61−64], ESCA [65−71], Auger electron spectroscopy [65,72], electron microprobe [60,73−75], X-ray fluoroescence spectroscopy [56], ion scattering spectrometry (ISS), secondary ion mass spectrometry (SIMS) [68,76−86], and others [87−90]. Details of these techniques are outside the scope of this treatise. However, some salient features are summarized in Table 10.4.

REFERENCES

1. ASTM D907, Standard Definitions of Terms Relating to Adhesives, *Annual Book of ASTM Standards*, 22, 233 (1977).
2. R. J. Good, J. Adhes., 8, 1 (1976).
3. D. D. Eley and D. Tabor, in *Adhesion*, D. D. Eley, ed., Oxford University Press, London, 1961, pp. 1−18.
4. C. Kemball, in *Adhesion and Adhesives, Fundamentals and Practice*, J. E. Rutzler, Jr., and R. S. Savage, eds., Wiley, New York, 1954, pp. 69−71.
5. G. Salomon, in *Adhesion and Adhesives*, Vol. 1, 2nd ed., R. Houwink and G. Salomon, eds., Elsevier, New York, 1965, pp. 1−140.
6. K. L. Mittal, Polym. Eng. Sci., 17, 467 (1977).
7. J. J. Bikerman, *The Science of Adhesive Joints*, 2nd ed., Academic Press, New York, 1968.
8. N. A. de Bruyne, Aero Res. Tech. Notes, 179, 1 (1957); J. Sci. Instrum., 24, 29 (1947).
9. S. S. Voyutskii, *Autohesion and Adhesion of High Polymers*, translated by S. Kaganoff, Interscience, New York, 1963.
10. R. J. Good, in *Treatise on Adhesion and Adhesives*, Vol. 1, R. L. Patrick, ed., Marcel Dekker, New York, 1967, pp. 9−68.
11. J. L. Gardon, in *Treatise on Adhesion and Adhesives*, Vol. 1, R. L. Patrick, ed., Marcel Dekker, New York, 1967, pp. 269−324.
12. S. Wu, in *Polymer Blends*, Vol. 1, D. R. Paul and S. Newman, eds., Academic Press, New York, 1978, pp. 243−293.
13. L. H. Sharpe and H. Schonhorn, Adv. Chem. Soc., 43, 189 (1964).
14. W. A. Zisman, Adv. Chem. Ser., 43, 1 (1964).
15. J. R. Huntsberger, in *Treatise on Adhesion and Adhesives*, Vol. 1, R. L. Patrick, ed., Marcel Dekker, New York, 1967, pp. 119−149.
16. E. H. Andrews, *Fracture in Polymers*, American Elsevier, New York, 1968.
17. J. P. Berry, in *Fracture Processes in Polymeric Solids*, B. Rosen, ed., Interscience, New York, 1964, pp. 195−234.
18. S. Wu, J. Phys. Chem., 74, 632 (1970).
19. J. P. Berry, J. Appl. Phys., 34, 62 (1963).
20. R. E. Robertson, Adv. Chem. Ser., 154, 89 (1976).

21. G. R. Irwin, in *Structural Mechanics*, Proceedings of Symposium on Naval Structural Mechanics, J. N. Goodier and N. J. Hoff eds., Pergamon Press, New York, 1960, pp. 557–594.
22. E. Orowan, Welding J. Res. Suppl., 1–4 (March 1955).
23. A. A. Griffith, Phil. Trans. R. Soc. Lond., *A221*, 163 (1920).
24. M. L. Williams, J. Appl. Polym. Sci., *13*, 29 (1969).
25. *Modern Plastics Encyclopedia*, McGraw-Hill, New York, 1974.
26. D. L. G. Sturgeon and R. I. Lacy, *Handbook of Fillers and Reinforcements for Plastics*, H. S. Katz and J. V. Milewski, eds., Reinhold, New York, 1978, pp. 511–544.
27. A. H. Cottrell, Proc. R. Soc. Lond., *A282*, 2 (1964).
28. G. P. Anderson, S. J. Bennett, and K. L. DeVries, *Analysis and Testing of Adhesive Bonds*, Academic Press, New York, 1977.
29. G. C. Pimentell and A. L. McClellan, *The Hydrogen Bond*, W. H. Freeman, San Francisco, 1960.
30. L. Pauling, *The Nature of the Chemical Bond*, 3rd ed., Cornell University Press, Ithaca, N.Y., 1960.
31. Y. K. Syrkin and M. E. Dyatkina, *Structure of Molecules and the Chemical Bond*, Wiley-Interscience, New York, 1950.
32. N. F. Mott and H. Jones, *Theory of the Properties of Metals and Alloys*, Oxford University Press, New York, 1936.
33. J. E. Rutzler, Jr., Adhes. Age, *2*(7), 28 (1959).
34. F. W. Reinhart, in *Adhesion and Adhesives: Fundamentals and Practice*, J. E. Rutzler, Jr., and R. L. Savage, eds., Wiley, New York, 1954, pp. 9–15.
35. A. N. Gent and R. P. Petrich, Proc. R. Soc. Lond., *A310*, 433 (1969).
36. A. Ahagon and A. N. Gent, J. Polym. Sci., Polym. Phys. Ed., *13*, 1285 (1975).
37. W. M. Bright, in *Adhesion and Adhesives: Fundamentals and Practice*, J. Rutzler and R. L. Savage, eds., Wiley, New York, 1954, pp. 130–138.
38. D. H. Kaelble, J. Adhes., *1*, 102 (1969).
39. D. H. Kaelble and F. A. Hamm, Adhes. Age, *11*(7), 25 (1968).
40. E. H. Andrews and A. J. Kinloch, Proc. R. Soc. Lond., *A332*, 401 (1973).
41. G. A. Ilkka and R. L. Scott, in *Adhesion and Cohesion*, P. Weis, ed., Elsevier, Amsterdam, 1962, pp. 65–73.
42. L. Reegan and G. A. Ilkka, in *Adhesion and Cohesion*, P. Weiss, ed., Elsevier, Amsterdam, 1962, pp. 159–75.
43. C. L. Weidner, Adhes. Age, *6*(7), 30 (1963).
44. J. R. Huntsberger, J. Polym. Sci., *A1*, 1339 (1963).
45. R. J. Good, in *Recent Advances in Adhesion*, L. H. Lee, ed., Gordon and Breach, New York, 1973; pp. 357–378; also in ASTM STP *640*, 18–29 (1978).

46. C. A. Dahlquist, in *Aspects of Adhesion* Vol. 5, D. J. Alner, ed., CRC Press, Cleveland, Ohio, 1969, pp. 183–201.
47. J. J. Bikerman, J. Colloid Sci., *2*, 163 (1947).
48. J. J. Bikerman, in *Recent Advances in Adhesion*, L. H. Lee, ed., Gordon and Breach, New York, 1973, pp. 351–356.
49. J. J. Bikerman, Ind. Eng. Chem., *59*(9), 41 (1967).
50. H. Schonhorn, in *Adhesion Fundamentals and Practice*, Gordon and Breach, New York, 1969, pp. 12–21.
51. H. Schonhorn and R. H. Hansen, in *Adhesion, Fundamentals and Practice*, Gordon and Breach, New York, 1969, pp. 22–28.
52. R. E. Robertson, J. Adhes., *7*, 121 (1975).
53. N. J. DeLollis and O. Montoya, J. Adhes., *3*, 57 (1971).
54. S. Wu, J. Adhes., *5*, 39 (1973).
55. W. Lang, *Nomarsky Differential Interference Contrast Microscopy*, S-41-210.2-5-3, Carl Zeiss, 7082 Oberkochen, West Germany.
56. D. W. Dwight, M. E. Counts, and J. P. Wightman, in *Colloid and Interface Science*, Vol. 3, M. Kerker, ed., Academic Press, New York, 1976, pp. 143–156.
57. R. L. Patrick, W. G. Gehman, L. Dunbar, and J. A. Brown, J. Adhes., *3*, 165 (1971).
58. R. L. Patrick, J. A. Brown, N. M. Cameron, and W. G. Gehman, Appl. Polym. Symp., *16*, 87 (1971).
59. R. L. Patrick, in *Treatise on Adhesion and Adhesives*, Vol. 3, R. L. Patrick, ed., Marcel Dekker, New York, 1973, pp. 163–230.
60. J. I. Goldstein and H. Yakowitz, ed., *Practical Scanning Electron Microscopy*, Plenum Press, New York, 1975.
61. N. J. Harrick, *Internal Reflection Spectroscopy*, Wiley, New York, 1967.
62. N. J. Harrick, in *Characterization of Metal and Polymer Surfaces*, Vol. 2, L. H. Lee, ed., Academic Press, New York, 1977, pp. 193–206.
63. J. K. Barr and P. A. Flournoy, in *Physical Methods in Macromolecular Chemistry*, Vol. 1, B. Carroll, ed., Marcel Dekker, New York, 1969, pp. 109–164.
64. D. L. Allara, in *Characterization of Metal and Polymer Surfaces*, Vol. 2, L. H. Lee, ed., Academic Press, New York, 1977, pp. 193–206.
65. T. A. Carlson, *Photoelectron and Auger Spectroscopy*, Plenum Press, New York, 1975.
66. D. M. Hercules, in *Characterization of Metal and Polymer Surfaces*, Vol. 1, L. H. Lee, ed., Academic Press, New York, 1977, pp. 399–430.
67. D. T. Clark, in *Characterization of Metal and Polymer Surfaces*, Vol. 2, L. H. Lee, ed., Academic Press, New York, 1977, pp. 5–52.

References

68. W. L. Baun in *Adhesion Measurement of Thin Films, Thick Films and Bulk Coatings*, ASTM STP 640, K. L. Mittal, ed., American Society for Testing and Materials, Philadelphia, 1978, pp. 41–53.
69. R. W. Phillips, J. Colloid Interface Sci., **47**, 687 (1974).
70. H. Yasuda, H. C. Marsh, S. Brandt, and C. N. Reilley, J. Polym. Sci., Polym. Chem. Ed., **15**, 991 (1977).
71. T. Rhodin and C. Brucker, in *Characterization of Metal and Polymer Surfaces*, Vol. 1, L. H. Lee, ed., Academic Press, New York, 1977, pp. 431–466.
72. J. T. Grant, in *Characterization of Metal and Polymer Surfaces*, Vol. 1, L. H. Lee, ed., Academic Press, New York, 1977, pp. 133–154.
73. I. M. Stewart, in *Characterization of Metal and Polymer Surfaces*, Vol. 1, L. H. Lee, ed., Academic Press, New York, 1977, pp. 127–132.
74. A. J. Tousimis and L. Marton, eds., "Electron Probe Microanalysis," *Advances in Electronics and Electron Physics*, Suppl. 6, Academic Press, New York, 1969.
75. G. A. Hutchins, "Electron Probe Microanalysis," *Characterization of Solid Surfaces*, P. F. Kane and G. B. Larrabee, eds., Plenum Press, New York, 1974.
76. G. H. Morrison, in *Characterization of Metal and Polymer Surfaces*, Vol. 1, L. H. Lee, ed., Academic Press, New York, 1977, pp. 351–366.
77. I. M. Stewart, in *Characterization of Metal and Polymer Surfaces*, Vol. 1, Metal Surfaces, L. H. Lee, ed., Academic Press, New York, 1977, pp. 367–374.
78. J. A. Hugh, "Secondary Ion Mass Spectrometry," *Methods and Phenomena of Surface Analysis*, S. P. Wolsky and A. W. Czanderna, eds., Elsevier, Amsterdam, 1975.
79. C. A. Evans, Jr., Anal. Chem., **44**(13), 67A (1972).
80. G. Slodzian, Surf. Sci., **48**, 161 (1975).
81. R. E. Honig, in *Advances in Mass Spectrometry*, Vol. 6, A. R. West, ed., Applied Science Publishers, London, 1974, pp. 337–364.
82. G. H. Morrison and G. Slodzian, Anal. Chem., **47**(11), 932A (1975).
83. J. L. McCall and W. Mueller, eds., *Microstructural Analysis—Tools and Techniques*, Plenum Press, New York, 1973.
84. W. C. McCrone and J. G. Delly, eds., *The Particle Atlas*, 2nd ed., Vol. 2, Ann Arbor Science Publishers, Ann Arbor, Mich., 1973, pp. 168–178.
85. W. L. Baun, in *Characterization of Metal and Polymer Surfaces*, Vol. 1, L. H. Lee, ed., Academic Press, New York, 1977, pp. 375–390.

86. C. A. Evans, Jr., Anal. Chem., 47(9), 855A (1975).
87. G. W. Simmons and H. Leidheiser, Jr., in *Characterization of Metal and Polymer Surfaces*, Vol. 1, L. H. Lee, ed., Academic Press, New York, 1977, pp. 49–64.
88. E. W. Muller and S. V. Krishnaswamy, in *Characterization of Metal and Polymer Surfaces*, Vol. 1, L. H. Lee, ed., Academic Press, New York, 1977, pp. 21–48.
89. P. J. Estrup, in *Characterization of Metal and Polymer Surfaces*, Vol. 1, L. H. Lee, eds., Academic Press, New York, 1977, pp. 187–210.
90. R. D. Andrews and T. R. Hart, in *Characterization of Metal and Polymer Surfaces*, Vol. 2, L. H. Lee, ed., Academic Press, New York, 1977, pp. 207–240.

11
Formation of Adhesive Bond

11.1. ELEMENTARY PROCESSES IN ADHESIVE BOND FORMATION

The first step in the formation of an adhesive bond is the establishment of interfacial molecular contact by wetting. The molecules will then undergo motions toward preferred configurations to achieve the adsorptive equilibrium, diffuse across the interface to form a diffuse interfacial zone, and/or react chemically to form primary chemical bonds across the interface.

The van der Waals force diminishes rapidly with distance, varying with the inverse seventh power of distance between two molecules and with the inverse cube of distance between two macroscopic plates (Chapter 2). Appreciable attractions are obtained only when the distance of separation is at or near an equilibrium intermolecular distance on the order of 5Å.

The extent of molecular or local segmental diffusion across the interface determines the structure of the interfacial zone, which critically affects the mechanical strength of an adhesive bond. Negligible diffusion will give a sharp interface. In this case, if the interfacial attraction consists mainly of dispersion force, the adhesive strength will be low because of interfacial molecular slippage under applied stress. For a sharp interface, high adhesive strength can be obtained only when strong polar interactions or chemical bonds exist across the interface. On the other hand, if the interface is sufficiently diffuse as a result of extensive diffusion, dispersion force alone will give high adhesive strength.

11.2. WETTING AND ADHESION

Wetting can affect adhesion in two ways. First, incomplete wetting will produce interfacial defects and thereby lower the adhesive bond strength. Second, better wetting can increase the adhesive bond strength by increasing the work of adhesion, which is directly proportional to the fracture energy. These two views are analyzed below.

11.2.1. Wetting and Interfacial Defect

Stress concentration will occur around an unwetted interfacial defect as an adhesive bond is stressed [1,2]. When the local stress exceeds the local strength, the adhesive bond will fracture. In other words, interfacial defects may act as the critical flaw. Thus, wetting can affect the mechanical strength of an adhesive bond.

When an adhesive is applied on an adherend, many microscopic unwetted voids are formed at the interface, as real surfaces are almost never perfectly smooth. Random flaws may also arise from air entrapment or local material inhomogeneity. The driving force for the wetting of these interfacial voids is the spreading coefficient [3,4].

$$\lambda_{12} = \gamma_2 - \gamma_1 - \gamma_{12} \tag{11.1}$$

where λ_{12} is the spreading coefficient of phase 1 (the adhesive) on phase 2 (the adherend). The size of unwetted interfacial defect c is related to λ_{12} by [4]

$$c = c_0 \left(1 - \frac{\lambda_{12}}{\gamma_2}\right)^n \tag{11.2}$$

where c_0 is the size of the unwetted interfacial void when $\lambda_{12} = 0$ and n is a positive constant which has a value of 2 in most cases. Applying Eq. (11.2) in Eq. (10.5) gives [4]

$$\sigma_f = k \left(\frac{EG}{c_0}\right)^{1/2} \left(1 - \frac{\lambda_{12}}{\gamma_2}\right)^{-n/2} \tag{11.3}$$

where σ_f is the fracture strength, $k = (1/\pi)^{1/2}$ for plane stress or $[\pi(1 - \nu^2)]^{-1/2}$ for plane strain, and E and G are the elastic modulus and fracture energy of the system, respectively. Thus, wettability is directly related to the mechanical strength of an adhesive bond. The adhesive bond strength is affected not only by the wettability, but also by the mechanical properties.

The effect of wettability on adhesive bond strength can better be visualized by rewriting Eq. (11.3) as

$$\sigma_f = \frac{K_m}{1 - (\lambda_{12}/\gamma_2)} \tag{11.4}$$

$$= \frac{K_m \gamma_2}{\gamma_1 + \gamma_{12}} \tag{11.5}$$

where K_m is a function of the mechanical properties and we have used $n = 2$. The results discussed above can be tested experimentally as described below.

Case I: Different Adhesives on Different Adherends

According to Eq. (11.4), the fundamental wettability factor controlling the adhesive bond strength is the quantity $[1 - (\lambda_{12}/\gamma_2)]^{-1}$, or alternatively λ_{12}. Provided that the mechanical parameter K_m is more or less constant, a plot of σ_f versus $[1 - (\lambda_{12}/\gamma_2)]^{-1}$ for a series of adhesive bonds should give a straight line. This is shown for various adhesives on various adherends, bonded at 140°C, in Table 11.1 and

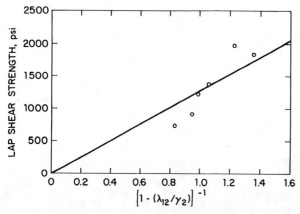

Figure 11.1. Lap shear strength versus $[1 - (\lambda_{12}/\gamma_2)]^{-1}$ for various polymer pairs; plotted from Table 11.1.

Table 11.1. Correlation Between Shear Strength and Spreading Coefficient for Various Polymer Pairs

Polymer pair Phase 1	Phase 2	Shear strength, psi	γ_1 at 140°C, dyne/cm	γ_2 at 140°C, dyne/cm	γ_{12} at 140°C, dyne/cm	λ_{12} at 140°C, erg/cm^2	$\left(1 - \dfrac{\lambda_{12}}{\gamma_2}\right)^{-1}$	W_a at 140°C, erg/cm^2
Polyethylene	Poly(methyl methacrylate)	750	28.8	32.0	9.7	−6.5	0.831	51.1
Poly(methyl methacrylate)	Polystyrene	950	32.0	32.1	1.7	−1.6	0.952	62.4
Poly(vinyl acetate)	Polystyrene	1250	28.6	32.1	3.7	−0.2	0.994	57.0
Poly(n-butyl methacrylate)	Poly(vinyl acetate)	1400	24.1	28.6	2.9	+1.6	1.059	49.8
Poly(n-butyl methacrylate)	Poly(methyl methacrylate)	2000	24.1	32.0	1.9	+6.0	1.230	54.2
Polydimethylsiloxane	Polychloroprene	1850	14.1	33.2	6.5	+12.0	1.565	40.8

Column 1: By definition, phase 1 has lower surface tension than phase 2.
Column 2: Lap joints bonded at 140°C for 15 min.
Columns 3, 4, 5, 6, and 8: λ_{12}, γ_{12}, and W_a are measured directly at 140°C for the polymer melts by the pendent drop method. The surface tensions accurate within 0.5 dyne/cm; the interfacial tensions within 0.1 dyne/cm.

Wetting and Adhesion

Figure 11.1. The scatter of the data probably arises from the variability of the K_m values. The interfacial and surface tensions used are accurately measured directly at the bonding temperature of 140°C by the pendent drop method [3–8]. On the other hand, no correlations exist between the bond strength and the interfacial tension or the work of adhesion, indicating that wetting of interfacial void is the controlling factor in these instances.

Case II: A Given Adhesive on a Series of Adherends

Here, γ_1 is constant. Since $\gamma_{12} \ll \gamma_1$ usually, we have, from Eq. (11.5),

$$\sigma_f \simeq k_1 \gamma_2 = k_1 W_a - k_2 \tag{11.6}$$

where k_1 and k_2 are constants. Furthermore, since $W_a = (1 + \cos\theta_{12})\gamma_1$, Eq. (11.6) can be recast as

$$\sigma_f = k_3(1 + \cos\theta_{12}) - k_2 \tag{11.7}$$

where θ_{12} is the contact angle of the adhesive (phase 1) on the adherend (phase 2). Thus, provided that the mechanical parameter K_m is constant, the bond strength should vary linearly with γ_2, W_a, or $\cos\theta_{12}$. Several experimental confirmations are given below.

Various polymer adherends are bonded with an epoxy adhesive ($\gamma_1 = 50$ dyne/cm). The tensile butt strength correlates with the (critical) surface tension of the adherends and the work of adhesion [9] (Table 11.2). The bond strength is directly proportional to the (critical) surface tension, consistent with Eq. (11.6) (Figure 11.2). Alternatively, the bond strength also varies linearly with W_a (Figure 11.3). The straight line intersects $\sigma_f = 0$ at W_a, consistent with Eq. (11.6).

The peel strength of a rubber adhesive (pressure-sensitive adhesive tape, $\gamma_1 = 20$ dyne/cm) on various adherends correlates with the work of adhesion [10,11] (Table 11.3 and Figure 11.4). The straight line intersects $\sigma_f = 0$ at W_a, as predicted by Eq. (11.6). The work of peel W_p, calculated as $W_p = (P/b)(1 - \cos\theta_p)$, where P is the peel force, b the specimen width, and θ_p the peel angle, is many orders of magnitude greater than the work of adhesion, because of large viscoelastic dissipation in deforming the adhesive.

Correlations between the shear strength and $\cos\theta_{12}$ for an epoxy adhesive on various surface-treated polyethylenes and aluminum alloys are shown in Table 11.4 and Figure 11.5 [12,13]. The nonpolar component of surface tension γ_2^d should be roughly identical for untreated and treated adherends. Plots of σ_f/γ_2^d versus $\cos\theta_{12}$ gives a single straight line for both the polyethylene adherends and the

Table 11.2. Correlation Between Tensile Butt Strength and Interfacial Properties for an Epoxy Adhesive on Various Substrates

Adherend	Bond strength, psi	γ_c, dyne/cm	γ_{12}, dyne/cm	Work of adhesion, W_a, erg/cm^2
Poly(ethylene terephthalate)	2580	43	3.45	89.6
Poly(1,4-cyclohexylene dimethyl terephthalate)	2600	43	3.45	89.6
Poly(vinylidene chloride)	1900	40	4.77	85.2
Poly(vinyl chloride)	1920	40	4.36	85.6
Poly(vinyl alcohol)	1650	37	6.71	80.3
Polystyrene	1100	33	7.67	75.3
Poly(vinyl fluoride)	1320	28	10.05	67.9
Polytetrafluoroethylene	350	18.5	13.9	54.6

Source: From Ref. 9.

Wetting and Adhesion 365

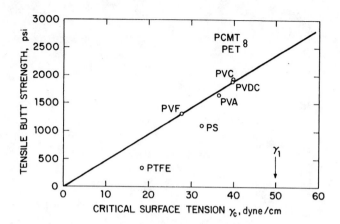

Figure 11.2. Tensile butt strength versus surface tension of adherend, bonded with an epoxy adhesive. (After Ref. 9.)

aluminum adherends [14], shown in Figure 11.5. The fundamental significance of this superposition is not known.

A correlation between the shear strength of an epoxy adhesive on various chromic acid-treated polyethylenes and the contact angle of water θ_w is reported [15]. The polar component γ_2^d can be calculated by the harmonic-mean equation [7],

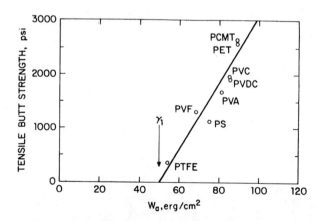

Figure 11.3. Tensile butt strength versus work of adhesion for an epoxy adhesive on various polymer adherends; plotted from Table 11.2.

Table 11.3. Correlation Between Peel Strength and Interfacial Properties for a Rubber Adhesive on Various Adherends [a]

Adherend	Peel strength, g/cm	Work of peel, erg/cm²	γ_c, dyne/cm	Work of adhesion, W_a, erg/cm²
Poly(ethylene terephthalate)	480	470,000	43	57.5
Poly(vinyl chloride)	457	450,000	39	54.7
Polystyrene	433	420,000	33	50.8
Polychlorotrifluoroethylene	370	360,000	31	46.3
Polyethylene	287	280,000	31	49.3
Polytetrafluoroethylene	228	220,000	18.5	37.7
Cellulose triacetate	669	660,000	48.8	66.8
Cellulose tricaprylate	358	360,000	20.9	39.8
Cellulose trilaurate	307	310,000	20.2	38.6
Cellulose tristearate	287	280,000	19.0	37.4

[a] Peel rate is 100 cm/min; γ_c of the rubber adhesive is 20 dyne/cm.
Source: From Refs. 10 and 11.

Table 11.4. Correlation of Shear Strength with Interfacial Properties for an Epoxy Adhesive on Surface-Treated Polyethylene and Aluminum [a]

Surface treatment	Lap shear strength, σ_f, psi	θ_{12}, deg	$\cos \theta_{12}$	W_a, erg/cm^2	σ_f / γ_2^d
Polyethylene, γ_2^d = 36 dyne/cm					
None	54	35.4	0.815	75.7	1.50
Acetone wipe (AW)	72	33.8	0.831	76.5	2.00
AW + 20 min 23°C acid	144	26.5	0.895	79.0	4.00
AW + 60 min 23°C acid	182	22.0	0.927	80.4	5.05
AW + 1 min 71°C acid	202	20.4	0.937	80.8	5.61
AW + 5 min 71°C acid	224	18.5	0.948	81.2	6.22
Aluminum alloy 2024, γ_2^d = 293 dyne/cm (for alumina)					
None	1252	27.0	0.891	78.9	4.26
Acetone wipe (AW)	1584	19.8	0.941	80.9	5.41
AW + 5 min 23°C acid	1610	18.8	0.947	81.2	5.50
AW + 15 min 23°C acid	1916	12.4	0.977	82.4	6.54
AW + 7 min 50°C acid	2234	6.5	0.994	83.1	7.62

[a] The epoxy adhesive is a diglycyl ether of bisphenol cured with a polyamide. The uncured liquid epoxy adhesive has a surface tension of 41.7 dyne/cm. θ_{12} is the contact angle of the adhesive on the adherend.

Source: From Refs. 12 and 13.

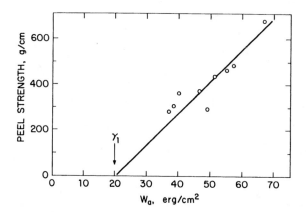

Figure 11.4. Peel strength versus work of adhesion for a rubber adhesive on various adherends; plotted from Table 11.3.

$$(1 + \cos \theta_w)\gamma_w = 4\left(\frac{\gamma_w^d \gamma_2^d}{\gamma_w^d + \gamma_2^d} + \frac{\gamma_w^p \gamma_2^p}{\gamma_w^p + \gamma_2^p}\right) \quad (11.8)$$

where $\gamma_w^d = 22.1$ dyne/cm, $\gamma_w^p = 50.7$ dyne/cm, and γ_2^d is assumed to be constant and equal to the value for the untreated polyethylene (36.0 dyne/cm). Thus, the increased shear strength is seen to arise from increased adherend polarity, shown in Table 11.5 and Figure 11.6.

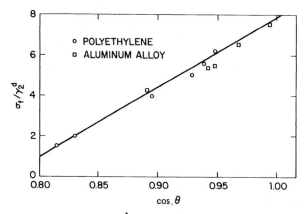

Figure 11.5. σ_f/γ_2^d versus $\cos \theta$ for an epoxy adhesive on surface-treated polyethylene and aluminum alloy; plotted from Table 11.4.

Table 11.5. Correlation Between Lap Shear Strength and Wettability for an Epoxy Adhesive on Chromic Acid-Treated Polyethylene

θ_{H_2O}	Shear strength, σ_f, psi	γ_2^d, dyne/cm	γ_2^p, dyne/cm	γ_2, dyne/cm
89 (Untreated)	10	36.0	5.31	41.3
55	175	36.0	20.9	56.3
51	187	36.0	23.5	59.5
50	240	36.0	23.5	59.5
46	290	36.0	25.7	61.7
42	440	36.0	28.0	64.0
38	575	36.0	29.9	65.9
38	600	36.0	29.9	65.9
36	700	36.0	31.2	67.2
35	675	36.0	31.2	67.2

Note: θ_{H_2O} is the contact angle of water on the untreated or chromic acid-treated polyethylene (treatment time up to 15 min). σ_f versus θ_{H_2O} data from Ref. 15.

The peel strengths of a styrene-butadiene rubber on various silicone rubbers have been found to correlate with the spreading coefficient. The release property of the silicones improves with decreasing spreading coefficient [16].

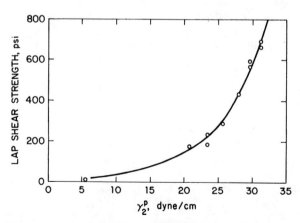

Figure 11.6. Lap shear strength versus polar component of the surface tension of treated polyethylene adherends; plotted from Table 11.5.

Case III: Various Adhesives on a Given Adherend

Here, γ_2 is constant, and Eq. (11.4) becomes

$$\sigma_f = \frac{k_4}{\gamma_1 + \gamma_{12}} = \frac{k_4}{\gamma_1} \qquad (11.9)$$

where $k_4 = K_m \gamma_2$, a constant. Thus, the bond strength should vary inversely with the surface tension of the adhesive, provided that K_m is constant. Few such experiments have been reported.

11.2.2. Wetting and Fracture Energy

In the preceding section, the effect of wetting on adhesion is analyzed in terms of the effect of wetting on interfacial defect. However, wetting may also affect adhesion through its effect on fracture energy. As discussed in Section 14.4.3, the fracture energy is directly proportional to the work of adhesion. From Eq. (14.51), the fracture energy for interfacial separation is given by

$$G = W_a \psi(Ra_T) \qquad (11.10)$$

where G is the fracture energy, W_a the work of adhesion, $\psi(Ra_T)$ the viscoelastic dissipation function, and Ra_T the effective rate. At zero rate, viscoelastic effects will be absent, and the separation process will be reversible. Therefore, the equilibrium fracture energy is equal to the work of adhesion, that is, $\psi(Ra_T) = 1$, when $Ra_T = 0$.

Thus, the adhesive bond strength should increase with increasing work of adhesion. For a given adhesive on a series of different rigid adherends, the viscoelastic function $\psi(Ra_T)$ should be nearly constant, and the adhesive bond strength should therefore be porportional to the work of adhesion. This is indeed true, as shown for the butt strength of an epoxy adhesive on various adherends (Table 11.2 and Figure 11.3) and for the peel strength of a rubber adhesive on various adherends (Table 11.3 and Figure 11.4). The importance of work of adhesion has also been emphasized elsewhere [17].

However, the fracture energy and the interfacial defect models are not always compatible. For instance, the nonreciprocal behavior, sometimes observed and discussed in Section 11.2.4, is predicted by the interfacial defect model but not by the fracture energy model. Apparently, different mechanisms may predominate in different situations.

Wetting and Adhesion

11.2.3. The Optimum Wettability Condition for Adhesion

Both the interfacial defect model and the fracture energy model predict that the optimum condition for wetting and adhesion is the matching of the polarities of the two phases. Consider that we have a given adherend, having a given surface tension γ_2 and polarity x_2^p. What surface properties should the adhesive have so that the wetting and adhesion are maximized?

The interfacial defect model indicates that the spreading coefficient should be maximized; the fracture energy model indicates that the work of adhesion should be maximized. The spreading coefficient and the work of adhesion are given by

$$\lambda_{12} = 2\phi(\gamma_1\gamma_2)^{1/2} - 2\gamma_1 \tag{11.11a}$$

$$W_a = 2\phi(\gamma_1\gamma_2)^{1/2} \tag{11.11b}$$

both of which are maximized when the interaction parameter ϕ has its maximum value. Letting

$$\frac{\partial \phi}{\partial x_1^p} = 0$$

at constant γ_1/γ_2 and x_2^p, it can be shown that the ϕ has the maximum value of $\phi_{max} = 2(g_1 + g_2)^{-1}$ for the harmonic-mean equation or $\phi_{max} = 1$ for the geometric-mean equation, when the polarities of the two phases are exactly matched (see Section 3.2.2), that is,

$$x_1^p = x_2^p \quad \text{(condition I)} \tag{11.12a}$$

Thus, the optimum condition for adhesion based on wettability is that the polarities of the adhesive and the adherend are equal [3]. This explains de Bruyne's empirical "rule" that polar/nonpolar pairs "never" form strong bonds.

On the other hand, what is the optimum surface tension for the adhesive? To answer this, consider that we have a given adherend having given surface tension γ_2 and polarity x_2^p, and the interaction parameter ϕ is constant. Increasing the adhesive surface tension will decrease the spreading coefficient but increase the work of adhesion. Thus, the interfacial defect model and the fracture energy model give antagonistic predictions. The optimum condition must then be a compromise between the two mechanisms. This optimum condition should correspond to the situation where spontaneous spreading occurs, that is, $\lambda_{12} = 0$, when

$$\gamma_1 = \phi^2 \gamma_2 \quad \text{(condition II)} \tag{11.12b}$$

and the corresponding work of adhesion is $W_a = 2\phi^2 \gamma_2$. Thus, Eqs. (11.12a) and (11.12b) define the optimum wettability condition for adhesion.

Several experimental observations appear to support the foregoing predictions. Two such examples are given in Figure 11.7, one for poly(vinyl chloride-vinyl acetate) adhesive on various surface-modified steel adherends, and the other for a polyurethane adhesive on various polymer adherends [18]. Plots of bond strength versus surface tension of adherends give curves with maximum bond strengths occurring when the surface tensions of the adhesive and the adherend are nearly equal, that is, $\gamma_1 \sim \gamma_2$. Other examples are known [19–24].

It is interesting to note that the optimum wettability conditions given by Eqs. (11.12a) and (11.12b) are similar to the optimum condition predicted by consideration of the diffusion process of bond formation [25,26], discussed in Section 11.3.5.

Interfacial tension between adhesive and adherend has also been sug-suggested as the most important wettability criterion for adhesion [9,19,20]. This should, however, be regarded as a special case of the more general criteria discussed above. It is also interesting to note that Table 11.1 shows that no correlation exists between the interfacial tension and bond strength in the particular cases examined.

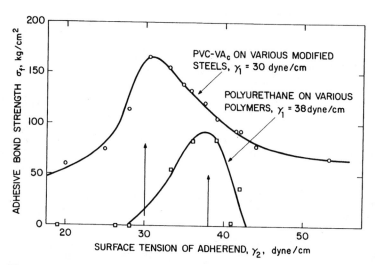

Figure 11.7. Adhesive bond strength versus surface tension of adherend for a poly(vinyl chloride-vinyl acetate) adhesive on various surface-treated steels and a polyurethane adhesive on various polymer adherends. Plotted from Ref. 18.

Wetting and Adhesion

11.2.4. Nonreciprocal Behavior

The interfacial defect model predicts a nonreciprocal adhesive behavior; that is, the adhesion between two phases depends on which phase is used as the adhesive and which phase is used as the adherend. The nonreciprocal behavior is sometimes observed in bonding polymer pairs, [27,28]. In contrast, the fracture energy model does not predict such a behavior.

It can be shown that if phase a spreads on phase b, phase b will not spread on phase a. Therefore, the interfacial defect model predicts that if phase a (used as the adhesive) adheres strongly to phase b (used as the adherend), phase b (used as the adhesive) will adhere only weakly to phase a (used as the adherend). Let σ_{fij} be the bond strength for phase i (adhesive) on phase j (adherend); then we have

$$\frac{\sigma_{fab}}{\sigma_{fba}} = \frac{\gamma_b(\gamma_b + \gamma_{ab})}{\gamma_a(\gamma_a + \gamma_{ab})} \sim \frac{\gamma_b}{\gamma_a} \qquad (11.13)$$

which follows from Eq. (11.5). The viscoelastic term K_m exactly cancels out when adhesive joints of the same configuration are used in both cases.

Two examples of nonreciprocal behavior are given in Table 11.6. In the first example, polyethylene (mp 130°C) is first used as the adhesive, and hot-pressed at 150°C between cross-linked epoxy-amine supported on aluminum adherends to form lap joints. Next, uncured liquid epoxy-amine is used as the adhesive supported on aluminum adherends, and cured by hot-pressing at 124°C against solid polyethylene film [28]. In the second example, poly(methyl methacrylate) is first used as the adhesive, spread from a solution onto cross-linked alkyd resin supported on aluminum adherends, and hot-pressing at 150°C [27]. The ratio of the bond strength $\sigma_{fab}/\sigma_{fba}$ agrees used as the adhesive, spread from a solution onto poly(methyl methacrylate) supported on aluminum adherends, and then cured by hot pressing at 150°C. The ratio of the bond strength $\sigma_{fab}/\sigma_{fba}$ agrees with the ratio of the surface tension γ_b/γ_a, as evident in Table 11.6, lending support to the interfacial defect model. Other polymer pairs exhibiting nonreciprocal behavior include poly(vinyl acetate) and polystyrene, poly(vinyl acetate) and poly(n-butyl methacrylate), and so on. In these examples, the first polymer adheres to the second, but not vice versa. The nonreciprocal behavior is, on the contrary, not predicted by the fracture energy model.

However, not all polymer pairs exhibit nonreciprocal behavior. That is, many polymer pairs exhibit similar bond strength regardless of which polymer is used as the adhesive and which polymer is used as the adherend. Examples of reciprocal polymer pairs include poly(methyl methacrylate) and poly(vinyl acetate), poly(methyl meth-

Table 11.6. Comparison of Experimental and Predicted (by the Interfacial Void Model) Nonreciprocal Adhesion Behavior

Phase a	Phase b	Shear strength of adhesive joint, psi		γ_a, dyne/cm	γ_b, dyne/cm	$\left(\dfrac{\gamma_b}{\gamma_a}\right)^2$	$\dfrac{\sigma_{fab}}{\sigma_{fba}}$
		σ_{fab}	σ_{fba}				
Polyethylene	Epoxy/amine	2400	750	31	47	2.3	3.2
Alkyd resin	Poly(methyl methacrylate)	1500	700	30	41	1.9	2.1

Source: From Ref. 27 and 28.

Wetting and Adhesion

acrylate) and polychloroprene, poly(vinyl acetate) and nitrocellulose, poly(methyl methacrylate) and poly(butyl methacrylate), and so on.

11.2.5. Kinetics of Bond Formation by Wetting

The rate of spontaneous wetting is given by [29-31]

$$\frac{d \cos \theta}{dt} = \frac{\gamma_1}{\eta_1 L}(\cos \theta_\infty - \cos \theta) \tag{11.14}$$

where γ_1 is the surface tension of the adhesive, η_1 the viscosity of the adhesive, L the kinetic length characteristic of the adhesive/adherend pair, θ_∞ the contact angle at infinite time, and θ the instantaneous contact angle at time t (Section 8.2.3). The quantity $(\gamma_1/\eta_1 L)$ is thus a wetting kinetic parameter, which sometimes controls the bond strength [32].

The rate of wetting of interfacial void has been given as [4]

$$c = c_\infty (1 - \alpha e^{-t/\tau})^{-2} \tag{11.15}$$

where c is the size of the interfacial void at time t, c_∞ that at infinite time, and α and τ are constants. Applying Eq. (11.15) in (11.4) gives

$$\frac{\sigma_{f\infty} - \sigma_{f0}}{\sigma_{f\infty} - \sigma_f} = \exp\left(\frac{t}{\tau}\right) \tag{11.16}$$

where σ_f is the adhesive bond strength at time t, $\sigma_{f\infty}$ that at infinite time, σ_{f0} that at time zero, and τ the retardation constant. Thus, bond strength develops by first-order kinetics. The rate constant k is the reciprocal of the retardation constant, that is, $k = 1/\tau$. The first-order kinetic plot for the bonding of a rubber adhesive tape on gold at 35°C is given in Figure 11.8 [33]. The peel strength increases rapidly with bonding time, then levels off to a plateau. The $(\sigma_{f\infty} - \sigma_{f0})/(\sigma_{f\infty} - \sigma_f)$ versus t plot is a straight line, conforming to Eq. (11.16), with a retardation constant of 51.3 min. Other examples include the bonding of poly(vinyl chloride) on steel, poly(vinyl butyral) on aluminum and glass, and benzyl cellulose on aluminum [34].

The retardation constant varies with temperature according to the Arrhenius relation [33,34],

$$\tau = B \exp\left(\frac{E_a}{RT}\right) \tag{11.17}$$

376   11 / Formation of Adhesive Bond

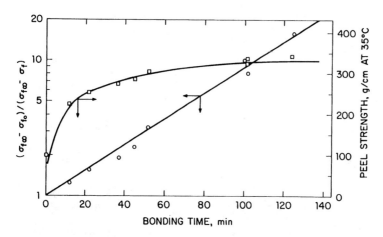

Figure 11.8. First-order kinetic plot for bonding of a rubber adhesive on gold at 35°C. (Data taken from Ref. 33.)

where B is a constant, E_a the activation energy for the adhesive bonding process, R the gas constant, and T the absolute temperature. Arrhenius plots for the bonding of poly(vinyl chloride) on steel, poly(vinyl chloride) plasticized with 34% dioctyl phthalate on steel, and a rubber adhesive on gold are shown in Figure 11.9. The E_a val-

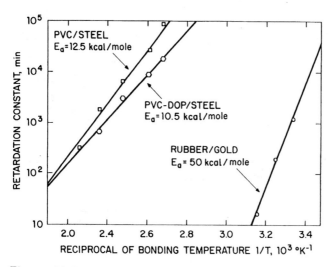

Figure 11.9. Retardation constant versus reciprocal of bonding temperature. (Data taken from Refs. 33 and 34.)

ues obtained are 12.5 kcal/mole for PVC/steel, 10.5 kcal/mole for PVC/steel, and 50 kcal/mole for rubber/gold. These are similar to those for the flow of polymers (Table 11.7), indicating that these bonding processes are controlled by wetting flow of the adhesives.

To a good approximation, the viscosity η of a polymer follows [35]:

$$\eta = C \exp\left(\frac{E_f}{RT}\right) \quad (11.18)$$

where C is a constant and E_f the flow activation energy, which is usually 5–50 kcal/mole. E_f is independent of shear stress and shear rate when evaluated with η at constant shear stress. But it tends to increase with decreasing shear rate when evaluated with η at a constant shear rate. E_f tends to be greater for large flow unit, bulky side groups, and rigid chains. Polydimethylsiloxane has the lowest known E_f (4 kcal/mole), because of its great chain flexibility [39,40].

Combining Eqs. (11.16) and (11.17) gives

$$\ln \frac{\sigma_{f_\infty} - \sigma_{f0}}{\sigma_{f_\infty} - \sigma_f} = \frac{t}{B} \exp\left(\frac{-E_a}{RT}\right) \quad (11.19)$$

which relates the adhesive bond strength σ_f (measured at a given temperature) to the bonding time t and the bonding temperature T. Thus, a plot of $\ln t - \ln \{\ln [(\sigma_{f_\infty} - \sigma_{f0})/(\sigma_{f_\infty} - \sigma_f)]\}$ versus $1/T$ should give a straight line. For short times $(\sigma_f - \sigma_{f_\infty}) \leq 0.4$, it simplifies to

$$\frac{\sigma_f}{\sigma_{f_\infty}} \sim \frac{t}{B} \exp\left(\frac{-E_a}{RT}\right) \quad (11.20)$$

Thus, a plot of log σ_f versus $1/T$ should give a straight line for short times, illustrated for a poly(vinyl chloride) adhesive on steel in Figure 11.10. The butt joints were prepared by sandwiching the adhesive film between the flat ends of two steel cylinders. The bonding time was 10 min at bonding temperatures between 130 and 170°C. The tensile shear strengths were measured at the room temperature [34].

Kinetic data should be interpreted with caution. Many different physical and chemical processes can occur simultaneously during adhesive bonding, including wetting, diffusion, chemisorption, crystallization, cross-linking, thermal degradation, and relaxation or buildup of internal stress. All can affect the bond strength. Activation energy alone does not provide an unambiguous clue to the predominant bonding process, as many different processes have similar activation energies. For instance, diffusion also has activation energies of 5–50 kcal/mol, just as for melt flow [41,42].

Table 11.7. Activation Energies for Polymer Flow and Adhesive Bonding

Polymer	Activation energy, kcal/mole	
	Polymer flow, E_f	Adhesive bonding, E_a
Poly(vinyl chloride)	25	12.5 (on steel)
Poly(vinyl chloride) plasticized with 34% dioctyl phthalate	33	10.5 (on steel)
Rubber adhesive (natural rubber, rosin, plasticizer and zinc oxide filler blend)	—	50 (on gold)
Polydimethylsiloxane	4	—
Polyethylene high density	6-7	—
low density	12	—

Polypropylene	9–10	—
Polybutadiene (cis)	5–8	—
Polyisobutylene	12–15	—
Poly(butadiene-styrene 75/25) GR-S or Buna S rubber	21	—
Poly(acrylonitrile-butadiene styrene), ABS 30% rubber	24	—
Polystyrene	25	—
Poly(α-methyl styrene)	32	—
Poly(vinyl butyral)	26	—
Poly(ethylene terephthalate)	19	—
Polycarbonate	26–30	—

Source: Data collected from Refs. 33–40.

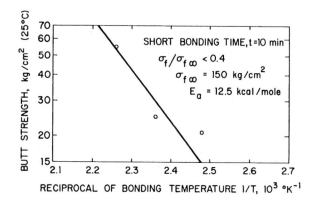

Figure 11.10. Kinetics of adhesive bond formation at short times according to Eq. (11.20) for a poly(vinyl chloride) adhesive on steel. (Data from Ref. 34.)

11.3. DIFFUSION AND ADHESION

11.3.1. Wetting-Diffusion Process

As two phases attain molecular contact by wetting, segments of the macromolecules will diffuse across the interface to various extents, depending on material properties and bonding condition. Consider the autohesion of simple liquids and polymers (Figure 11.11). For simple liquids, wetting and diffusion occur simultaneously; the interface disappears instantaneously upon interfacial contact. For polymers, however, autohesion is a two-stage process; wetting is followed by interdiffusion. The autohesion will be complete only after extensive interdiffusion of chain segments across the interface to reestablish the entangled network.

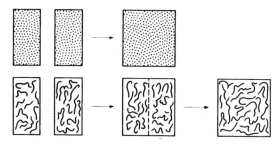

Figure 11.11. Schematics of autohesion of simple liquids (top) and polymers (bottom). (After Ref. 43.)

Diffusion and Adhesion

Dissimilar polymers are usually incompatible, and their diffusion coefficients quite small. Thus, interdiffusion of entire macromolecules across the interface is unlikely between dissimilar polymers However, both theoretically and experimentally, local segmental diffusion is known to occur readily, forming a diffuse interfacial layer of 10–1000 Å between two incompatible polymers. Such local segmental diffusion is favored thermodynamically as interfacial free enrgy is minimized by limited interdiffusion, and kinetically possible segmental movement is confined locally. Interdiffusion is an important process in adhesive bonding, since the diffuseness of the interface greatly affects the mechanical strength of the interface. Some experimental evidence of the wetting-diffusion process in autohesion and adhesion is given below.

Effect of Bonding Temperature on Autohesion of Poly(methyl methacrylate)

The effect of bonding temperature on the autohesion of poly(methyl methacrylate) appears to illustrate the two-stage wetting-diffusion process [43–45] (Figure 11.12). Two pieces of plasticized or unplasticized poly(methyl methacrylate) are pressed together at various bonding temperatures for a constant bonding time of 15 min and a bonding pressure of 40 kg/cm^2. The bond strengths of the autohesive bonds are then measured at the room temperature. The bond strength versus bonding temperature plots show two plateaus. The bonds fail

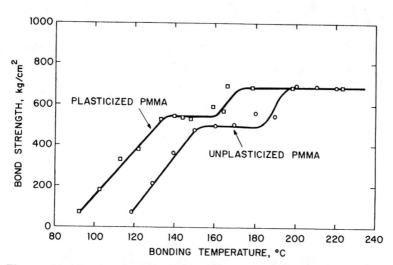

Figure 11.12. Autohesion of plasticized and unplasticized poly(methyl methacrylate) as a function of bonding temperature at a constant bonding time of 15 min and bonding pressure of 40 kg/cm^2. (After Refs. 43–45.)

interfacially on or below the first plateau, and fail cohesively on the second plateau. Thus, apparently, the first plateau corresponds to the attainment of complete wetting, and the second plateau to the attainment of equilibrium interdiffusion.

Effect of Bonding Time on Autohesion of Butadiene-Acrylonitrile Rubber

The bond strength continues to increase significantly with bonding time after the interfacial contact is complete in the autohesion of SKN-40 butadiene-acrylonitrile rubber [46] (Figure 11.13). The large continued increase of the bond strength is attributed to the interdiffusion of molecular segments, which further increases the interfacial strength. An optical method (based on the distortion of internal reflection of light at the interface) is used to measure the interfacial contact area [47–49]. The resolution of such a method, however, does not exceed half the wavelength of visible light (the resolution is about 2000–4000 Å).

Effect of Bonding Time on Adhesion of Butadiene-Acrylonitrile Rubber on Poly(vinyl alcohol)

The adhesive bond strength for SKN-40 butadiene-acrylonitrile rubber on a carbon black-filled poly(vinyl alcohol) continues to increase significantly with contact time after the interfacial contact is complete [46]

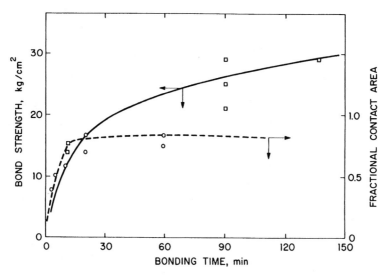

Figure 11.13. Autohesion of (SKN-40) butadiene-acrylonitrile rubber. (After Ref. 46.)

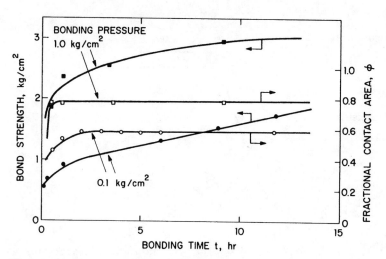

Figure 11.14. Bond strength and contact area versus bonding time for the adhesion of (SKN-40) butadiene-acrylonitrile rubber to carbon black-filled poly(vinyl alcohol). (After Ref. 46.)

(Figure 11.14). The rubber adhesive is pressed onto the adherend under a constant pressure of 0.1 or 1.0 kg/cm^2 at the room temperature. The interfacial contact area is measured by an optical distortion method on a device which also allows simultaneous measurement of the shear strength of the adhesive bond. The continued rise of the adhesive bond strength with contact time after the completion of the wetting process is attributed to the segmental interdiffusion of the adhesive and the adherend.

The rate of interfacial contact follows [50]

$$\phi + \ln(1 - \phi) = -\frac{Pt}{\eta} \qquad (11.21)$$

where ϕ is the fractional interfacial contact area, P the applied pressure, t the time, and η the viscosity of the adhesive. The rate of increase of adhesive bond strength conforms to first-order kinetics.

Effect of Bonding Temperature on the Adhesion of
Polyolefins to Butyl Rubber

Polyolefins (such as polyethylene and polypropylene) adhere weakly to a butyl rubber with apparent interfacial failure when bonded below the melting points of the polyolefins (T_m = 135°C for PE and 175°C for PP), but adhere strongly to the butyl rubber with cohesive failure when bonded above the melting points [51] (Figure 11.15). The peel

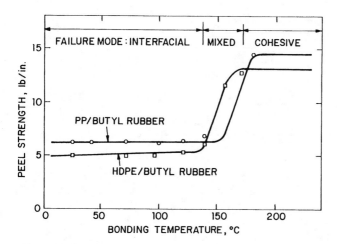

Figure 11.15. Peel strength (23°C) versus bonding temperature for polypropylene/butyl rubber and polyethylene/butyl rubber joints. (From Ref. 50.)

strength increases sharply near the respective melting point, accompanied by transition of failure mode from apparent interfacial failure to cohesive failure. Investigations of the interfaces by Fourier-transform internal reflection IR spectroscopy and interference microscopy show that the sharp increase in peel strength at the melting points of the polyolefins is associated with the formation of a diffuse interface. Similar behavior has also been observed with autohesive and adhesive bonding of several plasma-treated polymers (Section 9.3.6).

11.3.2. Structure of Diffuse Interface

Dissimilar polymers are usually incompatible, because of their long-chain nature [52–55]. In addition, their viscosities are so high that their diffusion coefficients are quite small. Consequently, diffusion of an entire macromolecule across the interface is unlikely. However, both theory and experiment show that local-segmental interdiffusion can occur across the interface. Such local-segmental interdiffusion is favored thermodynamically as interfacial free energy is lowered by limited interdiffusion, and is kinetically possible as the transport of an entire macromolecule is not involved.

The condition for molecular mixing is $\Delta G_m = \Delta H_m - T \Delta S_m \leq 0$, where ΔG_m is the Gibbs free energy of mixing. The enthalpy of mixing ΔH_m is usually positive (endothermic) in the absence of strong specific interactions. The combinatorial entropy ΔS_m favors mixing

Diffusion and Adhesion

but is usually too small for macromolecules because of limited chain configurations. Thus, ΔG_m tends to be positive, and mixing of entire macromolecules cannot occur. However, local segmental interdiffusion across the interface is possible, and favored thermodynamically. This is analyzed below by several theoretical approaches.

Analysis of Interdiffusion by Flory-Huggins Theory

The entropy of mixing is given by [56–59]

$$\Delta S_m = -R(N_1 \ln \phi_1 + N_2 \ln \phi_2) \tag{11.22}$$

where N is the number of moles and ϕ is the volume fraction. The enthalpy of mixing is given by

$$\Delta H_m = \frac{RTV}{\bar{v}_r} \chi_{12} \phi_1 \phi_2 \tag{11.23}$$

where V is the total volume, $\bar{v}_r$ the molar volume of a repeat unit, and χ_{12} the interaction parameter.

Assume equal molecular weight, molar volume, and density ρ. The free energy of mixing becomes [60–66]

$$\Delta G_m = \frac{\rho RTV}{M_c} \left[\frac{M_c}{M} (\phi_1 \ln \phi_1 + \phi_2 \ln \phi_2) + 2\phi_1 \phi_2 \right] \tag{11.24}$$

where

$$M_c = \frac{2\rho \bar{v}_r}{\chi_{12}} \tag{11.25}$$

Stable one-phase mixture can exist only when $(\partial^2 \Delta G_m / \partial \phi^2)_{T,P} > 0$. M_c turns out to be the critical molecular weight for miscibility. When $M < M_c$, stable one-phase mixture can exist for all compositions. When $M > M_c$, mixing can occur only at compositional extremities. These compatible zones become smaller as M/M_c increases. This is illustrated in Figure 11.16, where $\Delta \tilde{G}_m = (M_c/\rho RTV)\Delta G_m$, $\Delta \tilde{H}_m = (M_c/\rho RTV)\Delta H_m$, and $\Delta \tilde{S}_M = (M_c/\rho RTV)\Delta S_m$. The χ_{12} is given by [60,61]

$$\chi_{12} = \frac{\bar{v}_r}{RT} (\delta_1 - \delta_2)^2 \tag{11.26}$$

where δ is the solubility parameter. An improved relation is given by [63]

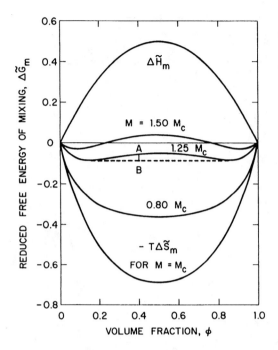

Figure 11.16. Reduced free energy of mixing versus composition at different molecular weights. (After Ref. 62.)

$$\chi_{12} = \frac{\overline{v}_r}{RT}[(\delta_1^d - \delta_2^d)^2 + (\delta_1^p - \delta_2^p)^2] \qquad (11.27)$$

Thus,

$$M_c = \frac{2\rho RT}{(\delta_1^d - \delta_2^d)^2 + (\delta_1^p - \delta_2^p)^2} \qquad (11.28)$$

The compatibility limit can be estimated by Eq. (11.28). For instance, polyethylene has $\delta^d = 7.82$ cal$^{1/2}$/cm$^{3/2}$ and $\delta^p = 0$, and poly(n-butyl methacrylate) has $\delta^d = 8.00$ and $\delta^p = 3.50$ cal$^{1/2}$/cm$^{3/2}$ [4,42,64]. Let $\rho = 1$ g/cm^3 and $T = 300$ K. Thus, Eq. (11.28) gives $M_c = 98$, indicating that interdiffusion of molecular segments of about 10 carbon atoms can occur across the interface at equilibrium in this system.

Analysis by Statistical Thermodynamics

Mean-field and lattice theories [67–72] show that the interfacial regions between two incompatible polymers are formed by limited segmental

Diffusion and Adhesion

interdiffusion (see Sections 3.2.8 and 3.2.9). For instance, Helfand [70] gave the interfacial thickness a_I explicitly as

$$a_I = \left[\frac{2(B_1^2 + B_2^2)}{\alpha}\right]^{1/2} \tag{11.29}$$

where

$$B_k = \frac{\rho_{0k}}{6}\left(\frac{\langle R^2 \rangle_k}{r_k}\right) \tag{11.30}$$

$$\alpha = (\rho_{01}\rho_{02})^{1/2}\chi_{12} \tag{11.31}$$

$$= \frac{1}{kT}(\delta_1 - \delta_2)^2 \tag{11.32}$$

where ρ_{0k} is the number density of the repeat unit of component k and $\langle R^2 \rangle_k$ is the mean-square end-to-end distance of a molecule of k. The relations above have been used to calculate the interfacial thickness between incompatible polymers (Table 3.20). The interfacial thickness at equilibrium is of the order of 10 ∿ 1000 Å.

For instance, the interfacial thickness between polyethylene and poly(n-butyl methacrylate) is 21 Å; that between poly(methyl methacrylate) and polystyrene is 160 Å. These are consistent with experimental evidence, discussed next.

Classical Thermodynamic Analysis

This has been discussed by Kammer (see Section 3.2.10). Interfacial composition and thickness are explicitly expressed in terms of thermodynamic parameters. The results are consistent with statistical mechanical theories.

11.3.3. Experimental Evidence of Interdiffusion

Indirect evidence includes the effects of contact time, temperature, pressure, and molecular structure (molecular weight, chain flexibility, side group, polarity, double bond, and compatibility) on the strength of autohesive and adhesive bonds [43] (see also Section 11.3.1). Direct evidence includes measurement of diffusion coefficient [73–76], and observation of interfacial structure by electron microscopy [47, 77–79], radiothermoluminiscence technique [80], paramagnetic probe [81], UV luminescence [82,83], and optical microscopy [84].

Diffusion coefficients of oligomers and polymers have been measured and found to be of the order of 10^{-13} to 10^{-7} cm^2/sec [73–76].

Table 11.8. Diffusion Coefficient and Activation Energy for Some Polymers and Organic Liquids

Diffusant	Medium
Linear polyethylene, 98% deuterated $M_w = 3600$, $M_w/M_n = 2.25$ 4600 2.2 11000 3.4 17000 2.2 23000 1.8	Molten linear polyethylene, $M_w = 1.6 \times 10^5$ $M_w/M_n = 16$
Behenyl behenate, $CH_3(CH_2)_{20}COO(CH_2)_{21}CH_3$	Same as above
	Branched polyethylene $T_m = 107°C$ $M_w = 10^5$, $M_w/M_n = 5$
Stearamide, $CH_3(CH_2)_{16}CONH_2$	Same as above
Polybutadiene MW = 1600	Polybutadiene ($M_w = 410,000$, $M_n = 90,000$) Natural rubber Styrene-butadiene rubber Ethylene-propylene rubber ($M_w = 170,000$) $M_n = 70,000$ Polydimethylsiloxane (MW 200,000)
Polyisoprene MW = 110,000	Unvulcanized natural rubber Vulcanized natural rubber
Octadecane	Polyisoprene Polyethylene Polychloroprene Polybutadiene
Octadecyl stearate	Polybutadiene

Diffusion coefficient, D, cm²/sec	E_d,[a] kcal/mole	E_f,[b] kcal/mole	Reference
0.367 $M_w^{-2.0}$ (176°C, melt)	–	–	73
1.10 × 10⁻⁶ (150°C, melt)	6.6	–	73
1.98 × 10⁻⁸ (60°C, solid)	22.4 (below 80°C)	–	73
1.05 × 10⁻⁶ (120°C, melt)	7.2 (above 80°C)	–	73
2.24 × 10⁻⁶ (120°C)	6.7 (above T_m)	–	73
3.2 × 10⁻⁷ (80°C)			
1.8 × 10⁻⁷ (25°C)	–	–	76
3.3 × 10⁻⁸ (25°C)	–	–	76
3.4 × 10⁻⁸ (25°C)	–	–	76
1.8 × 10⁻⁸ (25°C)	–	–	76
9.4 × 10⁻⁷ (25°C)	–	–	76
3.4 × 10⁻¹³ (100°C)	8.7	–	75
1.4 × 10⁻¹³ (100°C)	–	–	75
4.4 × 10⁻⁷ (60°C)	8.8	–	90
2.1 × 10⁻⁷ (80°C)	12.2	–	90
1.1 × 10⁻⁷ (60°C)	12.7	–	90
1.6 × 10⁻⁶ (80°C)	9.3	–	90
1.8 × 10⁻⁷ (60°C)	6.7	–	90

Table 11.8. (Continued)

Diffusant	Medium
Poly(n-butyl methacrylate) (MW 180,000)	Self-diffusion
Polyisobutylene (MW 1600)	Self-diffusion
Polyethylene (MW 4100)	Self-diffusion
Polyethylene (MW 5800)	Self-diffusion
Polydimethylsiloxane (MW 680)	Self-diffusion

[a] E_d is the activation energy for diffusion.
[b] E_f is the activation energy for viscous flow.

Some typical values are listed in Table 11.8. Experimantal methods are discussed in Section 11.3.4. These values indicate that interdiffusion across the interface of 10–1000 Å can occur within minutes to hours.

Transmission electron microscopy shows that the interface between two polymers appears as a band in which the electron density varies gradually from one phase to the other, indicating intermixing of the two components in the interfacial zone [47,77–79]. Typically, the interfacial zones are 10–1000 Å thick, consistent with theoretical predictions. Figure 11.17 shows the electron photomicrograph of the interfacial zone between poly(methyl methacrylate) and poly(vinyl chloride) bonded at 215°C. Figure 11.18 shows the graduation of optical density across the interfacial zone. As can be seen, the interfacial zone is about 3000 Å thick in this system. The thickness increases with contact time at a given temperature, and eventually reaches an equilibrium value between incompatible polymers or disappears between two compatible polymers [77,78,83]. The rate of growth of the interfacial thickness follows Fick's second law of diffusion in the initial stage, and then diminishes as the equilibrium is approached [84]. Table 11.9 lists the interfacial thickness between some polymer pairs.

11.3.4. Diffusion Coefficient

Diffusion coefficient D is the proportionality constant in Fick's first law, $dq/dt = -D(dc/dx)$ or in Fick's second law, $dc/dt = D(d^2c/dx^2)$, where dq is the flux in time dt with a concentration gradient dc/dx in

Diffusion coefficient, D, cm^2/sec	E_d,[a] kcal/mole	E_f,[b] kcal/mole	Reference
2.3×10^{-9} (47°C)	13.2	13.8	74
1.1×10^{-6} (181°C)	5.5	12–15	87
6.0×10^{-8} (30°C)			
1.6×10^{-7} (150°C)	5.3	6–7	88
1.1×10^{-7} (150°C)	–	–	88
1.9×10^{-6} (25°C)	4.0	4.0	89

the x direction [85,86]. The diffusion coefficients of gases and simple liquids through polymers have been measured widely [41,85,86]. However, those for polymers through polymers have not been as widely known.

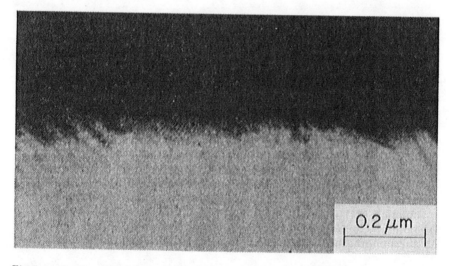

Figure 11.17. Electron photomicrograph of the interfacial region between poly(methyl methacrylate) and poly(vinyl chloride) bonded at 210–220°C. (After Ref. 77.)

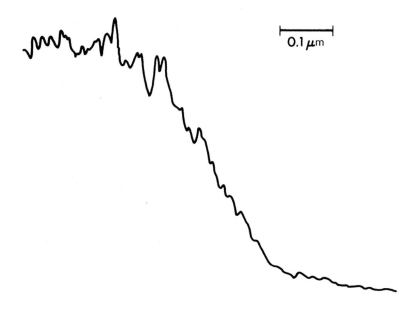

Figure 11.18. Variation of optical density across the interfacial region between poly(methyl methacrylate) and poly(vinyl chloride) welded at 210–220°C as measured from the electron photomicrograph of Figure 11.16. (After Ref. 77.)

Klein and Briscoe [73] measured the diffusion coefficients of deuterated polyethylenes, stearamide, and behenyl behenate through linear and branched polyethylenes by the IR absorption technique. Bueche and coworkers [74] measured the self-diffusion coefficients of ^{14}C tagged polystyrene and poly(n-butyl acrylate). Bresler and coworkers [75] measured the tritium-tagged polyisoprene into natural rubber. Ferry and coworkers [76] measured the diffusion coefficients of ^{14}C-tagged polybutadiene through natural rubber, styrene-butadiene rubber, ethylene-propylene rubber, and others. McCall and coworkers [87–89] measured the self-diffusion coefficients of polyisobutylene, polyethylene, and polydimethylsiloxane by the NMR spin-echo technique. Some diffusion coefficients are listed in Table 11.9.

Einstein [91] showed that $D = kT/f_0$, where f_0 is the molecular frictional factor (that is, the force needed to pull a molecule through its surrounding at unit speed). Bueche [92] thus obtained

$$D\eta = \frac{RT}{36} \frac{<R>^2}{M} \rho \qquad (11.33)$$

Table 11.9. Interfacial Thickness Between Polymers [a]

Polymer pair	Bonding condition	Interfacial thickness, Å	Technique
Polystyrene/poly(methyl methacrylate)	128-225°C	50	TEM
Butadiene-acrylonitrile rubber (MW 62,000)/polychloroprene (MW 480,000)	100°C, 1 hr	1,800 (eq)	TEM
Butadiene-acrylonitrile rubber (high MW)/polychloroprene (high MW)	100°C, 1 hr	500	TEM
Polystyrene (MW 490,000)/poly(methyl methacrylate) (high MW)	140°C, 10 hr	300 (eq)	TEM
Poly(vinyl chloride)/poly(methyl methacrylate)	165°C 215°C	1,200 3,000	TEM TEM
Poly(vinyl chloride)/poly(butyl methacrylate)	165°C 215°C	0 1,000	TEM TEM
Poly(vinyl chloride) (MW 53,000)/polyethylene (MW 24,000)	160°C	90,000 (eq)	OM
Isotactic polypropylene (MW 130,000)/polyethylene (MW 24,000)	160°C	28,000 (eq)	OM
Polystyrene (MW 102,000)/polystyrene (MW 490,000)	140°C 5 min 30 min 60 min	0 2,000 8	TEM TEM TEM

[a] TEM, transmission electron microscopy; OM, optical microscopy; (eq) denotes equilibrium value.
Source: Data collected from Refs. 77-79 and 84.

where η is the viscosity, $<R^2>$ the mean-square end-to-end distance of the macromolecule, M the molecular weight, and ρ the density. Equation (11.33) was reported to predict adequately the self-diffusion coefficient of poly(n-butyl acrylate) [74]. Ordinarily, $D\eta \sim 5 \times 10^{-8}$ dyne, a constant.

Molecular-Weight Dependence

Bueche's entangle model views the molecular entanglement as increasing the drag and hence the resistance to motion of a diffusing chain. The diffusion coefficient is predicted to decrease with increasing molecular weight, expressible by [93,94]

$$D = KM^{-\alpha} \tag{11.34}$$

where K and α are constants. Since D = constant, Bueche [73] suggested $\alpha = 1$ below M_e and $\alpha = 3.5$ above M_e, where M_e is the critical entanglement molecular weight. However, α has been found experimentally to have values generally between 1 and 5/3 for polymers above M_e [74,88,92].

Recently, the entangle model was recognized as being inadequate. A reptation model was then proposed by de Gennes [95,96], which views the translational motion of a macromolecule to occur by snakelike (reptile) wriggling along its length. This model predicts that $\alpha = 2$. Recently, Klein and Briscoe [73] accurately measured the diffusion coefficients of fractions of deuterated polyethylene (M_w = 3600--23,000) in a linear polyethylene ($M_w = 1.6 \times 10^5$) and found that

$$D = KM_w^{-2.0 \pm 0.1} \tag{11.35}$$

which elegantly verified the reptation model (Figure 11.18).

Vasenin [97] considered the molecular friction in terms of the exchange of momentum between the groups on the moving molecule and its surroundings, and obtained

$$D = \frac{D_1(2n - b)}{a^u n(2n - 2b)} \tag{11.36}$$

where D_1 is the diffusion coefficient of the first member of the homologous series to which the diffusant belongs, n the number of carbon atoms in the molecule, b the number of double bonds, a the relative cross-sectional area compared with normal paraffin, and u a constant characteristic of the diffusion medium. Where side groups exist, as

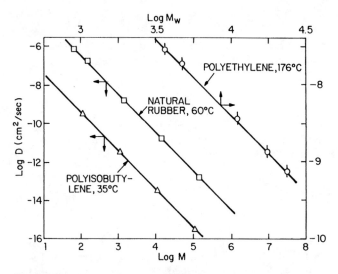

Figure 11.19. Diffusion coefficients of deuterated polyethylenes through a linear polyethylene ($M_w = 1.6 \times 10^5$, $M_w/M_n = 16$) and self-diffusion coefficients of natural rubber and polyisobutylene versus molecular weight. (Data collected from Refs. 73, 93, 94, and 97–99.)

in natural rubber and polyisobutylene, Eq. (11.36) predicts that $\alpha = 2$, also shown in Figure 11.19.

Temperature Dependence

Diffusion coefficient varies with temperature by [41, 73–75, 87–90]

$$D = D_o \exp\left(\frac{-E_d}{RT}\right) \qquad (11.37)$$

where E_d is the activation energy for diffusion and D_0 is a constant. E_d are often similar to E_f, shown in Table 11.8, making the distinction between diffusion and viscous flow difficult.

Concentration Dependence

Diffusion coefficient is independent of concentration for the diffusion of gases through polymers or simple liquids through simple liquids. It is, however, greatly dependent on diffusant concentration for the diffusion of organic liquids or polymers through polymers. These diffusants tend to swell or create holes in the polymers, and thus facilitate the diffusion. Various equations have been used to express the concentration dependence. A linear relation is given by [41, 93, 100]

$$D = D_0(1 + kc) \tag{11.38}$$

where c is the diffusant concentration and D_0 and k are positive constants. An exponential relation applicable over a wider range is [41, 101–103]

$$D = D_0 \exp kc \tag{11.39}$$

Vasenin [104] considered the effect of diffusant penetration on the energy of hole formation, and derived

$$D = D_0 (D_1/D_0)^{1-\exp[-\beta\phi/(1-\phi)]} \tag{11.40}$$

where D_1 is the diffusion coefficient at $\phi = 1$, ϕ the volume fraction of the diffusant in the medium, and β a positive constant having values between 1 and 5. This equation has been shown to be more accurate than Eq. (11.39). Vasenin [104] showed that an increase of 0.1 in ϕ could increase D by 1000-fold in some cases.

11.3.5. Diffusion and Bond Strength

Fracture Energy Analysis

In this model, interpenetration of molecular segments at the interface reduces molecular slippage, increases adhesive fracture energy, and thus increases the bond strength. The adhesive fracture enrgy Ga is given by

$$G_a = (1 - s)(G_1 G_2)^{1/2} \tag{11.41}$$

where s is the slippage factor and G_1 and G_2 are the cohesive fracture energies of phases 1 and 2, respectively. The slippage factor should decay with increased interfacial molecular penetration, and thus

$$s = 1 - \exp\left(\frac{-m}{a_I}\right) \tag{11.42}$$

where a_I is the interfacial thickness and m is a positive constant. Thus, we have

$$G_a = (G_1 G_2)^{1/2} \exp\left(\frac{-m}{a_I}\right) \tag{11.43}$$

Diffusion and Adhesion

Interfacial thickness has been given by mean-field theory. Using Eqs. (11.29) and (11.32) in Eq. (11.43) gives

$$G_a = (G_1 G_2)^{1/2} \exp\left[\frac{-m}{2^{3/2}(kT)^{1/2}} \cdot \frac{\delta_1 - \delta_2}{(B_1^2 + B_2^2)^{1/2}}\right] \quad (11.44)$$

where we define $(\delta_1 - \delta_2) \geq 0$. Thus, G_a increases with increasing compatibility of the two phases, and is greatest when $\delta_1 = \delta_2$.

Using Eq. (11.44) in Eq. (10.5), Griffith's adhesive fracture criterion, gives

$$\sigma_f = \left[\frac{(E_1 E_2)^{1/2}(G_1 G_2)^{1/2}}{\pi c_{12}}\right]^{1/2}$$

$$\cdot \exp\left[\frac{-m}{2^{3/2}(kT)^{1/2}} \cdot \frac{\delta_1 - \delta_2}{(B_1^2 + B_2^2)^{1/2}}\right] \quad (11.45)$$

where σ_f is the fracture strength and we let $E_{12} = (E_1 E_2)^{1/2}$. Equation (11.45) may be rewritten as

$$\sigma_f = p \exp[-q(\delta_1 - \delta_2)] \quad (11.46)$$

Thus, a plot of log σ_f versus $(\delta_1 - \delta_2)$ should give a straight line with a slope $-(q/2.303)(\delta_1 - \delta_2)$, provided that p and q are constant for the series of adhesive bonds. The adhesive strength should increase with compatibility. These values are confirmed experimentally below.

The peel strengths of various adhesives on poly(ethylene terephthalate) are given in Table 11.10. When the peel strength is plotted against the solubility parameter of the adhesive, a curve is obtained having a maximum where the solubility parameters of the adhesive and the poly(ethylene terephthalate) are identical (Figure 11.20). The adhesive strength increases with increasing compatibility of the adhesive and the adherend. These data can be used to test Eq. (11.46). Indeed, when log (peel strength) is plotted against $(\delta_1 - \delta_2)$, a straight line is obtained (Figure 11.21).

The peel strength between various polymers and rubbers are also used to test Eq. (11.46). When log (peel strength) is plotted against $(\delta_1 - \delta_2)$, straight lines are indeed obtained (Figure 11.22). This again confirms the validity of the present analysis.

Deryagin and coworkers [105] showed a qualitative correlation between adhesive strength and compatibility. Furthermore, in some cases, the plot of bond strength versus surface tension of the ad-

Table 11.10. Bond Strength and Solubility Parameter for the Adhesion of Various Polymers to Poly(ethylene terephthalate) [a]

Polymer adhesive	Peel strength, g/cm	Solubility parameter δ, $(cal/ml)^{1/2}$	Failure locus
Polyethylene	52	7.80	i
Polyisoprene	77	8.07	i
Poly(butadiene-styrene)	336	8.42	i
Poly(butadiene-styrene-2-Vinylpyridene)	231	8.91	i
Polychloroprene	479	8.94	i
Chlorosulfonated polyethylene	536	9.20	i
Acrylic copolymer	577	9.31	i
Poly(ethylene-vinyl acetate	699	9.20	i
Poly(vinyl chloride)	869	9.41	i
Poly(butadiene-acrylonitrile)	929	9.48	i
Polyurethane	1532	10.00	c
Poly(vinyl acetate-dibutyl maleate)	1812	10.46	c
Epoxy/isocyanate adhesive	1633	10.58	c
Poly(vinyl acetate)	817	10.99	i
Poly(vinylidene chloride-acrylonitrile)	280	12.08	i
Alkoxy alkylated nylon	127	12.64	i

[a] Peel strength measured at 90° peel angle and 5 cm/min rate at room temperature. The adherend is poly(ethylene terephthalate) having $\delta = 10.30$ $(cal/ml)^{1/2}$. The failure locus i = interfacial failure; c = cohesive failure.

Source: From Ref. 25.

Diffusion and Adhesion

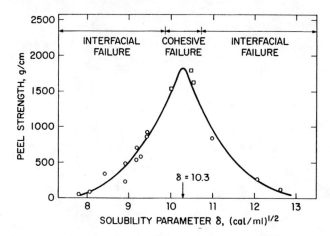

Figure 11.20. Peel strength versus solubility parameter for various adhesives on poly(ethylene terephthalate); plotted from Table 11.11. (After Ref. 25.)

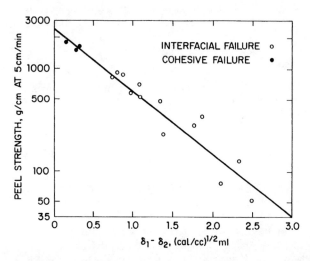

Figure 11.21. Semilogarithmic plot of peel strength versus solubility parameter difference for various adhesives on poly(ethylene terephthalate) adherend.

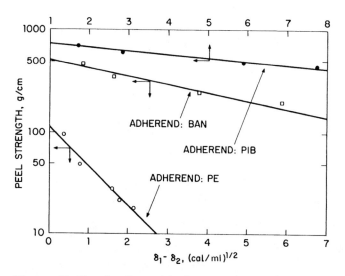

Figure 11.22. Semilogarithmic plot of peel strength versus solubility parameter difference for various adhesives on adherends PIB (polyisobutylene), BAN (SKN-40 butadiene-acrylonitrile rubber), and PE (polyethylene), respectively.

hesive is curved, the maximum occurring when the surface tensions of the adhesive and the adherend are equal, such as shown in Figure 11.7, (see Section 11.2.3). Such a plot is equivalent to a σ_f versus $\Delta \delta$ plot, and the observed behavior suggests that the adhesive strength increases with compatibility.

Vasenin's Kinetic Theory

Vasenin [106–108] considered the adhesive bond strength in terms of momentum exchange in pulling off interfacial molecules. The force f required to pull a molecule is

$$f = \left(\frac{mv}{2}\right)\nu n \tag{11.47}$$

where m is the mass of atomic group, v the pull rate, ν the frequency of collision of the groups (which can be taken as the vibrational frequency of a —CH_2— group), and n the number of groups. The number n of groups which have diffused into the adherend in time t is then obtained by applying proper boundary conditions to Fick's second law. With certain simplifying assumptions, Vasenin obtained

$$n = \left[\frac{\pi}{2L \cos(\theta/2)}\right]^{1/2} K_D^{1/2} t^{1/4} \tag{11.48}$$

Diffusion and Adhesion

where L is the length of C—C bond, θ the C—C valence angle, and K_D the diffusion coefficient at t = 1 sec. By using proper numerical values and combining Eqs. (11.47) and (11.48), the adhesive strength is obtained as

$$\sigma_f = 5.55\nu \left[\left(\frac{2\rho_1}{M_1}\right)^{2/3} K_{D1}^{1/2} + \left(\frac{2\rho_2}{M_2}\right)^{2/3} K_{D2}^{1/2}\right] vt^{1/4} \quad (11.49)$$

where ρ_j is the density of phase j and M_j is the molecular weight of phase j.

Thus, the theory predicts a $t^{1/4}$ dependence of adhesive strength. This is confirmed experimentally. Figure 11.23 plots the data of Forbes and McLeod [109] for the butt joint strength between natural rubber and a butadiene-acrylonitrile rubber, showing a good fit with $t^{1/4}$. Figures 11.24 and 11.25 plot the data of Voyutskii and Shtarkh [110] for the autohesive peel strength of polyisobutylene, again showing a good fit with $t^{1/4}$ but not with $t^{1/3}$ or $t^{1/2}$. In fact, the rate of adhesive strength development in many systems obeys [4]

$$\sigma_f = at^b \quad (11.50)$$

where a is a constant and b is usually between 1/4 and 1/5. Thus, Vasenin's theory adequately account for the kinetic process only within a practical time period.

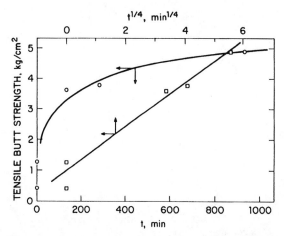

Figure 11.23. Tensile butt strength of joints between natural rubber and a butadiene-acrylonitrile rubber versus bonding time. (After Ref. 109.)

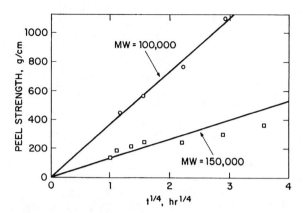

Figure 11.24. Kinetics of autohesion of polyisobutylenes. (Data from Ref. 110.)

However, the theory is inadequate for predicting the final strength of adhesive bond. Equation (11.47) is based on equilibrium consideration and, therefore, cannot describe the fracture strength, which is a nonequilibrium property. Equation (11.49) predicts that the adhesive strength increases with decreasing molecular weight (that is, $M^{-2/3}$). This has been shown to be true in some cases [43], but is not generally valid. In many cases, the fracture strength increases with molecular weight, because of improved mechanical properties.

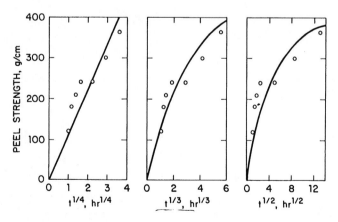

Figure 11.25. Kinetics of autohesion of polyisobutylene (MW 150,000). (Data from Ref. 110.)

Diffusion and Adhesion

11.3.6. Remarks on Some Experimental Results

Voyutskii [43,111,112] has summarized experimental observations on the effects of bonding conditions and molecular structures on autohesive and adhesive bond strengths. The bond strength increases with longer contact time, higher bonding temperature, higher bonding pressure, lower molecular weight, higher chain flexibility, absence of bulky short side groups, and lower degree of cross-linking. These were cited as proof of the importance of diffusion in adhesion. However, these also favor wetting process.

Effect of Bonding Pressure

Adhesive and autohesive strengths increases with increasing applied pressure during bonding [43,111,112] (Figure 11.26). Since diffusion is known to be independent of pressure, the phenomenon observed must arise from increased forced flow (see also Section 11.3.1).

Effect of Molecular Weight

Increased molecular weight will increase the cohesive strength of materials, and therefore should favor high bond strength. But the increased viscosity tends to retard both wetting and diffusion. Thus, the effect of molecular weight on bond strength will be variable, depending on material properties. This is consistent with experimental facts. The bond strength between various polyisobutylene adhesives

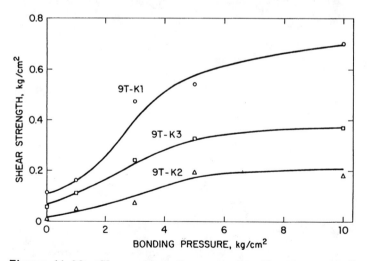

Figure 11.26. Shear strength versus bonding pressure for the autohesion of some unvolcanized carcass rubbers. (After Ref. 113.)

and a cellophane adherend increases with decreasing molecular weight of the adhesives [43,112]. The autohesive strength of polyisobutylene increases with decreasing molecular weight [110]. The bond strength between a polyamide adherend and various butadiene-acrylonitrile adhesives increases with decreasing molecular weight of the adhesive [111] (Figure 11.27). On the contrary, the bond strength of acrylic copolymers on steel increases with increasing molecular weight [113]. The bond strengths of poly(vinyl acetate) adhesives on aluminum and cellulose adherends increase with increasing molecular weight of the adhesives [114] (Figure 11.28). Similarly, the bond strength increases with increasing molecular weight of poly(vinyl acetate) on steels [115].

Temperature Dependence

The autohesive bond strength of rubbers increases with increasing bonding temperature, obeying [110,116]

$$\sigma_f = \sigma_{f_0} \exp\left(\frac{-E_a}{RT}\right) \tag{11.51}$$

where E_a is an activation energy. This equation is similar to Eq. (11.20) derived from wetting kinetics. Equation (11.51) is expected in view of Eqs. (11.18) and (11.33). If diffusion is the controlling process in the formation of adhesive or autohesive bond, then $E_a = E_d$. An Arrhenius plot for the autohesion of polyisobutylene is shown in

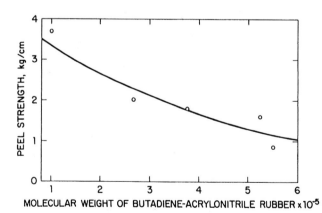

Figure 11.27. Effect of molecular weight on adhesive bond strength between butadiene-acrylonitrile rubbers (of various molecular weights) and a polyamide at a bonding temperature of 200°C. (Data from Ref. 111.)

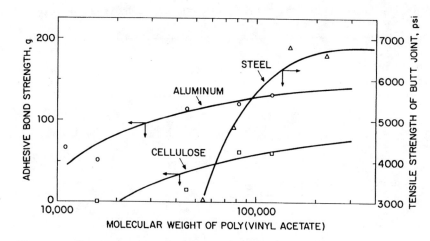

Figure 11.28. Effect of molecular weight of poly(vinyl acetate) adhesive on joint strength for aluminum, cellulose, and steel adherends, respectively. (Data from Refs. 115 and 116.)

Figure 11.29. The activation energy E_a is found to be 2.6 kcal/mole. In comparison, E_d is 5.5 kcal/mole and E_f is 12-15 kcal/mole for polyisobutylene (see Table 11.8). Thus, the value of 2.6 kcal/mole for E_a is rather small, but is closer to E_d than to E_f. Therefore, in this case, diffusion appears to be the controlling process.

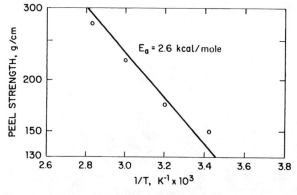

Figure 11.29. Arrhenius plot for the autohesion of polyisobutylene. (Data from Ref. 116.)

Table 11.11. Effect of Nitrile Content on the Peel Strength Between Butadiene-Nitrile Rubber and Cellophane

Acrylonitrile content in rubber, %	Peel strength, g/cm	Failure mode
18.4	1638	Cohesive
28.6	1382	Mixed
37.7	132	Adhesive

Source: From Ref. 43.

Effect of Molecular Structure

Increased chain rigidity tends to, but does not always, decrease the bond strength. Chain rigidity will increase the cohesive strength of the material, but will also increase the viscosity and retard the diffusion. Thus, the effect of chain rigidity on adhesion can be variable. However, usually the bond strength tends to decrease with increasing chain rigidity [117,118].

For instance, incorporation of rigid side groups such as methyl, t-butyl, and phenyl groups into a polymer generally lowers its adhesion [111,112]. Cross-linked polymers are generally quite difficult to adhere to, because of decreased molecular diffusivity. Table 11.11 shows the effect of the nitrile content of butadiene-nitrile rubber on its adhesion to cellophane [43]. Increased nitrile content should make the rubber and the cellophane more compatible. The bond strength should probably increase. But, in fact, it decreases with increasing nitrile content, probably because nitrile groups increase the chain rigidity and viscosity, and decrease the diffusivity.

11.4. CHEMICAL ADHESION

11.4.1. Effect of Interfacial Chemical Bonding on Fracture Energy

Interfacial chemical bonding can increase the adhesive bond strength by (1) preventing molecular slippage at a sharp interface during fracture, and (2) increasing the fracture energy by increasing the interfacial attraction.

Interfaces are classified into two types: sharp interface and diffuse interface. A sharp interface is obtained when little or no interfacial diffusion occurs; a diffuse interface is obtained when sufficient

Chemical Adhesion

interfacial diffusion occurs. If interfacial attraction involves mainly dispersion force, interfacial molecular slippage can occur during fracture at a sharp interface, resulting in low energy absorption and low mechanical strength. Dispersion forces between polymers (1--5 kcal/mole) are too weak to prevent molecular slippage at a sharp interface. However, those between a polymer and a high-energy material are higher, and may be sufficient to prevent slippage at a sharp interface. On the other hand, chameical bonds have much higher energies (50--250 kcal/mole) and can prevent molecular slippage at a sharp interface. Dispersion forces should, however, be sufficient to prevent slippage at a diffuse interface.

As adherends, cross-linked polymers and vulcanized rubbers have rather low bondability, because interfacial diffusion is difficult and the interface is sharp. In such cases, chemical bonding or specific interactions can greatly increase the bond strength.

Fracture energy G has been given as $G = G_0 \Psi(Ra_T)$, where G_0 is the equilibrium fracture energy and $\Psi(Ra_T)$ is a viscoelastic dissipation function (Section 14.4.3). If the interfacial attraction involves only physical interactions, G_0 will equal the work of adhesion in the case of interfacial failure, or equal the work of cohesion in the case of cohesive failure. On the other hand, if the interfacial attraction involves chemical bonding, G_0 is written as W_b, which includes chemical bond energy. Typically, covalent bond energy is 50 to 100 kcal/mole, and van der Waals energy is 2.5 to 5 kcal/mole. Therefore, the ratio W_b/W_a could be as much as 25 to 40. This is demonstrated experimentally below.

The 180°-peel strength of polybutadiene rubber and ethylene-propylene rubber chemically bonded to glass was investigated [119, 120]. Interfacial chemical bonding sites of varying densities are obtained by treating the glass with mixtures of vinylsilane and ethylsilane in varying proportions. Vinylsilane provides chemical bonding sites. A layer of the rubber is then applied and cross-linked in situ with dicumyl peroxide. The silanes covalently bond to the glass through ether linkages (Section 11.4.3). The rubber chemically bonds to the vinyl groups of the vinylsilane, as shown by the disappearance of the near-infrared absorption bands of terminal vinyl groups at 1.6 and 2.1 μm after cross-linking of the rubber on the vinylsilane-treated glass. The fracture energy is measured over the temperature range −80 to +80°C and a rate range of 10^{-6} to 10^{-2} m/sec, and found to vary with the rate R and the temperature T, but can be superposed by using the effective Ra_T, where a_T is the universal form of the WLF shift factor [121] given in Eq. (14.58).

The fracture energy, plotted as a function of the effective rate, increases with increasing interfacial chemical bond density (Figure 11.30). The fracture energy at the lowest effective rate of 10^{-20} m/sec is taken to be the equilibrium value G_0, which is plotted versus the amount of vinylsilane in the treatment mixture (Figure 11.31). A

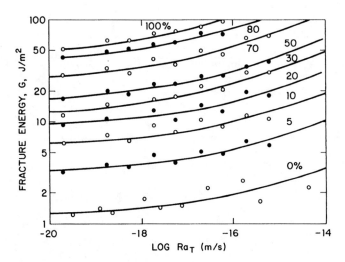

Figure 11.30. Adhesive fracture energy G (determined by 180°-peel test) for a polybutadiene cross-linked with 0.2% dicumyl peroxide against glass surfaces that are treated with various mixtures of vinylsilane and ethylsilane. The numbers on the curves are percent by weight of vinylsilane in the silane mixtures. (After Ref. 119.)

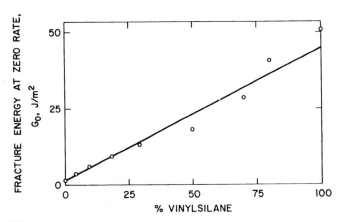

Figure 11.31. Adhesive fracture energy G_0 (determined by 180°-peel test) at zero rate versus percent by weight of vinylsilane in the vinylsilane-ethylsilane treating mixture. (After Ref. 119.)

Chemical Adhesion

linear relation is obtained, suggesting that the joint strength is directly proportional to the density of interfacial chemical bond. The G_0 for 100% vinylsilane-treated system is 50 J/m^2; that for 100% ethylsilane-treated system is 1.4 J/m^2. These are greater than the theoretical equilibrium values by a factor of about 25, indicating a significant rate effect even at an effective rate as low as 10^{-20} m/sec. However, the ratio of G_0 for vinylsilane to that for ethylsilane is about 35, which is comparable to the ratio of covalent bond energy (about 80 kcal/mole) to dispersion energy (about 2.5 kcal/mole). When polybutadiene is not cross-linked in situ, no interfacial chemical bonding occurs and the fracture energies on both the vinylsilane- and ethylsilane-treated glasses are identical at all effective rates.

A styrene-butadiene rubber is bonded to and cross-linked in situ on sodium-etched FEP adherend (Section 9.1.1). The equilibrium fracture energy is found to be about 2000 erg/cm^2, compared with the work of adhesion of about 70 erg/cm^2 (Table 9.1). The ratio of the two is about 30. Apparently, the styrene-butadiene rubber is covalently bonded to the sodium-etched FEP by cross-linking with the abundant C=C bonds on the sodium-etched FEP. On the other hand, when untreated FEP is used as the adherend, the equilibrium fracture energy is found to be only about 22 erg/cm^2, which is similar to the work of adhesion of about 48 erg/cm^2.

11.4.2. Evidence of Interfacial Chemical Bonding

Chemical bonding at the interface is difficult to detect, because of the thinness of the interface. However, in some systems, interfacial chemical bonding has been shown to occur, and greatly contribute to the adhesive strength (discussed below).

Epoxy/Cellulose Interface

The epoxy groups of an epoxy resin react with the hydroxyl groups of a cellulose at the interface, shown by internal reflection IR spectroscopy [122]. The chemical reaction causes the OH stretching band at 3350 cm^{-1} and the C—O stretching bands at 1000–1500 cm^{-1} of the cellulose to diminish, and the epoxide band at 915 cm^{-1} and the antisymmetric O-bridge stretching band at 1160 cm^{-1} of the epoxy resin to disappear. The chemical bonding accounts for the strong adhesion between the epoxy resin and the cellulose.

Phenolic Resin/Lignocellulose Interface

The strong adhesion between phenol-formaldehyde resin and lignocellulose has been shown to arise at least partially from chemical bonding between the phenol-formaldehyde prepolymer and the guaiacyl group of lignocellulose [123]. 3,5-Dibromo-4-hydroxybenzyl alcohol (DBHBA) is used as a model compound for phenol-formaldehyde prepolymer. This model compound cannot self-condense, and therefore will not be entrapped in the lignocellulose fiber by physical entanglement. An extractable-free lignocellulose fiber is used as the adherend. The extent of reaction is determined by bromine analysis by neutron activation after conversion to ^{82}Br and exhaustive extraction. As much as 1.6% by weight of bromine remains on the lignocellulose fiber, showing extensive interfacial chemical bonding.

Amine/Alkyd Interface

Many polymers adhere weakly to cross-linked polymeric adherends, failing interfacially. Incorporation of a small amount (0.01 mole%) of certain nitrogen-containing groups greatly increases the adhesion [124]. The failure becomes cohesive in either the adhesive polymer or the adherend. An example is the strong adhesion of an amino polymer to a cross-linked alkyd resin, due to interfacial amine-ester interchange reaction between the two phases to form amide bonds.

The amine-ester interchange reaction can readily be seen by using butylamine as a model compound for the amino polymer. When this amine is added to a toluene solution of an uncured alkyd resin, it readily reacts with the latter at room temperature to form dibutyl phthalamide, which crystallizes and precipitates out. The reaction occurs at the phthalic linkages.

A mixture of an amino polymer and an uncured alkyd resin is examined by FTIR spectroscopy. After baking the mixture, the amino absorption band decreases with concurrent appearance of the amide absorption band, suggesting that amine-ester interchange reaction can indeed occur at the interface.

Resin/Silane/Glass Interfaces

Organosilanes are widely used as primers on glass fibers to promote the adhesion between the resin and the glass in fiberglass-reinforced plastics [125,126]. They are also used as primers or integral blends to promote adhesion of resins to minerals, metals, and plastics [127, 128]. These organosilanes have the general structure R—Si—X$_3$, where R is an alkyl or a functional alkyl group and X is a hydrolyzable group such as alkoxy group or chlorine. Some commonly available silanes are listed in Table 11.12.

Various theories of silane adhesion promotion have been proposed (Section 12.6.3). The chemical bonding theory remains the most viable. Essentially, during application, (1) the X groups hydrolyze to silanol groups, which then react with the silanol groups on the glass surface to form ether linkages; (2) the hydrolyzed silanes also self-condense to form polysiloxanes; and (3) the functional groups on the R group can react with appropriate chemical groups in the resin and thus chemically coupling the resin to the glass. Silanes are thus known as coupling agents. The X groups serve only to provide hydrolytic sites. Their chemical nature does not affect the coupling action. An exception is silyl peroxide, such as vinyltri(peroxy-t-butyl) silane, whose peroxy groups act as in situ free radical initiator and can promote resin/silane reaction. The R group is chosen for its chemical reactivity with the resin. As discussed below, actual reactions are more complicated than the simplified picture given here. Investigations using IR spectroscopy, Raman spectroscopy, radioisotope tracer, ellipsometry, and electron microscopy confirm the chemical coupling mechanism.

Silanes can promote both dry and wet adhesions. The effect is particularly pronounced in improving the stability of adhesive bond against hydrolytic degradation in a wet environment (Sections 11.4.4 and 15.4.2).

Structure of Silane Treatment Solution. Silanes are commonly applied to glass from fresh aqueous solutions. Neutral silanes (where R is a vinyl, glycidyl, or methacryloxy group) as usually prepared in dilute acetic acid solution of pH about 4 hydrolyze rapidly to silane triols, and then slowly condense to oligomeric siloxanols [129],

$$R-Si-(OCH_3)_3 + 3 H_2O \xrightarrow[\text{FAST}]{\text{HYDROLYSIS}} R-Si-(OH)_3 + 3 CH_3OH$$

$$R-Si-(OH)_3 \xrightarrow[\text{SLOW}]{\text{CONDENSATION}} R-\underset{\underset{OH}{|}}{\overset{\overset{OH}{|}}{Si}}-\left[O-\underset{\underset{OH}{|}}{\overset{\overset{OH}{|}}{Si}}\right]_n-R$$

The monomeric and lower oligomeric siloxanols are soluble in water, but higher oligomers are insoluble. Aqueous silane solutions therefore, have a limited stability, and must be used in a few hours. A fresh silane solution contains mainly monomers and dimers. For instance, a fresh aqueous solution of vinyltrimethoxysilane contains 82% monomer, 15% dimer, and 3% trimer. But after aging until precipitation just starts, it contains 34% monomer, 23% dimer, 30% trimer, and 13% tetramer [129]. Silanes will totally lose coupling activity if they completely condense to polysiloxanes before use [130].

Aminosilanes having a nitrogen on the third carbon atom in the R group have unique solution properties. These γ-aminoalkoxysilanes

Table 11.12. Structure of Some Silane Coupling Agents

Name	Structure
Ethyltriethoxysilane	$CH_3CH_2Si(OC_2H_5)_3$
γ-Chloropropyltrimethoxysilane	$ClCH_2CH_2CH_2Si(OCH_3)_3$
Vinyltriethoxysilane	$CH_2\!=\!CHSi(OCH_2CH_3)_3$
Vinyltrichlorosilane	$CH_2\!=\!CHSiCl_3$
Vinyltriacetoxysilane	$CH_2\!=\!CHSi(OOCCH_3)_3$
Vinyltri(methoxyethoxy) silane	$CH_2\!=\!CHSi(OCH_2CH_2OCH_3)_3$
Vinyltri(peroxy-t-butyl) silane	$CH_2\!=\!CHSi(OOtBu)_3$
γ-Methacryloxypropyltrimethoxysilane	$CH_2\!=\!\overset{\displaystyle CH_3}{C}\!-\!\overset{\displaystyle O}{\overset{\|}{C}}\!-\!O(CH_2)_3\!-\!Si(OCH_3)_3$

Name	Structure
β-(3,4-Epoxycyclohexyl)ethyltrimethoxysilane	![epoxycyclohexyl]—CH$_2$CH$_2$Si(OCH$_3$)$_3$
γ-Glycidoxypropyltrimethoxysilane	CH$_2$—CHCH$_2$—O(CH$_2$)$_3$—Si(OCH$_3$)$_3$ (with epoxide O bridging CH$_2$-CH)
γ-Mercaptopropyltrimethoxysilane	HSCH$_2$CH$_2$Si(OCH$_3$)$_3$
γ-Aminopropyltriethoxysilane	NH$_2$(CH$_2$)$_3$Si(OC$_2$H$_5$)$_3$
N-β-(aminoethyl)-γ-aminopropyltrimethoxysilane	NH$_2$(CH$_2$)$_2$NH(CH$_2$)$_3$Si(OCH$_3$)$_3$
N-bis(β-hydroxyethyl)-γ-aminopropyltriethoxysilane	(HOCH$_2$CH$_2$)N(CH$_2$)$_3$Si(OC$_2$H$_5$)$_3$
Vinylbenzyl amino silane	CH$_2$=CH-CH$_2$-C$_6$H$_4$-CH$_2$CH$_2$NH(CH$_2$)$_3$-Si(OCH$_3$)$_3$
Styryl diamino silane	CH$_2$=CH-C$_6$H$_4$-CH$_2$NH(CH$_2$)$_2$NH(CH$_2$)$_3$-Si(OCH$_3$)$_3$

Table 11.13. Thickness of Adsorbed Silane Layer

Silane	Solid surface	Solution concentration and solvent
γ-Aminopropyl triethoxysilane	Borosilicate glass	1%, benzene
	Glass	1%, water
		5%, water
	Chrome	1%, water
Vinyltrimethoxysilane	Glass	0.1-1.0% water
		1%, methyl ethyl ketone
Vinyltriethoxysilane	E glass	0.1%, water
		1.0%, water
		1.5%, water
		2.0%, water

Source: Collected from Refs. 132-134.

hydrolyze almost immediately in water, forming stable aqueous solutions. Generally, stable solutions are obtained in hydrogen-bonding solvents such as water and alcohols. Solutions in nonpolar solvents such as toluene will precipitate upon contact with moist air. This stability is attributed to the formation of internal cyclic zwitterion, that is,

Aminosilanes having nitrogen on other than a β- or γ-carbon atom do not form stable aqueous solutions, because cyclization cannot occur.

Amino groups can also catalyze the condensation reaction between the silane and the glass surface [131]. Therefore, aminosilanes are self-catalytic and do not need external catalysts. On the other hand, amines such as n-propylamine can be used to catalyze the silane/glass coupling reaction for neutral silanes.

Silane/Glass Interface. Mechanism of Coupling: Hydrolyzed silanes first physically adsorb onto glass surface by hydrogen-bonding with

Amount of silane adsorbed, mg/cm^2	Film thickness, Å	Monolayer equivalent	Technique
-	2700	270	^{14}C radioisotope
5.7×10^{-4}	60	6	Ellipsometry
14.1×10^{-4}	150	15	Ellipsometry
0.9×10^{-4}	10	1	Ellipsometry
5.1×10^{-4} to 16.5×10^{-4}	50-160	5-16	Ellipsometry
0.8×10^{-4}	8	0.8	Ellipsometry
-	240	24	FT-IR spectroscopy
-	1790	179	FT-IR spectroscopy
-	1980	198	FT-IR spectroscopy
-	3250	325	FT-IR spectroscopy

the silanol groups on the glass, forming monolayer or multilayers (up to about 300 monolayer equivalents) depending on the concentration of the treating solution and the treating condition. The thickness of the adsorbed layer increases with silane concentration and treatment time [131-135]. Table 11.13 gives the thicknesses of some adsorbed silane layers. Adsorption from a nonaqueous solution requires a sufficient amount of water on the glass surface to hydrolyze the silane, and tends to give thinner layers.

Figure 11.32 shows the kinetics of silane adsorption onto glass [133]. Rapid adsorption occurs in the first 0.5-1 min, followed by slow bulidup [133,135]. Vinylsilane continues to adsorb slowly even after 400 min, but γ-aminopropylsilane reaches the saturation level within the first minute. Aminosilane adsorbs much faster than does vinylsilane, because of the autocatalytic effect of the amino group in the former. The continued adsorption of vinylsilane may be due to slow polymerization of the vinyl groups. Figure 11.33 shows the effect of silane concentration on the surface coverage [135]. The data for methacryloxypropyltrimethoxysilane and epoxypropyltrimethoxysilane coincide with each other. The continued slow buildup for these two silanes is again probably due to slow polymerization.

During drying or prolonged immersion at room temperature, two chemical reactions occur simultaneously: (1) condensation between the silanol group of silane and the silanol group on the glass to form ether linkage between the silane and the glass in the first monolayer (that is, silane/glass coupling reaction), and (2) condensation reaction between the silanol groups of neighboring silane molecules to form

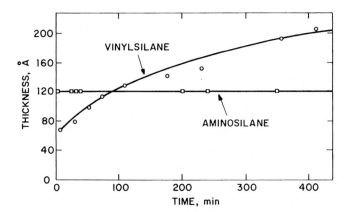

Figure 11.32. Thickness of silane layer adsorbed on glass. ○, Vinyl-tri(2-methoxyethoxy)silane adsorbed from 1% aqueous solution; □, γ-aminopropyltriethoxysilane adsorbed from 5% aqueous solution. (After Ref. 133.)

polysiloxane (that is, the self-condensation reaction of hydrolyzed silanes). Schematically,

The silane/glass coupling reaction occurs in the first monolayer. The formation of polysiloxane by self-condensation occur uniformly in the first monolayer, but only in isolated patches in the multilayers. Therefore, up to 98% of the adsorbed silane hydrolyzate in the multilayer is physically adsorbed, not chemically bound to the glass surface. These physically adsorbed silanes may degrade the resin/glass adhesion, and should be kept to a minimum.

Some Experimental Supporting Evidence. Radioisotope and Electron Microscopy: Schrader and coworkers [132] treated borosilicate glass with a 1% benzene solution of γ-aminopropyltriethoxysilane having a ^{14}C label on the carbon atom α to the amino group. The

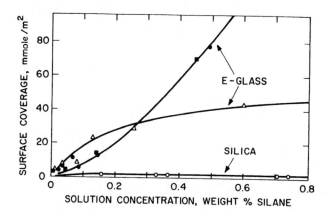

Figure 11.33. Adsorption of three different silanes from aqueous solutions. ●, γ-Aminopropyltriethoxysilane; Δ, γ-methacryloxypropyltrimethoxysilane; ■, 3-(2',3'-epoxypropoxy)propyltrimethoxysilane. (After Ref. 135.)

amount of silane deposited is determined by counting radioactivity. The treated borosilicate glass is then extracted in water at room temperature, and the radioactivity followed with extraction time. Next, the glass is extracted with boiling water and the residual radioactivity counted periodically. The extraction study reveals that the adsorbed silane consists of three fractions (Figure 11.34). The major fraction (fraction 1), which could be as much as 98% of the total and about 270 monolayer equivalents depending on the amount deposited, consists of physically adsorbed silane hydrolyzates which are insoluble in benzene but are rapidly removed by cold water rinse. Fraction 2 consists of about 10 monolayer equivalents of chemisorbed polysiloxanes, which require 3-4 hr of extraction with boiling water for essentially complete removal. Electron photomicrographs of this fraction after 1 hr of boiling-water extraction show the presence of islands [132], which have also been observed by Sterman and Bradley [136]. After the complete tapering off of the radioactivity versus extraction time, electron photomicrographs now show a completely bare surface, the islands having all disappeared. However, some radioactivity equivalent to a monolayer remains on the apparently bare glass surface. This residue (fraction 3), which is much more tenaciously held to the surface than are fractions 1 and 2, is apparently attached to the glass surface by chemical bonding.

Raman Spectroscopy: Koenig and Shih [137] treated E-glass and silica fibers with vinyltriethoxysilane. The fibers are immersed in 2-3% aqueous solutions of the silane for 30 min at room temperature,

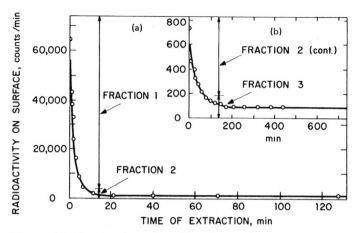

Figure 11.34. Amount of radioactive aminosilane remaining on polished borosilicate glass surface after extraction with water. (a) Extraction at 25°C; (b) extraction continued at 100°C. (After Ref. 132.)

dried for 5 hr at room temperature, and baked at 104°C for 10 min. The silane-treated fibers are then washed by immersion in toluene for 24 hr, followed by rinsing with boiling water for 2 hr. Raman spectra show that the vinyl lines at 1600, 1410, 1276, and 967 cm^{-1} remain, although reduced, after the extraction. The stability against extraction suggests chemical bonding between the silane and the fiber. Chlorination of the silane-treated fiber indicates that not all vinyl groups are available for chlorination, as the vinyl line at 1600 cm^{-1} is not completely removed, whereas the C——Cl stretching mode at 725 cm^{-1} increases with the extent of chlorination. This is consistent with the existence of multilayers. Apparently, the vinyl groups which are buried deep inside the multilayers are not available for chlorination.

Infrared spectroscopy. Kaas and Karods [131] treated silica powder (surface area 150 m^2/g) with a 1% acetone solution of γ-aminotriethoxysilane at room temperature. After the treatment, the free hydroxyl band at 3745 cm^{-1} of the silica surface decreases, the hydrogen-bonded hydroxyl band at 3500 cm^{-1} of the silica surface increases, and the C——H stretching bands at 2940 and 2880 cm^{-1} of the silane appear, indicating adsorption of the silane by hydrogen bonding with the silanol groups on the silica. After evacuation of the treated silica at 3×10^{-5} torr and 460°C for 5 hr, the C——H stretching bands remain, although decreased, indicating that the physically adsorbed silanes are removed by the evacuation but that the remaining fractions are chemically bound to the silica surface. In a parallel experiment using a 1:1 mole ratio of ethyltriethoxysilane and n-propylamine, n-propyl amine is found to catalyze the condensation reaction between the silane

Chemical Adhesion

and the silica. Amino groups can hydrogen-bond with the silanol groups on the silica surface, but the hydrogen bonds are readily reversible under evacuation. The adsorbed γ-aminopropylsilane appears to exist as dimers [139],

Ishida and Koenig [137,138] used Fourier-transform IR spectroscopy to study the bonding between vinylsilane and silica or E-glass surface. Substraction spectra of an oligomeric polyvinylsiloxanol (deposited on silica surface) before and after heat treatment (200°C for 3 hr) show that the OH stretching bands at 893 and 970 cm^{-1} of the silicasilanol groups disappear, and the oligomer-Si-O-Si-silica antisymmetric stretching bands at 1170 and 1080 cm^{-1} appear, indicating chemical bonding between the silane oligomer and the silica surface. Similar results have also been obtained in other IR spectroscopic studies [140–143].

Resin/Silane Interface. Chemical bonding between resin and silane has been shown by several different techniques. Sterman and Marsden [144] polymerized styrene and methyl methacrylate, respectively, in the presence of methacryloxysilane-treated glass fibers. After exhaustive extraction with benzene and trichloroethylene, respectively, residual carbon analysis showed that a thin layer of polymer remains on the glass.

Johannson and coworkers [135] polymerized ^{14}C-labeled methyl methacrylate and styrene mixture on untreated and silane-treated E-glass. Radioactivity measurement shows that a sizable residue of the resin remains on the silane-treated glass, but not on the untreated glass, after extraction.

Koenig and Shih [137] polymerized methyl methacrylate in the presence of vinylsilane-coated E-glass and silica fibers, using an azo free radical initiator. After exhaustive extraction with methyl

ethyl ketone, Raman spectroscopy shows that significant amount of poly(methyl methacrylate) retains on the surface, and that about 30–40% of the vinyl siloxane groups on the glass surface have reacted with the acrylic monomer.

Ahagon and coworkers [120] used the characteristic IR absorption peaks of terminal vinyl groups at 1.6 and 2.1 μm to follow the chemical bonding between vinylsilane-coated silica and polybutadiene. The rubbery polymer is mixed with an equal weight of the silane-coated silica powder, 2.7% by weight of dicumyl peroxide and 0.32% by weight of sulfur, and heated in a press at 160°C for 1 hr. After this treatment, the terminal vinyl peak decreases by about 40% in intensity. However, when a similar sample without the cross-linking agents is used, the terminal vinyl peak before and after the heat treatment stays substantially the same. This indicates that the vinyl groups of the silane molecules covalently bond to the polymer during cross-linking of the polymer.

11.4.3. Adhesion Promoting Functional Groups

Small amounts (0.001–0.1 mole fraction) of appropriate reactive functional groups, when incorporated into polymers, can greatly increase the adhesive bond strength. This has been widely practiced, and is perhaps the most fruitful way of promoting adhesion [145]. In some cases as little as 0.001–0.01 mole fractions of functional groups can dramatically increase the adhesive strength. At such low amounts, polymer bulk properties and wettability are practically unchanged. Furthermore, the effectiveness of functional groups in adhesion promotion is quite specific with respect to surface chemical composition. These findings suggest that the improved adhesion results from interfacial chemical bonding. Excess amounts of functional groups should be avoided, as they may degrade the bulk properties and thus adversely affect the joint strength.

Concentration Effect

Adhesive bond strength often increases with increasing concentration of the functional group in the adhesive by [115,146–152]

$$\sigma_f = \sigma_{f_0} + kC^n \tag{11.52}$$

where σ_f is the adhesive strength when the concentration of the functional group in the adhesive is C, σ_{f_0} the adhesive strength when the functional group is absent, and k and n are constants. Usually, n is about 0.6–1.0.

Figure 11.35 shows that n = 2/3 for the adhesion of an epoxy adhesive to aluminum, and of polyvinylformal adhesive to steel [147,148].

Chemical Adhesion

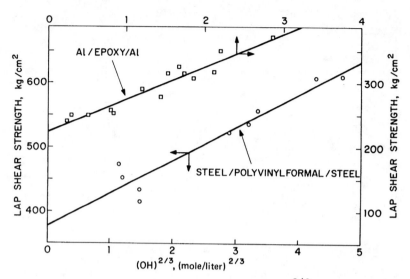

Figure 11.35. Lap shear strength versus $(OH)^{2/3}$ fo the adhesive for aluminum strips bonded with epoxy adhesives and cold-rolled steel strips bonded with poly(vinyl formal) adhesives. (Data from Refs. 147 and 148.)

Regression analysis gives $\sigma_f = 51.0(OH)^{2/3} + 224$ for the lap shear strength of aluminum/epoxy/aluminum joints, and $\sigma_f = 50.8(OH)^{2/3} + 378$ for the lap shear strength of steel/polyvinylformal/steel joints, where σ_f is in kg/cm^2 and (OH) in mole/liter [148]. Similar relation has also been found for the peel strength of laminates of poly(vinyl chloride-maleic acid) to cellulose [115,147]. In this case, $\sigma_f = 795 (COOH)^{2/3} + 16.8$, where σ_f is in g/cm at 20°C and (COOH) in mole/1000 g of polymer. The $C^{2/3}$ dependence of σ_f appears to suggest that the adhesive strength is proportional to the surface concentration of functional groups.

In other cases, joint strength increases rapidly with concentration and levels off at a plateau. The curve follows the Langmuir adsorption isotherm,

$$\sigma_f = \sigma_{f,m} \theta = \sigma_{f,m}\left(\frac{bc}{1+bc}\right) \qquad (11.53)$$

where $\sigma_{f,m}$ is the fracture strength at complete surface coverage, θ the fractional surface coverage, c the concentration of functional groups in the bulk, and b is a constant. Such behavior is exemplified by the adhesion of a carboxylated adhesive (R-COOH) on epoxy adherend (Figure 11.36). The plateau in the bond strength may arise either from the saturation adsorption of the functional groups at the

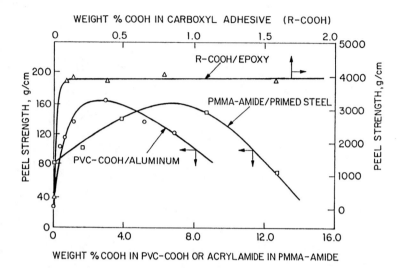

Figure 11.36. Peel strength versus amount of functional group. (After Refs. 115, 150 and 151.)

interface, or from cohesive failure of adhesive or adherend. In the case cited, the saturation adsorption appears to be the mechanism.

In some cases, on the contrary, excessive amounts of functional groups will decrease the adhesive strength [115,150-153]. This behavior is shown for poly(vinyl chloride-maleic acid)/aluminum joints [115], and for poly(methyl methacrylate-acrylamide)/epoxy-ester-primed steel joints [150] in Figure 11.36. Excessive functional groups can decrease the bond strength by degrading the bulk properties of the adhesive, corroding the adherend, or introducing deleterious impurities into the adhesive. A more vivid example is shown in the strengths of butt joints and lap joints of polypropylene radiation-grafted with bis(2-chloroethyl)vinyl phosphonate (BCVP) sandwiched between two aluminum adherends [153] (Figure 11.37). BCVP can promote adhesion to aluminum, steel, and copper, probably by complex formation with the metals (Table 11.14). Too much BCVP will, however, decrease the joint strength by weakening the cohesive strength of the adhesive.

Specificity of Functional Groups

The effectiveness of functional groups in promoting adhesion shows considerable specificity with respect to the adherend. Table 11.14 illustrates such specificity for various functional groups radiation-grafted onto polypropylene used as adhesive on various metals (alu-

Table 11.14. Effect of Various Functional Groups on the Tensile Butt Strength of Polypropylene to Metal Adhesive Bonds

Grafted monomer	Amount grafted, wt %	Apparent reactive group	Tensile butt strength, psi[a]		
			Aluminum	Steel	Copper
$CH_2=CH-C_6H_4-CH_2SO_3H$	0.5	Sulfonic acid	1200 (c)	1100 (c)	1000 (c)
$HOOC-CH=CH-COOH$	0.5	Fumaric acid	600 (i)	1900 (c)	1100 (c/i)
$CH_2=CH-C_6H_4-B(OH)_2$	3.0	Boronic acid	1200 (c)	2000 (c)	1300 (c)
$CH_2=CH-P(=O)(OH)_2$	3.5	Phosphonic acid	700 (i)	—	800 (i)
$CH_2=CH-P(=O)(OCH_2CH_2Cl)_2$	0.75	Chlorophosphonate (BCVP)	3800 (c)	3100 (c)	3300 (c)
$CH_2=CH-P(=O)(OCH_2CH_3)_2$	4.0	Phosphonate	550 (i)	450 (i)	100 (i)
$CH_2=CH-CH_2OH$	0.5	Hydroxyl	500 (i)	1200 (c/i)	700 (i/c)
$CH_2=CH-CH_2C(=O)OCH-CH_2$ (epoxide)	4.5	Epoxide	800 (i/c)	1800 (c)	750 (i/c)
$CH_2=CHSi(OCH_2CH_3)_3$	0.5	Silane	600 (i)	1200 (c/i)	200 (i)
$CH_2=CHCH_2NH_2$	0.5	Primary amine	650 (i/c)	1500 (c)	550 (i)
$CH_2=CHCH_2N(CH_3)_2$	0.5	Tertiary amine	900 (i/c)	800 (i/c)	300 (i)
$CH_2=CH-CN$	3.2	Nitrile	500 (i)	1100 (c/i)	1200 (c/i)
Polypropylene (control)	—	—	300 (i)	600 (i)	100 (i)

[a] Failure locus i = interfacial failure; c = cohesive failure.
Source: From Ref. 153.

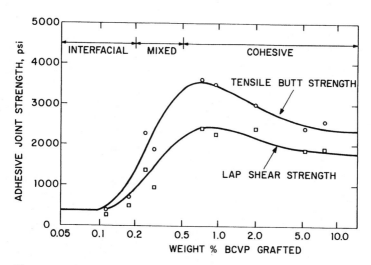

Figure 11.37. Adhesive joint strength at room temperature versus amount of bis(2-chloroethyl)vinyl phosphonate radiation-grafted onto polypropylene. The adherends are aluminum. (After Ref. 153.)

minum, steel, and copper). For instance, the epoxide group effectively promotes adhesion of polypropylene to steel, but only marginally to aluminum and copper. On the other hand, the chlorophosphonate group effectively promotes adhesion to all three adherends, probably by complex formation. Phosphonic acid and ethyl phosphonate groups are very marginal in promoting adhesion, probably because they do not form complexes.

If functional groups promote adhesion by changing the bulk properties, their effectiveness should be nonspecific. Functional groups are polar, and may thus improve the wettability and interfacial diffusion. Undoubtedly, such effects operate in some cases. However, in most cases, the specificity of the functional group arises from chemical bonding, or hydrogen bonding.

Selection of Functional Groups

To promote adhesion, functional groups should be selected for interfacial chemical bonding or specific interaction such as hydrogen bonding. Chemical bonding is much more effective than hydrogen bonding. Some useful functional groups are discussed below. Many functional monomers are quite toxic. Therefore, proper caution should be exercised in their use.

Chemical Adhesion

Carboxyl groups promote adhesion to metals, glass, and polymers. They can be incorporated into adhesives by copolymerization with carboxyl monomers, or using additives containing carboxyl groups. For instance, acrylic acid in ethylene-acrylic acid copolymers promotes adhesion to aluminum, steel, and copper [152]. Methacrylic acid copolymerized in butadiene rubbers (such as polybutadiene, butadiene-styrene rubber, and butadiene-acrylonitrile rubber) promotes adhesion to aluminum, steel, glass, rayon, nylon, and polyester [151]. Maleic acid copolymerized in vinyl chloride-vinyl acetate copolymers promotes adhesion to cellulose and aluminum [115,147]. α,β-Unstaruated carboxylic acids and their salts grafted onto polyolefins by melt blending promote adhesion to metals [154]. Carboxylated polyesters are useful as additives to promote adhesion of various polymers (such as cellulose acetate butyrate, vinyl chloride-vinyl acetate copolymers, poly(methyl methacrylate), polystyrene, and polycarbonate) to various adherends (such as steel, brass, copper, chromium, aluminum, nylon, and polyester) [155]. Useful carboxyl monomers include acrylic acid, methacrylic acid, maleic acid, and itaconic acid [156].

Nitrogen-containing groups promote adhesion to various organic and inorganic adherends [124,157,158]; see also Section 11.4.2.C. Useful amino monomers include aminoethyl methacrylate, aminoethyl vinyl ether, 2-(1-aziridinyl)ethyl methacrylate, t-butylaminoethyl methacrylate, dimethylaminoethyl methactylate, vinyl pyridine derivatives, ethylene-urea derivatives, oxazolidine derivatives, oxazoline derivatives, oxazine derivatives, piperidine derivatives, piperazine derivatives and morpholine derivatives [124,157,158].

Hydroxyl and methylol groups promote adhesion to various organic and inorganic adherends. For instance, hydroxyl groups can promote adhesion of polyethylene to glass. Methylol groups can promote adhesion to polyamides, probably through reaction with the active hydrogen on the amide nitrogen [159]. Some useful monomers are hydroxyethyl acrylate and methacrylate, and methylol acrylamide and methacrylamide [160,161].

Epoxide groups promote adhesion to various organic and inorganic surfaces having active hydrogens such as hydroxyl, carboxyl, amino, and amide groups [157,162,163]. Epoxy resin is used to prime aramid fibers before RFL dip to achieve adhesion to the fibers [164]. Useful epoxide monomers include glycidyl methacrylate, allylglycidyl ether, and glycidylsilane [157].

Isocyanate groups are quite reactive, and promote adhesion to surfaces having hydroxyl, carboxyl, amino, or amide groups [159, 162,163]. For instance, isocyanate groups can promote adhesion to a polyamide surface, probably by reacting with the amide hydrogen [145].

Phosphoric acid group can be incorporated into a polymer by reacting phosphoric acid with a polymer containing an epoxide group [165]. Such a polymer has been reported to improved adhesion to metals [165]. A *sulfonic acid group* can be incorporated into a polymer by copolymerizing with sulfoethyl methacrylate [166]. Such copolymers show improved adhesion to a variety of adherends.

11.4.4. Coupling Agents

Several classes of compounds are known to promote adhesion apparently by chemically coupling the adhesive to the adherends [127]. These include silanes; esters of phosphoric, phosphonic, or phosphorous acid; chromium complexes; and titanates.

Silanes

Silane coupling mechanisms have been discussed in Section 11.4.2. Mechanisms of wet strength promotion are discussed in Section 15.4.2. The history of silane development has been reviewed elsewhere [167-169]. Here, the use of silanes as adhesion promoters is discussed.

Silanes having reactive alkyl groups (vinyl, amino, epoxide, mercaptan, hydroxyl) can chemically couple the adhesive and the adherend by reacting with appropriate groups, and thus promote adhesion. Silanes having nonreactive alkyl groups have no chemical coupling activity, and thus do not promote adhesion. This is illustrated in Table 11.15 [170]. However, a few exceptions are known [171]. In such cases, adhesion appears to arise from improved interfacial compatibility.

Silanes may be used as pretreatment for adherends [170,172-178], as an additive in integral blends [128,172-175] or copolymerized with the resin. For instance, vinylsilane and methacryloxysilane can be copolymerized with vinyl and acrylic resins. Mercaptosilane can be incorporated into vinyl or acrylic polymers by chain transfer in the vinyl polymerization. Silanes have been used to promote adhesion to hydrophilic adherends, such as glass [170,172-178], aluminum, and steel [174,175,178], and mineral fillers such as silica, alumina, clay, talc, and calcium carbonate [170,178,179]. The effectiveness is highest with silica and aluminum, moderate with TiO_2, clay, and talc, and least with calcium carbonate [178,179], consistent with mineral surface reactivity. Calcium carbonate lacks reactive surface groups [178, 179].

Chromium Complexes

Trivalent chromium complexes with carboxyl ligands are useful as coupling agents [127,180].

Table 11.15. Adhesion of Elastomers Vulcanized on Silane-Treated Glass Plates

Alkyl group of silane	Peel strength, g/cm					
	Natural rubber	Styrene butadiene rubber	Ethylene/ propylene diene rubber	Nitrile rubber	Neoprene	Chloro- sulfonated polyethylene
No silane	20	0	200	0	0	210
Nonreactive silanes						
Propyl	-	0	120	-	-	-
3-Chloropropyl	110	20	200	90	20	210
Phenyl	-	120	160	-	-	-
Chlorophenyl	-	70	230	-	-	-
Reactive silanes						
Vinyl	120	160	390	20	20	320
Methacryloxypropyl	90	20	820	20	20	360
Aminoethylamino propyl	790	210	620	1400	200	3600 (c)[a]
Styryl diamino	860	360	3600 (c)[a]	980	390	3600 (c)[a]
Mercaptopropyl	360	540	1900	210	720	460

[a]Failure locus c = cohesive failure; all other specimens failed interfacially.
Source: From Ref. 170.

Dilution with water (pH 5–7) will start hydrolysis, in which some chloride ions are replaced by hydroxyl groups [180],

As hydrolysis continues, the complex begins to polymerize by olation [180],

The complex may react with the adherend at any time during olation at negatively charged surface sites. Thus, acidic SiOH sites on glass will react with the complex, if the pH is high enough to ionize to SiO-.

Table 11.16. Structures of Titanate Coupling Agents

Name	Structure
Tetralkoxy type	
Tetraisopropyl titanate	$Ti(OCH-CH_3)_4$ with CH_3 branch
Isopropyl tri(lauryl-myristyl) titanate	$CH_3CH(CH_3)-O-Ti-[O(CH_2)_nCH_3]_3$
Carboxy monoalkoxy type	
Isopropyl triisostearoyl titanate	$CH_3CH(CH_3)-O-Ti-[OC(=O)(CH_2)_{14}CH(CH_3)-CH_3]_3$
Isopropyl isostearoyl dimethacryl titanate	$CH_3CH(CH_3)-O-Ti(OC(=O)(CH_2)_{14}CH(CH_3)CH_3)(OC(=O)C(CH_3)=CH_2)_2$

Table 11.16. (Continued)

Name	Structure
Isopropyl isostearoyl di(4-aminobenzoyl) titanate	$\text{CH}_3\text{CH(CH}_3\text{)-O-Ti}[\text{O-C(=O)(CH}_2)_{14}\text{CH(CH}_3\text{)CH}_3][\text{O-C(=O)-C}_6\text{H}_4\text{-NH}_2]_2$
Sulfonyl monoalkoxy type	
Isopropyl tri(dodecylbenzene-sulfonyl) titanate	$\text{CH}_3\text{CH(CH}_3\text{)-O-Ti}[\text{O-S(=O)}_2\text{-C}_6\text{H}_4\text{-CH(CH}_3\text{)(CH}_2)_9\text{CH}_3]_3$
Isopropyl di(dodeceylbenzene-sulfonyl)-4-aminobenzene-sulfonyl titanate	$\text{CH}_3\text{CH(CH}_3\text{)-O-Ti}[\text{O-S(=O)}_2\text{-C}_6\text{H}_4\text{-CH(CH}_3\text{)(CH}_2)_9\text{CH}_3]_2[\text{O-S(=O)}_2\text{-C}_6\text{H}_4\text{-NH}_2]$

Chemical Adhesion

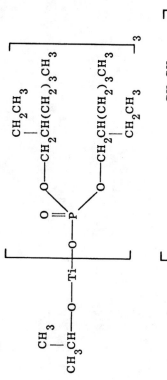

Phosphato monoalkoxy type

Isopropyl tri(diisooctyl-phosphato) titanate

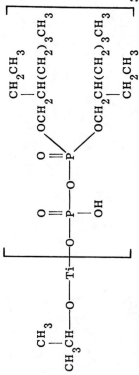

Pyrophosphato monoalkoxy type

Isopropyl tri(dioctyl-pyrophosphato) titanate

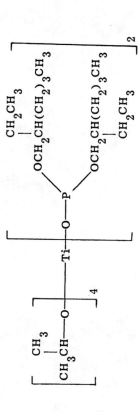

Phosphito coordinate type

Tetraisopropyl di(dioctyl-phosphito) titanate

Table 11.16 (continued)

Name	Structure
Chelate type	
Titanium isostearate methacrylate oxyacetate	(structure: Ti chelate with O–C(=O)–(CH$_2$)$_{14}$–CH(CH$_3$)–CH$_3$, O–C(=CH$_2$)–OCH$_3$, and O–CH$_2$–C(=O)–O ring ligands)
Ethylene isostearate methacrylate titanate	(structure: Ti chelate with O–C(=O)–(CH$_2$)$_{14}$–CH(CH$_3$)–CH$_3$, O–C(=CH$_2$)–OCH$_3$, and O–CH$_2$–CH$_2$–O ring ligands)

The preferred pH varies with the nature of the glass surface. The glass-complex bond is given as [182],

```
        R                    R
        |                    |
        C                    C
       ╱ ╲                  ╱ ╲
      O   O   H            O   O
      |   |   |   H        |   |
      |   H   O   |        |   H
      |       |   O        |       
     Cr◄──── Cr            Cr◄──── Cr      COMPLEX
      |       |             |       |
      O       O             O       O
   ═══╪═══════╪═════════════╪═══════╪═══
      |       |             |       |
     Si      Si            Si      Si      GLASS
       ╲   ╱   ╲          ╱  ╲   ╱
         O       O       O      O
```

The chromium complex can react with the resin (adhesive) in three ways: (1) coordinate covalent bonding with carboxyl, hydroxyl, or amino groups; (2) reaction with CrCl or CrOH groups to form Cr—O—R linkages with the elimination of either HCl or H_2O; and (3) copolymerization between the vinyl group in R and the vinyl groups in the resin [180].

Type (1) and (3) reactions depend on the nature of the R group. Reactive R groups contain vinyl, carboxyl, amino, or hydroxyl groups capable of chemical bonding with the resin matrix. Examples are methacrylic acid, acrylic acid, phthalic acid, succinic acid, and fumaric acid [127,181–183]. Stearic acid chromium complex is an example of the complex with a nonreactive R group, used primarily to impart a hydrophobic finish on a hydrophilic surface [127].

Titanates

Alkyl ortho titanates, $Ti(OR)_4$, are useful as coupling agents for glass, metals, and polymers [183,184]. The reactivity depends on the nature of the alkyl group R. Isopropoxy group is the most reactive and the most often used. Some typical titanate coupling agents are listed in Table 11.16.

These titanates can chemically bond to surfaces containing hydroxyl groups through condensation reaction between the isopropoxy, oxyacetate, or dioxyethylene group of the titanate and the hydroxyl group of the surface. The reactivity is affected by the functional group adjacent to the titanium, and increases in the order: (least reactive) sulfonyl, phosphato, pyrophosphato, phosphito, carboxy, alkoxy (most reactive). The reaction gives a monolayer coverage, as opposed to a multilayer coverage for silanes. Titanates do not polymerize by self-condensation in solution or in monolayer, except for the lower tetralkoxy titanates, such as tetraisopropyl titanate. The chelated titanates react preferentially with surface hydroxyls rather than with water, and are therefore useful in aqueous systems.

The chemically bound titanates on the surface can interact with the resin by several different mechanisms: (1) increased compatibility (for instance, stearyl group can improve compatibility with polyolefins); (2) transesterification of the —O—X—R group with carboxyl- or hydroxyl-containing resins [the transesterification reactivity increases in the order: (least reactive) carboxyl, alkoxy, phosphato, sulfonyl (most reactive); pyrophosphato and phosphito (coordinate type) groups are inactive]; (3) covalent bonding between the R groups and the resin (the R groups may contain reactive functional groups such as a vinyl, hydroxyl, mercapto, or amino group, which can react with appropriate groups in the resin).

11.5. MECHANICAL ADHESION

Mechanical anchoring is utilized widely; for instance, in metal plating of plastics, in adhesion to surface-treated metals, and in adhesion to porous substrates (paper, leather, and fabrics).

11.5.1. Surface Roughness and Bond Strength

Surface roughness may increase the adhesive bond strength by increasing the surface area, promoting wetting, or providing mechanical anchoring sites. Furthermore, weak boundary layers may be removed during the roughening process. All can contribute to increased bond strength. However, mechanical interlocking should be the most effective.

Roughness tends to lower the contact angle when the intrinsic angle θ_0 is less than 90°, or increase it when the intrinsic angle is greater than 90° (Section 1.10). Spontaneous wicking will occur when the Wenzel's roughness factor reaches a critical value r_c, that is,

$$r_c = \frac{1}{\cos \theta_0} \tag{11.54}$$

The bond strength tends to increase with adherend roughness [185], as illustrated in Figure 11.38.

Mechanical Adhesion

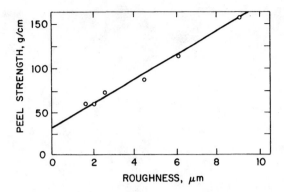

Figure 11.38. Effect of adherend (nickel) roughness on 180°-peel strength using a polyurethane adhesive. (After Ref. 185.)

11.5.2. Micromechanical Interlocking

Mechanical interlocking can produce strong adhesive bonds that are resistant to hydrolytic and thermal degradation. To be effective, the adherend surface must have sufficient numbers of microscopic undercutting or rootlike cavities. Some examples are discussed below.

Adhesion of Metal Platings to Plastics

The adhesion of metal platings to polypropylene and ABS plastics has been shown to arise from mechanical interlocking [186-188]. The metal plating process involves first treating the plastic surface to produce numerous cavities capable of mechanical interlocking, sensitized with stannous chloride solution, activated by depositing Pd^0 from Pd^{2+} solution, depositing electroless nickel or copper, and then electroplating the desired metal, such as chromium [186]. Strong adhesion of the metal plating to the plastics is obtained only when the plastics has been pretreated to produce numerous interlocking cavities (Section 9.1.2 and Figures 9.8 and 9.9).

Adhesion to Pretreated Metals

Various methods for pretreating metal surfaces have been reviewed elsewhere [189--192]. Many of these pretreatments may not only change the surface chemical compositions, but also produce interlocking surface sites. Mechanical interlocking contributes greatly, if not critically, to adhesion to such surfaces.

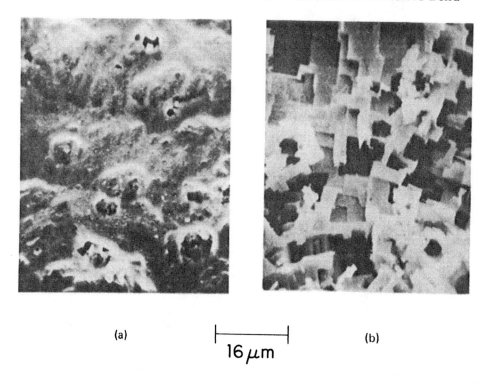

(a) |—— 16 μm ——| (b)

Figure 11.39. SEM photomicrographs of etched aluminum surface. Etched with 6% hydrochloric acid. (a) 2024 aluminum; (b) 3003 aluminum.

Figure 11.39 shows the surface topography of two different aluminum adherends etched with hydrochloric acid. The etched 2024 aluminum gives a relatively structureless surface with few isolated pits. On the otherhand, the etched 3003 aluminum gives numerous cubic crystal structures with tunneling and undercutting cavities. Polymers can strongly anchor to the latter, but not to the former.

Figure 11.40 shows the surface topography of an original and phosphate-treated cold-rolled steel. The original surface is relatively featureless and smooth. After phosphating, numerous intermeshing platelets of iron phosphate crystals can be seen on the surface. The interplatelet spaces provide numerous interlocking sites.

Figure 11.41 shows the surface topography of two polyethylene surfaces which have been sintered on two different anodized aluminum substrates; the aluminum was dissolved away with a dilute sodium hydrox-

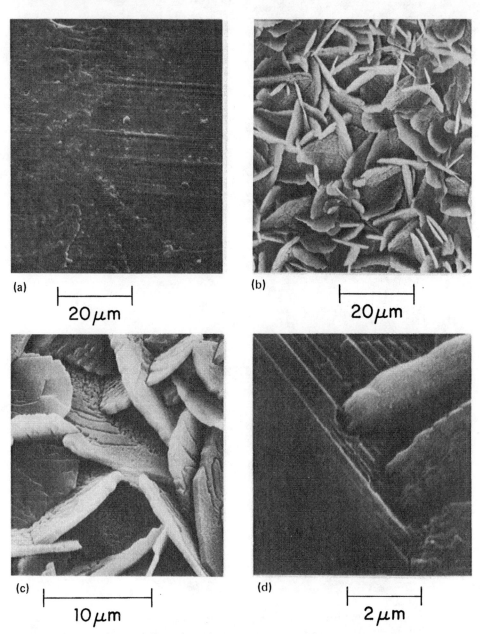

Figure 11.40. (a) Original cold-rolled steel; (b) phosphate conversion coating; (c) and (d) higher magnifications of (b).

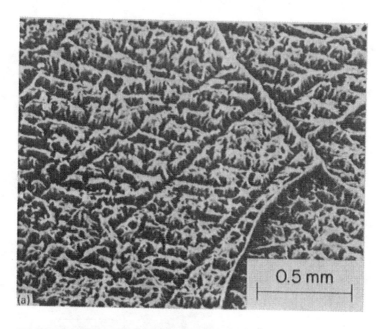
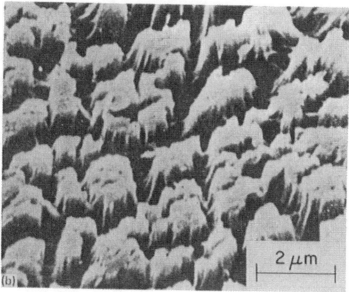

Figure 11.41. Scanning electron photomicrographs of two polyethylene surfaces bonded onto two anodized aluminum adherends which were later removed by dissolving with aqueous sodium hydroxide solution. (a) Aluminum anodized in 15% sulfuric acid at 20 V for 60 min; (b) aluminum anodized in 4% phosphoric acid at 20 V for 60 min. (After Ref. 193.)

Table 11.17. Effect of Surface Topography of Electroformed Copper (or Nickel) Foils on the Peel Strength of Copper (or Nickel) Laminates Bonded with an Epoxy Adhesive

Surface topography of copper foil		Peel strength, g/cm
Flat		670
0.3-μm dendrites		680
0.3-μm dendrites + oxides		785
3-μm Pyramids		1055
2-μm Low hills + 0.3-μm dendrites		1310
2-μm Low hills + 0.3-μm dendrites + oxides		1570
3-μm Pyramids + 0.3-μm dendrites + oxides		2420
Nickel foil with knobbed nodules		2330

Source: From Ref. 197.

ide solution. The polyethylene surfaces reproduce the topography of the anodized aluminum adherends, having numerous tufts and pores [193-195]. Mechanical interlocking accounts for the strong adhesion of polyethylene to such surfaces. Other examples have been reported elsewhere [196,197].

Table 11.17 shows the correlation of surface topography of electroformed copper with the peel strength of copper foil/epoxide laminates [197]. The surface topography of the electroformed copper foil is varied by varying the plating bath composition and plating conditions. The adhesive strength increases with mechanical interlocking.

REFERENCES

1. N. A. de Bruyne, Aero Res. Tech. Notes, 168, 1 (1956).
2. C. Mylonas, Exp. Stress Anal., 12, 129 (1955).
3. S. Wu, J. Adhes., 5, 39 (1973); also in *Recent Advances in Adhesion*, L. H. Lee, ed., Gordon and Breach, New York, 1973, pp. 45–63.
4. S. Wu, in *Polymer Blends*, Vol. 1, D. R. Paul and S. Newman, eds., Academic Press, New York, 1978, pp. 243–293.
5. S. Wu. J. Colloid Interface Sci., 31, 153 (1969).
6. S. Wu, J. Phys. Chem., 74, 632 (1970).
7. S. Wu, J. Polym. Sci., C34, 19 (1971).
8. S. Wu, J. Macromol. Sci., C10, 1 (1974).
9. M. Levine, G. Ilkka, and P. Weiss, J. Polym. Sci. Polym. Lett. Ed., 2, 915 (1964).
10. C. A. Dahlquist, in *Aspects of Adhesion*, Vol. 5, D. J. Alner, ed., CRC Press, Cleveland, Ohio, 1969, pp. 183–201.
11. C. A. Dahlquist, ASTM STP 360, 46 (1963).
12. M. J. Barbarisi, Nature, 215, 383 (1967).
13. M. J. Barbarisi, *Wetting of High and Low Energy Surfaces by Liquid Adhesives and Its Relation to Bond Strength*, Tech. Rep. 3456, Picatiny Arsenal, Dover, N.J., 1967 (AD651093 National Technical Information Service, Dept. of Commerce, Springfield, Va.).
14. D. H. Kaelble, *Physical Chemistry of Adhesion*, Wiley-Interscience, New York, 1971.
15. N. A. de Bruyne, Nature, 180, 262 (1957).
16. D. J. Gordon and J. A. Calquhoun, Adhes. Age, 19(6), 21 (1976).
17. W. A. Zisman, in *Adhesion and Cohesion*, P. Weiss, ed., Elsevier, Amsterdam, 1962, pp. 176–208.
18. G. A. Dyckerhoff and P. J. Sell, Angew. Makromol. Chem., 21(312), 169 (1972).
19. K. L. Mittal, Polym. Eng. Sci., 17, 467 (1977).
20. A. W. Neumann, Staub-Reinhalt. Luft, 28(11), 24 (1968).
21. M. Toyama, T. Ito and H. Moriguchi, J. Appl. Polym. Sci., 14, 2295 (1970); 17, 3495 (1973).
22. M. Toyama and T. Ito, Polyn.-Plast. Technol. Eng., 2 (2). 161 (1973).

23. M. Toyama, Y. Kitazaki, and A. Watanabe, J. Adhes. Soc. Jpn., 6, 356 (1970).
24. M. Toyama, T. Ito, and H. Moriguchi, J. Appl. Polym. Sci., 14, 2039 (1970).
25. Y. Iyengar and D. E. Erickson, J. Appl. Polym. Sci., 11, 2311 (1967).
26. S. S. Vogutskii, S. M. Yagnyatinskaya, L. Ya. Kaplunova, and N. L. Garetovskaya, Rubber Age, 37 (February 1973).
27. S. Wu, Org. Coat. Plast. Chem., 31(2), 27 (1971).
28. L. H. Sharpe and H. Schonhorn, Adv. Chem. Sci., 43, 189 (1964).
29. B. W. Cherry and C. M. Holmes, J. Colloid Interface Sci., 29, 174 (1969).
30. T. D. Blake and J. M. Haynes, J. Colloid Interface Sci., 30, 421 (1969).
31. S. Newman, J. Colloid Interface Sci., 26, 209 (1968).
32. B. W. Cherry and S. E. Muddarris, J. Adhes., 2, 42 (1970).
33. W. M. Bright, in *Adhesion and Adhesives, Fundamentals and Practice*, J. E. Rutzler, Jr., and R. L. Savage, eds., Wiley, New York, 1954, pp. 130–138.
34. K. Kanamaru, Kolloid Z. Z. Polym., 192, 51 (1963).
35. L. E. Nielsen, *Polymer Rheology*, Marcel Dekker, New York, 1977, pp. 31–46.
36. A. B. Bestul and H. V. Belcher, J. Appl. Phys., 24, 696 (1953).
37. W. Philippoff and F. H. Gaskins, J. Polym. Sci., 21, 205 (1956).
38. C. L. Sieglaff, SPE Trans., 4(2), 1 (1964).
39. H. Schott, J. Appl. Polym. Sci., 6(3), S-29 (1962).
40. R. S. Porter and J. F. Johnson, J. Polym. Sci., C15, 373 (1966).
41. G. J. van Amerongen, Rubber Chem. Technol., 37, 1065 (1964).
42. D. W. van Krevelen, *Properties of Polymers: Their Estimation and Correlation with Chemical Structure*, Elsevier, Amsterdam, 1976, pp. 410–425.
43. S. S. Voyutskii, *Autohesion and Adhesion of High Polymers*, Interscience, New York, 1963.
44. S. S. Vogutskii, Adhes. Age, 5(4), 30 (1962).
45. N. A. Grishin and S. S. Voyutskii, Vysokomol. Soedin., 1, 1778 (1959).
46. N. S. Korenevskaya, V. V. Lavrentyev, S. M. Yagnyatinskaya, V. G. Raevskii, and S. S. Voyutskii, Vysokomol. Soedin., 8, 1247 (1966); Polym. Sci. USSR, 8, 1372 (1966).
47. S. S. Voyutskii, J. Adhes., 3, 69 (1971).
48. I. V. Kragelskii, Trenie Iznos [Friction and Wear], Mash. (1962).

49. V. V. Lavrentyev, Vysokomol. Soedin., *4*, 1151 (1962); Dokl. Akad. Nauk SSSR, *175*(1), 125 (1967); Proc. Acad. Sci. USSR, *175*, 495 (1967).
50. L. F. Plisko, V. V. Lavrentyev, V. L. Vakula, and S. S. Voyutskii, Vysokomol. Soedin., *A14*, 2131 (1972); Polym. Sci. USSR, *14*, 2501 (1972).
51. N. H. Sung, Polym. Eng. Sci., *19*, 810 (1979).
52. S. Krause, J. Macromol. Sci., *C7*, 251 (1972).
53. S. Krause, in *Polymer Blends*, Vol. 1, D. R. Paul and S. Newman, eds., Academic Press, New York, 1978, pp. 15–113.
54. I. C. Sanchez, in *Polymer Blends*, Vol. 1, D. R. Paul and S. Newman, eds., Academic Press, New York, 1978. pp. 115–139.
55. T. K. Kwei and T. T. Wang, in *Polymer Blends*, Vol. 1, D. R. Paul and S. Newman, eds., Academic Press, New York, 1978, pp. 141–184.
56. P. J. Flory, J. Chem. Phys., *9*, 660 (1941); *10*, 51 (1942).
57. P. J. Flory, *Principles of Polymer Chemistry*, Cornell University Press, Ithaca, N.Y., 1953.
58. M. L. Huggins, J. Chem. Phys., *10*, 51 (1942).
59. M. L. Huggins, Ann. N.Y. Acad. Sci., *43*, 1 (1942).
60. J. H. Hildebrand and R. L. Scott, *The Solubility of Nonelectrolytes*, 3rd ed., Reinhold, New York, 1950.
61. J. H. Hildebrand, *Regular Solutions*, Prentice-Hall, Englewood, Cliffs, N.J. 1962.
62. D. R. Paul, in *Polymer Blends*, Vol. 1, D. R. Paul and S. Newman, eds., Academic Press, New York, 1978, pp. 1–14.
63. S. A. Chen, J. Appl. Polym. Sci., *15*, 1247 (1971).
64. C. M. Hansen and A. Beerbower, in *Kirk-Othmer Encyclopedia of Chemical Technology*, 2nd ed., Suppl., Wiley, New York, 1971, pp. 889–910.
65. R. L. Scott, J. Chem. Phys., *17*, 279 (1949).
66. H. Tompa, Trans. Faraday Soc., *45*, 1142 (1949).
67. E. Helfand and Y. Tagami, J. Polym. Sci., *B9*, 741 (1971).
68. E. Helfand and Y. Tagami, J. Chem. Phys., *56*, 3592 (1972).
69. E. Helfand and Y. Tagami, J. Chem. Phys., *57*, 1812 (1972).
70. E. Helfand and A. M. Spase, J. Chem. Phys., *62*, 1327 (1975).
71. E. Halfand, J. Chem. Phys., *63*, 2192 (1975).
72. R. J. Roe, J. Chem. Phys., *62*, 490 (1975).
73. J. Klein and B. J. Briscoe, Proc. R. Soc. Lond., *A365*, 53 (1979).
74. F. Bueche, W. M. Cashin, and P. Debye, J. Chem. Phys., *20*, 1956 (1952).
75. S. E. Bresler, G. M. Zakharov, and S. V. Kirillov, Vysokomol. Soedin., *3*, 1072 (1961); Polym. Sci. USSR, *3*, 832 (1962).
76. C. K. Rhee, J. D. Ferry, and L. J. Fetters, J. Appl. Polym. Sci., *21*, 783 (1977).

77. A. N. Kamenskii, N. M. Fodiman, and S. S. Voyutskii, Vysokomol. Soedin., 7, 696 (1965); Polym. Sci. USSR, 7, 769 (1965).
78. A. N. Kamenskii, N. M. Fodiman, and S. S. Voyutskii, Vysokomol. Soedin., A11, 394 (1969); Polym. Sci. USSR, 11, 442 (1969).
79. H. Van Oene and H. K. Plummer, ACS Org. Coatings Plast. Prep., 37(2), 498 (1977).
80. L. Yu Zlatkevich, V. G. Nikolskii, and V. G. Raevskii, Dokl. Akad. Nauk SSSR, 176, 1100 (1967); Proc. Acad. Sci. USSR, Phys. Chem. Sect., 176, 748 (1967).
81. L. Ya. Kapulunova, A. L. Kovarskii, T. S. Fedoseeva, A. M. Vasserman, A. S. Kuzminskii, and S. S. Voyutskii, Dokl. Akad. Nauk SSSR, 204, 1151 (1972); Proc. Acad. Sci. USSR, Phys. Chem. Sect., 204, 486 (1972).
82. L. P. Morozova and N. A. Krotova, Dokl. Akad. Nauk SSSR, 115, 747 (1957); Kolloid. Zh., 20, 59 (1958).
83. N. A. Krotova, L. P. Morozova, A. M. Polyakov, G. A. Sokolina, and N. N. Spefanovich, Kolloid, Zh., 26, 207 (1964).
84. J. Letz, J. Polym. Sci., A-2, 7, 1987 (1969).
85. R. M. Barrer, *Diffusion in and Through Solids*, 2nd ed., Cambridge University Press, London, 1951.
86. J. Crank and G. S. Park, eds., *Diffusion in Polymers*, Academic Press, New York, 1968.
87. D. W. McCall, D. C. Douglass, and E. W. Anderson, J. Polym. Sci., A1, 1709 (1963).
88. D. W. McCall, D. C. Douglass, and E. W. Anderson, J. Chem. Phys., 30, 771 (1959).
89. D. W. McCall, E. W. Anderson, and C. M. Huggin, J. Chem. Phys., 34, 804 (1961).
90. I. Auerbach, W. R. Miller, W. C. Kuryla, and S. D. Gehman, J. Polym. Sci., 28, 129 (1958).
91. A. Einstein, Ann. Phys., 17, 549 (1905).
92. F. Bueche, J. Chem. Phys., 20, 1959 (1952); also *Physical Properties of Polymers*, Interscience, New York, 1962.
93. A. Aitken and R. M. Barrer, Trans. Faraday Soc., 51, 116 (1955).
94. R. M. Barrer and G. Skirrow, J. Polym. Sci., 3, 549 (1948).
95. P. G. de Gennes, J. Chem. Phys., 55, 572 (1971).
96. P. G. de Gennes, Macromolecules, 9, 587 (1976).
97. R. M. Vasenin, Vysokomol. Soedin., 2, 857 (1960).
98. S. Prager, E. Bagley, and F. A. Long, J. Am. Chem. Soc., 75, 1255 (1953).
99. S. Prager and F. A. Long, J. Am. Chem. Soc., 73, 4072 (1951).
100. M. J. Hayes and G. S. Park, Trans. Faraday Soc., 51, 1134 (1955); 52, 949 (1956).

101. D. W. McCall, J. Polym. Sci., *26*, 151 (1959).
102. G. S. Park, Trans. Faraday Soc., *46*, 686 (1950).
103. H. Fujita, in *Diffusion in Polymers*, J. Crank and G. S. Park, eds., Academic Press, New York, 1968, pp. 75–105.
104. R. M. Vasenin, Vysokomol. Soedin., *2*, 851 (1960).
105. B. V. Deryagin, S. K. Zherebkov, and A. M. Medvedeva, Kolloid. Zh., *18*, 404 (1956); Colloid J. USSR, *18*, 399 (1956).
106. R. M. Vasenin, Vysokomol. Soedin., *3*, 679 (1961); Polym. Sci. USSR, *3*, 608 (1961).
107. R. M. Vasenin, Adhes. Age, *8*(5), 18 (1965).
108. R. M. Vasenin, Adhes. Age, *8*(6), 30 (1965).
109. W. G. Forbes and L. A. McLeod, Trans. Inst. Rubber Ind., *34*, 154 (1958).
110. S. S. Voyutskii and B. V. Shtarkh, Kolloid. Zh., *16*, 3 (1954); also in Rubber Chem. Technol., *30*, 548 (1957).
111. S. S. Voyutskii and V. L. Vakula, J. Appl. Polym. Sci., *7*, 475 (1963).
112. S. S. Voyutskii, Vysokomol. Soedin., *1*, 230 (1959); also in Rubber Chem. Technol., *33*, 748 (1960).
113. V. A. Pinegin, S. A. Vasileva, and L. M. Kepersha, quoted in Ref. 43, p. 152.
114. S. Gusman, Off. Dig., 884–905 (August 1962).
115. A. D. McLaren and C. J. Seiler, J. Polym. Sci., *4*, 63 (1949).
116. S. W. Lasoski, Jr., and G. Kraus, J. Polym. Sci., *18*, 359 (1955).
117. S. S. Voyutskii and V. M. Zamazii, Dokl. Akad. Nauk USSR, *81*, 63 (1951); also in Rubber Chem. Technol., *30* 544 (1957).
118. P. E. Cassidy, J. M. Johnson, and C. E. Locke, J. Adhes., *4*, 183 (1972).
119. A. Ahagon and A. N. Gent, J. Polym. Sci., Polym. Phys., Ed., *13*, 1285 (1975).
120. A. Ahagon, A. N. Gent, and E. C. Hsu, in *Adhesion Science and Technology*, Vol. 9A, L. H. Lee, ed., Plenum Press, New York, 1975, pp. 281–288.
121. J. D. Ferry, *Viscoelastic Properties of Polymers*, 2nd ed., Wiley, New York, 1970.
122. R. N. O'Brien and K. Hartman, J. Polym. Sci., *C34*, 293 (1971).
123. G. G. Allan and A. N. Neogi, J. Adhes., *3*, 13 (1971).
124. (a) K. Uno, M. Makita, S. Ooi and Y. Iwakura, J. Polym. Sci., A-1, *6*, 257 (1968); (b) U.S. Patents 2,940,872, 2,940,950, 3,248,397, 3,290,416, 3,290,417, 3,325,443, 3,307,006, 3,730,883, 3,338,885, 3,365,519.
125. E. P. Plueddemann, in *Interfaces in Polymer Matrix Composites*, E. P. Plueddemann, ed., Academic Press, New York, 1974, pp. 173–216.
126. E. P. Plueddemann, J. Adhes., *2*, 184 (1970).
127. P. E. Cassidy and B. J. Yager, Rev. Polym. Technol., *1*, 1 (1972).

References

128. P. E. Cassidy, J. M. Johnson, and G. C. Rolls, Ind. Eng. Chem., Prod. Res. Dev., *11*(2), 170 (1972).
129. E. P. Pleuddemann, Proc. 24th SPI Conf. Reinforced Plast. Div., Sect. 19-A, 1969,
130. H. A. Clark and E. P. Plueddemann, Mod. Plast., *40*(6), 133 (1963).
131. R. L. Kaas and J. K. Kardos, Polym. Eng. Sci., *11*, 11 (1971).
132. M. E. Schrader, I. Lerner, and F. J. D'Oria, Mod. Plast., *45*(1), 195 (1967).
133. D. J. Tutas, R. Stromberg, and E. Passaglia, SPE Trans., *4*(10), 256 (1964).
134. H. Ishida and J. L. Koenig, J. Colloid Interface Sci., *64*, 565 (1978).
135. O. K. Johannson, F. O. Stark, G. E. Vogel, and R. M. Fleischmann, J. Compos. Mater. *1*, 278 (1967).
136. S. Sterman and H. B. Bradley, SPE Trans., *1*, 224 (1961).
137. J. L. Koenig and P. T. K. Shih, J. Colloid Interface Sci., *36*, 247 (1971).
138. H. Ishida and J. L. Koenig, J. Colloid Interface Sci., *64*, 555 (1978).
139. M. E. Schrader and A. Block, J. Polym. Sci., *C34*, 281 (1971).
140. T. E. White, SPI Reinforced Plast. Div., 20th ANTEC, 3-B (1965).
141. L. I. Gulubenova, A. N. Shabadash, S. N. Nokonova, and M. S. Akutin, Polym. Sci., USSR, *4*, 422 (1963).
142. J. G. Koelling and K. E. Kolb, Chem. Commun. *1*, 6 (1965).
143. V. I. Alksne, V. Z. Kronberg, and Y. A. Edius, Mekh. Polim., *4*, 182 (1968); *3*, 601 (1967).
144. S. Sterman and J. G. Marsden, Proc. 21st PSI Conf. Reinforced Plast. Div., 1966.
145. W. C. Wake, in *Recent Advances in Adhesion*, L. H. Lee, ed., Gordon and Breach, New York, 1973, pp. 285–293.
146. C. H. Hofrichter, Jr., and A. D. McLaren, Ind. Eng. Chem., *40*, 329 (1948).
147. T. Nakabayashi, M. Kayumi, I. Kitani, and J. Tsukamoto, Kobunshi Kagaku, *10*(96), 156 (1953).
148. N. A. de Bruyne, J. Appl. Chem., *6*, 303 (1956).
149. D. D. Eley, in *Adhesion*, D. D. Eley, ed., Oxford University Press, London, 1961, pp. 266–279.
150. T. J. Mao and S. L. Reegen, in *Adhesion and Cohesion*, P. Weiss, ed., Elsevier, Amsterdam, 1962, pp. 209–216.
151. H. P. Brown and J. F. Anderson, in *Handbook of Adhesives*, I. Skeist, ed., Reinhold, New York, 1962, pp. 255–267.
152. W. H. Smarook and S. Bonotto, Polym. Eng. Sci., *8*, 41(1968).
153. A. F. Lewis and L. J. Forrestal, ASTM STP *360*, 59 (1963).
154. W. F. Busse and J. A. Boxler, U.S. Patent 2,838,437 (1958).
155. W. J. Jackson, Jr., and J. R. Caldwell, Adv. Chem. Ser., *99*, 562 (1971).

156. L. S. Luskin, in *Functional Monomers: Their Preparation, Polymerization and Application*, Vol. 2, R. H. Yocum and E. B. Nyquist, eds., Marcel Dekker, New York, 1974, pp. 357–554.
157. D. A. Tomalia, in *Functional Monomers: Their Preparation, Polymerization and Application*, Vol. 2, R. H. Yocum and E. B. Nyquist, eds., Marcel Dekker, New York, 1974, pp. 1–355.
158. L. S. Luskin, in *Functional Monomers: Their Preparation, Polymerization and Application*, Vol. 2, R. H. Yocum and E. B. Nyquist, eds., Marcel Dekker, New York, 1974, pp. 555–739.
159. F. H. Sexsmith and E. L. Polaski, in *Adhesion Science and Technology*, Vol. 9A, L. H. Lee, ed., Plenum Press, New York, 1975, pp. 259–279.
160. E. B. Nyquist, in *Functional Monomers: Their Preparation Polymerization and Application*, Vol. 1, R. H. Yocum and E. B. Nyquist, eds., Marcel Dekker, New York, 1974, pp. 299–488.
161. D. C. MacWilliams, in *Functional Monomers: Their Preparation, Polymerization and Application*, Vol. 1, R. H. Yocum and E. B. Nyquist, eds., Marcel Dekker, New York, 1974, pp. 1–198.
162. R. B. Dean, Off. Dig., *36*(473), 664 (1964).
163. J. E. Rutzler, Jr., Adhes. Age, *2*(6), 39 (1959).
164. Y. Iyengar, J. Appl. Polym. Sci., *22*, 801 (1978).
165. J. A. Robertson, Off. Dig., *36*(469), 138 (1964).
166. D. A. Kangas, in *Functional Monomers: Their Preparation, Polymerization and Application*, Vol. 1, R. H. Yocum and E. B. Nyquist, eds., Marcel Dekker, New York, 1974, pp. 489–640.
167. P. W. Erickson, J. Adhesion, *2*, 131 (1970).
168. P. W. Erickson and E. P. Plueddemann, in *Interfaces in Polymer Matrix Composites*, E. P. Plueddemann, ed., Academic Press, New York, 1974, pp. 1–31.
169. J. Bjorksten, J. Colloid Interface Sci., *67*, 552 (1978).
170. E. Plueddemann and W. T. Collins, in *Adhesion Science and Technology*, Vol. 9A, L. H. Lee, ed., Plenum Press, New York, 1975, pp. 329–338.
171. W. D. Bascom, Adv. Chem. Ser., *87*, 38 (1968).
172. S. Sterman and J. G. Marsden, Mod. Plast., *40*(11), 125 (1963).
173. S. Sterman and J. G. Marsden, Ind. Eng. Chem., *58*(3), 33 (1966).
174. M . C. Polniaszek and R. H. Schaufelberger, Adhes. Age, *11*(7), 25 (1968).
175. S. Sterman and J. B. Toogood, Adhes. Age, *8*(7), 34 (1965).

References

176. E. P. Plueddemann, J. Paint Technol., 42(550), 600 (1970).
177. E. P. Plueddemann, J. Paint Technol., 40(516), 1 (1968).
178. E. P. Plueddemann, Adhes. Age, 18(6), 36 (1975).
179. B. M. Vanderbilt and J. J. Jaruzelski, Ind. Eng. Chem., Prod. Res. Dev., 1, 188 (1962).
180. P. C. Yates and J. W. Trebilcock, SPE Trans., 199 (October 1961).
181. F. B. Hauserman, Adv. Chem. Ser., 23, 338 (1960).
182. J. A. Robertson and J. W. Trebilcock, Tappi, 58(4), 106 (1975).
183. S. J. Monte and P. F. Bruins, Mod. Plast., 68–72 (December 1974).
184. S. J. Monte and G. Sugarman, in *Additives for Plastics*, Vol. 1, R. B. Seymour, ed., Academic Press, New York, 1978, pp. 169–192.
185. S. L. Reegen and G. A. Ilkka, in *Adhesion and Cohesion*, P. Weiss, ed., Elsevier, Amsterdam, 1962, pp. 159–171.
186. I. A. Abu-Isa, Polym.-Plast. Technol. Eng., 2(1), 29 (1973).
187. D. R. Fitchmun, S. Newman, and R. Wiggle, J. Appl. Polym. Sci., 14, 2411, 2457 (1970).
188. J. S. Mijovic and J. A. Koutsky, Polym.-Plast. Technol. Eng., 9(2), 139 (1977).
189. C. V. Cagle, *Adhesive Bonding: Techniques and Applications*, McGraw-Hill, New York, 1968.
190. C. V. Cagle, ed., *Handbook of Adhesive Bonding*, McGraw-Hill, New York, 1973.
191. R. C. Snogren, *Handbook of Surface Preparation*, Palmerton, New York, 1974.
192. N. J. DeLollis, *Adhesives for Metals: Theory and Technology*, Industrial Press, New York, 1970.
193. K. Bright, B. W. Malpass, and D. E. Packham, Nature, 223, 1360 (1969).
194. D. E. Packham, K. Bright, and B. W. Malpass, J. Appl. Polym. Sci., 18, 3237 (1974).
195. B. W. Malpass, D. E. Packham, and K. Bright, J. Appl. Polym. Sci., 18, 3249 (1974).
196. A. J. Siegmund and P. Kukanskis, Adhes. Age, 19(2), 29 (1976).
197. D. J. Arrowsmith, Trans. Inst. Met. Finish., 48, 88 (1970).

12
Weak Boundary Layers

The fracture strength of an adhesive bond is determined not only by the interfacial structure and attraction, but also by the mechanical properties of the components and the interfacial zone, discussed in this and the following three chapters.

12.1. WEAK-BOUNDARY-LAYER THEORY
(See also Section 10.3)

An adhesive bond will fracture at its weakest link. Thus, if a cohesively weak layer exists at the interface, the adhesive bond may fracture within this weak boundary layer at low applied stress. Such a weak boundary layer is often very thin, and the fracture may be mistaken for an interfacial failure.

Bikerman [1-3] emphasized the role of the weak boundary layer and proposed that practically all cases of weak adhesive strength and all cases of apparent interfacial failure are due to the existence of weak boundary layers. Such a broad claim is not justified, however. Undoubtedly, weak boundary layers are the cause of weak bond strength in many cases. However, this fact alone cannot be logically generalized to all cases.

The deleterious effect of the weak boundary layer is nonetheless real and deserves proper attention. The weak boundary layers may be formed from migration of cohesively weak low-molecular-weight fractions, lubricatns or additives, adsorption of oily contaminants from the environment, or the formation of weak oxide layers on metals [1–3]. Examples of weak boundary layers are given next.

12.2. EXAMPLES OF WEAK BOUNDARY LAYERS

Many examples of weak boundary layers are discussed below. In most cases, the existence of a weak boundary layer is, however, inferred indirectly. Some examples are convincing; others have alternative explanations. Critiques and alternatives for some examples are discussed in Section 12.3.

Bikerman [1] suggested that "if a material does not stick to anything and nothing sticks to it, it is highly probable that this material is a carrier of a weak boundary layer, that is, it contains a soft or a nearly liquid ingredient which tends to accumulate at the specimen surface and to form there a soft or nearly liquid stratum. The focus of the least strength will be in this layer and rupture will proceed in it." Commercial polyethylene appears to be such an example. Most commercial polyethylenes contain low-molecular-weight fractions which are compatible with the rest in the molten state but are rejected to the surface when the melt freezes and crystallizes. Molecular-weight fractionation is known to occur in spherulite growth; the lower-molecular-weight species tend to concentrate in the outer regions of the spherulite [4].

Several experiments appear to support this view. Several polyethylenes incapable of strong adhesion are purified by dissolving in boiling xylene and precipitated with acetone or butanone [5]. The

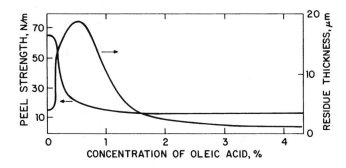

Figure 12.1. Peel strength versus concentration of oleic acid. (After Ref. 6.)

filtrate contains a greaselike material, amounting to about 1% of the original sample. The purified polymer is otherwise practically identical to the original sample; that is, tensile strength, modulus, and wettability are unchanged. The purified samples adhere strongly to every adherend that has no weak boundary layer of its own.

The effect of purification can be reversed by adding small amounts of suitable low-molecular-weight compounds to the purified adhesionable polyethylenes [6]. Oleic acid is such a compound, which can accumulate between polyethylene and metal. An aluminum ribbon is attached to glass with a molten mixture of adhesionable polyethylene and oleic acid. The joint is cooled and peeled at room temperature. The peel strength is 65 N/m when the oleic acid is less than 0.1%, and decreases rapidly to about 10 N/m when the oleic acid is greater, as shown in Figure 12.1. The solubility limit of oleic acid in the solid appears to be near 0.1%.

Figure 12.1 also shows the effect of oleic acid on the thickness of polyethylene residue remaining on the aluminum ribbon after peeling. The residual thickness of purified polyethylene is about 4 μm. As oleic acid is gradually added, the residual thickness increases to about 19 μm at about 0.4%. Thereafter, the residual thickness decreases to below 1 μm as the amount of oleic acid increases further. Such behavior is explained as follows. As the oleic acid exceeds its solubility limit, the excess acid accumulates at the polyethylene-aluminum and polyethylene-glass boundaries. Rupture occurs near the center of the weak boundary layer. As the oleic acid exceeds 0.4%, the weak boundary layer becomes so thick (over 0.1 μm) that the locus of rupture is determined by the position of the greatest stress, which is near the aluminum-polymer interface. The accumulation of oleic acid at the boundaries has also been demonstrated by electrical surface resistivity measurement [7].

Accumulation of oleic acid at the phase boundary can be prevented if the acid is solubilized in the polyethylene by suitable additives. Ethyl palmitate is a suitable compound [8], illustrated in Table 12.1. Ethyl palmitate appears to be a common solvent for polyethylene and

Table 12.1. Effect of Common Solvent Ethyl Palmitate on the Tensile Butt Strength of Steel/Polyethylene/Steel Joint

Adhesive	Steel/steel tensile butt strength, MPa
Purified adhesionable polyethylene (PAP)	9.2
PA + 1% oleic acid	0.1
PAP + 1% oleic acid + 5% ethyl palmitate	7.3–9.4

Source: From Ref. 8.

oleic acid. Addition of 1% oleic acid in the purified polyethylene decreases the tensile butt strength of steel/polyethylene/steel joint by a factor of about 100. However, the addition of only 5% ehtyl palmitate into the oleic acid-contaminated polyethylene can nearly completely recover its original strength.

Thus, weak boundary layers can be removed by purification or by the addition of a common solvent. Alternatively, they may be removed by increasing their cohesive strength, for instance, by cross-linking or transcrystallization. Schonhorn and Hansen [9—12] found that exposure of polyethylene to inert gas (helium, argon, nitrogen, or hydrogen) plasma greatly increases the bondability of polyethylene. They reported that the wettability of the treated polymer is not changed and polar groups are not detected by internal reflection IR spectroscopy. The only significant change is that a thin cross-linked skin about 1000 Å thick is formed on the polymer surface. Vivid evidence of the existence of a cross-linked skin on inert-gas-plasma-treated polyethylene is shown elsewhere [9]. They thus concluded that the improved bondability is due to strengthening of the weak boundary layer by cross-linking. This view is, however, not entirely convincing, as discussed in Section 12.3 (see also Sections 9.3.3—9.3.6).

Inert-gas-plasma treatment has also been found to greatly increase the bondability of polytetrafluoroethylene [12], poly(vinyl fluoride) [12], perfluorinated ethylene-propylene copolymer [13], and polycaprolactam [13]. The weak boundary layers on these polymers are presumably cross-linked by the plasma. On the other hand, the bondability of polypropylene is improved by exposure to oxygen or nitrous oxide plasma, but not by inert-gas (helium, argon, nitrogen, or hydrogen) plasmas [14]. This was thought to be because inert-gas plasma causes mainly chain scission and ablation of polypropylene without cross-linking, whereas oxygen or nitrous oxide plasma causes cross-linking (forming a cross-linking skin of about 300 Å) as well as oxidation (incorporating polar groups and increasing the wettability). Schonhorn and coworkers [14] argued that the increased bondability of polypropylene in the later case results from the surface cross-linking, not from the increased polarity and wettability.

Schonhorn and coworkers [15,16] also found that a polyethylene surface can be cross-linked by exposure to UV radiation or by fluorination. Exposure of polyethylene to UV radiation in helium causes surface cross-linking without surface oxidation, and greatly increases its bondability [15]. On the other hand, exposure of polyethylene to fluorine gas causes hydrogen abstraction, leading to fluorination and cross-linking [16]. The fluorination decreases the critical surface tension from 31 dyne/cm for the untreated surface to 20 dyne/cm after the treatment, and greatly increases its bondability. The increased bondability presumably results from surface cross-linking (see also Sections 9.1.4, 9.3.6 and 9.4).

Examples of Weak Boundary Layers

Schonhorn and coworkers [9,17–22] reported that the weak boundary layers on semicrystalline polymers such as polyethylene, polypropylene, polychlorotrifluoroethylene, poly(4-methylpentene-1), perfluorinated ethylene-propylene copolymer, and nylon 66 can also be strengthened by inducing transcrystallization in the surface layers. The semicrystalline polymers are molded against high-energy surfaces such as metals and metal oxides, which are then removed by chemical dissolution. They suggested that only high-energy surfaces have nucleation activity and can induce transcrystallization, whereas low-energy surfaces have no nucleation activity and do not induce transcrystallization. However, Kargin and coworkers [23], Fitchmun and Newman [24,25] and others [26] have already shown that nucleating activities of mold surfaces have nothing to do with their surface energies, that is, some low-energy surfaces can also induce surface transcrystallization, and not all high-energy surfaces are effective in inducing surface transcrystallization. A transcrystalline surface layer on polyethylene (molded against gold which has been removed by chemical dissolution) has been shown in Figure 5.13. Because of increased surface crystallinity, such surfaces have greatly increased wettability, cohesive strength, and bondability (see Tables 5.12, 5.13 and 9.10). The transcrystalline surface layers obtained by molding polyethylene or polypropylene against copper films are typically about 12 μm thick. The dynamic moduli of these transcrystalline layers are two to four times greater than those of the bulk materials, shown in Table 9.10 [20,27,28]. Although the transcrystalline surface layers have greatly increased wettability, Schonhorn and coworkers [9,17–22] suggested that their greatly increased bondability results mainly from strengthening of the weak boundary layer. This view is, however, not convincing; other mechanisms have also been proposed by others, as discussed in Section 12.3 (see also Section 9.3.6).

Polycaprolactam contains monomers that prevent strong adhesion of elastomers such as polyisobutylene [29]. After purification of the polycaprolactam by extraction, strong joints could be formed. Deposition of decanoic acid onto steel greatly decreases the tensile butt strength of a steel/poly(vinyl acetate)/steel joint [30]. As little as 0.14 μg/cm^2 or one monolayer of decanoic acid will drastically decrease the joint strength, as shown in Figure 12.2. Similarly, deposition of stearic acid onto steel adherends drastically decreases the joint strength of steel/epoxy/paint/steel/epoxy/steel joints [31]. However, petroleum jelly and lanolin spread on steel did not decrease the joint strength, probably because these materials are soluble in the paint.

Various polyethylenes were treated in aqueous ammonium persulfate. Their bondability to an epoxy adhesive increased markedly with treatment time [32]. Although ammonium persulfate is a strong oxidizing agent, very little evidence of polymer oxidation is found by either internal reflection IR spectroscopy or contact angle measurement. Rather, the treatment mainly produced a cross-linked surface layer,

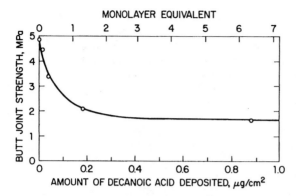

Figure 12.2. Butt joint strength versus amount of decanoic acid deposited. (After Ref. 30.)

as manifested by an insoluble residue amounting to about 1% by weight after solvent extraction of the treated polymers. Surface cross-linking of the polyethylenes was then proposed as the cause of increased bondability; see also Section 9.3.6. A critique of this view is discussed in Section 12.3.

All metals except gold are known to have an oxide film on their surface. Pure metals can be produced and preserved only under ultrahigh vacuum, where the oxygen partial pressure is below 10^{-9} torr. These oxide films, when formed under certain conditions, may be cohesively weak. Furthermore, metals and metal oxides have very high surface energies, of the order of 1000 dyne/cm. Consequently, low-energy contaminants such as mill oils, airborne grease, and moisture tend to absorb onto them, forming a weak boundary layer. Metals are, therefore, usually etched or treated with a chemical conversion coating to remove these contaminants and/or form a cohesively strong surface layer to replace any weak oxide film before adhesive bonding [33–37].

Oxide films can be classified into two types: porous oxide films and compact oxide filsm [38]. If the oxide formed occupies less volume than the metal that is replaced, the film is porous and nonprotective. If the oxide formed occupies a greater volume than the metal it replaces, the film is compact and protective. The critical density ratio can be defined as Md/mD, where M is the molecular weight of oxide, m the atomic weight of metal, D the oxide density, and d the metal density. If the ratio is less than unity, porous oxide films are produced; if the ratio is greater than unity, compact oxide films are produced. Alkali and alkaline earth metals give porous films, whereas aluminum and transition metals give compact films [38].

Porous oxide films grow mainly by diffusion of oxygen through the films, and tend to be cohesively weak, owing to their porosity. Compact oxide films grow by outward diffusion of the metal, such as with iron (FeO and Fe_3O_4). The migration of the metal ions from the surface leaves vacancies which will aggregate to form cavities. Because of volume expansion, further oxidation will create tensile stress. Cracking and flaking may now occur, exposing the metal to further oxidation [39]. Cracked and flaked oxide films may be cohesively weak. The thickness and structure of the oxide film will depend on the nature of the metal and the oxidation condition. For example, iron forms mainly FeO at the metal-oxide interface and mainly Fe_2O_4 at the oxide-air interface.

Aluminum oxide may be a weak boundary layer when formed under certain conditions [40]. Aluminum adherends were abraded with silicon carbide and then aged in either dry argon or a mixture of 80% argon, 20% oxygen, and enough moisture to make the relative humidity 50%. Aluminum/epoxy adhesive/aluminum lap joints were then prepared and tested. Aging in the humid argon-oxygen mixture greatly decreased the joint strength, probably because of the formation of weak oxide layer. In contrast, aging in dry argon did not appreciably change the joint strength (Table 12.2).

Bonding of sulfur-vulcanized rubber to brass (or to brass-plated steel) involves the formation of copper sulfide on the brass surface and covalent bonding of the rubber to the copper through sulfur linkage [41]. Strong bonding cannot be obtained, however, if the copper sulfide layer formed is powdery, which will act as a weak boundary layer [41,42].

During heat aging, weak boundary layers may be formed in metal-adhesive bonds, as the adhesives are degraded by oxidation catalyzed by the metal [43]. For instance, two aluminum strips are bonded with an adhesive consisting of a phenolic resin (eight parts) and an epoxy resin (one part). Its lap shear strength at room temperature is about 12.2 MPa. Heat aging of the joint in air at 288°C for 100 hr decreases the shear strength (at room temperature) to about 6.5 MPa. If stainless steel is bonded, instead of aluminum, the initial strength is

Table 12.2. Effect of Aging of Abraded Aluminum in Different Atmospheres on the Joint Strength of Aluminum/Epoxy Adhesive/ Aluminum Lap Joints

	Lap shear strength, MPa	
Aging time, min	Dry argon	Humid argon-oxygen
0	20.5	20.0
60	19.0	10.5

Source: From Ref. 40.

Table 12.3. Catalytic Effect of Metal Adherends During Heat Aging of Adhesive Joint on the Lap Shear Strength [a]

Metal	Lap shear strength, MPa	
	Before heating	After heating
Aluminum	12.2	6.5
Manganese	11.8	7.2
Chromium	11.2	6.3
Iron	9.9	3.8
Nickel	11.1	4.7
Zinc	12.7	7.8
Copper	9.6	0
Silver	14.6	8.0
Cerium	13.0	8.2

[a] The adhesive used consists of a phenolic resin (eight parts) and an epoxy resin (1 part). Heat aging at 288°C in air for 100 hr.
Source: From Ref. 43.

about the same, but its strength after heat aging as above is reduced to almost zero. When the heat aging is performed in nitrogen, only a moderate decrease in strength is noted. Apparently, a weak boundary layer is formed by oxidative degradation of the adhesive catalyzed by an ingredient on the metal surface. Table 12.3 illustrates the catalytic effects of various thin metal films deposited on an aluminum surface by displacement from an acid solution [43]. Heat aging was again in air at 288°C. As can be seen, copper is the most powerful oxidation catalyst. This effect is, however, not general; apparently, the oxidation occurs mainly with the epoxy resin. Heat againg in air has only a moderate effect on adhesive joints made with nylon or butadiene-acrylonitrile copolymers, regardless of the metal [1,43].

12.3. CRITIQUE OF WEAK-BOUNDARY-LAYER THEORY
(See also Section 10.3)

Undoubtedly, weak boundary layers can cause low bond strength. However, in most cases, the existence of a weak boundary layer has been inferred indirectly; alternative explanations are possible. Moreover, the weak boundary layer has not been isolated and characterized. It is not known how the properties of the weak boundary layer affect the fracture locus and the bond strength.

Critique of Weak-Boundary-Layer Theory

Recently, Robertson [44] elegantly showed that the mere existence of a weak boundary layer need not always result in weak bond strength. Weak boundary layers should not be indiscriminately attributed as the cause of weak adhesion. Two plates of poly(methyl methacrylate) having a molecular weight of about 3,000,000 and a cleavage fracture energy of about 140 J/m^2 are bonded with a thin interlayer of a polystyrene of narrow and low molecular weight of 10,300 and a cleavage fracture energy of about 0.60 J/m^2 [44-46]. The low-molecular-weight polystyrene interlayer serves as the weak boundary layer. The fracture toughness for crack initiation (K_{Ic}) and for crack arrest (K_{Ia}) of the PMMA/PS/PMMA sandwich assemblies are then determined by cleavage. The fracture toughness for both crack initiation and crack arrest are found to remain roughly constant and to correspond to those for the bulk poly(methyl methacrylate), when the weak polystyrene interlayer is thinner than about 1 µm. Only when the interlayer thickness is greater than about 1 µm does the fracture toughness drop precipitously to values as low as 15% of those for the bulk poly(methyl methacrylate), shown in Figure 12.3. Thus, a boundary layer of weak, low-molecular-weight polystyrene does not manifest its weakness until its thickness exceeds 1 µm, which is rather thick and effectively macroscopic. When the polystyrene interlayer is thin, the fracture path is found to meander between the polystyrene interlayer and the poly (methyl methacrylate) bulk. Although the polystyrene interlayer should be an easy path for crack growth, the crack often passes through the interlayer; the interlayer is found to have little

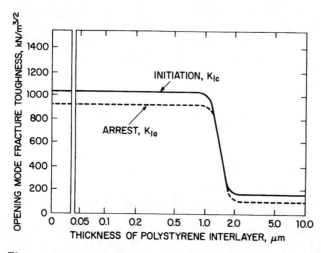

Figure 12.3. Opening mode fracture toughness versus thickness of "weak boundary layer" (low-molecular-weight polystyrene). (After Ref. 44.)

tendency to channel the crack. The critical thickness of a weak boundary layer would depend on the molecular weight, falling to less than 100 Å for a nonpolymeric layer.

Schonhorn and Hansen [9–12] suggested that the increased bondability of polyethylene by treatment with inert-gas plasma results from strengthening of the weak boundary layer by formation of a crosslinked skin; no polar groups are introduced and the wettability is not changed, as discussed in Section 12.2. Although cross-linked skins are indeed formed on polyethylene and perfluorinated ethylene-propylene copolymer upon plasma treatment [47–49], many workers [47–61] have conclusively shown that treatment with inert-gas plasmas does simultaneously incorporate polar groups onto polymer surfaces and greatly increase the wettability. Helium and argon plasmas have been shown by ESCA to introduce polar oxygen groups; and nitrogen plasma, both oxygen and nitrogen groups [49–51] (see Sections 9.3.4 and 9.3.6). In fact, only when polar groups are introduced and the wettability is increased does the bondability increase significantly [47–61].

Such a discrepancy appears to arise from inadequacy of the analytical techniques used in wettability and surface chemical analyses. Internal reflection IR spectroscopy used by Schonhorn and Hansen has a penetration depth of about 5 μm and is therefore not sufficiently sensitive for the detection of surface groups. Later workers such as Clark and Dilks [49], Yasuda and coworkers [50], and Blythe and coworkers [51] used ESCA, which has a penetration depth of about 10–20 Å, and can therefore give reliable surface analysis. These ESCA results show definitely that in every case inert-gas plasma introduces polar groups onto polymers (see also Sections 9.3.4–9.3.6).

Schonhorn and Ryan [15] also claimed that UV treatment of polyethylene in helium causes surface cross-linking and increased bondability without introducing polar groups or changing the wettability. However, Sowell [62] and DeLollis and coworkers [52,53] showed that the treatment does greatly increase the wettability of the polymers, and challenged the weak-boundary-layer concept. In evaluating the wettability, both the advancing and the receding contact angles should be measured. Unfortunately, most workers usually measure only the advancing contact angle, which often reflects only the low-energy portions and is often unaffected by the polar portions of the heterogeneous surface. This may account for the discrepancy in the reported wettability in some cases. Thus, strengthening of the presumed weak boundary layer may not solely account for the increased bondability in the cases described above; incorporation of polar groups and increased wettability may play a key role.

On the other hand, plasma treatments mainly make the polymers bondable at lower temperatures. The untreated polymers, in most cases, can give strong adhesive bonds if bonded at sufficiently high temperatures, for instance, above the crystalline melting points. Polar groups are invariably incorporated onto the treated surfaces, and the wettabilities are increased. Our preferred view is that the increased bondability by plasma treatment results mainly from softening of the polymer surface because of the production of polymeric scission products which promote interfacial flow and interdiffusion. The increased wettability also facilitates the formation of interfacial contact. These are discussed in more detail in Section 9.3.6.

DeLollis and Montoya [54] showed that weak boundary layers cannot explain the poor adhesion of epoxy adhesives to cured polydimethylsiloxane rubber, nor its improved adhesion after plasma treatment. Cured silicon rubbers are unbondable with usual adhesives, such as epoxies. They contain low-molecular-weight fractions of linear and cyclic polysiloxanes. Thus, the silicone rubbers fit the classical description of adherends containing weak boundary layers. Treatment of the silicone rubber with oxygen plasma was found to increase greatly its wettability and bondability. Analyses by both solvent extraction and internal reflection IR spectroscopy showed that the low-molecular-weight polysiloxane fractions are present on the surfaces of cured silicone rubbers both before and after the treatment. Strong bonds are formed even in the presence of what would be called weak boundary layers. Thus, the increased bondability by plasma treatment cannot be attributed to removal of the weak boundary layer, but rather seems to arise from incorporation of polar groups and enhanced wettabiltiy.

Schonhorn and coworkers [9,17—22] suggested that the bondability of semicrystalline polymers such as polyethylene, perfluorinated ethylene-propylene copolymer, polycaprolactam, and others can be greatly increased by inducing transcrystallization to strengthen the weak boundary layers that are presumably present on the surfaces of these polymers. They proposed that only high-energy mold surfaces can induce transcrystallization, which thereby improves bondability. In all the examples where the bondability is increased by transcrystallization, the polymers are molded against high-energy surfaces such as metals and metal oxides. When molded against low-energy surfaces, the bondability is not improved.

However, it is known that nucleation activity has nothing to do with surface energy; that is, many low-energy surfaces can also induce transcrystallization, and not all high-energy surfaces are effective [23—26] (Section 9.5). For instance, surface layers of polyethylene can be transcrystallized by molding against low-energy surfaces such as poly(ethylene terephthalate), polytetrafluoroethylene, perfluorinated ethylene-propylene copolymer, and others [23—26]. The bondability of such surfaces is not improved. Thus, strengthening of the weak boundary layer by transcrystallization cannot account for increased bondability.

Furthermore, transcrystallized polymer surfaces have dramatically increased surface tension and wettability (Sections 5.4 and 9.5). The increased bondability can be explained adequately by the greatly increased wettability, discussed in Section 9.5.

The increased surface tension and wettability of a transcrystallized polymer surface can be quantitatively explained in terms of the surface crystallinity (Section 5.4). However, Dwight and Riggs [63] and Briggs and coworkers [64] suggested that the increased wettability arises from the existence of a thin layer (20—60 Å) of polar, oxygenated residues on the transcrystallized surfaces, as detected by ESCA. For instance, oxygenated hydrocarbons are detected on perfluorinated ethylene-propylene copolymer transcrystallized against gold [63], and on polyethylene transcrystallized against aluminum [64]. Briggs and coworkers [64] suggested that the polar groups are formed on the polyethylene by thermal oxidation catalyzed by the metal surface during compression molding of the polymer melt.

Sharpe [26] seriously doubted the existence of weak boundary layers on semicrystalline polymers, which have often been cited as the cause of poor bondability of these polymers. He stated: "The evidence for their existence is purely inferential and related to joint strength alone. The reasoning goes that weak boundary layers are present because the joint is weak and, conversely, the joint is weak because weak boundary layers are present. There has been no evidence, independent of joint strength and related measurements, to demonstrate that they exist." He then proposed that the poor bondability of semicrystalline polymers such as polyethylene (modulus 30—60 kpsi) with high-modulus adhesives such as an epoxy adhesive (modulus 150—220 kpsi) is due to disparity in the modulus of the two phases, which gives rise to large stress concentration at the edges of the joint. Cross-linking of surface layers by plasma treatment or transcrystallization of surface layers of semicrystalline polymers improves their bondability by increasing the modulus of the surface layers and thus minimizing the stress concentration. However, no proofs have been offered, and the details of such effects are unknown.

REFERENCES

1. J. J. Bikerman, *The Science of Adhesive Joints*, 2nd ed., Academic Press, New York, 1968.
2. J. J. Bikerman, Ind. Eng. Chem., 59(9), 41 (1967).
3. J. J. Bikerman, in *Recent Advances in Adhesion*, L. H. Lee, ed., Gordon and Breach, New York, 1973, pp. 351—356.
4. H. D. Keith and F. J. Padden, J. Appl. Phys., 35, 1270 (1964).
5. J. J. Bikerman, Adhes. Age, 2(2), 23 (1959).
6. J. J. Bikerman, J. Appl. Chem., 11, 81 (1961).
7. J. J. Bikerman, SPE Trans., 2, 213 (1962).
8. J. J. Bikerman and D. W. Marshall, J. Appl. Polym. Sci., 7, 1031 (1963).

References

9. H. Schonhorn, in *Adhesion, Fundamentals and Practice*, The Ministry of Technology, U.K., Kingdom, Gordon and Breach, New York, 1969, pp. 12−21.
10. H. Schonhorn and R. H. Hansen, in *Adhesion, Fundamentals and Practice*, The Ministry of Technology, U.K., Gordon and Breach, New York, 1969, pp. 22−28.
11. R. H. Hansen and H. Schonhorn, J. Polym. Sci., *B4*, 203 (1966).
12. H. Schonhorn and R. H. Hansen, J. Appl. Polym. Sci., *11*, 1461 (1967).
13. H. Schonhorn and F. W. Ryan, J. Polym. Sci., A-2, *7*, 105 (1969).
14. H. Schonhorn, F. W. Ryan, and R. H. Hansen, J. Adhes., *2*, 93 (1970).
15. H. Schonhorn and F. W. Ryan, J. Appl. Polym. Sci., *18*, 235 (1974).
16. H. Schonhorn and R. H. Hansen, J. Appl. Polym. Sci., *12*, 1231 (1968).
17. H. Schonhorn and F. W. Ryan, J. Polym. Sci., A-2, *6*, 231 (1968).
18. H. Schonhorn and F. W. Ryan, Adv. Chem. Ser., *87*, 140 (1968).
19. K. Hara and H. Schonhorn, J. Adhes., *2*, 100 (1970).
20. H. L. Frisch, H. Schonhorn, and T. K. Kwei, J. Elastoplast., *3*, 214 (1971).
21. H. Schonhorn, J. Polym. Sci., Polym. Lett., Ed., *5*, 919 (1967).
22. H. Schonhorn, Macromolecules, *1*, 145 (1968).
23. V. A. Kargin, T. I. Sogolova, and T. K. Shaposhuikova, Dokl. Akad. Nauk SSSR, *180*, 901 (1968); Trans. Acad. Sci. USSR, *180*, 406 (1968).
24. D. R. Fitchmun and S. Newman, J. Polym. Sci., A-2, *8*, 1545 (1970).
25. D. R. Fitchmun and S. Newman, Polym. Lett., *7*, 301 (1969).
26. L. H. Sharpe, in *Recent Advances in Adhesion*, L. H. Lee, ed., Gordon and Breach, New York, 1973, pp. 437−450.
27. T. K. Kwei, H. Schonhorn, and H. L. Frisch, J. Appl. Phys., *38*, 2512 (1967).
28. S. Matsuoka, J. H. Daane, H. E. Bair, and T. K. Kwei, J. Polym. Sci., Polym. Lett. Ed., *6*, 87 (1968).
29. V. G. Raevskii, S. S. Voyutskii, V. E. Gul, A. N. Kamenskii, and I. Moneva, Izv. Vyssh. Uchebn. Zaved., Khim. Khim. Tekhnol., *8*, 305 (1965).
30. S. W. Lasoski and G. Kraus, J. Polym. Sci., *18*, 359 (1955).
31. T. R. Bullet and J. L. Prosser, Trans. Inst. Met. Finish, *41*, 112 (1964).
32. C. E. M. Morris, J. Appl. Polym. Sci., *14*, 2171 (1970).
33. S. Spring, *Preparation of Metals for Painting*, Reinhold, New York, 1965.

34. S. Spring, *Metal Cleaning*, Reinhold, New York, 1963.
35. C. V. Cagle, *Adhesive Bonding, Techniques and Applications*, McGraw-Hill, New York, 1968.
36. N. J. DeLollis, *Adhesives for Metals, Theory and Technology*, Industrial Press, New York, 1970.
37. R. C. Snogren, *Handbook of Surface Preparation*, Palmerton, New York, 1974.
38. N. B. Pilling and R. E. Bedworth, J. Inst. Met., *29*, 529 (1923).
39. A. F. Lewis and L. J. Forrestal, in *Treatise on Coatings*, Vol. 2, Part 1, R. R. Myers and J. S. Long, eds., Marcel Dekker, New York, 1969, pp. 57–98.
40. R. F. Wegman, Adhes. Age, *10*(1), 20 (1967).
41. S. Buchan, *Rubber to Metal Bonding*, 2nd ed., Crosby Lockwood, London, 1959.
42. Yu. I. Markin, V. M. Gorchakova, V. E. Gul, and S. S. Voyutskii, Izv. Vyssh. Uchebn. Zaved. Khim. Khim. Tekhnol., *5*, 810 (1962); Chem Abstr., *58*, 12693 (1962).
43. J. M. Black and R. F. Blomquist, Ind. Eng. Chem., *50*, 918 (1958).
44. R. E. Robertson, J. Adhes., *7*, 121 (1975).
45. R. E. Robertson, Adv. Chem. Ser., *154*, 89 (1976).
46. R. P. Kusy and M. J. Katz, Polymer, *19*, 1345 (1978).
47. M. Hudis and L. E. Prescott, J. Appl. Polym. Sci., *19*, 451 (1975).
48. M. Hudis and L. E. Prescott, J. Polym. Sci., *B10*, 179 (1972).
49. D. T. Clark and A. Dilks, in *Characterization of Metal and Polymer Surfaces*, Vol. 2, L. H. Lee, ed., Academic Press, New York, 1977, pp. 101–132.
50. H. Yasuda, H. C. Marsh, S. Brandt, and C. N. Reilley, J. Polym. Sci., Polym. Chem. Ed., *15*, 991 (1977).
51. A. R. Blythe, D. Briggs, C. R. Kendall, D. G. Rance, and V. J. I. Zichy, Polymer, *19*, 1273 (1978).
52. N. J. DeLollis, Rubber Chem. Technol., *46*, 549 (1973).
53. R. R. Sowell, N. J. DeLollis, H. J. Gregory, and O. Montoya, J. Adhes., *4*, 15 (1972).
54. N. J. DeLollis and O. Montoya, J. Adhes., *3*, 57 (1971).
55. E. H. Andrews and A. J. Kinloch, Proc. R. Soc. Lond., *A332*, 385 (1973).
56. B. W. Malpass and K. Bright, in *Aspects of Adhesion*, Vol. 5, D. J. Alner, ed., CRC Press, Cleveland, Ohio, 1969, pp. 214–225.
57. J. R. Hall, C. A. L. Westerdahl, A. T. Devine, and M. J. Bodnar, J. Appl. Polym. Sci., *13*, 2085 (1969).
58. J. R. Hall, C. A. L. Westerdahl, M. J. Bodnar, and D. W. Levi, J. Appl. Polym. Sci., *16*, 1465 (1972).
59. C. A. L. Westerdahl, J. R. Hall, E. C. Schramm, and D. W. Levi, J. Colloid Interface Sec., *47*, 610 (1974).

References

60. E. L. Lawton, J. Appl. Polym. Sci., *18*, 1557 (1974).
61. C. Y. Kim, G. Suranyi, and D. A. I. Goring, J. Polym. Sci., *C30*, 533 (1970).
62. R. R. Sowell, as quoted in Ref. 52.
63. D. W. Dwight and W. M. Riggs, J. Colloid Interface Sci., *47*, 650 (1974).
64. D. Briggs, D. M. Brewis, and M. B. Konieczdo, J. Mater. Sci., *12*, 429 (1977).

13
Effect of Internal Stress on Bond Strength

During setting, adhesives usually shrink as a result of solidification by loss of solvent, cross-linking, or cooling from the bonding temperature. As the adhesive-adherend contact area is constrained by adhesion, the shrinkage will induce internal stresses which can reduce the fracture strength of the adhesive bond. If the internal stress is σ_i, the actual fracture strength σ is then

$$\sigma = \sigma_o - \sigma_i \qquad (13.1)$$

where σ_o is the fracture strength for the internal-stress-free adhesive bond.

Internal stress will relax as long as the adhesive remains sufficiently fluid, but will start to build up as soon as the adhesive reaches the solidification point, where the adhesive viscosity becomes sufficiently high. Therefore, the magnitude of internal stress is often independent of the initial solvent concentration, baking temperature, or adhesive thickness [1-5]. Solvent-cast coatings of poly(isobutyl methacrylate) on rigid substrates have been found to develop internal stresses of about 4 MPa on drying; those of polystyrene about 10 MPa [1-3]. Poly(methyl methacrylate) coatings on rigid substrates have been found to develop internal stresses of about 6 MPa on cooling from a baking

Table 13.1. Effect of Internal Stress on the Lap Shear Strength of Aluminum Powder-Filled Polystyrene [a]

Filler concentration, % by weight	Inherent residual stress, MPa	Lap shear strength, MPa	
		Unannealed	Annealed
0	15.2	1.0 (i)	6.1 (i/c)
20	11.0	1.9 (c)	6.5 (c)
30	6.2	2.5 (c)	7.5 (c)
40	4.8	3.0 (c)	8.9 (c)

[a] Aluminum/polymer/aluminum lap shear joints; i, interfacial failure; c, cohesive failure.
Source: From Figure 13.1 and Ref. 7.

temperature of 120°C [4]. Internal stresses of about 18 MPa have been measured with strain gauges embedded in an unsaturated polyester cured at 120°C [6].

Internal stresses as a function of drying time in cast polystyrene films containing various amounts of aluminum powder have been measured [7]. The greatest internal-stress buildup occurs in the unfilled polymer below the glass transition temperature; fillers generally reduce the internal stress, as shown in Figure 13.1. Lap shear joints were prepared with these aluminum powder-filled polystyrenes. Table 13.1 gives the joint strength for unannealed specimens and those annealed at 130°C (about 30°C above the glass transition temperature of

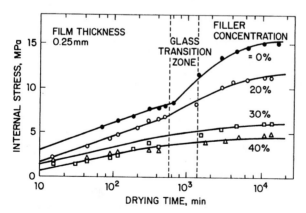

Figure 13.1. Effect of internal stress on lap shear strength of aluminum powder-filled polystyrene adhesive. (After Ref. 7.)

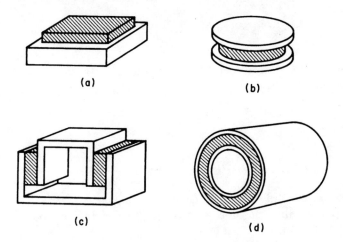

Figure 13.2. (a) Shrinkage of a coating; (b) free thickness contraction in a butt joint; (c) no thickness contraction in a sandwich structure; (d) no thickness contraction in radial direction.

polystyrene). Relaxation of the internal stress by annealing or use of fillers generally increases the joint strength, suggesting that internal stress has significant effect on joint strength.

Four shrinkage geometries are analyzed below: (1) shrinkage of a coating, (2) free contraction of thickness in a butt joint, (3) no contraction of thickness in a sandwich structure, and (4) no contraction of thickness in radial direction, as shown in Figure 13.2.

13.1. SHRINKAGE OF A COATING ADHERING TO A SUBSTRATE

In this case, the coating-substrate contact area is constrained, but the thickness can shrink. Consider the isothermal drying of a coating by loss of solvent. The volume of the solvent lost from the coating after the solidification point, ΔV, is

$$\Delta V = \phi_s V_s - \phi_d (V_s - \Delta V) \tag{13.2}$$

where ϕ_s is the volume fraction of the solvent at the solidification point, ϕ_d that of the retained solvent in the "dry" coating, and V_s the volume of the coating at the solidification point. Experimentally, the shrinkage stress of a coating on a rigid substrate is practically independent of the coating thickness, indicating that the internal stress is uniformly distributed within the coating [1-3,5]. Thus, the internal linear strain e_i is

$$e_i = \frac{\phi_s - \phi_d}{3(1 - \phi_d)} \qquad (13.3)$$

Assuming Hookean behavior for the coating, the internal stress in the biaxial plane stress condition is

$$\sigma_i = \frac{E}{1 - \nu} \left[\frac{\phi_s - \phi_d}{3(1 - \phi_d)} \right] \qquad (13.4)$$

where E is Young's modulus and ν is the Poisson's ratio for the dry coating.

On the other hand, consider the cooling of a coating at constant composition. Let the solidification temperature be T_s and the temperature of "cooled" coating be T. Then, the shrinkage stress is

$$\sigma_i = E(T_s - T) \Delta \alpha \qquad (13.5)$$

where $\Delta \alpha$ is the difference in the linear thermal expansion coefficients of the coating and the substrate, again assuming uniform stress distribution.

Shrinkage stress in a coating can be determined by measuring the deflection of a cantilever substrate (such as steel blades of 0.1–0.25 mm thickness) coated on one side [1,4,5,8]. As the coating shrinks, the substrate is bent with the coating on the concave side (Figure 13.3). Assume that (1) the elastic properties of the coating and the substrate are isotropic, (2) the elastic limits are not exceeded in the bending, (3) the coating adheres to the substrate, and (4) the shrinkage stress is constant through the coating thickness. As the coating shrinks, the substrate is bent to a spherical section. This is true for small curvatures found in practice, except for a small error due

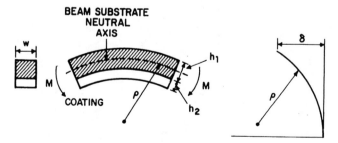

Figure 13.3. Bending of a beam. Subscripts: 1, substrate; 2, coating.

Shrinkage of a Coating Adhering to a Substrate

to the end clamp. The radius of curvature ρ and the end deflection δ are related by

$$\frac{1}{\rho} = \frac{2\delta}{L^2} \tag{13.6}$$

where L is the length of the substrate. The internal stress σ_i is given by [8]

$$\sigma_i = \frac{E_1 h_1^3}{6 h_2 \rho (h_1 + h_2)(1 - \nu_1)} + \frac{E_2(h_1 + h_2)}{2\rho(1 - \nu_2)} \tag{13.7}$$

where E is Young's modulus, h the thickness, ν is Poisson's ratio, and the subscripts 1 and 2 refer to the substrate and the coating, respectively. In terms of end deflection δ, the expression is given by

$$\sigma_i = \frac{\delta E_1 h_1^3}{3 h_2 L^2 (h_1 + h_2)(1 - \nu_1)} + \frac{\delta E_2 (h_1 + h_2)}{L^2 (1 - \nu_2)} \tag{13.8}$$

The last term in Eq. (13.7) or (13.8) is a correction term, accounting for the additional stress that would be induced into the coating if the bent substrate is straightened. When $h_1/h_2 \geq 10$ and $E_1/E_2 \geq 100$, the last term amounts to less than about 1% of the total internal stress, and thus can usually be neglected [8].

A more rigorous analysis gives [5]

$$\sigma_i = \frac{C_1 h_1^3}{6 h_2 \rho (h_1 + h_2)} \left(1 + \frac{C_2 h_2}{C_1 h_1}\right)\left(1 + \frac{C_2 h_2^3}{C_1 h_1^3}\right) + \frac{C_2(h_1 + h_2)}{2\rho} \tag{13.9}$$

where $C_i = E_i/(1 - \nu_i)$. Equation (13.9) is identical to Eq. (12.6) except for the correction terms, which may usually be neglected.

The modulus of the substrate may be determined by end loading it as a cantilever according to the relation

$$E_1 = \frac{4PL^3}{\delta w h_1^2} \tag{13.10}$$

where P is the end load and w the width of the substrate.

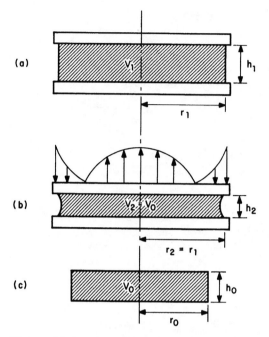

Figure 13.4. Analysis of internal stress in a sandwich structure.

13.2. FREE THICKNESS CONTRACTION IN A BUTT JOINT

Consider a circular adhesive layer bonded between two adherends (Figure 13.4). Initially, the adhesive layer has a volume V_1, radius r_1, and thickness h_1. After shrinkage in which the adhesive-adherend interface is constrained but the adhesive thickness can contract freely, the adhesive has a volume V_2, radius $r_2 = r_1$, and thickness h_2. If the adhesive could freely contract in all directions, the adhesive would have a volume $V_0 = V_2$, radius r_0, and thickness h_0 after the shrinkage. The constrained contraction can be regarded as a superposition of two deformations: a simple compression and then a constrained extension. The normal interfacial stress is obtained as [9,10]

$$\frac{\sigma}{2Ee_0} = \frac{1 + (r_0^2/h_0^2)[1 - (x^2/r_0^2)]}{1 + (r_0^2/2h_0^2)} - 1 \qquad (13.11)$$

where σ is the normal interfacial stress (positive for tension and negative for compression), E is Young's modulus of the adhesive, x the radial distance from the center, and

Free Thickness Contraction in a Butt Joint

$$e_0 = \frac{r_1}{r_0} - 1 = \frac{h_1}{h_0} - 1 \tag{13.12}$$

The maximum tensile stress occurs at the center of the bonded surface, having a value of

$$\frac{\sigma_{max}}{Ee_0} = \frac{r_0^2}{2h_0^2}\left(1 + \frac{r_0^2}{2h_0^2}\right)^{-1} \tag{13.13}$$

A maximum compressive stress of equal magnitude occurs at the edge. Both the tensile and the compressive stresses decrease to zero at $x = r_0/\sqrt{2}$. For thick adhesives ($h_0 \gg r_0$), $\sigma_{max} = 0$, and the shrinkage stress is unimportant. However, for thin adhesives ($r_0 \gg h_0$), $\sigma_{max} = 2Ee_0$, which can be quite significant. Practically, e_0 can be of the order of 2.5%. Thus, σ_{max} is quite small, only about 1 kg/cm^2, for conventional rubber vulcanizates with E about 20 kg/cm^2, but can be as high as 500 kg/cm^2 for conventional thermoplastics with E about 10,000 kg/cm^2.

The tangential interfacial stress τ is given by [9]

$$\tau = -\frac{h_0}{2}\left(\frac{\partial \sigma}{\partial x}\right) \tag{13.14}$$

Hence, from Eq. (13.11),

$$\frac{\tau}{2Ee_0} = \frac{x/h_0}{1 + (r_0^2/2h_0^2)} \tag{13.15}$$

The maximum tangential stress occurs at the edges of the interface, that is, $x = r_0$, having a value of

$$\frac{\tau_{max}}{2Ee_0} = \frac{r_0/h_0}{1 + (r_0^2/2h_0^2)} \tag{13.16}$$

which is smaller than σ_{max} for both thick and thin adhesives. Note that τ is highest when $r_0 = \sqrt{2h_0}$. For conventional rubber vulcanizates of E about 20 kg/cm^2, τ_{max} amounts to only about 0.7 kg/cm^2, but for conventional thermoplastics of E about 10,000 kg/cm^2, τ_{max} is as high as 350 kg/cm^2.

13.3. NO THICKNESS CONTRACTION IN A SANDWICH STRUCTURE

When the adherends are held at a constant separation h, where $h = h_0(1 + e_0)$, the stresses developed during shrinkage are very much more severe. The normal interfacial stress is in tension everywhere, given by [9]

$$\frac{\sigma}{Ee_0} = 1 + 3\left(\frac{r_0^2}{h_0^2}\right)\left(1 - \frac{x^2}{r_0^2}\right) \qquad (13.17)$$

Its maximum value occurs at the center of the bonded surfaces, given by

$$\frac{\sigma_{max}}{Ee_0} = 1 + \frac{3r_0^2}{h_0^2} \qquad (13.18)$$

When the adhesive layer is thin ($r_0 \gg h_0$), σ_{max} will be extremely large. For instance, when $r_0 = 10h_0$, the maximum tensile stress is 7.5E for a linear shrinkage of 2.5%. Such a stress would inevitably cause failure either at the interface or in the adhesive layer.

Using Eq. (13.14), the tangential interfacial stress is obtained as

$$\frac{\tau}{Ee_0} = \frac{3x}{h_0} \qquad (13.19)$$

which has a maximum value occurring at the edges,

$$\tau_{max} = \frac{3Ee_0 r_0}{h_0} \qquad (13.20)$$

Thus, when $r_0 = 10h_0$ and with a linear shrinkage of 2.5%, the interfacial shear stress developed is as high as 0.75E, which could cause spontaneous delamination.

13.4. NO THICKNESS CONTRACTION IN A LONG, NARROW ADHESIVE LAYER IN AN ANNULUS

This condition prevails in rubber tubes bonded on their curved surfaces which are free to contract only in the axial direction. The normal interfacial stress is [9]

$$\frac{\sigma}{Ee_0} = 2 + \frac{6}{h_0^2}\left(\frac{w_0^2}{4} - x^2\right) \qquad (13.21)$$

where x now denotes the distance in the radial direction from the center of the adhesive layer and w_0 is the axial radial thickness of the adhesive. The tangential interfacial stress is

$$\frac{\tau}{Ee_0} = \frac{6x}{h_0} \qquad (13.22)$$

Thus, the maximum tensile stress occurs at the center of the adhesive layer, given by

$$\frac{\sigma_{max}}{Ee_0} = 2 + 3\left(\frac{w_0^2}{2h_0^2}\right) \qquad (13.23)$$

The maximum shear stress occurs at the edges, given by

$$\frac{\tau_{max}}{Ee_0} = \frac{3w_0}{h_0} \qquad (13.24)$$

REFERENCES

1. S. G. Croll, J. Coatings Technol., 50(638), 33 (1978).
2. S. G. Croll, J. Coatings Technol., 51(648), 64 (1979).
3. S. G. Croll, J. Appl. Polym. Sci., 23, 847 (1979).
4. S. Gusman, Off. Dig., 34(451), 884 (1962).
5. T. S. Chow, C. A. Liu, and R. C. Penwell, J. Polym. Sci., Polym. Phys. Ed., 14, 1311 (1976).
6. A. J. Bush, Mod. Plast., 35, 143 (February 1958).
7. A. F. Lewis and L. J. Forrestal, in *Treatise on Coatings*, Vol. 2, Part 1, R. R. Meyers and J. S. Long, eds., Marcel Dekker, New York, 1969, pp. 57–98.
8. E. M. Corcoran, J. Paint Technol., 41(538), 635 (1969).
9. A. N. Gent, Rubber Chem. Technol., 47, 202 (1974).
10. A. N. Gent and E. A. Meinecke, Polym. Eng. Sci., 10, 48 (1970).

14
Fracture of Adhesive Bond

This chapter contains two parts. In Part I, the fundamentals of fracture mechanics are discussed. In Part II, the analysis and testing of adhesive bonds are discussed.

PART I. FUNDAMENTALS OF FRACTURE MECHANICS
(See also Chapter 10)

The real strength of a material is usually much lower than its ideal (theoretical) strength, because the material contains defects or cracks, causing the local stress to exceed the local strength. In addition, a material may deform plastically at a low stress. Thus, the fracture strength of a material depends on the presence of flaws and fracture mechanisms. In 1920, Griffith [1] found that the fracture strength of a brittle, elastic material is determined quantitatively by preexisting cracks. In 1948, Irwin [2-6] reformulated the Griffith crack theory and proposed the term *fracture mechanics*. Many have since contributed to its progress [7-17].

Fracture mechanics has been applied to adhesive fracture only recently. Williams [18-21] showed that the Griffith-Irwin theory of cohesive fracture applies equally to adhesive (interfacial) fracture. In cohesive fracture, the fracture energy is expended to create two similar surfaces, whereas in adhesive fracture, the fracture energy is expended to create two dissimilar surfaces. Gent [22] developed general fracture criteria for adhesive joint, applicable to either brittle-elastic or viscoelastic adhesives.

14.1. LINEAR ELASTIC FRACTURE MECHANICS

The effect of a preexisting crack on the fracture strength of a linear-elastic material can be analyzed by two equivalent approaches: the critical stress criterion and the critical energy criterion. Two critical stress analyses and one critical energy analysis are given below.

14.1.1. Stress Concentration Analysis

Inglis [23] analyzed the local stresses around an elliptical hole in a semi-infinite plate (Figure 14.1), and found that the maximum local stress occurs at the ends of the major axis of the elliptical hole, given by

$$\frac{\sigma_{max}}{\sigma} = 1 + \frac{2a}{b} \tag{14.1}$$

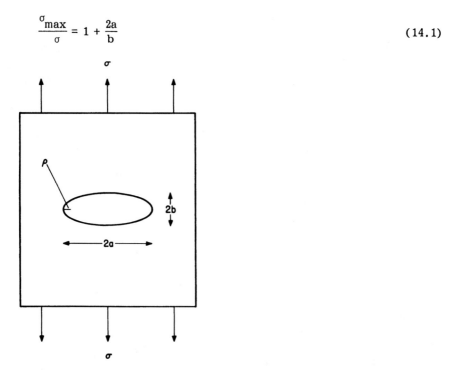

Figure 14.1. Elliptical hole in a uniformly loaded semi-infinite plate.

Linear Elastic Fracture Mechanics

where σ_{max} is the maximum local stress, σ the applied stress, a the half major axis, and b the half minor axis of the ellipse. The radius of curvature ρ at the end of the ellipse is

$$\rho = \frac{b^2}{a} \tag{14.2}$$

Thus, Eq. (14.1) becomes

$$\frac{\sigma_{max}}{\sigma} = 1 + 2\left(\frac{a}{\rho}\right)^{1/2} \tag{14.3}$$

In most cases $a \gg \rho$; therefore,

$$\frac{\sigma_{max}}{\sigma} \sim 2\left(\frac{a}{\rho}\right)^{1/2} \tag{14.4}$$

The quantity $2(a/\rho)^{1/2}$ is thus termed the *stress concentration factor*. Stress concentrations in a wide variety of crack configurations have been analyzed and listed [24–26]. When the maximum local stress equals the ideal (local) cohesive strength, fracture will occur. Therefore, Eq. (14.4) becomes

$$\sigma_f \sim \frac{\sigma_0}{2}\left(\frac{\rho}{a}\right)^{1/2} \tag{14.5}$$

where σ_f is the fracture strength and σ_0 the ideal (local) cohesive strength. Relation (14.5) relates the real fracture strength to the ideal strength and the crack geometry, and shows that the real fracture strength is usually very much lower than the ideal cohesive strength.

The energy G required to cleave a unit interfacial area is approximately

$$G = \frac{\sigma_0^2 z_0}{2E} \tag{14.6}$$

where z_0 is the equilibrium interatomic distance and E is Young's modulus. Applying relation (14.6) in Eq. (14.5) gives [7]

$$\sigma_f \sim \left(\frac{EG}{2a}\right)^{1/2} \tag{14.7}$$

provided that $\rho \sim z_0$, as it must be for cleavage fracture. Apart from the numerical factor, Eq. (14.7) is identical to the Griffith equation derived from consideration of energy balance, discussed next. Thus, the critical stress criterion and the critical energy criterion are equivalent.

14.1.2. Energy Balance Analysis: Griffith Theory

Equation (14.4) indicates that the local stress will tend to infinity as the crack tip radius approaches zero. In reality, however, the crack tip radius cannot be smaller than the atomic radius, and is usually blunted by local plastic flow. The crack tip radius is therefore generally unknown, and the critical stress criterions is difficult to utilize.

However, Griffith [1] noted that the total energy expended to create fracture surfaces is finite and should equal the release of potential energy (that is, the elastic strain energy and the work done by the movement of external loads). In other words, fracture will proceed when the release of potential energy due to the growth of crack is sufficient to supply the energy needed to create new fracture surfaces. The fracture criterion is thus given by

$$-dU \geq dW \tag{14.8}$$

where $-dU$ is the decrease of potential energy and dW is the increase of surface energy as the interfacial area of the crack extends by dA. Let us define the fracture energy G as the energy required to separate one unit interfacial area, that is,

$$G \equiv \frac{dW}{dA} \tag{14.9}$$

Note that one unit of interfacial area consists of two units of surface area. The Griffith criterion can then be restated as

$$-\frac{\partial U}{\partial A} \geq G \tag{14.10}$$

for fracture to occur.

Griffith [1] identified the fracture energy with the surface free energy, writing $G = 2\gamma_s$, where γ_s is the surface tension of the solid. However, the energy input required to create fracture surface includes not only surface free energy but also local plastic work and other energy dissipation processes, such as light emission, acoustic emission, and electrostatic charging. Therefore, generally, G should be identified as a characteristic fracture energy which includes all forms of energy dissipation processes.

Linear Elastic Fracture Mechanics

For the case of constant load displacement and a plane stress condition, Griffith [1] found that the elastic strain energy lost by the introduction of an Inglis-type elliptical crack is

$$U - U_0 = -\frac{\pi \sigma^2 a^2 h}{E} \tag{14.11}$$

where U is the elastic strain energy of the body with the crack, U_0 that without the crack, σ applied stress, a half crack length, h the thickness, and E the elastic modulus. The increase in surface energy is

$$W - W_0 = 2ahG \tag{14.12}$$

Since $\sigma = \sigma_f$ at fracture, the well-known Griffith equation is obtained as

$$\sigma_f = \left(\frac{EG}{\pi a}\right)^{1/2} \tag{14.13}$$

for a plane stress condition. On the other hand, for a plane strain condition, the relation becomes

$$\sigma_f = \left(\frac{EG}{\pi(1-\nu^2)a}\right)^{1/2} \tag{14.14}$$

where ν is Poisson's ratio.

Equation (14.10) represents an equilibrium condition. Since

$$\frac{\partial^2 U}{\partial A^2} = -\frac{\pi \sigma^2}{2Eh} < 0 \tag{14.15}$$

the equilibrium is unstable, and fracture will proceed to completion.

Griffith relation is derived for linear-elastic material containing a sharp crack. Although Eqs. (14.13) and (14.14) do not explicitly contain the crack tip radius ρ, as is the case for the stress concentration analysis in Eq. (14.5), the carck tip radius is nevertheless assumed to be very sharp, such that local stress will reach the critical value as the Griffith criterion is satisfied. Such a premise demands that the fracture energy G be dependent on the crack tip radius. Combining Eqs. (14.5) and (14.13) gives

$$G \simeq \frac{\pi}{4}\left(\frac{\sigma_0^2}{E}\right)\rho \tag{14.16}$$

for a plane stress condition, which explicitly shows the dependence of G on ρ. The G will approach zero as ρ becomes infinitely small. However, this does not occur in reality, since a very sharp crack tip will be blunted by local plastic flow, so ρ will always be finite.

The Griffith relation has been verified for brittle solids such as glass, poly(methyl methacrylate), and polystyrene [1,27,28]. The relation $\sigma_f \propto a^{1/2}$ is obeyed. The fracture energy G, equals the work of cohesion for ideally brittle glass, but is several orders of magnitude greater than the work of cohesion for nonideally brittle solids such as poly(methyl methacrylate) and polystyrene. This is because in the fracture of an ideally brittle solid, no plastic yielding occurs and the fracture energy is expended primarily to supply the surface free energy. But in the fracture of a nonideally brittle solid, extensive plastic yielding occurs around the crack tip, which absorbs energies many orders of magnitude greater than the surface free energy; Table 10.1.

14.1.3. Stress Intensity Analysis: Griffith-Irwin Theory

Irwin [5,6] analyzed fracture in terms of the crack tip stress field. There are three major loading modes (Figure 14.2).

Mode I: Opening mode or tensile mode, where the crack surfaces move directly apart

Mode II: Sliding mode in in-plane shear mode, where the crack surfaces slide over one another in a direction perpendicular to the leading edge of the crack

Mode III: Tearing or antiplane shear mode, where the crack surfaces move relative to one another and parallel to the leading edge of the crack

Mode I is encountered in overwhelming majority of practical situations, and has therefore received the greatest attention both theoretically and experimentally. Modes II and III are encountered less frequently. Mode I has the lowest fracture energy for isotropic materials, and therefore should be an adequate criterion. Mixed-mode (I and II) fractures are usually controlled by the mode I fracture [29–32]. Recently, however, the mixed-mode (I and II) fracture of a rubber-modified epoxy resin has been found to have a fracture energy lower than that for the mode I fracture by a factor of 10. This is probably due to the effect of rubber particles on the micromechanical processes of fracture [33]. Thus, mode I fracture alone may not adequately characterize the fracture properties of certain composite materials.

Using the crack tip stress functions of Westergaard [34], Irwin [4–6] gave the following relations:

Linear Elastic Fracture Mechanics

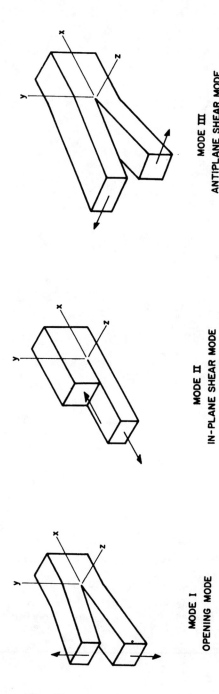

Figure 14.2. Three basic loading modes. (After Ref. 10.)

Mode I Loading

$$\sigma_x = \frac{K_I}{\sqrt{2\pi r}} \cos \frac{\theta}{2} \left(1 - \sin \frac{\theta}{2} \sin \frac{3\theta}{2}\right) \tag{14.17}$$

$$\sigma_y = \frac{K_I}{\sqrt{2\pi r}} \cos \frac{\theta}{2} \left(1 + \sin \frac{\theta}{2} \sin \frac{3\theta}{2}\right) \tag{14.18}$$

$$\tau_{xy} = \frac{K_I}{\sqrt{2\pi r}} \left(\sin \frac{\theta}{2} \cos \frac{\theta}{2} \cos \frac{3\theta}{2}\right) \tag{14.19}$$

Mode II Loading

$$\sigma_x = -\frac{K_{II}}{\sqrt{2\pi r}} \sin \frac{\theta}{2} \left(2 + \cos \frac{\theta}{2} \cos \frac{3\theta}{2}\right) \tag{14.20}$$

$$\sigma_y = \frac{K_{II}}{\sqrt{2\pi r}} \left(\sin \frac{\theta}{2} \cos \frac{\theta}{2} \cos \frac{3\theta}{2}\right) \tag{14.21}$$

$$\tau_{xy} = \frac{K_{II}}{\sqrt{2\pi r}} \cos \frac{\theta}{2} \left(1 - \sin \frac{\theta}{2} \sin \frac{3\theta}{2}\right) \tag{14.22}$$

Mode III Loading

$$\tau_{xz} = -\frac{K_{III}}{\sqrt{2\pi r}} \sin \frac{\theta}{2} \tag{14.23}$$

$$\tau_{yz} = \frac{K_{III}}{\sqrt{2\pi r}} \cos \frac{\theta}{2} \tag{14.24}$$

where σ is the normal stress, τ the shear stress, θ the polar angle, and r the radial coordinate (Figure 14.3). Thus, the constants K_I, K_{II}, and K_{III} are the only parameters that determine the magnitude of the local stress for a given specimen configuration and applied stress, and are termed the *stress intensity factors* for the three corresponding loading modes. In general $K = f(\sigma,a)$, where the function f depends on specimen configuration and loading mode. Stress intensity functions for various specimen configurations and loading modes have been listed [26,35,36].

Linear Elastic Fracture Mechanics

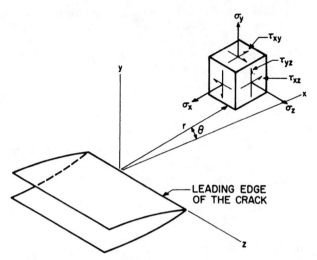

Figure 14.3. Coordinates measured from the leading edge of a crack and stress components in the crack tip stress field. (After Ref. 10.)

Since mode I is the most frequently encountered and usually has the lowest fracture energy, the subscript I will sometimes be omitted; that is, K will sometimes be used in place of K_I. For a large plate of width D containing a central slit of length 2a and loaded by a uniform tensile stress σ normal to the crack axis (that is, mode I loading), K is [4]

$$K = \sigma \left(D \tan \frac{\pi a}{D} \right)^{1/2} \tag{14.25}$$

When $D \gg a$, this becomes the case of an infinite plate and Eq. (14.25) becomes

$$K = \sigma \sqrt{\pi a} \tag{14.26}$$

The critical stress criterion states that fracture will occur when the local stress reaches the local strength. From Eq. (14.18), for a crack of tip radius ρ, the maximum local stress $\sigma_{y,max}$ occurs at $\theta = 0$ and $r = \rho$, and thus

$$K_c = \sigma_0 (2\pi\rho)^{1/2} \tag{14.27}$$

where σ_0 is the ideal (local) cohesive strength and K_c is the critical value of K at the onset of fracture. Thus, the critical stress criterion of fracture may be restated as: Fracture will occur when the stress intensity factor reaches a critical value K_c. Note that K_c is dependent on crack tip radius.

For an infinite plate containing a central crack of length 2a and loaded with a uniform tensile stress σ normal to the crack axis, the critical stress intensity factor is given by

$$K_c = \sigma_f \sqrt{\pi a} \tag{14.28}$$

which follows from Eq. (14.26). K_c has also been termed the *fracture toughness*, and has units of $N/m^{3/2}$ or $lb/in.^{3/2}$.

K and G are related by [2-6]

$$K^2 = EG \tag{14.29}$$

for a plane stress condition, or

$$K^2 = \frac{EG}{1 - \nu^2} \tag{14.30}$$

for a plane strain condition. These relations follow readily from Eqs. (14.13), (14.14), and (14.26).

Irwin [2-6] defined the quantity $-(\partial U/\partial A) \equiv \zeta$ as the *strain energy release rate*. Since it has a dimension of energy per unit interfacial area, which is numerically identical to force per unit interfacial lingth, it is also known as the *crack extension force*. The critical energy criterion can then be stated as: Fracture will occur when ζ reaches the critical value ζ_c. This ζ_c is identical to the fracture energy G defined in Eq. (14.9).

14.2. LOCAL PLASTIC FLOW

A zone of plastic flow will occur around the crack tip whenever the local stress exceeds the yield strength of the material. The local plastic flow will blunt the tip of a preexisting crack and truncate the local stress at the value of the yield strength. This will increase the apparent crack length, which can be assumed to be the actual crack length plus a certain fraction of the local plastic zone thickness. Two models for estimating the size of the local plastic zone are given below.

14.2.1. Irwin Plastic Zone Model

Consider the local elastic stress normal to the plane of crack extension where $\theta = 0$. The local elastic stress $\sigma_y = K/\sqrt{2\pi r}$ will exceed the yield

Local Plastic Flow

strength σ_P at some distance r from the carck tip, and thereby truncate the elastic stress at that value. The thickness of the plastic zone r_p is then given by [5,37]

$$r_P = \frac{1}{2\pi} \frac{K^2}{\sigma_P^2} \qquad (14.31)$$

for a plane stress condition. On the other hand, for a plane strain condition, the triaxial stress field tends to supress the plastic zone; the plane strain plastic zone thickness is thus smaller and given by [38]

$$r_P = \frac{1}{6\pi} \frac{K^2}{\sigma_P^2} \qquad (14.32)$$

The size of the plastic zone also varies with θ. When analyzed by the more general distortion energy theory, the thickness of the plastic zone is found to be [37]

$$r_P = \frac{K^2}{2\pi\sigma_P^2} \left(1 + 3\sin^2\frac{\theta}{2} \cos^2\frac{\theta}{2}\right) \qquad (14.33)$$

for a plane stress condition.

14.2.2. Dugdale Plastic Zone Model

Dugdale [39] proposed another model for the case of plane stress. The plastic zone is considered to be a narrow conical strip extending a distance r_P from each crack tip (Figure 14.4). The internal crack of length 2a is allowed to extend elastically to $2(a + r_P)$. An internal stress is applied around the plastic zone to reclose the crack. This internal stress must equal the yield strength of the material. Thus, r_P is given by

$$\frac{r_P}{a} = -1 + \sec\left(\frac{\pi\sigma}{2\sigma_P}\right) \qquad (14.34)$$

When applied stress σ is much lower than the yield strength, Eq. (14.34) reduces to

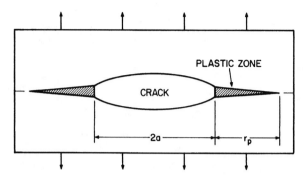

Figure 14.4. Dugdale plastic zone model for non-strain-hardening solids.

$$\frac{r_P}{a} = \frac{\pi^2}{8}\left(\frac{\sigma}{\sigma_P}\right)^2 \tag{14.35}$$

14.3. GENERALIZATION OF FLAW THEORY TO VISCOELASTIC MATERIALS: RIVLIN AND THOMAS THEORY

Rivlin and Thomas [40–47] showed that the Griffith criterion can be generalized to viscoelastic materials by reinterpreting the energy terms. They proposed

$$-\left(\frac{\partial U}{\partial A}\right)_L \geq G \tag{14.36}$$

for fracture to occur, where U is the elastically stored energy in the specimen as a whole, A the interfacial area, and G a characteristic fracture energy per unit interfacial area. The subscript L denotes that no external work is done on the system during the interchange of energy between the body of the specimen as a whole and the crack. The criterion is perfectly general, as it avoids any reference to the classical elasticity theory. The left-hand side of Eq. (12.58) may readily be expressed in terms of measurable quantities such as force and strain for a given specimen configuration. This is illustrated by the following two examples.

14.3.1. Tensile Specimen

Consider a long sheet of uniform thickness h, containing an edge crack of length a and loaded in tension (Figure 14.5, left). Since A = ah, Eq. (14.36) becomes [7]

$$-\frac{1}{h}\left(\frac{\partial U}{\partial a}\right)_L \geq G \qquad (14.37)$$

A similar specimen containing no crack has a uniform elastic-strain-energy density (that is, elastic-strain energy per unit volume) of Q. The insertion of the crack reduces this energy to zero within the shaded area of the sheet. Geometrical consideration suggests that this area be proportional to a^2. The loss of strain energy due to the crack will therefore be given by $\beta a^2 hQ$, where β is a constant that equals π for linear-elastic materials and ranges from 1 to 3 for elastomers [48]. The left-hand side of Eq. (14.37) becomes

$$-\left(\frac{\partial U}{\partial a}\right)_L = 2\beta ahQ \qquad (14.38)$$

Fracture will occur at a critical strain energy density Q_c, which is the strain energy per unit volume at break. Thus, the fracture criterion becomes

$$G = 2\beta a Q_c \qquad (14.39)$$

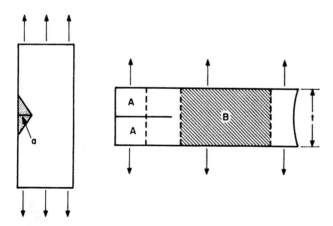

Figure 14.5. Left: Simple tension of a specimen with an edge crack. Right: Pure shear specimen: A, relaxed region, B, pure shear region.

If the introduction of the crack is assumed to have only a second-order effect on the strain energy remote from the crack, the critical strain energy density can be determined from the stress-strain curve of the specimen.

14.3.2. Pure Shear Specimen

Consider a sheet of uniform thickness h, containing a long crack parallel to and gripped in tension along the long edges (Figure 14.5, right). Provided that the specimen is long compared with the grip distance, the region B will be in a state of pure shear, and the region A along the crack will be stress free. The regions near the crack tip and at the crack-free end of the specimen will be in an undetermined stress state. The effect of extending the crack will thus be to remove a portion of Δa in length from the pure shear region and transfer it to the unstrained region. The loss of elastic energy is [7]

$$-\Delta U = Qh T \Delta a \qquad (14.40)$$

where Q is the strain energy density in the material in pure shear and t is the grip distance before application of the strain. Therefore,

$$-\frac{1}{h}\frac{\partial U}{\partial a} = Qt \qquad (14.41)$$

and the fracture criterion is

$$G = tQ_c \qquad (14.42)$$

The Rivlin and Thomas theory has been successful in treating materials such as elastomers which do not obey the classical linear elasticity [40–49]. The theory is general for viscoelastic materials, and has also been applied to viscoelastic liquids [50]. The essential feature of the theory is to identify $-(\partial U/\partial A)$ as the recoverable elastic-strain energy and G as a characteristic fracture energy.

14.4. PROPERTIES OF FRACTURE ENERGY

14.4.1. Components of Fracture Energy

Griffith [1] identified the fracture energy as the surface free energy of the newly created surfaces,

$$G = 2\gamma_s \qquad (14.43)$$

Properties of Fracture Energy

where γ_S is the surface tension of the solid. The factor 2 accounts for the fact that one unit of interfacial area gives two units of surface area. Table 10.1 lists typical fracture energies for some materials. As can be seen, for ideally brittle materials, (such as soda-lime glass), the fracture energy is indeed twice the surface tension.

However, for the vast majority of materials which are not ideally brittle (such as polymers and metals), the fracture energy is many orders of magnitude greater than the surface free energy. This is because fracture is usually not a reversible process, but rather consists of many irreversible dissipative processes, such as plastic yielding around the crack tip, cavitation, static electrification, light and acoustic emissions, and so on. Furthermore, fracture surfaces may be rough, so that the true surface area is greater than the planar geometrical area. However, the viscoelastic dissipation by plastic yielding is by far the most important. The energy dissipated in plastic yielding is many orders of magnitude greater than those in other processes. For instance, in the impact fracture of notched specimens of a rubber-toughened polymer, extensive cavitation occurs in the energy-absorbing zone around the crack tip, making this zone appear white. The energy absorbed per unit volume in this zone is 4.5 cal/ml by plastic yielding (stretching and alignment of polymer molecules), 0.7×10^{-2} cal/ml by cavitation, and 0.6×10^{-5} cal/ml due to the increased surface area of the rough fracture surface. Therefore, the fracture energy may be written as

$$G \simeq 2(\gamma_S + \gamma_P) = W_c + W_P \tag{14.44}$$

where γ_P is the plastic work per unit surface area, W_c the work of cohesion, and W_P the plastic work per unit interfacial area. Since $W_P \gg W_c$ is most cases, the fracture energy can be given as

$$G \simeq W_P \tag{14.45}$$

Since W_P is a viscoelastic quantity, the fracture energy is thus dependent on rate, temperature, and loading mode, just as would be for a viscoelastic property, although independent of specimen geometry.

14.4.2. Extension to Adhesive Fracture

In terms of energy concept, adhesive (interfacial) and cohesive fractures are similar. The essential difference lies in the interpretation of the fracture energy [18–21]. In cohesive fracture, two similar surfaces are created. In adhesive fracture, two dissimilar surfaces are created (Figure 14.6). The adhesive fracture energy G_a is thus given by

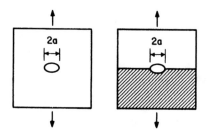

Figure 14.6. Griffith model for cohesive and adhesive fractures.

$$G_a = (\gamma_{S1} + \gamma_{P1}) + (\gamma_{S2} + \gamma_{P2}) - \gamma_{12} \tag{14.46}$$

where γ_{12} is the interfacial tension between phases 1 and 2. Equation (14.46) can be rewritten as

$$G_a = W_a + (\gamma_{P1} + \gamma_{P2}) = W_a + W_{P12} \tag{14.47}$$

where W_a is the work of adhesion and $W_{P12} = \gamma_{P1} + \gamma_{P2}$. The condition for adhesive fracture then becomes

$$-\left(\frac{\partial U}{\partial A}\right)_L \geq G_a \tag{14.48}$$

in complete analogy with that for cohesive fracture.

Consider the case where the adherend is infinitely rigid as compared with the adhesive. For end-bonded half-planes with a central crack of length 2a, the Griffith criterion for adhesive fracture is thus [94,97]

$$\sigma_f = \left(\frac{EG_a}{\pi a}\right)^{1/2} \tag{14.49}$$

for a plane stress condition. On the other hand, for a plane strain condition, the relation is

$$\sigma_f = \left[\frac{EG_a}{\pi(1-\nu^2)a}\right]^{1/2} \tag{14.50}$$

in analogy with cohesive fracture.

Properties of Fracture Energy

On the other hand, Gent [22,51,52] has extended the Rivlin-Thomas theory to the interfacial fracture of viscoelastic adhesives, in complete analogy with cohesive fracture discussed in Section 14.3. These are discussed further in Part II of this chapter.

14.4.3. Fracture Energy Function

The fracture energy has been given as the sum of thermodynamic work and plastic work in Eqs. (14.44)−(14.47). Recently, an alternative and very useful relation has been proposed [53,54],

$$G = G_0 \psi(R) \tag{14.51}$$

where G_0 is the fracture energy at zero rate (equilibrium fracture energy), R the rate, and $\psi(R)$ a rate-dependent viscoelastic function. This relation has been confirmed experimentally and theoretically [53, 54]. The rate and temperature effects are equivalent for a viscoelastic property, and can be superimposed by using an effective rate. Therefore, Eq. (14.51) can be rewritten as

$$G = G_0 \psi(Ra_T) \tag{14.52}$$

where a_T is an appropriate shift factor and Ra_T is the effective (or reduced) rate. At zero rate, viscoelastic effects are absent, and the separation process is reversible. Therefore,

$$\psi(Ra_T) = 1 \quad \text{when } Ra_T = 0 \tag{14.53}$$

and $G = G_0$. If only physical (van der Waals) attractions are present across the interface, the equilibrium fracture energy is the thermodynamic work, that is,

$$G_0 = W_c \quad \text{for cohesive fracture} \tag{14.54}$$

and

$$G_0 = W_a \quad \text{for adhesive fracture} \tag{14.55}$$

On the other hand, if chemical bonds are present across the interface, the equilibrium fracture energy will include the chemical bond energy, that is,

$$G_0 = W_b \tag{14.56}$$

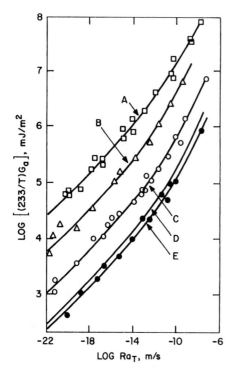

Figure 14.7. Adhesive fracture energy G_a versus reduced crack propagation rate Ra_T at $T_g = -40°C$ for butadiene-styrene rubber on various adherends as measured by 90°-peel test. (A) Fracture energy for bulk adhesive; (B) fluorinated ethylene-propylene copolymer, etched in sodium-naphthalene solution for 120 sec; (C) fluorinated ethylene-propylene copolymer, commercially etched; (D) poly(ethylene terephthalate); (E) nylon 11. (After Ref. 54.)

where W_b is the reversible work of separation, including the chemical bond energy. In such case, $W_b \gg W_a \simeq W_c$.

The situation described above has been demonstrated experimentally [54]. A lightly cross-linked styrene-butadiene rubber (40:60 weight ratio, $T_g = -40°C$) is bonded to various rigid adherends. The fracture energies are measured by the 90°-peel test over wide ranges of temperature (−35 to 100°C) and rate (1 nm/sec to 0.1 mm/sec), and superimposed on the effective rate using the universal form of the WLF equation (Figure 14.7). All curves (for different adherends) are parallel, suggesting the form of Eq. (14.52). The viscoelastic function is given by [53]

Table 14.1. Equilibrium Fracture Energy G_0 and Work of Adhesion W_a for a Lightly Cross-linked Styrene-Butadiene Rubber Adhering to Various Adherends

Adherend	G_0, mJ/m^2	W_a, mJ/m^2
Fluorinated ethylene-propylene copolymer (FEP)	21.9	48.4
Polychlorotrifluoroethylene	74.9	62.5
Nylon 11	70.8	71.4
Poly(ethylene terephthalate)	79.4	72.3
Plasma-treated FEP (helium)	68.5	56.8
Sodium naphthalene-treated FEP		
10 sec	851	68.0
20 sec	1170	70.2
60 sec	1290	69.8
90 sec	1620	71.1
120 sec	1780	71.1
500 sec	2420	72.2
1000 sec	1990	71.8

Source: From Ref. 54.

$$\psi(Ra_T) = 1 + 1.25 \times 10^3 (Ra_T)^{0.42} \qquad (14.57)$$

where Ra_T is in cm/sec. Equilibrium fracture energy G_0 and work of adhesion W_a for the rubber adhering to the various adherends are listed in Table 14.1. The G_0 and W_a values indeed agree quite well for the adherends, including untreated FEP (fluorinated ethylene-propylene copolymer), plasma-treated FEP, polychlorotrifluoroethylene, nylon 11, and poly(ethylene terephthalate), indicating that the adhesive bonds are formed by van der Waals attraction. On the other hand, the G_0 values are much greater than the W_a values for the various sodium naphthalene-treated FEP, indicating that chemical bonds are involved in forming the latter adhesive bonds.

14.4.4. Rate-Temperature Effect

The effects of rate and temperature on fracture energy are equivalent. The cohesive fracture energy of poly(methyl methacrylate), measured by cleavage [55], is given in Figure 14.8. The curves can be super-

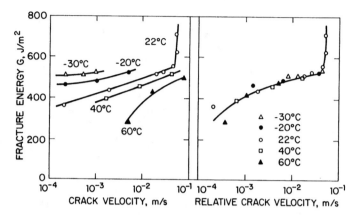

Figure 14.8. Temperature dependence of fracture energy for slow crack propagation in poly(methyl methacrylate). Left: Actual velocities. Right: Velocities shifted to match the curve at 22°C. (After Ref. 55.)

imposed to give a single master curve by shifting the rates to match the 22°C curve. The effects of rate and temperature on the adhesive fracture energy, measured by the 180°-peel test [56], for a cross-linked polybutadiene rubber bonded to a glass adherend (treated with a 50:50 mixture of ethylsilane and vinylsilane to give various covalent bond densities at the interface) are shown in Figure 14.9. All curves can be superimposed by plotting against the effective rate Ra_T, using the universal form of the WLF equation [57],

$$\log a_T = \frac{-17.4(T - T_g)}{51.6 + T - T_g} \qquad (14.58)$$

where the T_g of the polybutadiene rubber is $-90°C$. Another example has been given in Figure 14.7.

14.4.5. Effect of Loading Mode

Fracture energy is affected by the loading mode [58]. The fracture energies of a polyurethane adhesive bonded to an acrylic adherend tested under various loading modes are listed in Table 14.2. Mode I (opening mode) has the lowest fracture energy; mode II (in-plane shear) is intermediate; mode III (antiplane shear) has the highest fracture energy. As mentioned before, mode I usually has the lowest fracture energy and is encountered in most cases. Even in mixed-mode

Table 14.2. Fracture Energy of a Polyurethane Adhesive Bonded to Poly(Methyl Methacrylate) Adherend Tested Under Various Loading Modes

Mode	Test type	Adhesive fracture energy, G_a, in.-lb/in.^{2}
I	Blister test	0.17
	90°-peel test	0.18
II	0°-cone test	0.40
III	90°-cone-torsion test	0.58

Source: From Ref. 58.

fractures, mode I is usually the controlling factor. Recently, however, mixed-mode (I and II) fractures of some multiphase adhesives have been found to have fracture energies vastly lower than those of mode I [33] (Table 14.3). This probably arises from the effect of rubber particles on the micromechanical process of fracture.

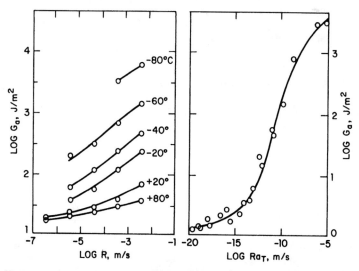

Figure 14.9. Adhesive fracture energy G_a versus rate R or effective rate Ra_T at various temperatures for a cross-linked polybutadiene rubber on glass treated with a 50:50 mixture of ethylsilane and vinylsilane as measured by 180°-peel test. (After Ref. 56.)

Table 14.3. Comparison of Opening Mode and Mixed-Mode Adhesive Fracture Energy

	Fracture energy, J/m^2	
Adhesive	Mode I cleavage	Mixed (I and II) mode cleavage + shear
Epoxy (unmodified) [a]	116	140
Epoxy-carboxyl rubber [b]		
10% rubber	3500	110
30% rubber	2200	110
Epoxy-rubber	2300	870
Nylon-epoxy	6100	750
Narmco ME-329 adhesive [c]	630	55
3M AF-243 adhesive [d]	860	220

[a] DGEBA (diglycidyl ether-bisphenol A) epoxy resin cured with hexahydrophthalic anhydride.
[b] DGEBA epoxy resin modified with a carboxy-terminated butadiene-acrylonitrile rubber, M_n = 3500 (B. F. Goodrich CTBN) and cured with piperidine.
[c] Narmco Materials, Inc.
[d] 3M Company.
Source: From Ref. 33.

14.4.6. Mixed Cohesive and Adhesive Fracture

In mixed cohesive and adhesive fracture, the fracture energy is given by linear additivity of the various modes [59],

$$G = x_1 G_1 + x_2 G_2 + x_3 G_3 \tag{14.59}$$

where x_1 is the fractional area where cohesive fracture of phase 1 occurs, x_2 that where cohesive fracture of phase 2 occurs, x_3 that where adhesive (interfacial) fracture occurs between phases 1 and 2, and G_1, G_2, and G_3 are the corresponding fracture energies. This relation has been demonstrated for a styrene-butadiene rubber bonded to various polymer adherends [59].

Tensile Tests (Butt Joints)

14.4.7. Effect of Crack Tip Radius

Combination of Eqs. (14.27) and (14.28) gives

$$G = 2\pi \left(\frac{\sigma_0^2}{E}\right)\rho \qquad (14.60)$$

which is identical to Eq. (14.16) except for the numerical factor. Equation (14.60) indicates that G is affected by the crack tip radius. However, whenever the local stress around the crack tip exceeds the yield strength, a region of plastic flow will occur. This tends to blunt the crack tip and increase the fracture energy. A similar situation has been discussed in Section 14.1.2.

PART II. ANALYSIS AND TESTING OF ADHESIVE BONDS

The mechanical strenfth of an adhesive joint is its most important property. Many tests have been developed to evaluate it. Fracture strength tests include tensile tests of butt joints, shear tests of lap joints, peel tests, and others. These tests may also give fracture energy when the test results are appropriately analyzed. Other fracture energy tests include cantilever beam tests, blister tests, and cone tests. Both fracture energy and fracture strength are dependent on rate, temperature, and loading mode. However, the fracture energy is independent of specimen geometry, whereas the fracture strength is dependent on specimen geometry. Therefore, fracture energy is a more fundamental measure of the quality of an adhesive joint. Tests should be chosen to approximate the service conditions.

In this section, various commonly utilized adhesive joint tests will be analyzed. Details of test specifications can be found in ASTM Standards and elsewhere [58,60–63].

14.5. TENSILE TESTS (BUTT JOINTS)

14.5.1. Description of Tensile Tests

Typical tensile tests of butt joints are described, for example, in ASTM D897, depicted in Figure 14.10. In ASTM D897, the specimen diameter is chosen to give a cross-sectionsl area of 1 in.^{2}. Therefore, this test is often called the *pi tensile test*. ASTM D2094 describes methods for preparation of bar (square cross section) and rod (circular cross section) specimens. ASTM D2095 describes testing techniques and specimen fixtures. These are recommended for use with either metal or plastics. ASTM D1344 describes cross-lap test

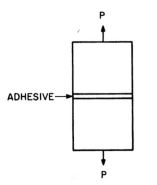

Figure 14.10. Tensile test on a butt joint.

for use with glass, wood, sandwich, and honeycomb structures, depicted in Figure 14.11.

The stresses induced in the adhesive layer in a butt joint under tension can be quite complicated. When the adhesive and the adherend have identical moduli, the tensile stress will be uniformly distributed within the adhesive layer. However, when the moduli are unequal, the tensile stress will be nonuniform with superimposed shear stress (Figure 14.12). In addition, stress concentrations can also occur along the three-phase boundaries caused by differential lateral contraction of the adhesive and the adherend under tension.

14.5.2. Geometrical Stress Concentration

Consider two cylindrical adherends joined base to base with an adhesive layer (Figure 14.13). Let the radius of the adherend be R and the original thickness of the adhesive be h. When an external tension is applied along the longitudinal axis of the cylinders, the adhesive layer will contract laterally along the air-adhesive interface, creating a side groove, while the adhesive thickness increases to h + Δh. If

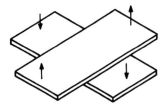

Figure 14.11. Cross-lap tensile test.

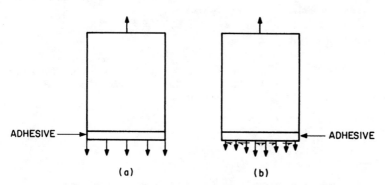

Figure 14.12. Schematic stress distributions in butt joints under tension for case (a), where the adherend and adhesive have identical moduli, and for case (b), where the moduli are different. (After Ref. 58.)

the profile of this groove is approximated as a half ellipse, the long-half axis will be $(1/2)(h + \Delta h)$ and the short half-axis $(2R/\pi)[h/(h + \Delta h)]$. Substituting in Eq. (14.1) gives [64]

$$\frac{\sigma_{max}}{\sigma} = 1 + \frac{8}{\pi} \left(\frac{R}{h}\right) \frac{e}{(1 + e)^2} \tag{14.61}$$

where e is the adhesive tensile strain ($e = \Delta h/h$). Thus, if $R/h = 100$ and e is 0.03, the stress concentration factor σ_{max}/σ will be as high as 9.

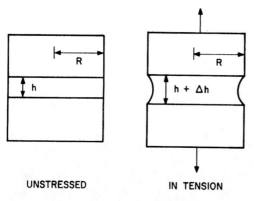

Figure 14.13. Geometrical stress concentration in a butt joint.

14.5.3. Poisson's Contraction and Shear Stress

Consider again two cylindrical adherends joined base to base with an adhesive. When a tensile stress of σ is applied, both the adherend and the adhesive will contract radially. The net radial strain $\lambda(r)$ at a radial position r is $\lambda(r) = (r/R)[(\nu_2/E_2) - (\nu_1/E_1)]\sigma$. The interfacial shear stress τ in the adhesive is thus

$$\tau = \frac{r}{R}\left[\left(\frac{\nu_2}{E_2}\right) - \left(\frac{\nu_1}{E_1}\right)\right]E_2\sigma \tag{14.62}$$

The maximum shear stress occurs at the periphery, where $r = R$. Let $E_1 \gg E_2$; then $\tau_{max} \simeq \nu_2\sigma$. If $\nu_2 = 0.5$, then $\tau_{max} \simeq 0.5\sigma$, which can be quite significant. More elaborate analyses of interfacial shear stresses in butt joints can be found elsewhere [65-67].

14.5.4. Photoelastic Analysis of Butt Joint

Stresses in butt joints have been investigated by photoelastic technique [68]. The results agree with theoretical predictions. The stress is not uniform in the butt joint. It varies slowly near the center of the adhesive layer, and increases rapidly toward the adhesive-air boundary. The stress is the greatest at the three-phase (adhesive-adherend-air) boundary, indicating that the greatest stress concentration is there.

14.5.5. Fracture Strength Analysis of Butt Joint

Complete stress analysis for a butt joint with an elastic adhesive is quite complicated. Instead, the case with a Bingham plastic adhesive has been analyzed [64,69]. Consider two flat plates bonded together with a Bingham plastic adhesive (Figure 14.14). The flat plates are very long in the direction normal to the plane of the drawing and have width w. The adhesive thickness is h_0. When an average tensile stress σ is applied parallel to the y axis, the local tensile stress σ_y in the y direction in the adhesive is independent of y. However, the tensile stress σ_x in the x direction will depend on both x and y. This stress will cause the adhesive to flow toward the plane $x = 0$. In addition, there are shear stresses acting along slip lines which cross each other at right angles. When the local shear stress along a slip line reaches the yield strength τ_0, yielding will occur. The applied tensile stress σ_f that will cause the fracture by yielding is thus given by [64]

Tensile Test (Butt Joints)

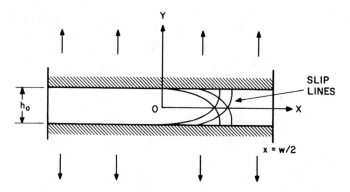

Figure 14.14. Slip lines in a Bingham plastic adhesive in a butt joint under tension.

$$\sigma_f = \frac{w \tau_0}{h_0} \quad (14.63)$$

Similarly, for two circular adherends of radius R, the relation is given by

$$\sigma_f = \frac{2}{3} \frac{R \tau_0}{h_0} \quad (14.64)$$

In either case, the theory predicts that σ_f would increase without limit as ever thinner adhesives are used, which is qualitatively consistent with experimental fact discussed later.

14.5.6. Fracture Energy Analysis of Butt Joint

Two extreme cases of tensile butt joints are considered: (1) a thin adhesive layer with a relatively long edge crack (that is, pure shear, Figure 14.15a), and (2) a thick adhesive layer with a relatively short edge crack (that is, simple tension, Figure 14.15b). In both cases, the fracture energy G_a per unit area of interface comes from the strain energy stored in the deformable adhesive.

For the first case, analogous to Eq. (14.42), the relation is given by [22]

$$G_a = h_0 Q_c \quad \text{(pure shear)} \quad (14.65)$$

where h_0 is the thickness of adhesive layer in the unstrained state and Q_c the critical strain energy density, that is, the strain energy per unit volume of the adhesive at fracture.

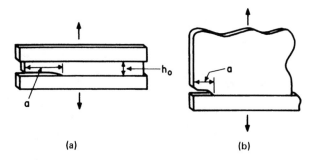

Figure 14.15. Pure shear (a) and simple tension (b) geometries. (After Ref. 22.)

For the second case, analogous to Eq. (14.39), the relation is given by [22,52]

$$G_a = \beta a Q_c \quad \text{(simple tension)} \tag{14.66}$$

where a is the length of the edge crack (or half of crack length if the crack lies well within the interior of the bond), and β is a numerical factor given to a good approximation by $\pi(1 + e_f)^{-1/2}$, where e_f is the tensile strain at fracture.

A useful empirical relation encompassing both extreme cases is given by [22]

$$G_a = h_0 Q_c \left[1 - \exp\left(-\frac{\beta a}{h_0}\right)\right] \tag{14.67}$$

valid for any elastic or viscoelastic adhesives.

If the overall strains are small and the adhesive is linearly elastic, the critical strain energy density is $Q_c = \sigma_f^2/2E$, where σ_f is the fracture strength and $\beta = \pi$. Thus, for the first case,

$$\sigma_f = \left(\frac{2EG_a}{h_0}\right)^{1/2} \quad \text{(pure shear)} \tag{14.68}$$

and, for the second case,

$$\sigma_f = \left(\frac{2EG_a}{\pi a}\right)^{1/2} \quad \text{(simple tension)} \tag{14.69}$$

Tensile Test (Butt Joints)

Therefore, when the thickness of adhesive layer is small compared with the crack (the first case), the thickness is the important dimension. On the other hand, when the thickness is large compared with the crack (the second case), the size of the crack becomes important. In both cases, the smaller the critical dimension, the greater is the fracture strength. Thus, for thin adhesives, the strength is predicted to increase with decreasing adhesive thickness, in agreement with experimental fact (discussed later).

For a more complete treatment of this effect, however, it is necessary to take into account stress concentration at the edges and non-uniform stress distribution in thin adhesive layers [51]. Consider a thin circular adhesive layer: failure will start at the edge of an interface where by chance a large flaw exists. As the local strains are larger near the edges of the adhesive, Q_c will exceed the average value Q_c' by a factor [22]

$$\alpha^2 \left(1 + \frac{3R^2}{h_0^2}\right)\left(1 + \frac{R^2}{2h_0^2}\right)^{-1}$$

where R is the radius of the adhesive layer and α is the stress concentration factor in the vicinity of the edges. The overall energy density Q_c' is given in terms of the average applied stress σ_f by

$$Q_c' = \frac{\sigma_f^2}{2E[1 + (R^2/2h_0^2)]}$$

Thus, the average fracture strength is given by

$$\sigma_f = \frac{1}{\alpha}(2EQ_c)^{1/2}\left(1 + \frac{R^2}{2h_0^2}\right)\left(1 + \frac{3R^2}{h_0^2}\right)^{-1/2} \qquad (14.70)$$

where Q_c is the critical strain energy in the vicinity of the edge for the crack to grow catastrophically. Q_c is related to the fracture energy G_a by Eq. (14.66) when the crack is initially much smaller than the thickness of the adhesive layer. When the adhesive thickness h_0 is much greater than the radius R, the fracture strength is independent of the thickness; when the thickness is much smaller than the radius, the fracture strength is inversely proportional to the thickness. These are generally observed.

14.5.7. Experimental Observations

Effect of Adhesive Thickness

Experimentally, the strength of a butt joint increases with decreasing adhesive thickness [70–77], shown in Figures 14.16 to 14.18. Thus, the thinner the adhesive, the stronger the butt joint. This behavior is, however, only specific with butt joints. The opposite is often found for lap joints and peel joints.

At very low adhesive thickness, the tensile strength of a butt joint may exceed that of the bulk adhesive. As the thickness increases, the joint strength tends to decrease and approach the ultimate tensile strength of the bulk adhesive, (Figure 14.18). The upper curve is for a high-molecular-weight polyethylene, the lower curve for a polyethylene wax of molecular weight 2750. The adherends are steel to steel for circles, and glass to glass for squares. The dashed lines on the right are the tensile strengths of the two bulk polyethylenes. The butt joint strength with the thinnest adhesive is about 2 to 2.5 times the ultimate tensile strength of the bulk adhesive.

A butt joint of two brass cylinders joined with a eutectic solder less than 0.025 cm thick has a tensile strength of 165 MPa, about three times the ultimate tensile strength of the bulk solder. A butt joint of two steel cylinders joined with a thin layer of paraffin wax has a tensile strength of 2.2 MPa, about 2.5 times the ultimate tensile strength of the bulk paraffin wax [72].

These observations are consistent with Eqs. (14.64) and (14.70). Derived for Bingham-plastic adhesives, Eq. (14.64) is not strictly applicable to elastic or viscoelastic adhesives. However, Eq. (14.70)

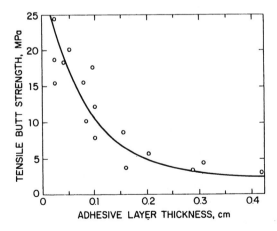

Figure 14.16. Effect of adhesive thickness on tensile butt strength for two steel cylinders bonded with poly(methyl methacrylate) at a loading rate of 44.4 MPa/sec. (After Ref. 71.)

Tensile Test (Butt Joints)

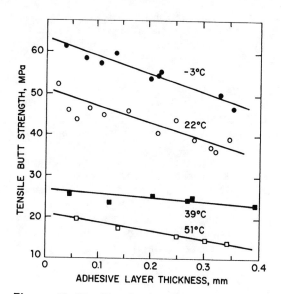

Figure 14.17. Tensile butt strength versus adhesive layer thickness for steel plugs bonded with poly(vinyl acetate) at 148°C. (After Ref. 75.)

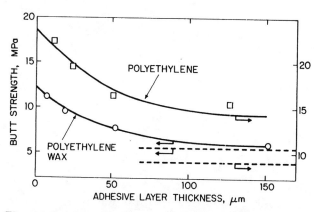

Figure 14.18. Tensile butt strength versus adhesive layer thickness: □, for glass to glass joined with a high-molecular-weight polyethylene; ○, for steel to steel joined with a polyethylene wax (MW about 2800). (After Ref. 74.)

should be generally applicable to viscoelastic adhesives. It predicts that the butt joint strength should be inversely proportional to the adhesive thickness for very thin adhesives, and independent of the adhesive thickness for very thick adhesives, in agreement with experiment.

Several other explanations have also been offered to explain the effect of adhesive thickness. Bikerman [73,74] suggested that as the adhesive thickness decreases, the probability of a large flaw occurring in the adhesive layer decreases, and therefore the butt joint strength increases. However, this does not explain why the strength of lap joint and peel joint often increases with increasing adhesive thickness. Gardon [78] suggested that since the rigid adherend inhibits the contraction of adhesive, the shear stresses induced in the adhesive are lower and the adhesive cross-sectional area at break is larger than in uniaxial tension. Meissner and Baldauf [72] suggested that as the adhesive thickness increases, internal stress arising from differential shrinkage of adhesive and adherend tends to increase, and

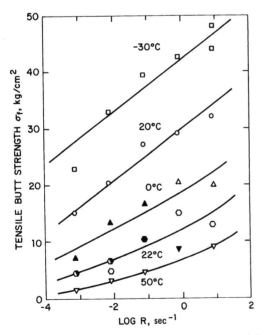

Figure 14.19. Tensile strength of butt joint with a thin adhesive layer (1 mm thickness) versus extension rate R. Open symbols are for interfacial failure; filled symbols are for cohesive failure. The adhesive is a butadiene-styrene rubber. The adherends are PET polyester films glued to steel cylinders with an epoxy adhesive. (After Ref. 51.)

Tensile Test (Butt Joints)

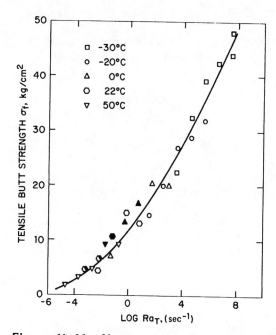

Figure 14.20. Master curve for tensile butt strength versus effective extension rate Ra_T at 23°C for the data of Figure 14.19. (After Ref. 51.)

therefore, the butt joint strength tends to decrease. However, this does not explain why the butt joint strength for very thin adhesives can exceed the ultimate tensile strength of the bulk adhesive.

Rate-Temperature Effect

The tensile strength of butt joints over wide ranges of temperature and rate have been investigated [51]. The joints are prepared by using a butadiene-styrene rubber (weight ratio 60:40, $T_g = -40°C$) as the adhesive, and poly(ethylene terephthalate) film as the adherend. The PET films are bonded to steel cylinders using an epoxy adhesive. The results are plotted in Figure 14.19. The results at various rates and temperatures can be superimposed to give a single master curve by using the universal form of the WLF shift factor, Eq. (14.58), shown in Figure 14.20. This shows that the joint strength is associated mainly with viscoelastic effects rather than with thermodynamic properties. The fact that the data for both interfacial and cohesive failures can be superimposed may, at first sight, suggest that the interfacial failure is actually a cohesive failure very near the interface. However, this is not true. The observed behavior arises from the fact that energy dissipative processes are similar in cohesive and interfacial fractures [51].

The effects of adhesives thickness on the strengths of these butt joints are shown in Figure 14.21. The joint strength for thin (0.1 cm) adhesive is greater than for thick (2.5 cm) adhesive at all rates. At low rates, the joint strength with thick adhesive coincides with the tensile strength of the bulk adhesive. At higher rates, however, the joint strength for both thin and thick adhesives falls below the tensile strength of the bulk adhesive.

Fracture Energy Measurements

Fracture energies of butt joints in simple tension and in pure shear have been measured [52]. Lightly cross-linked butadiene-styrene rubber (60:40 weight ratio, T_g = −40°C, cross-linked with dicumyl peroxide) is used as the adhesive, and PET films rigidly cemented to steel blocks are used as the adherends. The critical strain energy density Q_c of the adhesive is determined by integrating the tensile stress-strain curve for an adhesive sheet without crack.

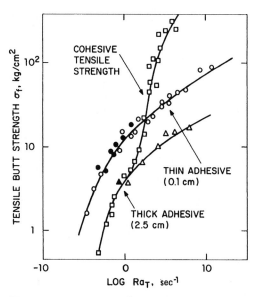

Figure 14.21. Tensile butt strength versus effective rate Ra_T at 23°C for thin and thick adhesive layers. Also shown is the cohesive tensile strength. Butadiene-styrene rubber adhesive on PET polyester adherends. (After Ref. 51.)

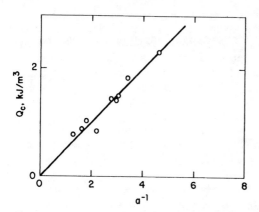

Figure 14.22. Experimental relation between critical strain energy density Q_c and half-length of a central crack a for simple extension specimen. Strain rate 1.6×10^{-4} sec^{-1}, temperature 20°C. (After Ref. 52.)

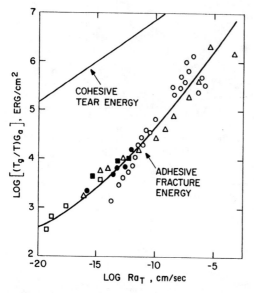

Figure 14.23. Adhesive fracture energy G_a versus effective crack propagation rate Ra_T for a cross-linked butadiene-styrene rubber on a PET polyester adherend. △, Simple tension at constant rate; □, simple tension at constant load; ■, pure shear at constant rate; ●, pure shear at constant load; ○, peel test. (After Ref. 52.)

In the case of simple tension, central cracks of variable length 2a are introduced at the adhesive-adherend interface. Figure 14.22 shows that the Q_c versus 1/a plots are linear, confirming the validity of Eq. (14.66). The fracture energies are thus determined over wide ranges of temperature and rate. Figure 14.23 shows that the fracture energies measured in simple tension and in pure shear can be superimposed to give a single master curve, again using the universal form of the WLF shift factor.

Fracture energies have also been measured using the 90°-peel test. The butadiene-styrene rubber is peeled from PET film adherend supported on a steel block (see Section 14.7.4). The fracture energies obtained by the peel test fall on the same master curve, shown in Figure 14.23. The fact that the fracture energies as determined by three different specimen and loading configurations (that is, simple tension, pure shear, and peeling) all superimpose to give a single master curve confirms the validity of the fracture energy concept and that the fracture energy is a material property independent of specimen geometry.

14.6. SHEAR TESTS (LAP JOINTS)

14.6.1. Description of Shear Tests

Single lap joints are the most widely used (Figure 14.24). Typical test methods are described, for example, in ASTM D1002 for metal to metal lap joints and ASTM D3163 for plastic-to-plastic lap joints. ASTM D2182 and ASTM D905 describe methods for testing adhesive shear strength by compression loading. ASTM D1759 describes how a lap joint can be prepared from sandwich laminates for shear testing, as illustrated in Figure 14.25.

The stresses induced in a single lap joint loaded in tension are quite nonuniform and complicated. Differential straining of the adherends in the overlap region produces shear stresses, whereas eccentric bending of the adherends produces tearing (peel) stresses. These two stresses are highest at or near the overlap ends. Thus, the applied load is borne mainly by the end zones of the overlap region.

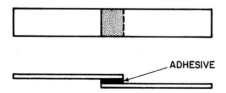

Figure 14.24. Single lap joint specimen.

Shear Tests (Lap Joints)

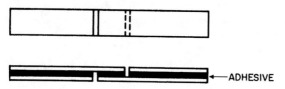

Figure 14.25. ASTM D3165 lap joint specimen prepared from sandwiched structure.

This is manifested in the effect of overlap length on the mean breaking stress of steel-to-steel lap joints bonded with a mixture of poly(vinyl formal) and phenolic resin [79] (Figure 14.26). The breaking load increases with overlap length, but the mean breaking stress decreases with the overlap length.

The effect of stress concentration near the two overlap ends in a single lap joint is further demonstrated strikingly below [80]. Lap joints with an overlap length of 2.5 cm fail at a loading of 1360 kg. When the joints are made in which the 2.5-cm overlap region is only half-filled with the adhesive (that is, 1.25 cm of the center zone of the overlap region is unfilled), the breaking load is 1130 kg. Thus, removal of the middle half of the adhesive layer lowers the breaking load by only 20%.

On the other hand, the breaking stress is generally independent of the width of the adherends, but is dependent on the thicknesses of the adherends and the adhesive layer.

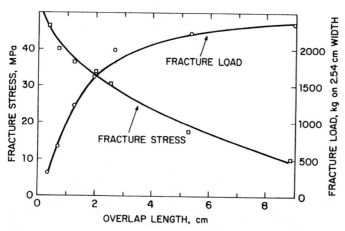

Figure 14.26. Effect of overlap length on the fracture load and fracture stress of steel overlap joint bonded with poly(vinyl formal)-phenolic adhesive. (After Ref. 79.)

Table 14.4. Comparison of Butt Joint Strength, Lap Joint Strength, and Bulk Ultimate Tensile Strength for Some Adhesives [a]

Adhesive	Lap joint shear strength, MPa	Butt joint tensile strength, MPa	Ultimate tensile strength of adhesive, MPa
Nylon 6	28.2	65.2	82.5
Nylon 66	26.8	66.8	72.2
Nylon 48	16.5	28.2	29.6
Nylon 610	21.3	57.2	48.2
Nylon 11	17.9	41.3	58.5
Polyethylene	7.6	17.2	31.0

[a] The joints are aluminum/adhesive/aluminum.
Source: From Ref. 81.

Some typical values for butt joint strength, lap joint strength, and ultimate tensile strength for some adhesives are compared in Table 14.4 [81]. Generally, the joint strength increases with increasing adhesive cohesive strength, provided that the failure is not interfacial. The tensile strength of a butt joint is similar to the tensile strength of the adhesive, but the shear strength of a lap joint is lower, roughly half of butt joint strength, probably because the shear strength of a polymer is often about half of its tensile strength.

14.6.2. Differential Straining

Shear stresses are generated as a result of differential straining of the adherends in the overlap region, depicted in Figure 14.27. In the unloaded state, each element is represented by a square. If the adherends are nondeformable, each adherend will move as a block without deforming its elements. The adhesive will deform to accommodate the displacement of the adherends, and thus experience a shear strain identical over the entire overlap region. Each adherend bears the full load just before the overlap region, and the load is gradually transmitted to the other adherend through the adhesive in the overlap region. The stress in the lower adherend is thus the highest at A and gradually diminishes toward B, where it is zero. The stress in the upper adherend is highest at B, and vanishes at A. On the other hand, if the adherends are deformable and linearly elastic, each element

Shear Tests (Lap Joints)

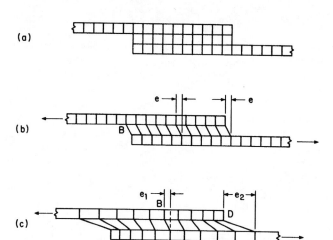

Figure 14.27. Schematics of shearing in the adhesive in a single lap joint: (a) unloaded joint; (b) loaded joint with inextensible adherends; (c) joint with elastic adherends. (After Ref. 82.)

of the adherend will extend in proportion to the existing stress at that element. The corresponding points B, C in the middle of the overlap region and D, E at an overlap end will be displaced by unequal amounts e_1 and e_2, respectively. The end displacement is much greater, giving a greater stress in the adhesive at each end of the overlap. Therefore, failure usually starts at the edge of the overlap region. Differential strains are much lower in tapered lap joints and scarfed joints (Figure 14.28), which account for their higher strengths.

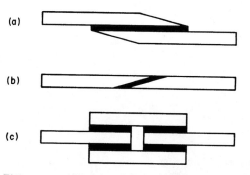

Figure 14.28. (a) Tapered lap joint; (b) scarfed joint; (c) double lap joint.

14.6.3. Tearing Stress Due to Eccentric Loading

The line of tension in a single lap joint is always oblique, since it is offset at least by the adherend thickness (Figure 14.29). This eccentric loading will generate a bending moment ($M = Pt/2$, where P is the tensile load and t the adherend thickness), which tends to pull the adherends apart and create tearing (peeling) stresses in the adhesive. The adherends will tend to bend to reduce such tearing stress. These tearing stresses are tensile and normal to the interface, and are generally confined to the area near the overlap ends. They are often quite significant, and are the critical factor in joint failure in many cases. Tearing stresses can be avoided, for instance, by using double lap joints (Figure 14.28).

Stress distributions in single lap joints have been analyzed by several workers [83–89]. Volkersen's analysis [83] considers only the shear stresses due to differential straining, and neglects the tearing stresses due to eccentric loading. The analysis of Goland and Reissner [84] considers both the shear stresses and the tearing stresses. Despite certain simplifications, this theory is the most rigorous and useful. The analysis of Plantema [85] combines some features of the above two theories. Lubkin [88] analyzed the scarfed joints.

14.6.4. Volkersen's Analysis of Lap Joint

Volkersen [83] considered the shear stresses in the adhesive due to differential straining of the adherends, but neglected the tearing

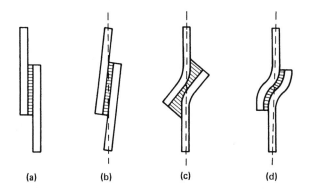

Figure 14.29. Bending and tearing in single lap joint. (a) before loading; (b) eccentric loading; (c) bending and tearing; (d) relief of tearing stress by bending (very flexible adherends).

Shear Tests (Lap Joints)

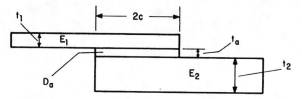

Figure 14.30. Single lap joint analyzed by Volkersen.

stresses due to eccentric loading. The theory is therefore inadequate in cases where bending moments are significant. Consider the single lap joint depicted in Figure 14.30. E_1 and E_2 are the tensile moduli of the two elastic adherends of thicknesses t_1 and t_2. The overlap length is $2c$. The shear modulus of the adhesive is D_a. The adhesive thickness is t_a.

The maximum shear stress is found at each overlap end. For linear elastic materials, the stress concentration factor α is found to be

$$\alpha = \frac{\tau_{max}}{\tau_m} = \frac{\delta}{\varepsilon}\left[\frac{2\varepsilon^2 - 1 + \cosh(2\varepsilon\delta)}{\sinh(2\varepsilon\delta)}\right] \tag{14.71}$$

where τ_{max} is the maximum shear stress, τ_m the mean shear stress, and δ and ε are dimensionless parameters defined by

$$\delta^2 = \frac{2c^2 D_a}{E_2 t_2 t_a} \tag{14.72}$$

and

$$\varepsilon^2 = \frac{E_1 t_1 + E_2 t_2}{2 E_1 t_1} \tag{14.73}$$

Figure 14.31 plots α versus δ^2 at several ε^2 values. When $\varepsilon\delta \sim 0$, $\alpha \sim 1$. The shear stresses are uniformly distributed. When $\varepsilon\delta$ is large, $n \simeq \delta/\varepsilon$.

On the other hand, for identical adherends, Eq. (14.71) becomes

$$\alpha = \delta \coth \delta \tag{14.74}$$

which indicates that α is a function of a single parameter δ. Thus, δ can be used as the scaling factor for comparison of lap joints.

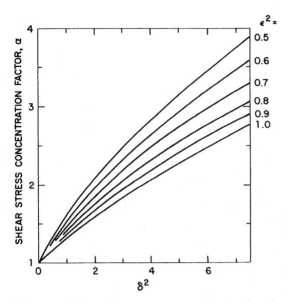

Figure 14.31. Shear stress concentration factor for single lap joints as a function of δ^2 and ϵ^2, according to Volkerson's theory.

14.6.5. Theory of Goland and Reissner

Despite some simplifying assumptions, the theory of Goland and Reissner [84] is the most rigorous and useful. Both the shear and the tearing stresses are considered. The deformation of the adherends in the overlap region is assumed to conform to cylindrically bent plates characterized by the dimensionless parameter k, which is the ratio of the bending moment just before the overlap region to the bending moment of the inflexible members, that is,

$$\frac{1}{k} = 1 + 2\sqrt{2} \tanh\left[\frac{c}{t}\sqrt{\frac{3}{2}(1-\nu^2)\frac{p}{E}}\right] \qquad (14.75)$$

where c is half the overlap length, t the adherend thickness, E the adherend tensile modulus, ν the adherend Poisson's ratio, and p the mean tensile stress in the adherend away from the overlap region (that is, p = F/bt, where F is the total tensile force acting on an adherend of width b and thickness t) (Figure 14.32). The k depends not only on the joint geometry (c and t) and the elastic properties (E and ν) of the adherends, but also on the applied tensile stress p. Therefore, the k changes with the load. The k is unity for undeformed adherends, because of either their stiffness or low applied load. As the adherend

Shear Tests (Lap Joints)

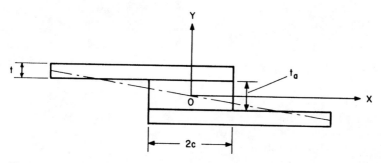

Figure 14.32. Single lap joint in unstrained state used in Goland and Reissner's analysis.

flexibility or the applied load increases, the k decreases toward zero as a limit. In practice, however, it usually remains above 0.35 (Figure 14.33).

The stress distributions in the overlap region are then calculated as a problem of plane strain for two extreme cases. In the first case, the adhesive layer is too thin and stiff to affect the flexibility of the joint. In the second case, the adhesive layer is thick and deformable and is the principal cause of joint flexibility. The first case (thin adhesive layer) is valid when

$$\frac{t_a}{E_a} \leq \frac{t}{10E} \quad \text{and} \quad \frac{t_a}{D_a} \leq \frac{t}{10D} \tag{14.76}$$

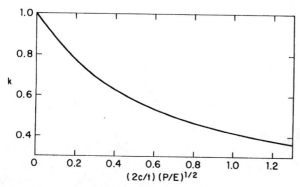

Figure 14.33. Bending factor k as a function of $(2c/t)(P/E)^{1/2}$. (After Ref. 84.)

such as for wood-to-wood joints. The second case (flexible adhesive layer) is valid when

$$\frac{t}{E} \leq \frac{t_a}{10E_a} \quad \text{and} \quad \frac{t}{D} \leq \frac{t_a}{10D_a} \tag{14.77}$$

such as for metal-to-metal joints. The symbols t_a, E_a, and D_a are thickness, tensile modulus, and shear modulus of the adhesive, respectively. The symbols t, E, and D are thickness, tensile modulus, and shear modulus of the adherends, respectively.

Thin Adhesive Layers

The existence of a thin adhesive layer is disregarded. The entire joint is considered to consist of a continuous, homogeneous material. The

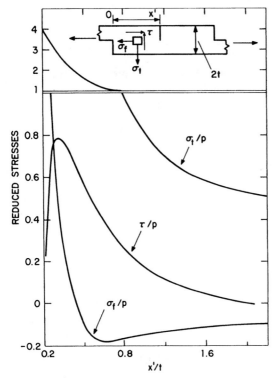

Figure 14.34. Longitudinal tensile stress σ_f, shear stress τ, and tearing stress σ_t as a function of the distance from the overlap edge x' in a single lap joint having k = 1. (After Ref. 84.)

Shear Tests (Lap Joints)

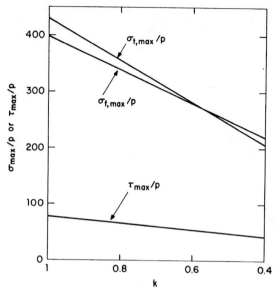

Figure 14.35. Reduction of stresses due to bending of adherends. (After Ref. 84.)

calculated stresses are those of the adherends along the glue line, and can be assumed to equal those in the adhesive layer. The shear stress τ, the tearing stress σ_t, and the longitudinal stress σ_f on the adherend fiber at the glue line are obtained as converging series. Their ratios to the mean tensile stress p along the glue line are shown for k = 1 (rigid adherends) in Figure 14.34. The highest stresses occur within a distance from the edge equal to about the adherend thickness. The σ_t and σ_f are highest (4.3p and 4.0p, respectively) at the very edge; τ is zero at the very edge but increases rapidly to a maximum of about 0.8p within a distance of 0.15t from the edge and then diminishes to zero at about 2t. These peak values are affected considerably by the bending of the adherends (Figure 14.35). As k decreases (increased bending), the maximum ratios τ/p, σ_t/p, and σ_f/p decrease.

Flexible Adhesive Layers

In joints with flexible adhesives, both the shearing and tearing stresses are maximum at the very edge of the overlap region. The shear stress τ is given by

$$\tau = -\frac{pt}{8c}\left[\frac{\beta c}{t}(1 + 3k)\frac{\cosh(\beta x/t)}{\sinh(\beta c/t)} + 3(1 - k)\right] \qquad (14.78)$$

where x is the coordinate parallel to the longitudinal axis of the adherend, $x = 0$ at the overlap center, $x = c$ at the overlap edge, and β is defined by

$$\beta^2 = \frac{8D_a t}{Et_a} \tag{14.79}$$

The maximum shear stress τ_{max} occurs at the overlap edge. Since the mean shear stress τ_m is given by $-(pt/2c)$, the shear stress concentration factor α_s is thus given by

$$\alpha_s = \frac{\tau_{max}}{\tau_m}$$

$$= \frac{1 + 3k}{4} \left(\frac{\beta c}{t} \right) \coth \left(\frac{\beta c}{t} \right) + \frac{4}{3} (1 - k) \tag{14.80}$$

which is plotted in Figure 14.36. When $\beta c/t \geq 2$, Eq. (14.80) simplifies to

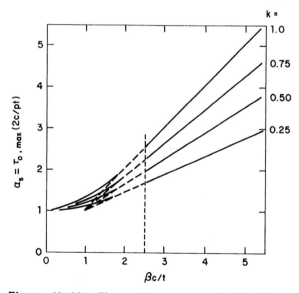

Figure 14.36. Shear stress concentration factor $\alpha_s = \tau_{max}/\tau_m$ as a function of $\beta c/t$ in a single lap joint. (After Ref. 82.)

Shear Tests (Lap Joints)

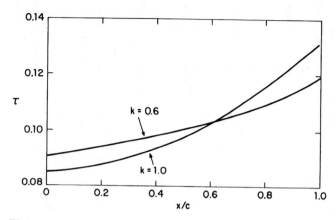

Figure 14.37. Variation of τ with x/c in a joint with $\beta c/t = 1$. (After Ref. 89.)

$$\alpha_s = \frac{1}{4}(1 + 3k)\frac{\beta c}{t} + \frac{3}{4}(1 - k) \tag{14.81}$$

Bending of the adherend (small k value) tends to reduce the shear stress. Figure 14.37 shows the variation of shear stress τ in a lap joint with $\beta c/t = 1$ at two different k values (1 and 0.6).

The tearing stress σ_t along the glue line is given by

$$\sigma_t = \frac{pt^2}{c^2 \Delta}\left\{\left[\frac{k}{2}\lambda^2 R_2 - k'\lambda \cosh \lambda \cos \lambda\right]\cosh\left(\frac{\lambda x}{c}\right)\cos\left(\frac{\lambda x}{c}\right) \right.$$
$$\left. + \left[\frac{k}{2}\lambda^2 R_1 - k'\lambda \sinh \lambda \sin \lambda\right]\sinh\left(\frac{\lambda x}{c}\right)\sin\left(\frac{\lambda x}{c}\right)\right\} \tag{14.82}$$

where

$$\lambda = \left(\frac{6E_a t}{Et_a}\right)^{1/4}\frac{c}{t} \tag{14.83}$$

$$\Delta = \frac{1}{2}(\sinh 2\lambda + \sin 2\lambda) \tag{14.84}$$

$$R_1 = \cosh \lambda \sin \lambda + \sinh \lambda \cos \lambda \tag{14.85}$$

$$R_2 = \sinh \lambda \cos \lambda - \cosh \lambda \sin \lambda \tag{14.86}$$

$$k' = \frac{kc}{t}\left[3(1-\nu^2)\frac{p}{E}\right]^{1/2} \tag{14.87}$$

and k has been defined in Eq. (14.75). The maximum tearing stress $\sigma_{t,max}$ is found to occur at the edge of the overlap, given by

$$\sigma_{t,max} = \frac{pt^2}{c^2\Delta}\left[\frac{k}{2}\lambda^2(\sinh 2\lambda - \sin 2\lambda)\right.$$

$$\left. - k'\lambda(\cosh 2\lambda + \cos 2\lambda)\right] \tag{14.88}$$

The tearing stress concentration factor α_t is then given by

$$\alpha_t = \frac{\sigma_{t,max}}{p} \tag{14.89}$$

Equation (14.89) can be given approximately by [82]

$$\alpha_t \simeq \frac{1}{2}k\lambda^2(t/c)^2 \tag{14.90}$$

which is accurate for k = 1, and has an error of about 4% for k = 0.4. On the other hand, for joints with long overlaps (that is, $\lambda \gtrsim 3$), Eq. (14.88) simplifies to

$$\sigma_{t,max} = \frac{pt^2}{2c^2}\left(\frac{k}{2}\lambda^2 - k'\lambda\right) \tag{14.91}$$

Figure 14.38 shows the variation of $\sigma_{t,max}$ with $\beta c/t$ at several k values in lap joints with relatively flexible adhesives. Figure 14.39 shows the variation of σ_t in a lap joint with $\lambda = 2.5$ at two different k values (1 and 0.6). The maximum tearing stress decreases with increasing overlap ratio $\beta c/t$. The tearing stress increases sharply near the overlap edge. This high tearing stress accounts for the fact that the failure of lap joint often starts by a splitting apart of the two adherends at the overlap edges.

The ratio of the maximum tearing stress to the mean shear stress is given by

$$\frac{\sigma_{t,max}}{2\tau_m} = (2)^{1/2}k\left(\frac{\beta c}{t}\right) \tag{14.92}$$

obtained from Eqs. (14.89) and (14.90), in which the relation $E_a = 2(1 + \nu_a)D_a$ is used and ν_a is assumed to be 1/3. Thus, k and $\beta c/t$ serve as the scaling factors for both the shear and tearing stresses.

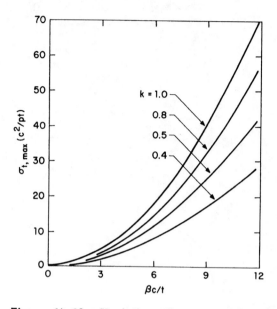

Figure 14.38. Variation of maximum tearing stress $\sigma_{t,max}$ with $\beta c/t$ and k for joints with relatively flexible adhesives. (After Ref. 89.)

Despite its simplifying approximations, the theory of Goland and Reissner is the most rigorous and useful analysis of stresses in lap joints. Recently, Wolley and Carver [90] reported a finite element analysis of stresses in a lap joint. Their results are in excellent agreement with those of the theory of Goland and Reissner, shown in Table 14.5.

14.6.6. Plantema's Modification

Plantema [85] gave an analysis that combines parts of the theory of Volkersen and the theory of Goland and Reissner. Volkersen's theory is used to calculate the differential strains and the stress distribution at the overlap edges. The bending moment and the resulting stress are then calculated, and the deformation of the adherends is calculated by the theory of Goland and Reissner. Plantema obtained the shear stress concentration factor α as

$$\alpha = w \coth w \qquad (14.93)$$

where

Table 14.5. Comparison of Tearing Stresses Calculated by Goland and Reissner Theory and by Finite Element Analysis for Single Lap Joint

Joint parameters			Maximum tearing stress concentration, $\sigma_{t,max}/p$	
E/E_a	$2c/t$	t_a/t	Finite element method	Theory of Goland and Reissner
10	5	0.06	1.43	1.44
50	5	0.06	0.65	0.64
100	5	0.06	0.59	0.63

Source: From Ref. 90.

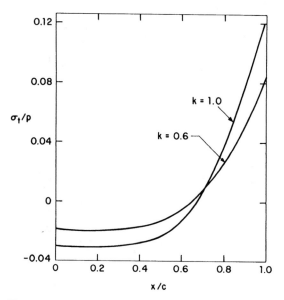

Figure 14.39. Variation of tearing stress σ_t with x/c in a typical joint with $t/c = 0.2$ and $\lambda = 2.5$. (After Ref. 89.)

Shear Tests (Lap Joints)

$$w = \frac{1}{2}(1 + 3k)^{1/2} \frac{\beta c}{t}$$

The tearing stress was not calculated.

14.6.7. Fracture Energy Analysis of Lap Joint

Anderson and coworkers [58] analyzed the fracture energy of a single lap joint using the stress distribution functions of Goland and Reissner. Fracture occurs in mixed I and II modes. The bending stress gives mode I loading; the shear stress gives mode II loading. Therefore, the fracture energy G_a is separated into two contributions,

$$G_a = \theta_b G_I + \theta_s G_{II} \qquad (14.94)$$

where G_I and G_{II} are mode I and mode II adhesive fracture energies, respectively, and θ_b and θ_s the fractional contributions for bending (mode I) and shearing (mode II), respectively; that is,

$$\theta_b = \frac{G_{a(b)}}{G_a} \qquad (14.95)$$

and

$$\theta_s = \frac{G_{a(s)}}{G_a} \qquad (14.96)$$

where $G_{a(b)}$ and $G_{a(s)}$ are the fracture energy components arising from bending and shearing, respectively. Bending produces tearing and is analogous to mode I. Shearing is analogous to mode II. These analogies are, however, only approximate, since bending will produce not only tearing but also shearing.

Experiments were conducted on single lap joints using aluminum adherends having a tensile modulus of 10.1 Mpsi and a Poisson's ratio of 0.34. A commercial epoxy adhesive was used as the adhesive. Two series of tests were conducted on samples prepared as outlined in Table 14.6. Joints were loaded to failure, and the fracture energy calculated, using the stress functions of Goland and Reissner. The results are shown in Figure 14.40. The fracture energy G_a is substantially independent of the overlap length 2c. The small slope apparently arises from the mixed-mode fracture. Figure 14.41 plots the fractional contributions of modes I and II fractures. As the overlap length increases, the contribution from mode I increases with a cor-

Table 14.6. Descriptions of Single Lap Joints Used in Fracture Energy Experiments

Preparation and geometry	Series A	Series B
Cure temperature, °C	68.3	57.2
Cure time, hr	12	5
Length of adherend (overlap region excluded), cm	16.5	8.9
Thickness of adherend, t, mm	3.2	4.8
Thickness of adhesive, t_a, mm	0.146	0.146
Width, cm	2.54	2.54
Overlap length, 2c, cm	1.5–3.6	1.3–3.0

Source: From Ref. 58.

responding decrease of mode II contribution. This accounts for the slight dependence of fracture energy on overlap length.

Gent [22] considered the shear test configurations shown in Figure 14.42. These are analogous to the tensile tests shown in Figure 14.15. The fracture energies are given by Eqs. (14.65) and (14.66), respectively, where Q_C now denotes the critical strain energy density due to shear deformation. For linear-elastic materials, the shear modulus $D = E/2(1 + \nu)$. Therefore, the fracture stress is given by

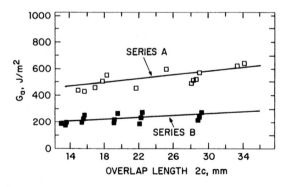

Figure 14.40. Adhesive fracture energy G_a versus overlap length 2c for a single lap joint. See Table 14.6 for materials and joint geometry. (After Ref. 58.)

Shear Tests (Lap Joints)

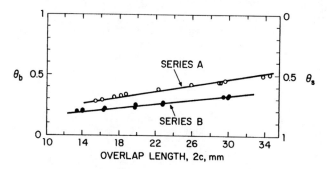

Figure 14.41. Fractional contributions to fracture energy by bending θ_b and by shearing θ_s versus overlap length for the results shown in Fig. 14.40. (After Ref. 58.)

$$\sigma_f = \frac{1}{\alpha} \left[\frac{E_a G_a}{t_a(1 + \nu_a)} \right]^{1/2} \tag{14.97}$$

for thin adhesives (having a relatively large debond length), and by

$$\sigma_f = \frac{1}{\alpha} \left[\frac{E_a G_a}{\pi(1 + \nu_a)a} \right]^{1/2} \tag{14.98}$$

for thick adhesives (having a relatively small debond length). The symbols α, E_a, ν_a, G_a, and a are stress concentration factor, adhesive tensile modulus, adhesive Poisson's ratio, adhesive fracture energy, and length of edge crack (or half length of central crack), respectively. Equation (14.98) should usually apply, since the debond zone should generally be much smaller than the adhesive thickness.

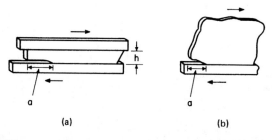

Figure 14.42. Shear tests with (a) thin and (b) thick adhesive layers. (After Ref. 22.)

14.6.8. Comparison with Experiments

Both the theory of Volkersen and the theory of Goland and Reissner agree reasonably with experiment; the latter being better than the former, as expected. Figure 14.43 shows a comparison of theoretical and experimental shear stresses in a lap joint with $2c = 7.5$ cm, $\beta c/t = 5.2$, $k = 1$, and $D_a = 280,000$ psi. As can be seen, the theory of Goland and Reissner agrees better with the experimental result than does the theory of Volkersen. Both theories correctly predict that the shear stress is highest at the overlap edges and decreases rapidly toward the middle of the overlap region.

As pointed out earlier, k and $\beta c/t$ are the two scaling factors for the stresses in lap joints. The parameter k characterizes the stresses arising from bending; the parameter $\beta c/t$ characterizes the stresses

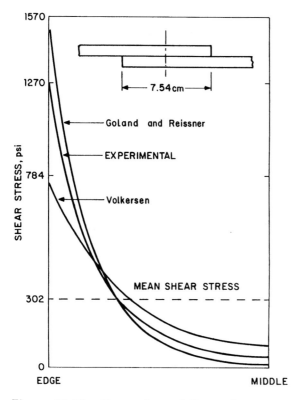

Figure 14.43. Comparison of theoretical and experimental shear stress distributions in a single lap joint with $2c = 7.54$ cm and $\beta c/t = 5.2$. (After Ref. 91.)

Shear Tests (Lap Joints)

arising from differential straining. Thus, lap joints having the same k and $\beta c/t$ values should have the same stress distribution and fracture strength [92,93].

In general, the fracture strength should increase with decreasing k (or increasing adherend flexibility) and decreasing $\beta c/t$. The fracture strength is given by

$$\sigma_f = \frac{F}{2cb} \qquad (14.99)$$

where F is the total load at fracture and b is the width of the adherend, that is, the total load divided by the area of the glued region. The k will decrease with decreasing E and t. The $\beta c/t$ will decrease with decreasing D_a and c, and increasing t, t_a, and E, since

$$\frac{\beta c}{t} = \left(\frac{8 D_a c^2}{E t t_a}\right)^{1/2} \qquad (14.100)$$

Thus, σ_f should decrease with increasing overlap length, although the total load increases, in agreement with the experiment shown in Figure 14.26. The theory also predicts that σ_f should be proportional to $t_a^{0.5}$ and $D_a^{-0.5}$. Experimentally, σ_f indeed often increases with increasing adhesive thickness and decreasing adhesive modulus, but is sometimes found to be independent of or contrarily to decrease with increasing adhesive thickness [64]. This discrepancy may arise from residual internal stresses in the adhesive or geometrical stress concentration at the three-phase boundary, which depends on the shape and curvature of the adhesive-air boundary and the adhesive thickness [82,89].

All theories assume constant stresses over the thickness of the adhesive layer. The highest shear stresses are predicted to develop at the overlap edges. However, the condition of equilibrium would limit the virtual stress to simple tangential tension or compression without any shear on the free adhesive-air boundary [82]. A longitudinal shear stress may exist where the boundary is oblique to this direction. A concave boundary will be in tension near the loaded adherend and in compression near the ending adherend, thus giving an intermediate stress-free point. With a square boundary, the areas of high tension and compression will be confined to the three-phase boundary, with the intermediate area under an almost uniform tearing stress. Such effects have been observed by photoelastic analysis and may account for the variable effects of adhesive thickness on fracture strength of lap joints [82].

14.7. PEEL TESTS (PEEL JOINTS)

14.7.1. Description of Peel Tests

Several typical peel tests are illustrated in Figure 14.44. Usually, the peel angle θ is kept constant during the test. Various mixed (modes I and II) loadings are obtained at different peel angles. Three commonly used configurations are the 90°-peel test (L-peel test), the 180°-peel test (U-peel test), and the T-peel test. ASTM D1876 describes the T-peel test. ASTM D903 describes the 180°-peel test. A test arrangement for the 90°-peel test was reported by Engel and Fitzwater [94]. ASTM D1781 describes a climbing drum peel test (Figure 14.45). ASTM D3167 describes a floating roller peel test, also shown in Figure 14.45.

14.7.2. Stress Analysis of Peel Test

Stresses induced in the adhesive layer during peeling have been analyzed [95–111]. In each case, both the adherend and the adhesive are assumed to be linearly elastic, and the stresses are assumed to be constant across the width and the thickness of the adhesive layer. Beam bending relations are used to obtain the stress distribution and the peel force.

The profile of a peeling specimen is shown schematically in Figure 14.46. The specimen is placed on the abscissa with its origin O at the point of detachment where the normal tensile stress in the adhesive is the greatest. In the region AO (curved region), the peeling strip

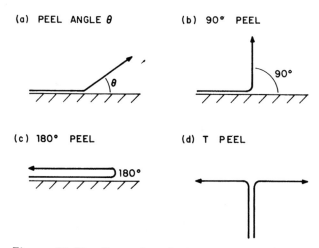

Figure 14.44. Several typical peel test configurations.

Peel Tests (Peel Joints)

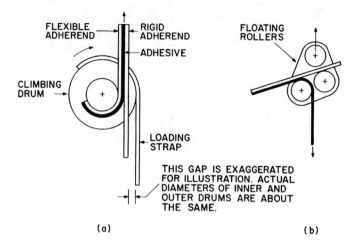

Figure 14.45. Schematics of (a) climbing drum peel test and (b) floating-roller peel test.

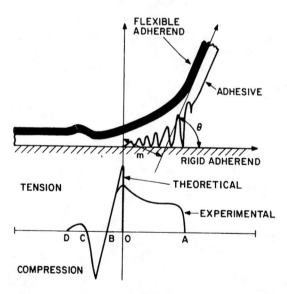

Figure 14.46. Schematics of peel profile and normal stresses in adhesive layer.

is curved, and some unbroken ligaments of the adhesive still bridge the peeling strip and the rigid adherend. If there are no unbroken ligaments, the normal stress should drop to zero at a distance infinitesimally to the right of the origin O. However, because of the unbroken ligaments, large tensile stresses persist in the region AO. In the region to the right of A (straight region), the peeling strip becomes straight and coincides with the direction of applied peel force P. The angle between the applied peel force and the rigid adherend is the peel angle θ. In the region OD (strained region), the cleavage stress is a highly damped harmonic function consisting of alternating zones of tension and compression. In the region to the left of D (unstrained region), the original adhesive bond is not disturbed.

Kaelble [98–100] gives the theoretical cleavage stress σ at a distance $-x$ from the point of rupture O ($x = 0$) as

$$\sigma = \sigma_o (\cos \beta x + K \sin \beta x) \exp (\beta x) \qquad (14.101)$$

where

$$\beta = \left(\frac{E_a b}{4EIt_a} \right)^{1/4} \qquad (14.102)$$

$$K = \frac{\beta m}{\beta m + \sin \theta} \qquad (14.103)$$

σ_o is the boundary cleavage stress at $x = 0$, E_a the elastic modulus of the adhesive, E the elastic modulus of the flexible adherend, t_a the adhesive layer thickness, b the bond width, I the moment of inertia of the peeling strip cross section, m the moment arm of the peel force, and θ the peel angle.

Equation (14.101) predicts that the cleavage stress is a highly damped harmonic function consisting of alternating zones of tension and compression. This prediction has been confirmed experimentally [100] (Figure 14.46). Theoretically, the cleavage stress falls abruptly to zero from the maximum value at the point of detachment ($x = 0$). Experimentally, however, large tensile stresses are found to persist well into the nominally detached region, because some unbroken ligaments of the adhesive remain.

The peel force P is given by [98–100]

$$\frac{P}{b} = \frac{t_a K^2 \sigma_o^2}{2E_a (1 - \cos \theta)} \qquad (14.104)$$

Peel Tests (Peel Joints)

which is applicable, however, only when $\beta m \gg \sin \theta$, and $K \sim 1$, as shown by Gent and Hamed [109]. Equation (14.104) predicts that the peel force is proportional to the stored strain energy per unit interfacial area $t_a \sigma_o^2 / 2E_a$ at the point of detachment, proportional to the adhesive thickness t_a and inversely proportional to $(1 - \cos \theta)$. These predictions agree qualitatively with experimental results, as discussed later.

Gardon [106,107] gives the theoretical peel force as [106,107]

$$\frac{P}{t_a^{0.25}} = n\sigma_o - \xi^{0.5} \left(\frac{P}{t_a}\right)(1 - \sin u)^{0.5} \qquad (14.105)$$

$$\tan u = 2 \left(\frac{\sigma_o \eta}{\xi}\right)^{0.75} - \left(\frac{1}{\xi}\right) P t_a^{0.5} \qquad (14.106)$$

$$\eta = 0.319 \, b \left(\frac{E}{E_a}\right)^{0.25} t^{0.75} \qquad (14.107)$$

$$\xi = 0.409 \, b(E_a E)^{0.5} t^{1.5} \qquad (14.108)$$

where u is the negative slope angle of the peeling strip at the point of detachment (that is, $x = 0$) and t is the thickness of the flexible adherend. The maximum failure stress σ_0 can be calculated from the dependence of peel force on the adhesive thickness by using the relations above. This maximum failure stress (at $x = 0$) is independent of the specimen geometry and, therefore, was suggested to be a better measure of the strength of a peel joint than the peel force.

Two limiting cases are noted below. For large t_a, Eq. (14.105) simplifies to

$$P = n\sigma_0 t_z^{0.25} \qquad (14.109)$$

For small t_a, Eq. (14.105) simplifies to

$$n\sigma_0 = \frac{P}{t_a^{0.25}} + \xi^{0.5} \left(\frac{P}{t_a}\right)^{0.5} \qquad (14.110)$$

Thus, a log P versus log t_a plot will give a curve having limiting slopes of 1 and 0.25 at small and large t_a values, respectively. In the intermediate ranges, P may be expressed as being proportional to a certain power of t_a having the power coefficient between 1 and 0.25. This has been demonstrated experimentally [107].

14.7.3. Fracture Energy Analysis of Peel Test

Several analyses of fracture energy in peel test have been reported [22,52,58,111–118]. Consider the energy balance for peeling of a viscoelastic adhesive layer (backed with a flexible adherend) from a rigid adherend (Figures 14.46 and 14.47). Peeling progresses steadily at constant peel force P and peel angle θ. The peeled portion is sufficiently long that the curvature and the strain in the curved region stay constant during peeling.

As the bond is peeled a distance c, the peel force moves a distance d in the direction of the applied force. The work of peel (that is, the work input) W is thus given by W = Pd. The load displacement d comes from the peeled length c and the additional tensile strain due to increased length of the straight region of the peeled strip. That is, d = (1 + e − cos θ) c = (λ − cos θ) c, where e is the tensile strain of the peeled strip in the straight region and λ the extension ratio, defined as λ = 1 + e. The work of peel is thus given by

$$W = P(\lambda - \cos \theta)c \qquad (14.111)$$

and the energy Γ expended in separating an interfacial area bc is given by

$$\Gamma = bcG \qquad (14.112)$$

where G is the fracture energy per unit interfacial area. The G has two components: the debond energy (that is, the energy required to debond the interface) and the plastic dissipation energy (that is, the energy dissipated irreversibly by yielding in a zone around the crack

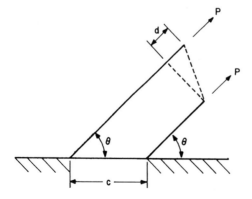

Figure 14.47. Schematics of peel work. Specimen width is b, adhesive thickness is t_a, and the flexible adherend thickness is t.

Peel Tests (Peel Joints)

tip). This dissipative zone will span the entire adhesive thickness for thin adhesive layers, but may be confined within a distance t_0 from the crack tip for thick adhesive layers. Thus,

$$G = G_0 + Ut_a \quad \text{for } t_a \leq t_0 \quad \text{(thin adhesive)} \tag{14.113}$$

and

$$G = G_0 + Ut_0 \quad \text{for } t_a \geq t_0 \quad \text{(thick adhesive)} \tag{14.114}$$

where G_0 is the debond energy, U the dissipation energy per unit volume of the adhesive, and t_a the adhesive thickness.

The energy is also expended in deforming the peeling strip in the straight region. This energy consists of the stored elastic energy and the plastically dissipated energy, that is,

$$V = S_a b c t_a + Sbct \tag{14.115}$$

where V is the deformation energy for peeling an interfacial area bc, S_a the deformation energy per unit volume of the adhesive, and S that of the flexible adherend. The deformation energy contains a stored elastic energy component (subscript e) and a plastically dissipated energy component (subscript p). Thus,

$$S_a = S_{ae} + S_{ap} = \int_0^e \sigma_a \, de \tag{14.116}$$

and

$$S = S_e + S_p = \int_0^e \sigma \, de \tag{14.117}$$

where σ_a is the tensile stress in the adhesive, σ the tensile stress in the flexible adherend, and e the tensile strain in the straight region of the peeled strip.

Conservation of energy requires that $W - \Gamma - V = 0$. Using Eqs. (14.111)–(14.117) gives

$$\frac{P}{b} = \frac{G + S_a t_a + St}{\lambda - \cos \theta} \tag{14.118}$$

which is generally valid for any viscoelastic adhesives. Note that the fracture energy G is dependent on adhesive thickness when $t_a \leq t_0$, and independent of adhesive thickness when $t_a \geq t_0$, as specified in Eqs. (14.113) and (14.114).

If the flexible adherend (backing) is flexible and inextensible, then $e = 0$, $\lambda = 1$ and $S_a = S = 0$. Equation (14.118) simplifies to

$$\frac{P}{b} = \frac{G}{1 - \cos \theta} \tag{14.119}$$

On the other hand, if the flexible adherend (backing) is flexible, but slightly extensible and linearly elastic, then $e \simeq 0$, $\lambda \simeq 1$, and $S \gg S_a$. Equation (14.118) then simplifies to

$$\frac{P}{b} = \frac{G}{(1/2)(1 + \lambda) - \cos \theta} \tag{14.120}$$

since $S_a \simeq 0$ and $S = P^2/2E(bt)^2$.

Alternatively, Gent [22] considered two extreme cases, shown in Figure 14.48. In the first case, the fracture energy comes from the bending energy stored in the detached portion, which is bent slightly by the peel force P, that is,

$$G = \frac{P^2 L^2}{2EIb} \tag{14.121}$$

(a)
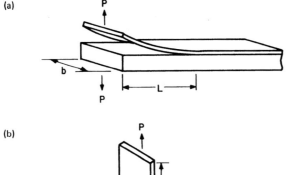

(b)
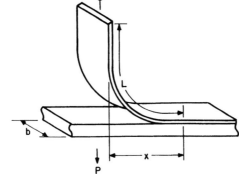

Figure 14.48. Cleavage (peel) tests with (a) stiff and (b) flexible detaching strip. (After Ref. 22.)

where L is the length of the peeled portion and EI is the bending stiffness of the bent strip, which may be a composite value consisting of contributions from the adhesive and the adherend. Equation (14.121) is valid only for small bending.

In the second case, the detaching strip is sufficiently flexible and thus bent through 90° by the peel force. If the detaching strip is flexible but inextensible, the fracture energy is simply given by

$$G = \frac{P}{b} \tag{14.122}$$

which follows from Eq. (14.119), since $\theta = 90°$. The two extreme cases can be encompassed by an empirical relation [22]

$$G = \frac{P}{b}\left(1 - \exp\frac{-L^2}{x^2}\right) \tag{14.123}$$

where x is the distance between the separation front and the perpendicular plane which the peeled portion attains when bent through 90°, and is given by

$$x^2 = \frac{2EI}{P} \tag{14.124}$$

which follows readily from a simple bending theory [22].

14.7.4. Experimental Observations

Fluctuation of Peel Force

Ideally, the peel force is constant at a constant peel rate. Practically, however, the peel force fluctuates, either randomly (Figure 14.49) or in a slip-stick fashion (Figure 14.50). The mean values are usually reported as the peel strength.

A random fluctuation of peel force is shown in Figure 14.49 for a T peel of two 2.5-cm-wide and 40-μm-thick cellophane films bonded with a 0.5-μm-thick layer of an acrylic (Rhoplex HA-8) adhesive peeled at 2 mm/sec. The fluctuating force conforms to Gaussian distribution [106]. The mean and the median of the peel force are identical and well defined. The variation of the peel force is thus due to random imperfections in the specimen. In random fluctuation, failure is initiated continuously, and propagates at the same rate as the testing rate.

A slip-stick fluctuation of peel force is shown in Figure 14.50 for a T peel of two 2.5-cm-wide and 40-μm-thick cellophane films bonded

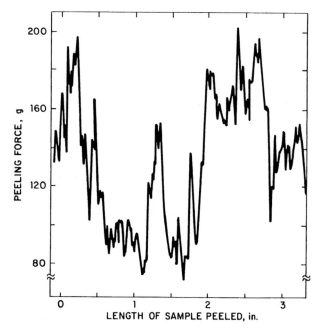

Figure 14.49. Steady-state peel force with continuous failure for T peel of two cellphane films (2.54 cm width and 40 μm thickness) bonded with an acrylic (Rhoplex HA-8) adhesive (.045 μ m thick) at a rate of 5 in./min. (After Ref. 106.)

with a 24-μm-thick layer of acrylic (Rhoplex B-60) adhesive peeled at 0.4 mm/sec. The peel force fluctuates between well-defined maxima and minima. The distance between the maximum and the minimum is independent of the testing rate [106]. In slip-stick rupture, the failure is initiated when the peel force reaches the maximum. The stored energy is then expended to propagate the crack at a rate faster than the testing rate. As the stored energy is depleted, the crack propagation ceases at the minimum peel force. As the peel force is increased again to the maximum, another failure cycle is repeated. Such slip-stick behavior has also been observed in tear [43], cleavage [119], and blister rupture [120,121].

The exact origin of slip-stick behavior is unknown. However, it is known to be affected by adhesive composition and adherend roughness. For example, slip-stick failure is observed when a carboxylated adhesive is peeled from an adherend. When the amount of pendent carboxyl group in the rubber adhesive is decreased, the failure mode changes to continuous failure. In another example, slip-stick failure is changed to continuous failure by roughening the adherend prior to

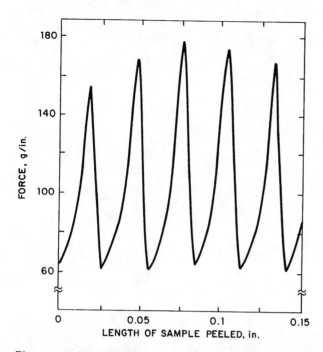

Figure 14.50. Steady-state peel force with slip-stick failure for T peel of two cellophane films (2.54 cm width and 40 μm thickness) bonded with an acrylic (Rhoplex B-60) adhesive (24 μ m thick) at a rate of 1 in./min. (After Ref. 106.)

adhesive bonding for cellophane films (adherends) bonded with an acrylic adhesive (T peel) [106]. However, the opposite effect is found with a semicrystalline poly(ethylene-vinyl acetate) adhesive peeled from a stainless steel adherend. In this case, slip-stick failure occurs on roughened adherend, whereas continuous failure occurs on smooth adherend [122].

Rate-Temperature Effect

Generally, the effects of rate and temperature on peel force are quite complicated, causing multiple transitions in peel force. These transitions usually arise from changes in the viscoelastic response of the bulk adhesive, and sometimes coincide with transitions of failure locus (from cohesive to interfacial failure) [123,124]. At a given rate, as the temperature increases, the peel force tends to decrease and a transition from interfacial to cohesive failure may occur [125,126]. At a given temperature, as the rate incrases, the peel force tends to increase, and then drops sharply with a transition from cohesive to inter-

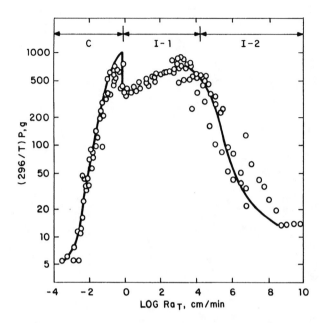

Figure 14.51. Rate-temperature superposition of 180°-peel force for an acrylic adhesive with a tape backing on a polystyrene adherend. Data are taken at 11 different temperatures from −38 to 70°C and at nine or more rates from 0.025 to 50 cm/min at each temperature. C stands for cohesive failure in the adhesive; I-1 stands for interfacial failure between adhesive and adherend; I-2 stands for interfacial failure between adhesive and tape baking. (After Ref. 124.)

facial failure [107,127−129]. The effects of rate and temperature on peel force are equivalent, and can be superimposed to give a single master curve by using an appropriate shift factor [123,124]. These are illustrated below.

Figures 14.51 and 14.52 show the superposition of reduced peel force (PT_0/T) versus reduced rate (Ra_T) for the 180° peel of an acrylic adhesive (equimolar copolymer of isoamyl acrylate and neopentyl acrylate, $T_g = 43°C$) on a tape backing from polystyrene and poly(tetrafluoroethylene-hexafluoropropylene) adherends, respectively [124]. The universal form of WLF shift factor is used with the reference temperature T_0 chosen to be 296 K, that is,

$$\log a_T = 9.8 - \frac{17.44(T - T_g)}{51.6 + T - T_g} \tag{14.125}$$

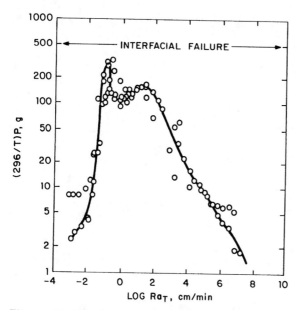

Figure 14.52. Rate-temperature superposition of 180°-peel force for an acrylic adhesive on fluorinated ethylene-propylene copolymer adherend. See Figure 14.51 for details. (After Ref. 124.)

so that $a_T = 1$ at 296 K. The peel rates are varied from 4×10^{-4} cm/sec to 0.83 cm/sec; the temperatures are varied from $-35°C$ to $70°C$.

As can be seen, the data superimpose quite well. At low rates, the peel force increases with increasing rate, reaches a plateau at intermediate rates, and then decreases at high rates. Such features are observed regardless of the adherends, indicating that the peel strength is mainly a manifestation of the viscoelastic properties of the bulk adhesive. In the case of polystyrene adherend, the rising region of the curve at low rates corresponds to the cohesive failure in the adhesive; the plateau region at intermediate rates corresponds to interfacial failure between the adhesive and the adherend; the falling region at high rates corresponds to separation of the adhesive from the tape backing. On the other hand, in the case of FEP adherend, however, interfacial failure between the adhesive and the adherend is observed at all rates covering the entire curve.

Viscoelastic and Failure-Locus Transitions

Gent and Petrich [123] contributed a great deal to the understanding of the peel strength behavior of adhesive joints. An un-crosslinked rubber (butadiene-styrene copolymer, 60:40 weight ratio,

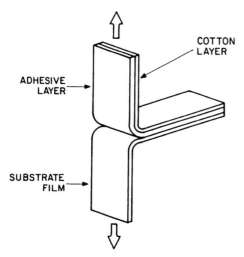

Figure 14.53. Schematics of T-peel test for an adhesive with cotton cloth backing and adhering to a substrate film. (After Ref. 123.)

M_n = 70,000 and T_g = −40°C) is used as the adhesive, supported on a woven cloth as the backing. The adherends used are PET polyester film (3 mils thick), cellophane film (0.9 mil thick), transcrystallized polyethylene film (5 mils thick), and polystyrene block, respectively. Test specimens are prepared by hot-pressing the adhesive between the cotton cloth backing and the adherend to give an adhesive

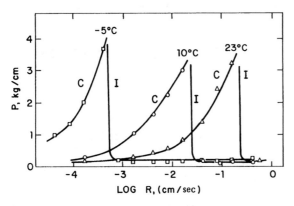

Figure 14.54. Peel force P versus peel rate R for an un-cross-linked butadiene-styrene rubber adhering to a PET polyester film. The symbols C and I denote cohesive failure and interfacial failure, respectively. (After Ref. 123.)

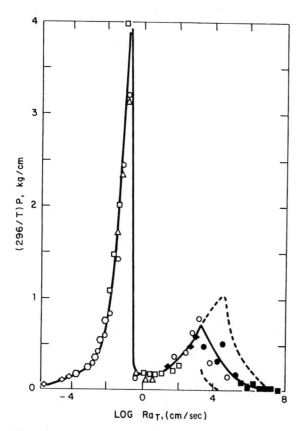

Figure 14.55. Master curve for peel force P versus peel rate Ra_T reduced to 23°C for an un-cross-linked butadiene-styrene rubber adhering to a PET polyester film. Dashed lines are extreme values for stick-slip behavior. (After Ref. 123.)

thickness of 0.5 mm. T-peel tests are performed at temperatures of −35 to 60°C and rates of 8.3×10^{-4} to 8.3 mm/sec (Figure 14.53). In the case of a polystyrene-block adherend, 90°-peel tests are used.

The peel forces on PET adherend versus peel rates at three different temperatures are presented in Figure 14.54. They show a common pattern: a steady rise in peel force with peel rate up to a critical rate, then an abrupt transition to much lower peel forces. At the same time, the locus of failure changes abruptly from cohesive failure of the adhesive layer to interfacial failure at the adhesive-adherend interface. This transition shifts to higher rates as the temperature is increased. The entire force-rate curves at different temperatures can

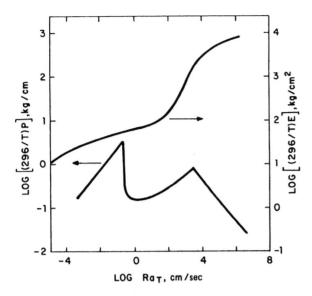

Figure 14.56. Peel force versus rate for an un-cross-linked butadiene-styrene rubber adhesive on PET polyester film, and tensile modulus versus rate for the bulk rubber adhesive. (After Ref. 123.)

be superimposed to give a single master curve by horizontal shift, using the universal form of the WLF shift factor, Eq. (14.125). Figure 14.55 shows that all curves combine successfully to a single master curve. The master curve obtained is quite complicated, having a high peak at low rates where the failure mode changes from cohesive to interfacial, and a low peak at high rates where the failure mode is entirely interfacial. The dashed lines at high rates show the upper and lower bounds of slip-stick peel forces.

The success of rate-temperature superposition indicates that the rate-temperature effects on peel strength reflects the viscoelastic properties of the adhesive. Comparison of the tensile modulus of the adhesive with the peel strength (Figure 14.56) shows that the high peak transition at low rates is associated with liquid-to-rubber transition, and the low peak transition at high rates is associated with the rubber-to-glass transition. The concurrent transition from cohesive to interfacial failure at the high peak is quite remarkable.

Two additional experiments were conducted to confirm that the high-peak transition at low rates is due to liquid to rubber transition [123]. In the first experiment, the butadiene-styrene adhesive is cross-linked with dicumyl peroxide. The peel strengths of un-cross-linked and cross-linked adhesives on PET film adherends are compared in Figure 14.57. The peel strengths are similar at high rates, but the

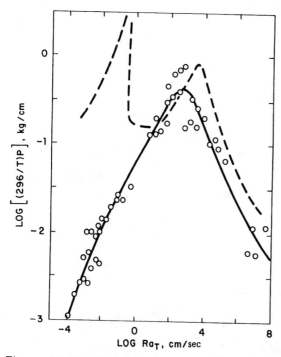

Figure 14.57. Peel force versus rate for cross-linked (full curve) and un-cross-linked (dashed curve) butadiene-styrene rubber adhering to PET polyester film. (After Ref. 123.)

cohesive-failure, high-strength region at low rates is absent for the cross-linked specimen. The failure in the cross-linked specimen is interfacial in the entire rate-temperature range tested. In the second experiment, a different butadiene-styrene adhesive is used, which contains a small high-molecular-weight fraction. Therefore, it is elastic as well as rubbery, but is otherwise identical to the original adhesive in chemical composition, number-average molecular weight, and glass temperature. Figure 14.58 shows the peel strength versus peel rate relation for the second adhesive on PET film adherend. The cohesive-failure, high-strength region at low rates is again absent, whereas the behavior at high rates is similar to that of the original adhesive. The failure is again entirely interfacial. If, however, this second adhesive is milled in a two-roll mill to break down the high-molecular-weight fraction and then used to prepare the peel specimen, the peel strength now shows cohesive failure at low rates and a transition from cohesive to interfacial failure at a critical rate (Figure 14.58).

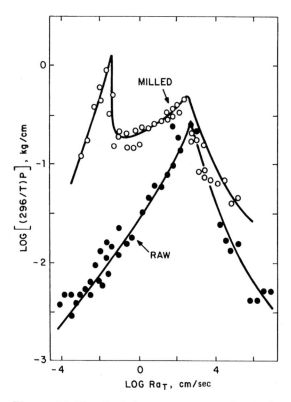

Figure 14.58. Peel force versus peel rate for the second rubber adhesive (see the text for a description) adhering to PET polyester film adherend. (After Ref. 123.)

These experiments clearly show that liquidlike flow is the cause of cohesive failure. The transition to interfacial failure occurs when the rate of deformation is so large that the adhesive cannot flow. In the absence of liquidlike flow, only a single broad peak is observed, and the failure occurs at the interface throughout. Liquidlike flow is not, however, the only mechanism for obtaining high-strength and cohesive failure at low rates; plastic yielding can give a similar effect.

Since the cotton cloth backing and the adherend can be regarded as flexible but inextensible in this case, the peel strength can be given by

$$\frac{P}{b} = G_0 + Ut_a \sim t_a \int_0^{e_m} \sigma_a \, de \qquad (14.126)$$

where e_m is the maximum tensile strain of the adhesive in the separation zone and σ_a is the tensile stress of the adhesive. For an uncross-linked rubber, as in this case, the integral represents the dissipated viscous energy, not the stored elastic energy. Equation (14.126) may be rewritten as

$$\frac{P}{b} = t_a \int_0^{f_m} \sigma_a \, de \qquad (14.127)$$

where f_m is the maximum stress attained. There are two possible rupture processes: (1) rupture of the adhesive when f_m equals the tensile strength of the adhesive f_a, and (2) separation of the adhesive from the adherend when f_m equals the maximum interfacial tensile stress f_i which the adhesive bond can withstand. This equation states that the peel strength is given by the deformation energy of the adhesive, represented by the area under the stress-strain curve of the adhesive up to the point where the tensile stress is large enough to either break the interfacial adhesive bond, $f_m = f_i < f_a$, or to rupture the adhesive cohesively, $f_m = f_a < f_i$.

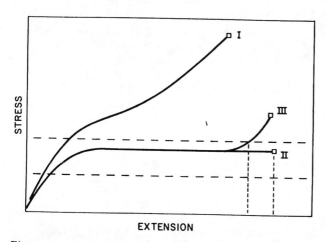

Figure 14.59. Schematic stress-strain curves for rubbery (I), liquid-like (II), and strain-hardening (III) materials. The horizontal dashed lines are possible levels of the maximum interfacial stress f_i. The vertical dotted lines denote the limiting extensions attainable in each case. (After Ref. 123.)

Several typical tensile stress-strain curves together with two possible levels of maximum interfacial bond strength f_i are depicted in Figure 14.59. At the lower level of f_i, the deformation energy is small and the peel strength is similar in all cases. At the upper level of f_i, major differences are found among the three cases. The rubberlike behavior (I) will lead to a low peel strength and interfacial failure. The plastic (yielding) behavior (II and III) will lead to a high peel strength and either cohesive failure (II) or interfacial failure (III). The peel strengths for curves II and III will be similarly high, but the failure is cohesive for II and interfacial for III. The curve III behavior is the basis for the clean detachment of some surgical tapes and bandages. The rubbery adhesive may undergo strain hardening by crystallization or molecular orientation, leading to the curve III behavior.

The original adhesive exhibits either type I or type II behavior at different rates. At high rates it shows type I behavior: the peel strength is low and the failure is interfacial. At low rates, it shows type II behavior: the peel strength is high, and the failure is cohesive. By treating f_i as an adjustable parameter, a complete theoretical peel strength versus rate curve can be constructed with Eq. (14.127), using the stress-strain curves of the adhesive at different rates. The results for the original adhesive on three different adherends (polystyrene, PET, and polyethylene) are given in Figure 14.60, where the solid lines are calculated theoretically. As can be seen, the theoretical curves agree reasonably with the experimental data, both in absolute magnitudes and abrupt transitions in peel strength. Thus, Eq. (14.127) with a single adjustable parameter f_i can quantitatively describe the peel strength behavior. The f_i values thus obtained are 3 kg/cm^2 for polyethylene adherend, 4 kg/cm^2 for PET and cellophane adherends, and 6 kg/cm^2 for polystyrene adherend. These may be considered as the intrinsic interfacial bond strength for the particular adhesive-adherend combinations [123].

As pointed out before, the low peak at high rates are associated with rubber-to-glass transition of the adhesive bulk property. This view is further supported by the fact that the transition occurs at essentially the same rate, regardless of the adherend used. On the other hand, a pronounced change in the specimen configuration occurs at rates near the high-rate peak. Instead of peeling the adhesive at 90°, the adherend is now peeled at 180° while the adhesive remains straight. This is because, at such high rates, the adhesive becomes glassy and stiff.

Two experiments are conducted to examine the effect of peel geometry on peel strength. In the first experiment (A), an adhesive layer is held horizontally by bonding its lower surface to a steel block. A PET film is then bonded to the upper surface of the adhesive, and peeled off at 90°. In this arrangement, the adhesive layer stays flat, and the PET adherend is bent through 90°. In the second experiment

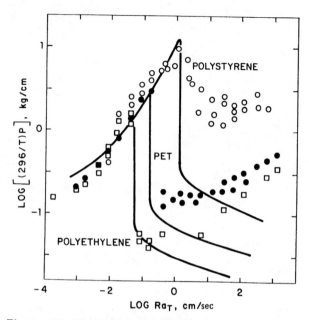

Figure 14.60. Peel force versus peel rate for an uncrosslinked butadient-styrene rubber adhering to three substrates. The acronym PET is poly(ethylene terephthalate). The full curves are calculated from stress-strain curves as described in the text. (After Ref. 123.)

(B), the PET film is bonded to a horizontal steel block, and the adhesive layer is peeled off at 90°. The peel strengths of these two arrangements are compared with that of the normal test specimen in Figure 14.61. All three give substantially the same results at low rates through the cohesive-interfacial transition region. But, at high rates, a marked difference occurs. Arrangement A gives similar results to those of the normal test specimen, whereas arrangement B does not show any transition at all. Thus, the sharp decline of peel strength at high rates observed for the normal test specimen is due to the increased stiffness of the adhesive layer. When the adhesive layer is prevented from bending (such as in arrangement A), the peel strength decreases in the same way. However, when the adhesive layer is forced to bend through 90° (such as in arrangement B), the peel strength does not decrease at all, but rather increases in the rubber-to-glass transition region, to reach a level about 100 times higher than that for arrangement A. This remarkable difference arises from little work done on the adhesive in the failure zone at high rates when the adhesive becomes stiff and inextensible.

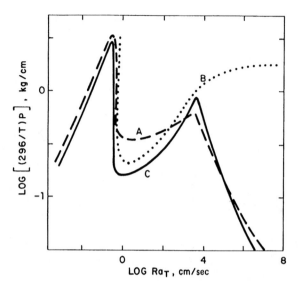

Figure 14.61. Peel force versus peel rate for an un-cross-linked butadiene-styrene rubber adhering to PET polyester film, determined with three different experimental arrangements: (A) PET film peeled off at 90° angle; (B) adhesive layer peeled off at 90° angle; (C) normal test piece (Figure 14.53). (After Ref. 123.)

The most favorable adhesive properties are those that resemble curves II and III in Figure 14.59, with the stresses adjusted so that the long horizontal portion lies close to, but below, the intrinsic interfacial bond strength f_i. The transition from cohesive to interfacial failure is not a necessary feature of the peel strength transition at low rates. For instance, a stress-strain curve resembling curve III would give a high peel strength, but the failure would be interfacial at low rates. The intrinsic interfacial bond strength f_i would affect the peel strength in two principal ways: higher f_i would shift the low-rate transition to higher rates, and higher f_i would generally give higher peel strength beyond this transition.

Fracture Energy

For a flexible but inextensible peeling strip, the fracture energy is related to the peel strength by

$$G = \frac{P}{b} \quad (90°\text{-peel test}) \tag{14.128}$$

and

Peel Tests (Peel Joints)   551

$$G = \frac{2P}{b} \quad (180°\text{-peel test}) \qquad (14.129)$$

which follows from Eq. (14.119) and can be used to calculate the fracture energy [52–54,59,130].

Figure 14.9 shows the rate and temperature effects on the fracture energy as measured by 180°-peel test for a cross-linked polybutadiene rubber (supported on a woven cotton cloth) coated on a glass adherend treated with a 50:50 mixture of ethylsilane and vinylsilane [56]. All the data can be superimposed to give a single master curve when plotted against the reduced rate Ra_T. The master curves for various interfacial-chemical-bond density are shown in Figure 11.30.

Figure 14.7 shows the reduced fracture energy $233G/T$ versus the reduced rate Ra_T for a lightly cross-linked rubber adhesive (butadiene-styrene copolymer 60:40 by weight, $T_g = -40°C$, $M_n = 70,000$ cross-linked with dicumyl peroxide) bonded to various adherends which are cemented rigidly to steel blocks, as measured by 90°-peel test. The universal form of WLF shift factor is again used for the superposition.

Fracture energies of the above lightly cross-linked butadiene-styrene rubber adhesive on PET adherends rigidly cemented to steel blocks have been measured for three different specimen configurations: simple tension, pure shear, and 90°-peel tests [52]. The fracture energies obtained for the three different specimen configurations superimpose to give a single master curve, shown in Figure 14.23. This remarkably demonstrates the validity of the energy criterion of adhesive bond fracture.

Effect of Peel Angle

Both the stress analysis and the energy analysis predict that the peel force varies inversely with $(1 - \cos \theta)$. Thus, the peel force should be the highest for 0° peel and the lowest for 180° peel. This agrees with experiment except near 0° and 180°, as illustrated in Figure 14.62. Here, a cellophane film and an aluminum foil are bonded with a rubber adhesive. The cellophane is rigidly clamped on a drum and the aluminum foil is peeled off at different angles and rates [99]. Additional examples can be found elsewhere [68,98,99,112,127,129,131].

Generally, two types of discrepancies are found. At low peel angles near 0°, the experimental peel force does not follow the theoretical relation. This probably arises from a change in the failure mode from cleavage to shear. At high peel angles near 180°, the peel force tends to be greater than at 90°. This probably arises from additional energy dissipation due to plastic yielding of the flexible adherend (or backing) as it bends through 180° [110]. This effect is quite pronounced, for example, for PET polyester film adhering to a thermoplastic rubber adhesive. When the polyester film is peeled in such a way to avoid plastic yielding, the peel energy is found to be 0.9 kJ/m^2 for both 90° and 180° peel angles. On the other hand, when the polyester

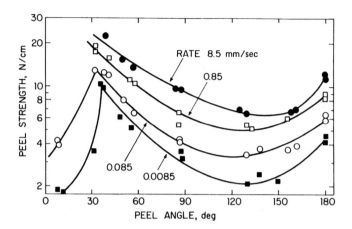

Figure 14.62. Effect of peel angle on peel strength for a cellophane film and an aluminum film bonded with a rubber adhesive. (After Ref. 99.)

film is allowed to bend freely, the peel energy is found to be quite different for the two angles, that is, 5.0 kJ/m² for 180° and 1.2 kJ/m² for 90°.

Effect of Adhesive Thickness

Experimentally, the peel force increases with increasing adhesive thickness for thin adhesives, and reaches a plateau for thick adhesives, shown in Figure 14.63 for a PET polyester film peeled from a Kraton rubber adhesive layer at a 180° peel angle [111].

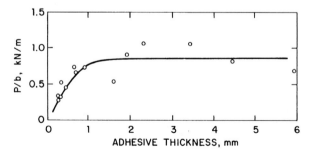

Figure 14.63. Effect of adhesive thickness on peel force per unit width (P/b) for a PET strip peeled from a KRATON rubber substrate at a 180° peel angle. (After Ref. 111.)

Peel Tests (Peel Joints)

The stress analysis (given in Section 14.7.2), predicts that the peel force should be directly proportional to the adhesive thickness. This discrepancy arises from the assumption used in the analysis that the boundary cleavage stress σ_0 is constant across the thickness of the adhesive. In fact, however, the boundary stress varies with the thickness, so that a larger mean stress is necessary to bring about the same detachment stress in a thin layer as in a thick layer [111].

On the other hand, energy analysis correctly predicts the observed effect of adhesive layer. For thin adhesives ($t_a \leq t_0$), we have

$$\frac{P}{b(1 - \cos\theta)} = G_0 + Ut_a \tag{14.130}$$

which predicts that the peel force increases with adhesive thickness. For thick adhesives ($t_a \geq t_0$), we have

$$\frac{P}{b(1 - \cos\theta)} = G_0 + Ut_0 \tag{14.131}$$

which predicts that the peel force is independent of adhesive thickness. See also Section 14.7.3.

Effect of Flexible Adherend Thickness

If the flexible adherend (or backing) is linearly elastic or inextensible, no energy is dissipated in bending it. However, if it is viscoelastic, bending will cause plastic deformation of the adherend, and thus increase the peel force and the fracture energy [68,110,111,132–134].

This additional peel force P_y due to yielding of the flexible adherent is greater for a thicker strip at a given bending curvature. The maximum P_y occurs when the strip is bent back on itself, given by [111]

$$P_{y,max} = \frac{t\sigma_y}{4} \tag{14.132}$$

where σ_y is the yield stress of the flexible adherend and t its thickness. Figure 14.64 shows the additional peel force P_y per unit width for PET strips of various thicknesses bent through 180° at various bend diameters D [111]. On the other hand, P_y drops sharply to zero as the bend diameter increases, and plastic yielding rapidly diminishes. The critical thickness t_c above which no plastic yielding occurs is given by [110]

$$t_c = \frac{12EP_0}{b\sigma_y^2} \tag{14.133}$$

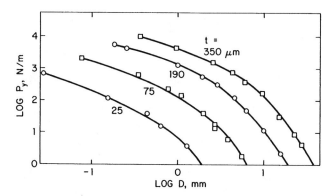

Figure 14.64. Additional peel force P_y per unit width due to propagating a 180° bend in PET strips of different thicknesses t. The diameter of the bend is D. (After Ref. 110.)

where P_0 is the peel force in the absence of plastic yielding. Thus, thick strips with weak adhesion will not yield and will have no plastic contribution to the peel force. However, in thin strips, the plastic contribution will be directly proportional to the thickness of the strip.

Experimentally, peel force versus backing thickness plot has been found to exhibit a maximum [110,129,135,136], in agreement with the analysis above. Figure 14.65 shows an example for 180° peel of a PET polyester film bonded with a rubber adhesive to a stainless steel at a peel rate of 5 mm/sec.

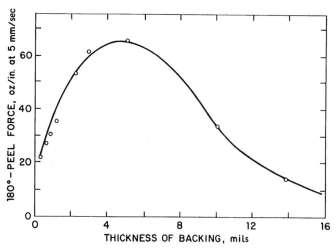

Figure 14.65. Effect of polyester backing thickness on the 180°-peel strength of an uncured synthetic adhesive. (After Ref. 135.)

14.8. CANTILEVER BEAM TESTS

Cantilever beam tests are analytically and experimentally simple, and can be used to measure the fracture energy. Figure 14.66 shows three typical cantilever-beam configurations useful in measuring adhesive fracture energy.

Consider the double-cantilever beam depicted in Figure 14.67. The specimen may be a monolithic material or an adhesive joint with the two halves above and below the crack joined with an adhesive. The materials are assumed to be linearly elastic. Also shown is the tensile load P versus displacement L for a crack length a. Let the crack extend slightly by δa at a critical load P_c under a fixed-grip condition. The stress-field energy released to extend the crack equals the area of the shaded triangular region. This energy release can be shown to be independent of the fixed-grip condition. Alternatively, let the crack

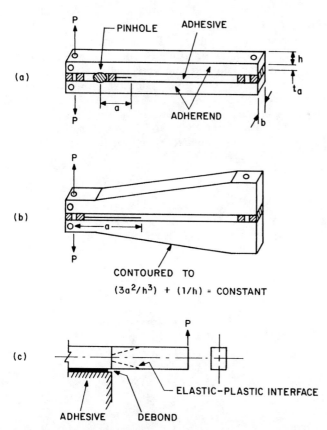

Figure 14.66. Cantilever beam specimens: (a) double-cantilever beam; (b) tapered-double-cantilever beam; (c) cantilever beam.

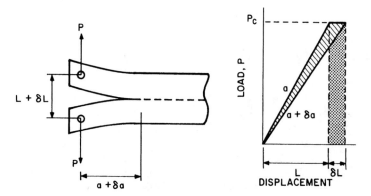

Figure 14.67. Schematics of cleavage of a double-cantilever beam, and load P versus displacement L curves.

extend δa under a fixed-load condition. The work done is equal to the area of the dotted rectangular region, which is just twice the area of the triangular region as δa approaches zero. The rectangular region contains both the recoverable and nonrecoverable stress-field energies. However, the nonrecoverable energy always equals to the area of the triangular region.

The fracture energy G_I is the (nonrecoverable) energy expended to extend the crack by unit interfacial area, and is equal to the area of the triangular region per unit interfacial area, that is,

$$G_I = \frac{1}{2} \frac{P_c}{b} \frac{dL}{da} \qquad (14.134)$$

where be is the specimen width. The compliance C of the specimen ($C = L/P_c$) is an increasing function of the crack length a. Thus,

$$G_I = \frac{1}{2} \frac{P_c^2}{b} \frac{dC}{da} \qquad (14.135)$$

Some load-displacement curves for an aluminum-epoxy-aluminum double-cantilever-beam specimen at room temperature are given in Figure 14.68. To obtain G, the critical load P_c is read from the diagram and the quanaity dC/da calculated by [137]

$$\frac{dC}{da} = \frac{8}{Eb}\left(\frac{3a^2}{h^3} + \frac{1}{h}\right) \qquad (14.136)$$

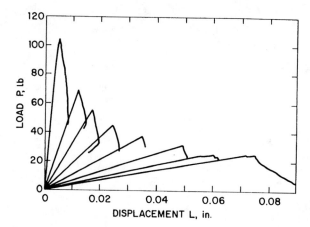

Figure 14.68. Load-displacement curves for an aluminum-epoxy-aluminum double-cantilever-beam specimen. (After Ref. 137.)

where h is the adherend thickness. The first term in the parentheses is the bending contribution, the second term the shear contribution. A specimen can be loaded to P_c to obtain one G value. After the crack jumps, the specimen can be reloaded to obtain a second G value. As the crack length increases, P_c will decrease to maintain a constant G. In this method, both P_c and a must be measured.

The method can be simplified by contouring the specimen so that the compliance changes linearly with a, as suggested by Ripling and coworkers [137]. In this case, dc/da is constant and only P_c needs to be monitored. A linear compliance specimen can be obtained if its height h is varied according to

$$\frac{3a^2}{h^3} + \frac{1}{h} = m \qquad (14.137)$$

where m is a constant. Specimens with m = 3 and 90 in.$^{-1}$, for instance, have been used by Mostovoy and coworkers [137-141]. Figure 14.69 shows some P-L curves for a tapered double-cantilever-beam specimen with m = 90 in.$^{-1}$. Two types of P-L curves are given, stable crack propagation and unstable crack propagation. In the latter case, the upper bound gives the crack initiation energy G_{Ic}; the lower bound gives the crack arrest energy G_{Ia}.

When the applied load P is below the critical load P_c, a quantity called the applied crack extenstion force (designated as G_j) can be defined as

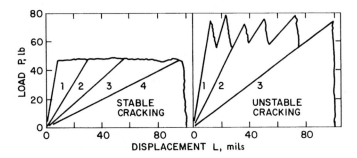

Figure 14.69. Load-displacement curves for linear compliance tapered-double-cantilever-beam specimens. The numbers denote unloading and reloading lines. In the unstable cracking, the upper limits correspond to crack initiation and the lower limits correspond to crack arrest. (After Ref. 137.)

$$G_j = \frac{1}{2} \frac{P^2}{b} \frac{dC}{da} \qquad (14.138)$$

which follows from Eq. (14.136). At the critical load, crack extension occurs, and the applied crack extension force becomes the critical crack extension force,

$$G_{Ic} = \frac{1}{2} \frac{P_c^2}{b} \frac{dC}{da} \qquad (14.139)$$

which is the fracture energy, and is identical to Eq. (14.135) except for the notation G_{Ic}. On the other hand, the fracture energy is also known as the critical strain energy release rate in Irwin's formulation [5,6].

A single cantilever beam is shown in Figure 14.66c. If plastic deformation occurs in the beam, the energy balance must include the energy dissipated as plastic work in those regions where the strain exceeds the elastic limits. Such an analysis has been given [142].

Double-cantilever beams (tapered and untapered) have been used extensively to investigate the fracture characteristics of adhesive joints by Mostovoy and coworkers [137–141,143–152] and others [32,33,153–155], and of bulk polymers [156,157]. The variables examined include adhesive thickness, crack velocity, temperature, humid environment, and cyclic loading.

Adhesive fracture energy versus adhesive thickness at 25°C for an epoxy resin (modified with 15% carboxyl-terminated butadiene-acrylonitrile rubber) using aluminum tapered-double-cantilever beams is shown

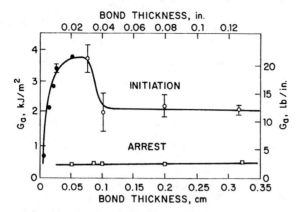

Figure 14.70. Adhesive fracture energy versus adhesive thickness at 25°C for an epoxy resin modified with 15% carboxyl-terminated butadiene-acrylonitrile rubber adhering to aluminum adherends. ●, stable propagation; ○, unstable propagation. (After Ref. 154.)

in Figure 14.70. At the thickness h_{max}, corresponding to the maximum G_{Ic}, there is a transition from stable to unstable crack propagation. Below h_{max}, the crack propagation is stable; above h_{max}, un-

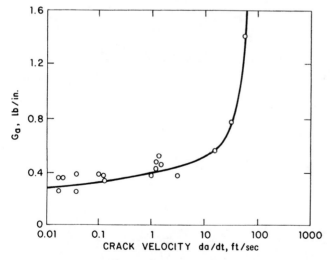

Figure 14.71. Effect of crack velocity on adhesive fracture energy for an amine-cured epoxy adhesive on aluminum adherends as measured with a tapered-double-cantilever beam. (After Ref. 145.)

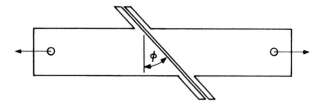

Figure 14.72. A scarfed specimen for a mixed modes I and II fracture test. (After Ref. 32.)

stable. G_{Ic} drops sharply below h_{max} and decreases to a constant value above h_{max}. However, the crack arrest energy G_{Ia} stays constant. Such behavior appears to arise from the presence of rubber particles, which modify the size of the plastic zone around the crack tip. The effect of temperature is much more complicated for adhesive joints than for bulk materials.

The effect of crack velocity on fracture energy for an amine-cured epoxy adhesive using tapered-double-cantilever beams is shown in Figure 14.71. As the crack velocity increases, G increases slowly, then rises sharply above a critical rate. The effects of chemical compositions and curing conditions on G have also been investigated in detail [145].

Stress corrosion cracking and fatigue of adhesive joints under cyclic loading have also been investigated with cantilever beams (see Chapter 15). Mixed-mode fracture of adhesive joints has been investigated, using "scarfed" specimens [32,33] (Figure 14.72). The results show that in some rubber-modified adhesives, the fracture energy of mixed-mode (I and II) fracture can be drastically (as much as 10 times) lower than that of mode I fracture, although mode I fracture has the lowest fracture energy and is predominant in most other cases, illustrated in Table 14.3. Tapered-double-cantilever beams have also been adapted to test the adhesive fracture energy of films and coatings.

14.9. OTHER FRACTURE ENERGY TESTS

These include blister tests and cone tests, and are discussed briefly below.

14.9.1. Blister Tests

This method was analyzed and developed by Williams [18–21] based on earlier methods proposed by Dannenberg [158] and by Malyshev

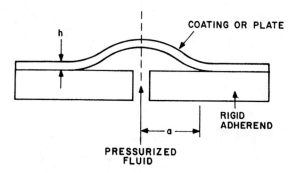

Figure 14.73. Configuration of a blister test.

and Salganik [159]. A plate or a coating is bonded to a rigid adherend except for a central region of radius a (Figure 14.73). A compressed fluid (such as air or mercury) is injected into the unbonded region to lift the plate, forming a blister whose radius stays fixed until a critical pressure P_c is reached. At the critical pressure, the radius of the blister increases, indicating separation of the bonded region. The critical pressure is given by [18]

$$P_c = \left[\frac{32}{3(1-\nu^2)}\left(\frac{h}{a}\right)^3\right]^{1/2}\left(\frac{EG_a}{a}\right)^{1/2} \qquad (14.140)$$

where h is the thickness of the plate, E its tensile modulus, ν its Poisson's ratio, and G_a the adhesive fracture energy. This method has been used to measure the fracture energy of dental adhesives on tooth, barnacle cement on various adherends, polyurethane coating on glass, explosively bonded metal plates [18–21,58,160], and poly-(methyl methacrylate)-steel joint [159].

Williams and coworkers [19–21] have also analyzed the case where the plate is bonded to the rigid adherend with a layer of soft adhesive. Mossakovskii and Rybka [161] analyzed the case where the plate is isotropic and infinitely thick. The blister method is particularly suitable for investigating coating adhesion.

14.9.2. Cone Tests

A cone test is shown in Figure 14.74. By using different cone angles, various mixed-mode loadings can be obtained. For instance, a 0° cone (a cylinder) pulled from a matrix or from a cylindrical hole is largely mode II loading. A 90° cone (button test) gives largely mode I loading when loaded in tension, and gives nearly pure mode III loading when

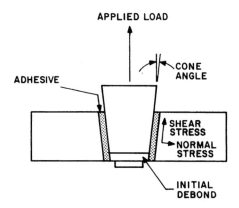

Figure 14.74. Cone test.

loaded in torsion. These tests have been analyzed and developed [58, 162], and are particularly useful for curved adherends, such as in the bonding of dowels, fibers, and tooth crowns.

14.9.3. Coating Adhesion Tests

Testing of coating adhesion presents a special problem, as a coating cannot be gripped directly. In principle, all the methods discussed above can be adapted to test coating adhesion, provided that a suitable method is devised to grip the coating. One method is to bond a backing (which provides a grip) to the coating with a suitable adhesive. This adhesive should adhere strongly to both the coating and the backing so that failure would not occur between them. The adhesive should also not swell or otherwise affect the cohesive and adhesive properties of the coating. In practice, such an adhesive is difficult to find.

A blister test is particularly useful for coatings, as the coatings are not gripped. Double-cantilever-beam tests have been adapted to investigate the adhesive fracture energy of thin coatings [155].

Consequently, many qualitative and semiquantitative methods have been devised and commonly used to evaluate coating adhesion [163]. These methods include various knife-cutting methods, scraping-scratching methods, crosscut-tape methods, inertia methods (such as ultrasonic vibration test and ultracentrifuge test), and impact tests (such as the gravel impact test and the bullet impact test). In all of these, quantitative interpretations in terms of fracture stress and energy are difficult. However, judicial applications of these qualitative tests often give useful results.

REFERENCES

1. A. A. Griffith, Phil. Trans. R. Soc. Lond., *A221*, 163 (1920).
2. G. R. Irwin, Trans. Am. Soc. Met., *40A*, 147 (1948).
3. G. R. Irwin and J. A. Kies, Welding J. Res. Suppl., 193s–198s (1954).
4. G. R. Irwin, J. Appl. Mech., *24*, 361 (1957).
5. G. R. Irwin, in *Encyclopedia of Physics*, Vol. 6, *Elasticity and Plasticity*, S. Flugge, ed., Springer-Verlag, Berlin, 1958, pp. 551–590.
6. G. R. Irwin, in *Structural Mechanics*, J. N. Goodier and N. J. Hoff, eds., Pergamon Press, New York, 1960, pp. 557–594.
7. E. H. Andrews, *Fracture in Polymers*, American Elsevier, New York, 1968.
8. J. F. Knott, *Fundamentals of Fracture Mechanics*, Wiley, New York, 1973.
9. A. S. Kobayashi, ed., *Experimental Techniques in Fracture Mechanics*, Iowa State University Press, Ames, Iowa, and Society for Experimental Stress Analysis, Westport, Conn., 1973.
10. P. C. Paris and G. C. Sih, in *Fracture Toughness Testing and Its Applications*, STP 381, American Society for Testing and Materials, Philadelphia, 1965, pp. 30–81.
11. J. G. Kaufman, ed., *Progress in Flaw Growth and Fracture Toughness Testing*, STP 536, American Society for Testing and Materials, Philadelphia, 1973.
12. P. C. Paris and G. R. Irwin, eds., *Fracture Analysis*, STP 560, American Society for Testing and Materials, Philadelphia, 1974.
13. J. R. Rice and P. C. Paris, eds., *Mechanics of Crack Growth*, STP 590, American Society for Testing and Materials, Philadelphia, 1976.
14. J. M. Barsom, ed., *Flaw Growth and Fracture*, STP 631, American Society for Testing and Materials, Philadelphia, 1977.
15. T. Yokobori, T. Kawasaki, and J. L. Swedlow, eds., *Proceedings of the First International Conference on Fracture, Sendai, Japan*, Vols. 1, 2, and 3, The Japanese Society for Strength and Fracture of Materials, 1966.
16. P. L. Pratt, ed., *Fracture*, Proceedings of the Second International Conference on Fracture (Brighton, 1969) Chapman & Hall, London, 1969.
17. D. M. R. Taplin, ed., *Advances in Research on the Strength and Fracture of Materials*, Proceedings of the Fourth International Conference on Fracture, Vol. 1, *An Overview;* Vol. 2A, *The Physical Metallurgy of Fracture;* Vol. 2B, *Fatigue;* Vol. 3A, *Analysis and Mechanics;* Vol. 3B, *Applications and Non-metals*, Vol. 4, *Fracture and Society;* Pergamon Press, New York, 1978.

18. M. L. Williams, J. Appl. Polym. Sci., *13*, 29 (1969).
19. M. L. Williams, J. Appl. Polym. Sci., *14*, 1121 (1970).
20. J. D. Burton, W. B. Jones and M. L. Williams, Trans. Soc. Rheol., *15*, 39 (1971).
21. M. L. Williams, in *Recent Advances in Adhesion*, L. H. Lee, ed., Gordon and Breach, New York, 1973, pp. 381–422.
22. A. N. Gent, Rubber Chem. Technol., *47*, 202 (1974).
23. C. E. Inglis, Proc. Inst. Nav. Archit., *55*, 219 (1913).
24. R. E. Peterson, *Stress Concentration Design Factors*, Wiley, New York, 1953.
25. H. Neuber, *Kerbspannungslehre,* Springer-Verlag, Berlin; English translation available from Edwards Bros., Ann Arbor, Mich., 1959.
26. H. Tada, P. C. Paris, and G. R. Irwin, *The Stress Analysis of Cracks Handbook*, Del Research Corp., 226 Woodbourne Dr., St. Louis, Mo., 1973.
27. J. P. Berry, in *Fracture Processes in Polymeric Solids*, B. Rosen, ed., Interscience, New York, 1964, pp. 195–234.
28. E. Orowan, *Fatigue and Fracture of Metals*, MIT Press, Cambridge, Mass., 1950.
29. G. G. Trantina, J. Comp. Mater., *6*, 371 (1972).
30. G. G. Trantina, *Combined Mode Crack Extension in Adhesive Joints*, TAM Rep. 352, University of Illinois, Urbana, Ill., 1971.
31. K. L. DeVries, M. L. Williams, and M. D. Chang, Exp. Mech., *14*, 89 (1974).
32. W. D. Bascom, C. O. Timmons, and R. L. Jones, J. Mater. Sci., *10*, 1037 (1975).
33. W. D. Bascom, R. L. Jones, and C. O. Timmons, in *Adhesion Science and Technology*, Vol. 9B, L. H. Lee, ed., Plenum Press, New York, 1975, pp. 501–511.
34. H. M. Westergaard, Trans. ASME, J. Appl. Mech., *61*, 49 (1939).
35. P. C. Paris and G. C. Shih, ASTM STP *381*, 30 (1965).
36. G. C. Shih, *Handbook of Stress Intensity Factors*, Lehigh University, Bethlehem, Pa., 1973.
37. R. W. Hertzberg, *Deformation and Fracture Mechanics of Engineering Materials*, Wiley, New York, 1976.
38. F. A. McClintock and G. R. Irwin, ASTM STP *381*, 84 (1965).
39. D. S. Dugdale, J. Mech. Phys. Solids, *8*, 100 (1960).
40. R. S. Rivlin and A. G. Thomas, J. Polym. Sci., *10*, 291 (1953).
41. A. G. Thomas, J. Polym. Sci., *18*, 177 (195).
42. H. W. Greensmith and A. G. Thomas, J. Polym. Sci., *18*, 189 (1955).
43. H. W. Greensmith, J. Polym. Sci., *21*, 175 (1956).
44. A. G. Thomas, J. Polym. Sci., *31*, 467 (1958).
45. A. G. Thomas, J. Appl. Polym. Sci., *3*, 168 (1960).
46. H. W. Greensmith, J. Appl. Polym. Sci., *3*, 175 (1960).

References

47. H. W. Greensmith, J. Appl. Polym. Sci., 3, 183 (1960).
48. H. W. Greensmith, J. Appl. Polym. Sci., 7, 993 (1963).
49. A. Ahagon, A. N. Gent, H. J. Kim, and Y. Kumagai, Rubber Chem. Technol., 48, 896 (1975).
50. J. F. Hutton, Proc. R. Soc. Lond., A287, 222 (1965).
51. A. N. Gent, J. Polym. Sci., A-2, 9, 283 (1971).
52. A. N. Gent and A. J. Kinloch, J. Polym. Sci., A-2, 9, 659 (1971).
53. A. N. Gent and J. Schultz, J. Adhesion, 3, 281 (1972); also in *Recent Advances in Adhesion*, L. H. Lee, ed., Gordon and Breach, New York, 1973, pp. 253–266.
54. E. H. Andrews and A. J. Kinloch, Proc. R. Soc. Lond., A332, 385 (1973).
55. L. J. Broutman and T. Kobayashi, International Conference on Dynamic Crack Propagation, Lehigh University, 1972, as quoted by A. R. Rosenfield and M. F. Kannien, J. Macromol. Sci., B7(4), 609 (1973).
56. A. Ahagon and A. N. Gent, J. Polym. Sci. Polym. Phys. Ed., 13, 1285 (1975).
57. J. D. Ferry, *Viscoelastic Properties of Polymers*, 2nd ed., Wiley, New York, 1970.
58. G. P. Anderson, S. J. Bennett, and K. L. DeVries, *Analysis and Testing of Adhesive Bonds*, Academic Press, New York, 1977,
59. E. H. Andrews and A. J. Kinloch, Proc. R. Soc. Lond., A332, 401 (1973).
60. C. V. Cagle, *Adhesive Bonding, Techniques and Applications*, McGraw-Hill, New York, 1968.
61. N. J. DeLollis, *Adhesives for Metals, Theory and Technology*, Industrial Press, New York, 1970.
62. S. Y. Elliott, in *Handbook of Adhesive Bonding*, C. V. Cagle, ed., McGraw-Hill, New York, 1973, pp. 31–3 to 31–24.
63. S. Semerdjiev, *Metal to Metal Adhesive Bonding*, Business Books Ltd., London, 1970.
64. J. J. Bikermann, *The Science of Adhesive Joints*, 2nd ed., Academic Press, New York, 1968.
65. S. Timoshenko, Phil. Mag., 6, 47, 1095 (1924).
66. Y. Kobatake and Y. Inoue, Appl. Sci. Res., A7, 100 (1958).
67. M. L. Williams, R. A. Schapery, A. Zak, and G. H. Lindsey, *The Triaxial Tensile Behavior of Viscoelastic Materials*, GALCIT Rep. SM63-6 (1963).
68. C. Mylonas, Proc. Soc. Exp. Stress Anal., 12(2), 129 (1955).
69. R. T. Shield, Q. Appl. Math., 15, 139 (1957).
70. J. W. McBain and W. B. Lee, J. Phys. Chem., 31, 1674 (1927).
71. H. P. Meissner and E. W. Merrill, ASTM Bull., 151, 80 (1948).
72. H. P. Meissner and G. H. Baldauf, Trans. ASME, 73, 697 (1951).

73. J. J. Bikerman, J. Soc. Chem. Ind. Trans., 60, 23 (1941).
74. J. J. Bikerman and C. R. Huang, Trans. Soc. Rheol., 3, 5 (1959).
75. S. W. Lasoski, Jr. and G. Kraus, J. Polym. Sci., 18, 359 (1955).
76. G. Kraus and J. E. Manson, J. Polym. Sci., 6, 625 (1951).
77. T. B. Crow, J. Soc. Chem. Ind. Trans., 43, 65 (1924).
78. J. L. Gardon, in *Treatise on Adhesion and Adhesives*, Vol. 1, R. L. Patrick, ed., Marcel Dekker, New York, 1967, pp. 269–324.
79. N. A. de Bruyne, Aircr. Eng., 16, 115, 140 (1944).
80. G. W. Koehn, in *Adhesion and Adhesives: Fundamentals and Practice*, J. E. Rutzler, Jr., and R. L. Savage, eds., Wiley, New York, 1954, pp. 120–126.
81. A. F. Lewis and G. A. Tanner, J. Appl. Polym. Sci., 6, S-35 (1962).
82. C. Mylonas and N. A. de Bruyne, in *Adhesion and Adhesives*, N. A. de Bruyne and R. Houwink, eds., Elsevier, New York, 1951, pp. 91–143.
83. O. Volkersen, Luftfahrt-Forsch., 15, 41 (1938).
84. M. Goland and E. Reissner, J. Appl. Mech., Trans. ASME, 66, 17 (1944).
85. F. J. Plantema, *De Schuitspanning in een Lijmnaad*, Rep. M 1181, Nat. Luchtvaartlaboratorium, Amsterdam, 1949.
86. R. W. Cornell, J. Appl. Mech., Trans. ASME, 75, 355 (1953).
87. J. L. Lubkin and E. Reissner, J. Appl. Mech., Trans. ASME, 78, 1213 (1956).
88. J. L. Lubkin, J. Appl. Mech., Trans. ASME, 24, 255 (1957).
89. I. N. Sneddon, in *Adhesion*, D. D. Eley, ed., Oxford University Press, London, 1961, pp. 207–253.
90. G. R. Wolley and D. R. Carver, J. Aircr., 8, 10 (1971).
91. B. Cooper, as quoted in Ref. 82.
92. A. Matting and U. Draugelates, Stahl Eisen, 84, 947 (1964).
93. G. Henning, Plaste Kautsch., 12, 459 (1965).
94. J. H. Engel and R. N. Fitzwater, in *Adhesion and Cohesion*, P. Weiss, ed., Elsevier, Amsterdam, 1962, pp. 89–100.
95. G. J. Spies, Aircr. Eng., 25, 64 (1953).
96. J. J. Bikerman, J. Appl. Phys., 28, 1484 (1957).
97. J. J. Bikerman, J. Appl. Polym. Sci., 2, 216 (1959).
98. D. H. Kaelble, Trans. Soc. Rheol., 3, 161 (1959).
99. D. H. Kaelble, Trans. Soc. Rheol., 4, 45 (1960).
100. D. H. Kaelbel, Trans. Soc. Rheol., 9, 135 (1965).
101. F. S. C. Chang, J. Appl. Phys., 30, 1839 (1959).
102. F. S. C. Chang, Trans. Soc. Rheol., 4, 75 (1960).
103. Y. Inoue and Y. Kobatake, Appl. Sci. Res., A8, 321 (1959).
104. C. Jouwersma, J. Polym. Sci., 45, 253 (1960).
105. S. Yurenka, J. Appl. Polym. Sci., 6, 136 (1962).

References

106. J. L. Gardon, J. Appl. Polym. Sci., 7, 625 (1963).
107. J. L. Gardon, J. Appl. Polym. Sci., 7, 643 (1963).
108. C. Mylonas, in Proc. 4th Int. Congr. Rheol., 2, 423 (1963).
109. A. N. Gent and G. R. Hamed, J. Adhes., 7, 91 (1975).
110. A. N. Gent and G. R. Hamed, J. Appl. Polym. Sci., 21, 2817 (1977).
111. A. N. Gent and G. R. Hamed, Polym. Eng. Sci., 17, 462 (1977).
112. T. Hata, M. Gamo, and Y. Doi, Kobunshi Kagaku, 22, 152 (1965).
113. P. B. Lindley, J. Inst. Rubber Ind., 5, 243 (1971).
114. T. Igarashi, J. Polym. Sci., 13, 2129 (1975).
115. T. Igarashi, J. Polym. Sci., 16, 407 (1978).
116. K. Kendall, J. Phys. D: Appl. Phys., 4, 1186 (1971).
117. K. Kendall, J. Adhes., 5, 179 (1973).
118. K. Kendall, J. Phys. D: Appl. Phys., 6, 1782 (1973).
119. A. L. Bailey, J. Appl. Phys., 32, 1407 (1961).
120. H. Dannenberg, J. Appl. Polym. Sci., 5, 124 (1961).
121. J. J. Bikerman, in "Testing of Polymers," Vol. 4, W. E. Brown, ed., Interscience, New York, 1969, pp. 213–235.
122. J. L. Racich and J. A. Koutsky, J. Appl. Polym. Sci., 19, 1479 (1975).
123. A. N. Gent and R. P. Petrich, Proc. R. Soc. Lond., A310, 433 (1969).
124. D. H. Kaelble, J. Adhes., 1, 102 (1969).
125. A. D. McLaren and C. J. Seiler, J. Polym. Sci., 4, 63 (1949).
126. W. M. Bright, in *Adhesion and Adhesives: Fundamentals and Practice*, J. E. Rutzler, Jr., and R. L. Savage, eds., Wiley, New York, 1954, pp. 130–138.
127. B. V. Deryagin, N. A. Krotova, and V. P. Smilga, *Adhesion of Solids*, translated by R. K. Johnston, Consultants Bureau, New York, 1978.
128. G. M. Hammond and R. C. W. Moakes, Trans. Inst. Rubber Ind., 25, 172 (1949).
129. D. W. Aubrey, G. N. Welding, and T. Wong, J. Appl. Polym. Sci., 13, 2193 (1969).
130. A. Ahagon, A. N. Gent, and E. C. Hsu, in *Adhesion Science and Technology*, Vol. 9A, L. H. Lee, ed., Plenum, New York, 1975, pp. 281–288.
131. D. H. Kaelble and C. L. Ho, Trans. Soc. Rheol., 18, 219 (1974).
132. A. J. Duke and R. P. Stanbridge, J. Appl. Polym. Sci., 12, 1487 (1968).
133. W. T. Chen and T. F. Flavin, IBM J. Res. Dev., 16, 203 (1972).
134. A. J. Duke, J. Appl. Polym. Sci., 18, 3019 (1974).
135. J. Johnston, Adhes. Age, 11(4), 20 (1968).
136. D. Satas and F. Egan, Adhes. Age, 9(8), 22 (1966).
137. E. J. Ripling, S. Mostovoy, and H. T. Corten, J. Adhes., 3, 107 (1971).

138. S. Mostovoy, E. J. Ripling, and C. F. Bersch, J. Adhes., 3, 125 (1971).
139. E. J. Ripling, S. Mostovoy, and C. F. Bersch, J. Adhes., 3, 145 (1971).
140. R. L. Patrick, W. G. Gehman, L. Dunbar, and J. A. Brown, J. Adhes., 3, 165 (1971).
141. R. L. Patrick, J. A. Brown, L. E. Verhooven, E. J. Ripling, and S. Mostovoy, J. Adhes., 1, 136 (1969).
142. M. D. Chang, K. L. DeVries, and M. L. Williams, J. Adhesion, 4, 221 (1972); also in *Recent Advances in Adhesion*, L. H. Lee, ed., Gordon and Breach, New York, 1973, pp. 423—436.
143. E. J. Ripling, S. Mostovoy, and R. L. Patrick, ASTM STP 360, 5 (1963).
144. E. J. Ripling, S. Mostovoy, and R. L. Patrick, Mater. Res. Stand., 4, 129 (1964).
145. S. Mostovoy and E. J. Ripling, J. Appl. Polym. Sci., 10, 1351 (1966).
146. S. Mostovoy and E. J. Ripling, J. Appl. Polym. Sci., 15, 641 (1971).
147. S. Mostovoy and E. J. Ripling, J. Appl. Polym. Sci., 13, 1083 (1969).
148. S. Mostovoy and E. J. Ripling, J. Appl. Polym. Sci., 15, 661 (1971).
149. S. Mostovoy, P. B. Crosley, and E. J. Ripling, J. Mater., 2, 661 (1967).
150. R. L. Patrick, J. A. Brown, N. M. Cameron, and W. G. Gehman, Appl. Polym. Symp., 16, 87 (1971).
151. S. Mostovoy and E. J. Ripling, in *Adhesion Science and Technology*, Vol. 9B, L. H. Lee, ed., Plenum, New York, 1975, pp. 513—562.
152. S. Mostovoy and E. J. Ripling, *Fracturing Characteristics of Adhesive Joints*, Materials Research Laboratory, Glenwood, Ill., Contract Nos. N00019-74-C-0274 (1975) and 00019-73-C-0163 (1974).
153. W. D. Bascom, R. L. Cottington, R. L. Jones, and P. Peyser, J. Appl. Polym. Sci., 19, 2545 (1975).
154. W. D. Bascom and R. L. Cottington, J. Adhes., 7, 333 (1976).
155. W. D. Bascom, P. F. Becher, J. L. Bitner, and J. S. Murday, ASTM STP 640, 63, (1978).
156. L. J. Broutman and F. J. McGarry J. Appl. Polym. Sci., 9, 589 (1965).
157. L. J. Broutman and F. J. McGarry, J. Appl. Polym. Sci., 9, 609 (1965).
158. H. Dannengerg, J. Appl. Polym. Sci., 5, 125 (1961).
159. B. M. Malyshev and R. L. Salganik, Int. J. Fracture Mech., 1, 114 (1965).
160. M. L. Williams, K. L. DeVries, and R. R. Despain, J. Dent. Res., 52(3), 517 (1973).

161. V. I. Mossakovskii and M. R. Rybka, PPM, *28*, 1061 (1964).
162. G. P. Anderson, K. L. DeVries, and M. L. Williams, Int. J. Fracture Mech., *10*(4), 565 (1974).
163. E. M. Corcoran, in *Paint Testing Manual*, G. G. Sward, ed., STP 500, American Society for Testing and Materials, Philadelphia, 1972, pp. 314–332.

15
Creep, Fatigue, and Environmental Effects

In Part I of this chapter, we discuss the creep and fatigue of adhesive joints. In Part II, we discuss the effects of wet environment on the short-term and the long-term (creep and fatigue) fractures of adhesive joints.

PART I. CREEP AND FATIGUE OF ADHESIVE JOINTS

Creep fracture occurs under prolonged static loading (usually tested under constant stress or strain), whereas fatigue fracture occurs under repeated cyclic loading in dynamic condition. In both cases, the maximum applied stress is lower than what will cause immediate catastrophic fracture. The crack growth rate is slow and stable [1,2]. The crack grows under an applied crack extension force G_j, which is therefore the fracture energy in creep or fatigue. The notation G_j emphasizes that this quantity is the applied crack extension force (see Chapter 14). When the applied load is below a threshold value ΔG_{Th}, the crack will not grow.

15.1. CREEP FRACTURE

15.1.1. Greensmith's Crack Growth Theory

Greensmith [3] proposed a theory applicable for noncrystallizing elastomers based purely on a propagation phenomenon. The rate of crack growth is given by

$$\frac{da}{dt} = qG^n \tag{15.1}$$

where a is crack length at time t, G the fracture energy, and q and n are numerical coefficients. For an edge crack, $G = \beta a Q_c$, which follows from Eq. (14.39). Therefore,

$$\frac{da}{dt} = q(\beta a Q_c)^n \tag{15.2}$$

Fracture is assumed to occur by growth of a preexisting flaw of effective length a_0. Integration of Eq. (15.2) gives the time to break under imposed conditions. Two cases are considered below.

Creep Fracture Test

The specimen is held at constant load or constant extension. Q_c and β are constant, and the time to break t_f is given by

$$t_f = \frac{1}{q} \int_{a_0}^{a} \frac{da}{(\beta a Q_c)^n}$$

$$= \frac{1}{(n-1)q\beta^n Q_c^n} \left(\frac{1}{a_0^{n-1}} - \frac{1}{a^{n-1}} \right) \tag{15.3}$$

Since $a \gg a_0$ at fracture, Eq. (15.3) becomes

$$t_f = \frac{1}{(n-1)q\beta^n Q_c^n a_0^{n-1}} \tag{15.4}$$

Equation (15.3) is true up to a certain value G_c, beyond which dc/dt increases rapidly. Assuming infinite velocity at $G > G_c$, Eq. (15.4) becomes

$$t_f = \frac{1 - (\beta a_0 Q_c / G_c)^{n-1}}{(n-1)q(\beta Q_c)^n a_0^{n-1}} \tag{15.5}$$

Creep Fracture

Tensile Test

The specimen is extended at constant strain rate v. The strain e at time t is

$$e = vt \qquad (15.6)$$

On the other hand, the stored energy density Q_c may be related to the strain by

$$Q_c = Be^p = B(vt)^p \qquad (15.7)$$

where B and p are adjustable parameters. Integration of Eq. (15.2) then gives

$$t_f = \frac{pn + 1}{(n - 1)q\beta^n Q_c^n a_0^{n-1}} \qquad (15.8)$$

where Q_c is the stored energy density at fracture. Taking into account the fact that the crack propagates catastrophically above G_c, Eq. (15.8) becomes

$$t_f = \frac{(pn + 1)[1 - (\beta a_0 Q_c/G_c)^{n-1}]}{(n - 1)q(\beta Q_c)^n a_0^{n-1}} \qquad (15.9)$$

The theory has been confirmed experimentally [3]. Figure 15.1 shows the results for specimens containing an artificial crack of length a_0. Theoretical predictions (solid lines) agree remarkably well with experimental data. Figure 15.2 shows the results for both tensile and creep tests on specimens in which a_0 is the intrinsic flaw size. Good agreement between the theoretical predictions (solid lines) and the experimental data can be seen.

15.1.2. Statistical Crack Growth Theories

Two theories that combine molecular and crack growth concepts are discussed below.

Halpin Theory

Slow propagation of existing cracks is considered in terms of the deformation and rupture of network chain at the crack tip [4]. The local stress σ_0 for rupture at the crack tip is related to the macroscopic average stress σ_f by

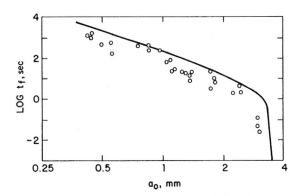

Figure 15.1. Relation between time to fracture t_f and artificial flaw size a_0 for a butadiene-styrene rubber. Points are experimental. The solid line is predicted by Greensmith's theory. (After Ref. 3.)

$$\sigma_0 = \alpha \sigma_f \tag{15.10}$$

where α is the stress concentration factor. The generalized creep compliance is defined as

$$C(t) = \frac{\phi(e)}{\sigma} \tag{15.11}$$

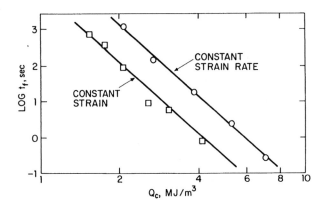

Figure 15.2. Time to fracture t_f versus critical strain energy density Q_c for butadiene-styrene rubber in actual tensile (○) and creep (□) tests (no artificial flaws). Solid lines are theoretical based on Eqs. (15.5) and (15.9). (After Ref. 3.)

Creep Fracture

where $\phi(e)$ is a function of strain e. The time scale t of molecular events at the crack tip is assumed to be related to the time t_f for macroscopic fracture by

$$t = \frac{t_f}{q} \qquad (15.12)$$

where q is a constant. A network chain at the crack tip is assumed to break at a critical strain e_c, giving the local breaking stress as

$$\sigma_0 = \frac{\phi(e_c)}{C(t_f/q)} \qquad (15.13)$$

The macroscopic fracture stress is thus given by

$$\sigma_f = \frac{\phi(e_c)}{\alpha C(t_f/q)} = \frac{K_0}{C(t_f/q)} \qquad (15.14)$$

and the macroscopic breaking strain e_f is such that

$$\phi(e_f) = K_0 \left[\frac{C(t_f)}{C(t_f/q)} \right] \qquad (15.15)$$

Figure 15.3 shows good agreement between the theoretical predictions (solid lines) and the experimental data.

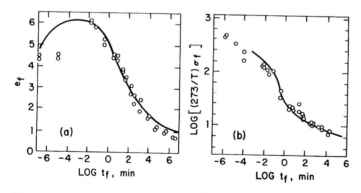

Figure 15.3. (a) Relation between fracture strain e_f and time to fracture t_f for styrene-butadiene rubber. Solid line shows the prediction of Eq. (15.14). (b) Relation between fracture stress σ_f and time to fracture in styrene-butadiene rubber. Solid line shows the prediction of Eq. (15.15). (After Ref. 4.)

Knauss Theory

The time required to generate a flaw of critical size, which then propagates catastrophically, has been considered [5]. The criterion for critical size, in accordance with Eq. (14.39), is given by

$$Q_c A^{1/2} \geq G \tag{15.16}$$

where Q_c is the stored energy density in the specimen, and A is the area of a circular crack of radius a, that is, $A = \pi a^2$. The crack grows to this critical size by activated breakage and reformation of network chains at the crack tip. The rates of bond breakage and reformation are given respectively by

$$\frac{dN_2}{dt} = -k_{21} N_2 + k_{12} N_1 \tag{15.17}$$

$$\frac{dN_1}{dt} = -k_{12} N_1 + k_{21} N_2 \tag{15.18}$$

where N_1 is the number of unbroken bonds, N_2 that of broken bonds, and k_{12} and k_{21} are the rate constants for the two opposing processes. $N_1 + N_2 = N$, where N is the total number of bonds. Assuming that the activation energies of the two processes are altered by the stored energy in the chain, Q_c/N_0, where N_0 is the number of chains per unit volume, the rate constants are then given by

$$k_{12} = B(T) \exp\left(\frac{-Q_c}{N_0 kT}\right) \tag{15.19}$$

$$k_{21} = B(T) \exp\left(\frac{Q_c}{N_0 kT}\right) \tag{15.20}$$

where $B(T)$ is a temperature-dependent coefficient. The rate of bond breakage and thus the rate of growth of crack is thus obtained by subtraction of Eqs. (15.17) and (15.18), giving

$$\frac{1}{2A_0 B(T)} \left(\frac{dA}{dt}\right) = -\left(\frac{N_2 - N_1}{N}\right) \cosh\left(\frac{Q_c}{N_0 kT}\right)$$

$$+ \sinh\left(\frac{Q_c}{N_0 kT}\right) \tag{15.21}$$

Creep Fracture

where A_0 is a constant with dimension of area. However, from Eq. (15.16), the critical condition is given by

$$\frac{dA}{dt} = -\frac{2G^2}{Q_c^3}\left(\frac{dQ_c}{dt}\right) \tag{15.22}$$

Thus, a relation among Q_c, dQ_c/dt, and T is established by Eqs. (15.21) and (15.22). A theoretical linear viscoelastic relation can be used to predict the dependence of Q_c on time and the time to fracture obtained by integration.

15.1.3. Absolute Rate Theories

The lifetime of a material under mechanical restraint was first considered as a thermally activated rate process by Tobolsky and Eyring [6]. The concept has been utilized to interpret the fatigue of polymers [7–9]. Coleman [7] assumed that when the mean displacement γ reaches a critical value γ_f, the breakdown process will proceed catastrophically. The time required for γ to reach γ_f is thus the lifetime (time to fracture), t_f. The rate of increase of γ is given by

$$\frac{d\gamma}{dt} = \frac{2\lambda kT}{h}\exp\left(-\frac{\Delta F^{\ddagger}}{RT}\right)\sinh\left[\frac{\sigma(t)\delta}{2kT}\right] \tag{15.23}$$

where $\Delta F^{\ddagger}$ is the free energy of activation for a jump, λ the jump distance, δ the displacement volume (that is, the product of the effective cross-sectional area per force center and the jump distance), T the absolute temperature, R the gas constant, k the Boltzmann constant, h the Planck constant, and $\sigma(t)$ the macroscopic mean stress at time t. Equation (15.23) has been integrated for several special cases, including the lifetime under dead load (creep under constant load), the lifetime under linearly increasing load, and the lifetime under cyclic loading (dynamic fatigue).

The lifetime under dead load is obtained as

$$\ln t_f = -B\sigma + \ln A \tag{15.24}$$

where

$$A = \frac{\gamma_f h}{\lambda kT}\exp\frac{\Delta F^{\ddagger}}{RT} \tag{15.25}$$

$$B = \frac{\delta}{2kT} \tag{15.26}$$

The lifetime under linearly increasing load is obtained as

$$\sigma_f = ut_f = \frac{1}{B} \ln (BuA) \qquad (15.27)$$

where σ_f is the macroscopic fracture stress and u is the rate of loading. The lifetime under cyclic loading is discussed in Section 15.2.

15.1.4. Applications to Adhesive Joints

Cherry and Holmes [9] investigated the yield strength of stainless steel lap joints bonded with polyethylene using various dead loads and strain rates. In the dead-load (constant-stress) experiment, log t_f varies linearly with applied stress. In the constant-strain-rate experiments, the yield stress varies linearly with the logarithm of strain. The failure strains are found to be the same in both experiments. The strength of the adhesive in the joint is found to be different from that in the unsupported state. This difference arises from a change in the defect structure of the polymer. A yield mechanism is proposed and analyzed in terms of absolute rate theory. The size of the flow unit can be calculated from the time dependence of yield strength. The ratio of these sizes for the polymer in the joint and in unsupported state is constant under all conditions examined.

The creep fracture of aluminum lap joints bonded with various adhesives was investigated at a constant stress rate (linearly increasing load) and a constant strain rate, respectively [10,11]. The Coleman and Knox theory was found to account satisfactorily for the constant-stress-rate experiments; the Cherry and Holmes theory, for the constant-strain-rate experiments.

The static fatigue of aluminum lap joints bonded with various adhesives, including nylon-epoxy, nitrile-phenolic, nitrile-epoxy, epoxy, and polyurethane adhesives, has been investigated [12]. The time to fracture under linearly increasing load is analyzed by

$$\frac{ut_f^2}{2} = K - (\sigma_0 - \sigma_{Th})t_f \qquad (15.28)$$

where t_f is the time to fracture, u the loading rate (the rate of linearly increasing load), K a material constant, σ_0 the stress at zero time, and σ_{Th} the threshold stress for static fatigue (static endurance limit). The relation was derived by assuming that the rate of crack growth is proportional to the amount by which the stress exceeds the threshold value [13,14]. The ratio σ_{Th}/σ_f, where σ_f is the static lap joint strength, is around 0.4 to 0.5 for most of the adhesives tested. Thus, for engineering purposes, the threshold stress may be assumed to be 0.25 of the short-term strength [12].

Fatigue Fracture

15.2. FATIGUE FRACTURE

Fatigue occurs under cyclic loading. The failure may arise from yielding, heat buildup, or crack propagation [2]. Here, the failure by crack propagation is considered. When the temperature rise in the specimen is too low to cause structural changes, fatigue fracture will result from initiation and propagation of fatigue cracks [2]. The total fatigue life N_f is given by

$$N_f = N_i + N_p \tag{15.29}$$

where N_i is the number of cycles required to initiate cracks and N_p that required to propagate the initiated cracks to final failure. If cracks capable of propagation under the applied stress are present initially, N_i is zero and the fatigue failure will result from propagation of preexisting cracks. This is often the case for rubbers and rubbery polymers such as low-density polyethylene. On the other hand, for metals and crystalline polymers, particularly synthetic fibers, cracks have to be created first under cyclic loading [2]. Unfortunately, fatigue crack initiation in polymers is little understood. Fatigue crack propagation is inherently slow and stable, contrary to catastrophic failure such as in the tensile test.

15.2.1. Power Law in Fatigue Fracture

The rate of crack growth in dynamic fatigue has been found to follow a power law [2,15-22]. Its physical basis has been discussed [2,19]. The growth of crack per cycle is given by

$$\frac{da}{dN} = \int_0^t \left(\frac{da}{dt}\right) dt \tag{15.30}$$

where a is the crack length, N is the number of cycles, and t the period per cycle. Applying Eq. (15.1) in Eq. (15.30) gives

$$\frac{da}{dN} = \frac{q}{\nu} G^n \tag{15.31}$$

where ν is the frequency. For simplicity, the specimen is assumed to be loaded and unloaded instantaneously at a constant load. Thus, G refers to the maximum stressed condition in the specimen. For a tensile specimen, integration gives

$$\frac{1}{n-1}\left(\frac{1}{a_0^{n-1}} - \frac{1}{a^{n-1}}\right) = \frac{q}{\nu} \beta^n Q_c^n N \tag{15.32}$$

Table 15.1. Crack Growth Constant n for Some Materials

Material	n Value
Polyethylene	3.5 (region I, brittle)
	2−2.5 (region III, ductile)
Metals	2.0
Poly(methyl methacrylate)	2.5−5.0
Natural rubber	2.0
Butadiene-styrene rubber	4.0

Source: From Ref. 19.

By letting $a \to \infty$ for complete failure, Eq. (15.32) becomes

$$N_f = (n-1)^{-1}\left(\frac{q}{\nu}\beta^n Q_c^n c_0^{n-1}\right)^{-1} \tag{15.33}$$

where N_f is the number of cycles to failure. This is termed *time-dependent fatigue* [2].

On the other hand, many materials are found to follow an even simpler power law relation,

$$\frac{da}{dN} = qG^n \tag{15.34}$$

where q and n are constants. For a tensile specimen, this becomes

$$\frac{da}{dN} = q(\beta a Q_c)^n \tag{15.35}$$

Equation (15.34) differs from Eq. (15.31) in the frequency term. Such behavior is termed *time-independent fatigue* and has been observed in rubbers [15−18], soft thermoplastics [19], glassy plastics [20,21], and metals [22]. The n value usually lies between 1 and 6. A value of about 2 holds for surprisingly wide variety of materials, including natural rubber and aluminum alloys. Table 15.1 lists the n values for some materials. Figure 15.4 shows the linear plot of a^{-1} versus N for natural rubber, where n = 2. Figure 15.5 shows the log (da/dN) versus log G plot for fatigue fracture of polyethylene. Region I has an n value of 3.5, corresponding to brittle failure. Region III has an n value of 2 to 2.5, corresponding to ductile failure. These two regions are connected by a transition zone (region II) in which crack growth is not uniquely defined by G. The transition occurs at dif-

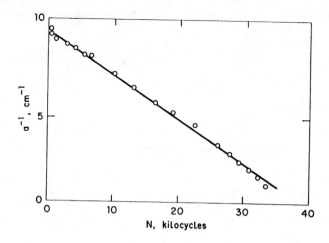

Figure 15.4. Growth of a fatigue crack in natural rubber. (After Ref. 16.)

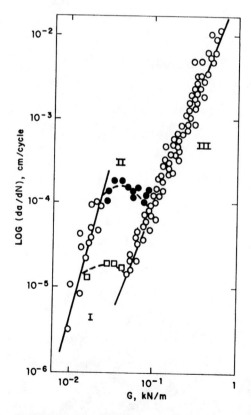

Figure 15.5. Dyanmic fatigue crack growth of a polyethylene. The transition region II depends on Q_c value. ●, $Q_c = 0.02$ MN/m^2; □, $Q_c = 0.23$ MN/m^2. (After Ref. 19.)

ferent da/dN for different values of Q_c, as illustrated for two different Q_c values.

15.2.2. Absolute Rate Theory of Fatigue

Coleman and Knox [8] applied the absolute rate theory to dynamic fatigue. For a sinusoidal loading, where the periodic stress $\sigma(t)$ is given by

$$\sigma(t) = p + c \sin(2\pi \nu t) \qquad (15.36)$$

with $p \gg q$, an asymptotic solution to Eq. (15.23) is

$$t_f = (2\pi Bq)^{1/2} A \exp[-B(p + c)] \qquad (15.37)$$

On the other hand, for a triangular loading with amplitude b and frequency ν, where

$$\sigma(t) = \begin{cases} 2b\nu t & \text{for} \quad 0 \le t < \dfrac{1}{2\nu} \\ 2b(1 - \nu t) & \text{for} \quad \dfrac{1}{2\nu} \le t < \dfrac{1}{\nu} \end{cases} \qquad (15.38)$$

and

$$\sigma(t) = \sigma\left(t - \dfrac{n}{\nu}\right) \qquad \text{for } t \ge \dfrac{1}{\nu} \qquad (15.39)$$

an asymptotic solution is given by

$$t_f = ABb \exp(-Bb) \qquad (15.40)$$

Equations (15.39) and (15.40) predict that for the fatigue experiments considered here, t_f is independent of frequency ν. Thus, whenever $\sigma(t)$ is periodic and $t_f \gg 1/\nu$, then t_f is independent of ν. For instance, in the isothermal fatigue of oriented fibers, it is the time to failure t_f, not the number of cycles to failure N_f, that is independent of the frequency, provided that $1/\nu$ is out of the range of relaxation time of the material tested.

Fatigue Fracture

15.2.3. Fatigue Threshold

In fatigue crack propagation, it is assumed that flaws exist for which $-(\partial U/\partial A)_L$ exceeds G at the stress level imposed in the specimen. However, since $-(\partial U/\partial A)_L = 2\beta a Q_c$, and provided that the preexisting flaw size a_0 is below a certain value, fatigue will not occur when the stored energy density is below its threshold value.

$$Q_{Th} = \frac{G}{2\beta a_0} \qquad (15.41)$$

where Q_{Th} is the fatigue threshold. For poly(methyl methacrylate), using $G = 3 \times 10^5$ erg/cm^3 and $a_0 = 10^{-2}$ cm gives a threshold value Q_{Th} of 5×10^6 erg/cm^3. Taking its tensile modulus as 3×10^3 kg/cm^2, a fatigue stress limit of 170 kg/cm^2 is predicted, which compared well with the experimental value of 150–300 kg/cm^2 [1].

15.2.4. Applications to Adhesive Joints

Adhesive joints that have high static strength may not have a correspondingly long fatigue life [23–27]. Generally, fatigue under cyclic loading is the most detrimental, that is, often even more so than stress corrosion cracking (Section 15.5.1). Interestingly, a wet environment sometimes does not shorten the cyclic fatigue life. However, fatigue life is significantly affected by stress state. For instance, the fatigue life of a scarf joint is much longer than that of a single-lap joint.

Mostovoy and Ripling [23,24] investigated the cyclic fatigue of adhesive joints using tapered-double-cantilever beams (Section 14.8). Several fracture energies are defined, that is, G_{Ic} is the fracture energy in a static (short term) test, and ΔG_{Th} is the fatigue threshold below which no crack grows. The crack growth rate can be given by

$$\frac{da}{dN} = q(\Delta G_j)^n \qquad (15.42)$$

where ΔG_j $(=G_{max} - G_{min})$ is the difference between the maximum and the minimum applied crack extension forces and n is typically about 3 to 6. In a cyclic test, the specimen is loaded and unloaded during each cycle. The crack extension force at the maximum load is G_{max} and that at the minimum load is G_{min} [Eq. (14.138)]. The crack growth rate in dynamic fatigue is related to ΔG, not to G_{max}. In most cases, however, tests are designed such that $G_{min} = 0$; hence $\Delta G_j = G_{max}$.

Four examples of a cyclic fatigue are given in Figures 15.6 to 15.9, where the crack growth rate da/dN is plotted against the applied crack

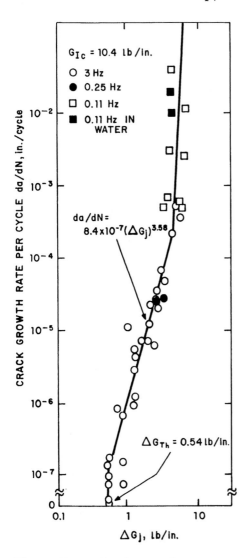

Figure 15.6. Crack growth rate per cycle da/dN versus ΔG_j for a modified epoxy adhesive (MRL-7) on sulfochromic acid-etched and unprimed aluminum adherends. (After Refs. 23 and 24.)

extension force ΔG_j. The crack does not grow up to 10^8 cycles below the threshold value ΔG_{Th}. The fatigue threshold ΔG_{Th} is usually one to two orders of magnitude lower than the corresponding static value G_{Ic}. However, no consistent trend exists between G_{Ic} and ΔG_{Th}. The fatigue threshold appears to be independent of cyclic frequency.

Fatigue Fracture

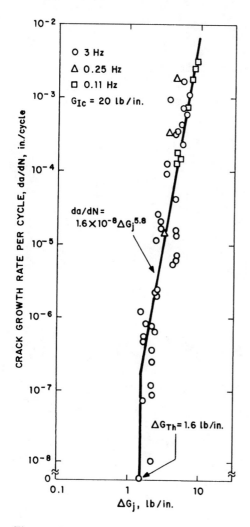

Figure 15.7. Crack growth rate per cycle da/dN versus ΔG_j for a nitrile-phenolic adhesive (MRL-3) with aluminum adherends (primed with EC 1593 primer) at room temperature and ambient humidity. (After Refs. 23 and 24.)

The crack growth rate da/dN becomes very rapid when loaded above ΔG_{Th}. Therefore, loading above ΔG_{Th} should be avoided in practical applications. The shape of da/dN curve is essentially independent of temperature, but the growth rate generally increases with temperature. Similar results have been reported by Marceau and coworkers [27].

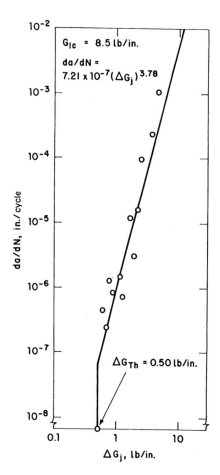

Figure 15.8. da/dN versus ΔG_j for a nitrile-epoxy adhesive (MRL-1) at room temperature and ambient humidity (15—35% RH) at 3 Hz. (After Ref. 23.)

The effects of a wet environment on the dynamic fatigue of adhesive joints are discussed in Section 15.5.2.

PART II. ENVIRONMENTAL EFFECTS

Adhesive joints tend to be weakened by exposure to a wet environment (water or humidity). In extreme cases, particularly when the adherends are hydrophilic, water or vapor may migrate to and accumulate at the interface, causing spontaneous separation of the adhesive from

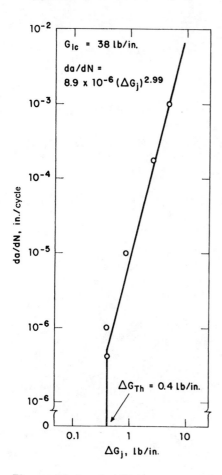

Figure 15.9. da/dN versus ΔG_j for a nylon-epoxy adhesive (MRL-2) at room temperature and ambient humidity (15–35% RH) at 3 Hz. (After Ref. 23.)

the adherend. Even in cases where spontaneous separation does not occur, water or vapor often markedly weakens the interfacial strength. These harmful effects are greatly amplified by combined actions of stress and environment, such as in static and dynamic fatigues.

Figure 15.10 shows the effects of continuous immersion in 24°C water (open symbols) and exposure to 94% relative humidity with temperature cycle between 30 and 65°C in 48 hr (solid symbols) on the shear strength of sulfochromate-treated aluminum lap joints bonded with two different structural adhesives (a nylon-epoxy adhesive and a nitrile rubber-

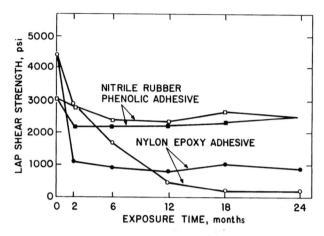

Figure 15.10. Effect of wet exposure on lap shear strength of aluminum lap joints. The joints are continuously exposed to wet environment, then removed for testing while still wet. Open symbols, continuous water immersion; solid symbols, 94% RH with 48 hr cycle from 86°F to 149°F. (After Ref. 28.)

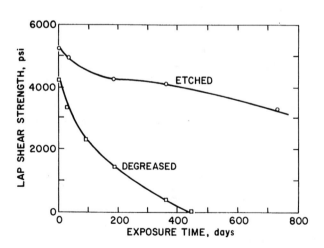

Figure 15.11. Effect of continuous immersion in room-temperature water on the shear strength of an aluminum single lap joint bonded with a nylon-modified epoxy adhesive. Case I: Aluminum etched with sulfochromic acid. Case II: Aluminum degreased only. (After Ref. 29.)

Spontaneous Separation

phenolic adhesive) [28]. The nitrile rubber-phenolic adhesive gives lower dry strength, but is rather resistant to water and humidity. The nylon-epoxy adhesive gives higher dry strength, but loses its joint strength rapidly on exposure to water and humidity, with cohesive failure in the dry state and interfacial failure in the wet state.

Figure 15.11 shows the effects of continuous immersion in 24°C water on the shear strength of aluminum lap joints bonded with a nylon-modified-epoxy adhesive [29]. When the aluminum is pretreated with sulfochromic acid, the joint strength decreases slowly with immersion time. However, when the aluminum is only degreased (and not pretreated otherwise), spontaneous separation occurs after about 450 days. More rapid degradation with water immersion has also been observed. For instance, aluminum lap joints bonded with polyethylene or poly(vinyl acetate) lose their strength completely in less than 1 week on water immersion. Additional examples have been reported elsewhere [30-55].

15.3. SPONTANEOUS SEPARATION

15.3.1. Thermodynamic Analysis

If the adherend has a greater affinity for a liquid in the environment than for the adhesive, the liquid will migrate to and accumulate at the interface, causing spontaneous separation of the adhesive from the adherend. This is likely for adhesive joints with high-energy adherends exposed to water. High-energy surfaces are highly hydrophilic. Their affinity for water is almost three times greater than for organic materials (Table 4.1).

Consider the process where an interface between the adhesive (phase 1) and the adherend (phase 2) is displaced by a liquid (phase 3) (Figure 15.12). The change in free energy is the reversible work of adhesion, $W_{a,L}$, where the subscript L denotes that the joint is immersed in a liquid medium (phase 3). For spontaneous separation to occur, the change in free energy for the process must be zero or negative, that is,

Figure 15.12. Interfacial displacement by a liquid.

$$W_{a,L} = \gamma_{13} + \gamma_{23} - \gamma_{12} \leq 0 \tag{15.43}$$

where γ_{ij} is the interfacial tension between phases i and j. This is the condition for spontaneous separation (thermodynamic failure). Conversely, the condition for stability is given by

$$W_{a,L} = \gamma_{13} + \gamma_{23} - \gamma_{12} > 0 \tag{15.44}$$

The preceding two relations can also be derived from an analysis of fracture energy (Section 15.3.2).

This thermodynamic criterion has been confirmed experimentally [56]. A terpolymer (vinylidene chloride/methyl acrylate/acrylic acid at weight ratio 76.9:19.2:3.9 with $\gamma^d = 38.9$, $\gamma^p = 14.7$, and $\gamma = 53.6$ dyne/cm) is coated on a flame-treated polypropylene ($\gamma^d = 33.5$, $\gamma^p = 4.1$, and $\gamma = 37.6$ dyne/cm) at a thickness of 5 μm. The specimens are then immersed in distilled water or various aqueous surfactant solutions. The coatings spontaneously separate from the adherend after 15 min of immersion in the liquids when $W_{a,L}$ values are negative, whereas no spontaneous separation occurs even after 6 months of immersion when $W_{a,L}$ values are positive (Table 15.2).

If the interaction between the liquid and the adherend (or the adhesive) is stronger than that between the adhesive and the adherend, γ_{13} (or γ_{23}) can be much smaller than γ_{12} such that $W_{a,L}$ is negative. In this case, spontaneous separation will occur upon immersion. This usually occurs when an adhesive joint with high-energy adherend is exposed to water or high humidity, unless the adhesive bonds to the adherend through hydrolytically stable chemical bonds. *Thus, interfacial chemical bonds are generally necessary to obtain stable adhesive bonds on high-energy adherends.*

On the other hand, adhesive bonds between two low-energy materials usually do not separate spontaneously upon water immersion. This is because low-energy materials are usually hydrophobic. They attract each other more strongly than water. However, spontaneous separation may still occur, such as when the low-energy materials contain sufficient amounts of hydrophilic groups or the interfaces are contaminated with hydrophilic impurities.

15.3.2. Fracture Energy Analysis of Wet Strength

The fracture energy G_L of an adhesive bond immersed in a liquid (or vapor) is given by

$$G_L = G_{0,L} \psi_L (Ra_T) \tag{15.45}$$

Table 15.2. Experimental Confirmation of Thermodynamic Criterion for Spontaneous Interfacial Separation of Adhesive Bonds Immersed in Liquids [a]

Immersion liquids	γ_{12}, dyne/cm	γ_{23}, dyne/cm	$W_{a,L}$, erg/cm^2	Immersion result
Dry, control	—	—	+87.7	—
Water	13.3	27.4	+37.3	N
Aqueous solutions of				
Sodium n-octyl sulfate	1.1	3.7	+1.4	N
Sodium n-decyl sulfate	1.0	2.1	−0.3	S
Sodium n-dodecyl sulfate	1.6	0.9	−0.9	S
Sodium n-tetradecyl sulfate	1.7	1.0	−0.7	S
Sodium n-hexadecyl sulfate	2.0	0.6	−0.8	S
Octylphenoxy polyethoxy ehtanol	0.7	2.6	−0.1	S
Sodium diisobutyl sulfosuccinate	6.4	12.8	+15.8	N
Sodium diisoamyl sulfosuccinate	7.9	1.2	+5.7	N
Sodium di(2-ethylhexyl) sulfosuccinate	9.3	1.6	+7.5	N

[a]The specimens are vinyl terpolymer coatings on flame-treated polypropylene; see the text. $\gamma_{12} = 3.5$ dyne/cm. The subscripts are 1 for coating, 2 for adherend, and 3 for immersion liquid. N, no spontaneous separation; S, spontaneous separation.
Source: From Ref. 56.

where the subscript L denotes the presence of a liquid. This follows from Eqs. (14.51)–(14.56). If physical interactions only are operative, then

$$G_L = W_{a,L}\psi_L(Ra_T) \quad \text{(interfacial failure)} \tag{15.46}$$

$$G_L = W_{c,L}\psi_L(Ra_T) \quad \text{(cohesive failure)} \tag{15.47}$$

The fracture criteria for several simple geometries are then given by

$$G_L = 2\beta a Q_{c,L} \quad \text{(simple tension)} \tag{15.48}$$

$$G_L = t_a Q_{c,L} \quad \text{(pure shear)} \tag{15.49}$$

$$G_L = \frac{P_L}{b}(1 - \cos\theta) \quad \text{(peeling)} \tag{15.50}$$

where a is the length of an edge crack (or half the length of a central crack), t_a the adhesive thickness, b the adhesive width, P_L the peel force in the liquid, θ the peel angle, and $Q_{c,L}$ the critical strain energy density of the adhesive immersed in the liquid, following from Eqs. (14.65), (14.66), and (14.119).

Consider interfacial failure of a linear-elastic adhesive on a rigid adherend. Appropriate combinations of the preceding equations give

$$\sigma_{f,L} = \left[\frac{E_L W_{a,L}\psi_L(Ra_T)}{\pi a}\right]^{1/2} \quad \text{(simple tension)} \tag{15.51}$$

$$\sigma_{f,L} = \left[\frac{2E_L W_{a,L}\psi_L(Ra_T)}{t_a}\right]^{1/2} \quad \text{(pure shear)} \tag{15.52}$$

$$\frac{P_L}{b} = \frac{W_{a,L}\psi_L(Ra_T)}{1 - \cos\theta} \quad \text{(peeling)} \tag{15.53}$$

where $\sigma_{f,L}$ (or P_L) is the fracture strength (or peel strength) in the presence of a liquid. Thus, when $W_{a,L} \leq 0$, then $\sigma_{f,L}$ (or P_L) will be zero, and spontaneous separation will occur.

On the other hand, if chemical bonds are formed across the interface, then

$$G_L = W_{b,L}\psi_L(Ra_T) \tag{15.54}$$

Effect of Interfacial Chemical Bonds

If these chemical bonds are stable toward the liquid (not completely hydrolyzable), $W_{b,L} > 0$ and spontaneous separation will not occur. This shows the importance of chemical bonds in retaining wet strength.

If the bulk properties are not affected by the liquid, $Q_{c,L} = Q_c$ and $\psi_L(Ra_T) = \psi(Ra_T)$. In this case, the ratio of the wet strength to dry strength is given by

$$\frac{\sigma_{f,L}}{\sigma_f} = \frac{W_{a,L}}{W_a} \qquad (15.55)$$

which has been demonstrated [57]. A styrene-butadiene rubber (40:60 weight ratio, $T_g = -40°C$) is backed with a cotton cloth and bonded to a film of poly(ethylene terephthalate). The rubber is then lightly cross-linked (with dicumyl peroxide and phenyl-2-naphthylamine at 150°C for 1 hr). T-peel strengths are measured while the specimens are immersed in various liquids, including water, alcohol/water mixtures, ethanol, butanol, glycerol, ethylene glycol, and formamide. Since $W_{a,L} > 0$, spontaneous separation does not occur. However, the peel strengths in liquids are invariably lower than in air, and quantitatively conform to Eq. (15.55).

15.4. EFFECT OF INTERFACIAL CHEMICAL BONDS

Interfacial chemical bonds are generally necessary for adhesive joints with high-energy adherends to achieve stability in water or high humidity. Several examples are discussed below.

15.4.1. Aluminum/Polyethylene Joints

Prolonged immersion of aluminum lap joints bonded with polyethylene causes spontaneous separation of the adhesive from the adherend [30, 58]. Considering the surface of the aluminum to consist of aluminum oxide, $W_{a,L}$ for the aluminum oxide-polyethylene interface is calculated to be -312 erg/cm^2, predicting a spontaneous separation, consistent with experiment. On the other hand, if a monolayer of stearic acid is chemisorbed onto the aluminum surface before adhesive bonding, the resulting lap joints have greatly improved water stability. In this case, spontaneous separation does not occur on prolonged water immersion, although the joint strength is weakened. The ionic bonds between the carboxyl group of the stearic acid and the aluminum oxide are not likely to be displaced by water. The $W_{a,L}$ value for the interface between the hydrocarbon group of the stearic acid and the polyethylene is calculated to be $+101.8$ erg/cm^2, predicting that polyethylene will not spontaneously separate from the chemisorbed stearic acid, consistent with experiment.

15.4.2. Silane Coupling Agents

Silane coupling agents can greatly increase the mechanical strength of adhesive joints (see also Chapter 11). The effects are particularly pronounced in retaining the wet strength of adhesive joints having hydrophilic adherends (such as glass, glass fibers, metals, and metal oxides). Chemical bonding of the adhesive to the adherend through silane molecules is the most likely mechanism. Several other mechanisms have also been proposed. Various theories are discussed.

Chemical Bonding Theory

This is the oldest and the most likely mechanism, discussed in Sections 11.4.2. and 11.4.4.

Wettability Theory

Good wetting minimizes interfacial voids, tending to improve joint strength. High-energy adherends tend to absorb moisture and greasy contaminates rapidly from the air, thus being converted to low-energy adherends. Organosilanes are thought to modify the high-energy surfaces by covering them with a uniform layer of absorbed organosilanes, preventing the absorption of moisture and greasy contaminates [59,60].

However, absorbed organosilanes have low surface energies (Table 15.3). Those organosilanes that are effective as adhesion promoters generally have rather low surface energies: γ less than 35 dyne/cm. The best coupling agent (a methacrylatosilane) for polyesters gives a γ of 44.8 dyne/cm. Chloropropylsilane (γ = 48.8 dyne/cm) and bromophenylsilane (γ = 49.4 dyne/cm) have higher surface energies, but are ineffective as coupling agents for polyesters [66]. Moreover, both ethylsilane and vinylsilanes have low surface energies, about 33.4 dyne/cm. Yet vinyl silane is an effective coupling agnet for butadiene-styrene rubber, whereas ethylsilane is not.

Laird and Nelson [67] found no correlation between wettability and adhesive bond life. The various glass surfaces investigated include a virgin E-glass surface formed by cleaving the glass immersed in the resin (an epoxy resin), melt-cast E-glass surface, chemically cleaned E-glass surfaces, alkali-deficient and alkali-rich E-glass surfaces, and contaminated, polished, and roughened E-glass surfaces. A virgin E-glass surface formed by cleaving the glass under the liquid epoxy resin gives poor bond life in water. Without silane treatment, all glass surfaces give poor bond life in water. Aminopropylsilane-treated glass surface is poorly wetted by an epoxy resin, yet the bond life in water is more than 200 times that without treatment. Plueddemann [68,69] investigated an extensive series of organosilanes as coupling agents for fiberglass-polyester laminates and found no correlation be-

Table 15.3. Surface Tension of Silane Coupling Agents Adsorbed onto Substrates from Aqueous Solutions[a]

Silane	Substrate	Surface tension at 20°C, dyne/cm
Alkyl and alkylene silanes		
Methyltrimethoxysilane, $CH_3Si(OCH_3)_3$	Soda-lime glass	28.0
Ethyltriethoxysilane, $CH_3CH_2Si(OC_2H_5)_3$	Silica (no catalyst)	36.7
	Silica (acetic acid catalyst)	34.5
	Silica (propionic acid catalyst)	27.8
	Silica (piperidine catalyst)	33.4
Phenyltrimethoxysilane, $C_6H_5Si(OCH_3)_3$	Soda-lime glass	43.0
Vinyltriethoxysilane, $CH_2{=}CHSi(OC_2H_5)_3$	Silica	33.4
Vinyltrimethoxysilane, $CH_2{=}CHSi(OCH_3)_3$	Soad-lime glass	28.6
γ-Methacryloxypropyltrimethoxysilane, $CH_2{=}C(CH_3){-}COO(CH_2)_3Si(OCH_3)_3$	Soda-lime glass	44.8

Table 15.3. (Continued)

Silane	Substrate	Surface tension at 20°C, dyne/cm
Amino silanes		
N-β-(Aminoethyl)-γ-aminopropyltrimethoxysilane, $NH_2(CH_2)_2NH(CH_2)_3Si(OCH_3)_3$	Soda-lime glass	33.7
γ-Aminopropyltriethoxysilane, $NH_2CH_2CH_2CH_2Si(OC_2H_5)_3$	Soda-lime glass	35.7
Epoxy silane		
Glycidoxypropyltrimethoxysilane, $H_2C\overset{\displaystyle{\diagdown}}{}\!\!\!\overset{O}{}\!\!\!\overset{\displaystyle{\diagup}}{}CHCH_2O(CH_2)_3Si(OCH_3)_3$	Soda-lime glass (no catalyst)	49.3
	Soda-lime glass (catalyzed with acetic acid, pH 4)	66.9
Halogenated alkyl silanes		
p-Chlorophenylethyltrimethoxysilane, $ClC_6H_4CH_2CH_2Si(OCH_3)_3$	Pyrex glass	51.0
	Silica	49.9
	Stainless steel	53.3
	α-alumina	49.9

Compound	Substrate	Value
γ-Chloropropyltrimethoxysilane, Cl(CH$_2$)$_3$Si(OCH$_3$)$_3$	Soda-lime glass	40.7
	Pyrex glass	48.8
	Stainless steel	49.9
p-Bromophenyltrimethoxysilane, BrC$_6$H$_4$Si(OCH$_3$)$_3$	Soda-lime glass	49.4
3-(1,1-Dihydroperfluorooctoxy)propyltriethoxysilane, CF$_3$(CF$_2$)$_6$CH$_2$O(CH$_2$)$_3$Si(OCH$_2$CH$_3$)$_3$	Pyrex glass	18.8
	Stainless steel	18.8
	Gold	21.5
γ-Perfluoroisopropoxypropyltrimethoxysilane, (CF$_3$)$_2$CFO(CH$_2$)$_3$Si(OCH$_3$)$_3$	Pyrex glass	24.2
	Silica	22.8
	Stainless steel	18.8
Mercapto silane		
γ-Mercaptopropyltrimethoxysilane, HS(CH$_2$)$_3$Si(OCH$_3$)$_3$	Soda-lime glass	41.9

[a] Surface tensions are determined from contact angles (partially from Refs. 61–65) of a series of testing liquids by the equation of state method (see Section 5.1.3).

tween wettability and effectiveness as adhesion promoters [68,69], although Lotz and coworkers [70] found some correlation. Thus, surface wettability is evidently not a primary mechanism of adhesion through organosilanes.

Deformable Layer Theory

Hooper [71] proposed the deformable layer theory, based on the observation that the fatigue life of fiberglass laminates is greatly improved by applying a finish layer on the fiberglass. He suggested that the finish layers are deformable and ductile and, therefore, can relieve the internal stresses. However, a typical silane layer of fiberglass is much too thin to be effective for stress relaxation. Erickson an coworkers [72,73] then proposed that the absorbed silane layer may induce a gradual change of chemical compositions and mechanical properties in an interfacial zone much thicker than 100 Å by deactivation or preferential absorption of resin matrix. Such a layer may be ductile and possess appropriate mechanical properties to provide effective stress relaxation and load transfer. However, this does not explain the improved bond life in the presence of water.

Restrained Layer Theory

According to this theory, silanes function by "tightening up" the resin structure in the boundary zone [74-76]. A boundary layer having modulus intermediate between those of the fiberglass and the resin may minimize stress concentration. However, this contradicts the need for a ductile boundary zone required in the deformable layer theory.

Dynamic Equilibrium Theory

Plueddemann [63,64,77] proposed that silanes improve the water resistance of adhesive bonds by a reversible hydrolyzable bond mechanism. Silanes must chemically couple the resin to the hydrophilic adherend. This chemical bonding alone is not sufficient to ensure resistance to degradation by water. The chemical bonds between the silanol groups of the organosilane layer and the hydrophilic surface may be either a hybrid ionic-covalent oxane bonds or hydrogen bonds. In either case, these interfacial bonds are constantly broken and re-formed by reversible hydrolysis if the resin is rigid (Figure 15.13). In a rigid system, reactive positions are held in such proximity that free silanols resulting from hydrolysis of an interfacial bond have no place to go out but eventually re-form the original bond or make a new bond with an adjacent group. As long as the interface is rigid, bond making and breaking is reversible in the presence of water. This dynamic equilibrium maintains interfacial chemical bonding, imparts ductility to the interface, and allows stress relaxation in the presence of water.

Effect of Interfacial Chemical Bonds

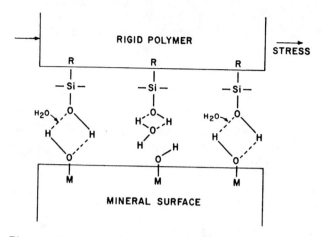

Figure 15.13. Bonding of a rigid polymer to a mineral surface through silane coupling, showing reversible hydrolysis. (After Refs. 64 and 77.)

On the other hand, if the resin is flexible, a hydrolyzed interfacial bond will be pushed away from the hydrophilic surface and will no longer be available for re-forming a new bond (Figure 15.14). Thus, flexible resins cannot form water-resistant bonds with organosilanes. Since flexibility is needed in many adhesives, a reactive rigid primer may be applied before the resin to obtain water resistance.

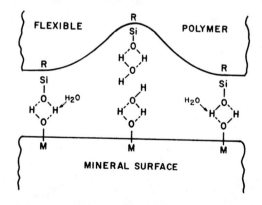

Figure 15.14. Bonding of a flexible polymer to a mineral surface through silane coupling, showing irreversible hydrolysis. (After Refs. 64 and 77.)

This theory combines the concepts of chemical bonding and stress relaxation. It predicts that silanes can improve water resistance only for rigid resins, and not for flexible resins. Its validity, however, remains to be proved.

15.4.3. Ionic Bonding at the Interface

Consider the adhesion to hydrophilic adherends such as metals. In ambient atmosphere, metal surfaces are covered with a layer of amphoteric metal hydroxides, which may undergo acid-base reactions with the adhesive. If the adhesive acts as an acid, then [78]

$$\log K_a = I_s + \log k_a \qquad (15.56)$$

where K_a is the equilibrium constant for the acidic reaction between the adhesive and the adherend, I_s the isoelectric point of the amphoteric adherend surface, and k_a the acidic dissociation constant of the adhesive. On the other hand, if the adhesive acts as a base, then

$$\log K_b = -\log k_b - I_s \qquad (15.57)$$

where K_b is the equilibrium constant for the basic reaction between the adhesive and the adherend and K_b is the basic dissociation constant of the adhesive.

The isoelectric points for a number of oxides and hydroxides have been collected by Parks [79] (Tables 15.4 and 15.5). The pk_a and pk_b for many organic compounds have been collected by Noller [80]. Using these values, the equilibrium constants are calculated for acidic and basic interfacial reactions between some typical adhesive functional groups and some typical oxide surfaces (Table 15.6).

Some observations can be made: (1) On acidic surfaces (such as SiO_2, $I_s = 2$), water will interact with the surface predominantly by the basic mode. The acidic mode of carboxylic acids and the basic mode of primary amines are stronger than that of water, but the interactions with alcohols and phenols are weaker than with water. Thus, on acidic surfaces, adhesives having amine or carboxyl groups should give more-water-resistant adhesive bonds. (2) On neutral surfaces (such as Al_2O_3, $I_s = 8$), water will interact by a hybrid acidic-basic mode. The acidic modes of phenols and carboxylic acids and the basic mode of primary amines are stronger than that of water. Thus, on neutral surfaces, adhesives having amine, carboxyl, or phenolic groups should give more-water-resistant adhesive bonds. (3) On basic surfaces (such as MgO, $I_s = 12$), water will interact mainly by acidic mode. The acidic modes of phenols and carboxylic acids and the basic mode of amines are stronger than that of water. Thus, on basic surfaces,

Effect of Interfacial Chemical Bonds

Table 15.4. Isoelectric Points for Some Oxides in Water

Oxides	Isoelectric point
α-Al_2O_3	6.6−9.2
γ-Al_2O_3	7.4−8.6
AlOOH (boehmite)	6.5−8.8
AlOOH (diaspore)	5.0−7.5
AlOOH (bayerite)	5.4−9.3
Amorphous Al_2O_3	7.5−8.0
$Fe(OH)_2$	11.5−12.5
Natural Fe_2O_3 (hematite)	5.4−6.9
Synthetic Fe_2O_3	6.5−8.6
Fe_2O_3	6.5−6.9
FeOOH (goethite)	6.5−6.9
FeOOH (lepidocrocite)	7.4
Hydrous Fe_2O_3 and amorphous hydroxides	8.5
Fe_3O_4	6.3−6.7
SiO_2 (quartz)	2.2
SiO_2 solutions and gels	1.8

Source: From Ref. 79.

Table 15.5. Effect of Hydration on Isoelectric Point

	Isoelectric point		
Oxide	Freshly calcined	After hydration	Difference
Fe_3O_4	6.7	8.6	1.9
Al_2O_3	6.7	9.2	2.5
TiO_2	4.7	6.2	2.5

Source: From Ref. 79.

Table 15.6. Interactions of Some Organic Compounds with Typical Oxide Surfaces

	Water, HOH	Alcohols, R-OH	Phenols, φ-OH	Carboxylic acids, R-COOH	Amines, R-NH$_2$
pk_a:	15.7	16	10	4.5	20
pk_b:	-1.7	-4	-7	-6	10
Acidic surface, SiO$_2$, I$_S$ = 2					
log K_a	-13.7	-14	-8	-2.5	-18
log K_b	-3.7	-6	-9	-8	+8
Neutral surface, Al$_2$O$_3$, I$_S$ = 8					
log K_a	-7.7	-8	-2	+3.5	-12
log K_b	-9.7	-12	-15	-14	+2
Basic surface, MgO, I$_S$ = 12					
log K_a	-3.7	-4	+2	+7.5	-8
log K_b	-13.7	-16	-19	-18	-2

Data partially from Ref. 78.

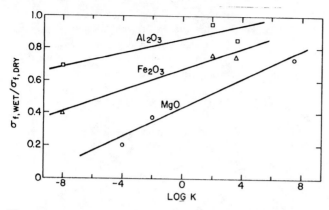

Figure 15.15. Correlation between log K and wet strength of lap joints after immersion in 25°C water for 7 days.

adhesives having phenolic, carboxyl, or amine groups should give more-water-resistant adhesive bonds. To summarize [78]:

On acidic surfaces (SiO_2, etc.):

    amines > carboxylic acids > alcohols > phenols

On neutral surfaces (Al_2O_3, Fe_2O_3, etc.):

    carboxylic acids $\geq$ amines > phenols > alcohols

On basic surfaces (MgO, etc.):

    carboxylic acids > phenols > amines > alcohols

These predictions are verified experimentally. Polyethylenes containing COOH, NH_2, and OH groups, respectively, are used as the adhesive to bond magnesium, aluminum, and cold-rolled steel, respectively. Tensile shear strengths of lap joints are measured in dry state and also after 7 days of immersion in water, and are plotted in Figure 15.15. The effectiveness of the functional groups in improving the wet strength retention is proportional to log K, consistent with the prediction.

15.5. STRESS CORROSION CRACKING AND FATIGUE IN WET ENVIRONMENT

Exposure of an adhesive joint to wet environment will often greatly degrade its mechanical strength, as water tends to invade the interface. Combined stress and wet environment are even more harmful. The fracture in wet environment under static stress is termed *stress corrosion cracking* (which may also be termed *wet static fatigue*). Generally, cyclic fatigue (in a dry or wet environment) is the most harmful to joint strength, and stress corrosion cracking is the next harmful, that is, $\Delta G_{Th} < G_{scc} \ll G_{Ic}$, where ΔG_{Th} is the threshold energy for cyclic fatigue, G_{scc} the threshold energy for stress corrosion cracking, and G_{Ic} the static fracture energy. Interestingly, however, a wet environment often does not increase the severity of cyclic fatigue as discussed later.

15.5.1. Stress Corrosion Cracking

Stress corrosion cracking can be induced by exposing statically stressed adhesive joint to wet environment. Three types of measurements may be made: (1) the time to failure versus applied stress, (2) the static strength of the joint versus applied stress and exposure time, or (3) the crack growth rate versus applied stress. ASTM D2919 describes a spring-loaded jig for applying any given stress to a joint [41]. An elegantly simple loading ring has been devised by Minford [29,53]. Mostovoy and coworkers [23,24] used double-cantilever beams. A qualitative wedge test was also devised [51,81]. Examples

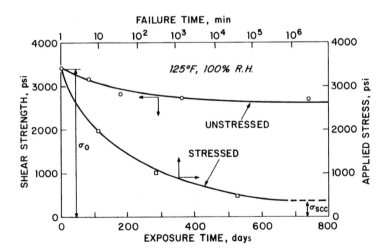

Figure 15.16. Wet strength of lap joints for acid-etched 6061-T6 aluminum bonded with a vinyl-phenolic adhesive under unstressed and stressed conditions at 125°F and 100% RH. (After Refs. 29 and 53.)

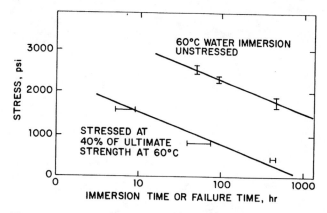

Figure 15.17. Shear strength of sulfochromate-etched aluminum bonded with a modified epoxy adhesive versus time of immersion in 60°C water under unstressed and stressed conditions. (After Ref. 49.)

of the three types of effects induced by stress corrosion cracking are given below.

Applied static stress and a wet environment can combine to greatly shorten the failure time of an adhesive joint. An example is given in Figure 15.16. The initial dry shear strength of the adhesive joint (aluminum lap joint bonded with a vinyl-phenolic adhesive) is 3400 psi. After exposure to 100% RH at 125°F for up to 700 days, the unstressed joint retains as much as 80% of its initial strength. However, this joint, stressed in tension at 2000 psi (about 60% of its initial strength) and exposed to the same wet environment, fails in only about 10 min. If the applied stress is reduced to 1000 psi (about 30% of its initial strength), the failure time increases to about 300 min. If the applied stress is only 500 psi, the failure time is about 40,000 min. The asymptotic stress value σ_{SCC} may be termed the *long-term durability limit* [53].

The unstressed wet strength curve is often parallel to the stressed failure curve [49]. An example for aluminum lap joints bonded with a modified epoxy adhesive is given in Figure 15.17. The upper line is a plot of the shear strength versus time of immersion in 60°C water. The lower line is a plot of the applied shear stress versus failure time for the same joint in the same wet environment, the applied stress being about 40% of the initial dry strength. The two lines are parallel.

Crack growth rates in stress corrosion cracking are investigated with double-cantilever beams [23,24,32,36,37,82]. The crack growth rates are measured at various crack extension forces G_j in wet environments (liquid water and high humidities) at various temperatures. Figure 15.18 shows the crack growth rate da/dt versus applied G_j for a modified epoxy adhesive (MRL-1) and sulfochromate-treated alumin-

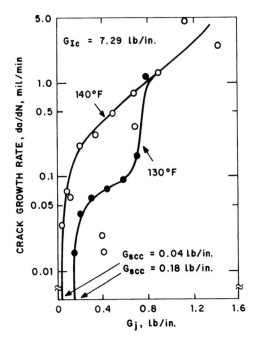

Figure 15.18. Stress corrosion cracking of nitrile-epoxy adhesive (MRL-1) with sulfochromic acid-etched and BR-127 primed aluminum adherends in liquid water at 130°F and 140°F as measured with tapered-double-cantilever beam. The dry G_{Ic} is relatively temperature insensitive and is 7.29 lb/in. between 0 and 200°F. (After Refs. 23 and 24.)

um (2024-T351) adherends immersed in liquid water at 130°F and 140°F. Figure 15.19 shows the same type of plot for a nylon-epoxy adhesive (MRL-2) on the same aluminum adherends immersed in 130°F water. In both cases, no crack growth is observed when G_j is below G_{scc}, which is thus the threshold value for stress corrosion cracking. Above the G_{scc}, the crack grows rapidly. The nylon-epoxy (MRL-2) adhesive has much higher short-term strength than does the modified epoxy (MRL-1) adhesive, but its long-term wet strength is not much better than that of the latter. When only short-term loading is anticipated, the G_{Ic} may be used as the engineering parameter, but when sustained loading in a wet environment is to be encountered, the G_{scc} is more appropriate.

Figure 15.20 shows the (short-term) fracture energy G_{Ic} and stress-corrosion-cracking threshold energy G_{scc} in liquid water for a mod-

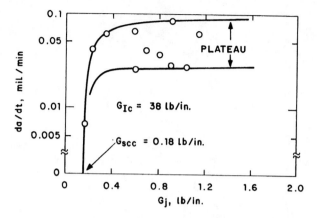

Figure 15.19. Stress corrosion cracking of a nylon-epoxy adhesive (MRL-2) on sulfochromic acid-etched and P-5 primed aluminum adherends in 130°F liquid water as measured with tapered-double-cantilever-beam specimen. G_{scc} is 200 times lower than G_{Ic}. (After Refs. 23 and 24.)

ified epoxy adhesive (MRL-7) on sulfochromate-treated aluminum adherends as a function of temperature. Below about 100°F, the G_{Ic} and G_{scc} drop precipitously, indicating that in this case the stress corrosion cracking sets in at about 100°F.

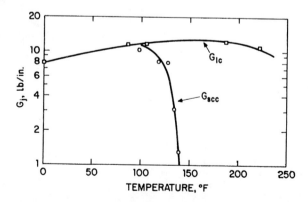

Figure 15.20. Comparison of G_{Ic} and G_{scc} in liquid water for a modified-epoxy adhesive (MRL-7) as a function of temperature. (After Ref. 23.)

15.5.2. Fatigue Fracture in Wet Environment

Wet dynamic fatigue of adhesive joints has been investigated with tapered-double-cantilever beams [23,24,27]. The crack growth rate per cycle da/dN follows a power law,

$$\frac{da}{dN} = q(\Delta G_j)^n \tag{15.58}$$

which is similar to the case for dry dynamic fatigue. A threshold energy ΔG_{Th} is also observed, below which the crack does not grow. Interestingly, however, the crack growth rates in dry and in wet environments in dynamic fatigue are similar. That is, wet environ-

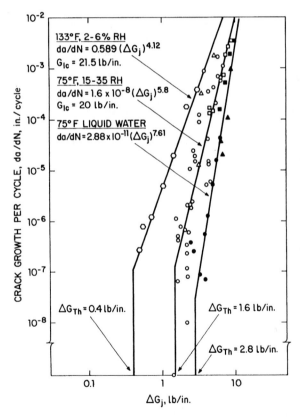

Figure 15.21. Crack growth rate da/dN versus applied crack extension force ΔG_j for a nitrile-phenolic adhesive (MRL-3) on aluminum adherends in the cyclic fatigue test. (After Ref. 23.)

ment often does not accelerate the crack growth rate over that in a dry environment. In some cases, a wet environment is found to slow down the crack growth rate, as shown in Figure 15.21 for a nitrile-phenolic adhesive (MRL-3) with aluminum adherends. Such anomalous behavior is sometimes observed.

Comparisons of the various fracture energies G_{Ic}, G_{scc}, and ΔG_{Th} can be made for a nitrile-epoxy adhesive (MRL-1) (Figures 15.8 and 15.18), for a nylon-epoxy adhesive (MRL-2) (Figures 15.9 and 15.19), and for a modified epoxy adhesive (MRL-7) (Figures 15.6 and 15.20). In all cases, $\Delta G_{Th}(dry) \sim G_{Th}(wet) \ll G_{scc} \ll G_{Ic}$. The effect of a wet environment on fatigue threshold energy is negligible. On the contrary, however, for a nitrile-phenolic adhesive (MRL-3), the dry ΔG_{Th} is lower than the wet ΔG_{Th} in liquid water, as mentioned before and shown in Figure 15.21. Such anomalous behavior is sometimes observed, but is not general.

REFERENCES

1. E. H. Andrews, *Fracture in Polymers*, American Elsevier, New York, 1973.
2. E. H. Andrews, in *Testing of Polymers*, Vol. 4, W. E. Brown, ed., Wiley-Interscience, New York, 1969, pp. 237–296.
3. H. W. Greensmith, J. Appl. Polym. Sci., *8*, 1113 (1964).
4. J. C. Halpin, J. Appl. Phys., *35*, 3133 (1964).
5. W. G. Knauss, as quoted in Ref. 1.
6. A. Tobolsky and H. Eyring, J. Chem. Phys., *11*, 125 (1943).
7. B. D. Coleman, J. Polym. Sci., *20*, 447 (1956).
8. B. D. Coleman and A. G. Knox, Text. Res. J., *27*, 393 (1957).
9. B. W. Cherry and C. M. Holmes, Br. J. Appl. Phys. (Ser. 2), *2*, 821 (1969).
10. E. McAbee, W. C. Tanner, and D. W. Levi, J. Adhes., *2*, 106 (1970).
11. E. McAbee, M. J. Bodnar, W. C. Tanner, and D. W. Levi, J. Adhes., *4*, 87 (1972).
12. A. F. Lewis, R. A. Kinmonth, and R. P. Kreahling, J. Adhes., *3*, 249 (1972).
13. H. S. Loveless, C. W. Deeley, and D. L. Swanson, SPE Trans., *2(2)*, 126 (1962).
14. H. S. Loveless and D. E. McWilliams, Polym. Eng. Sci., *10(3)*, 139 (1970).
15. A. G. Thomas, J. Polym. Sci., *31*, 467 (1958).
16. H. W. Greensmith, L. Mullins, and A. G. Thomas, in *Chemistry and Physics of Rubberlike Substances*, L. Bateman, ed., MacLaren and Sons Ltd., London, 1963, pp. 249–299.
17. A. N. Gent, P. B. Lindley, and A. G. Thomas, J. Appl. Polym. Sci., *8*, 455 (1964).

18. E. H. Andrews, J. Appl. Phys., 32, 542 (1961).
19. E. H. Andrews and B. J. Walker, Proc. R. Soc. Lond., A325, 57 (1971).
20. G. J. Lake and P. B. Lindley, in *Physical Basis of Yield and Fracture*, Institute of Physics and Physical Society, London, 1967, p. 176.
21. L. E. Culver, D. J. Burns, and H. F. Borduas, SPE ANTEC 23, 233 (1967).
22. P. C. Paris and F. Erdogan, J. Basic Eng., 85, 528 (1963).
23. S. Mostovoy and E. J. Ripling, in *Adhesion Science and Technology*, Vol. 9B, L. H. Lee, ed., Plenum Press, New York, 1975. pp. 513–562.
24. S. Mostovoy and E. J. Ripling, *Fracturing Characteristics of Adhesive Joints*, Materials Research Laboratory, Glenwood, Ill., Contract Nos. N00019-74-C-0274 (1975) and N00019-73-C-0163 (1974).
25. D. Y. Wang, Exp. Mech., 4, 173 (1964).
26. D. Y. Wang, Air Force Materials Laboratory, Rep. ASD-TDR-63-93 (1963).
27. J. A. Marceau, J. C. McMillan, and W. M. Scardino, Adhes. Age, 21(4), 37 (1978).
28. N. J. DeLollis and O. Montoya, J. Appl. Polym. Sci., 11, 983 (1967).
29. J. D. Minford, in *Treatise on Adhesion and Adhesives*, Vol. 3, R. L. Patrick, ed., Marcel Dekker, New York, 1973, pp. 79–122.
30. R. A. Gledhill and A. J. Kinloch, J. Adhes., 6, 315 (1974).
31. E. J. Ripling, S. Mostovoy, and H. T. Corten, J. Adhes., 3, 107 (1971).
32. E. J. Ripling, S. Mostovoy, and C. F. Bersch, J. Adhes., 3, 145 (1971).
33. R. L. Patrick, W. G. Gehman, L. Dunbar, and J. A. Brown, J. Adhes., 3, 165 (1971).
34. R. L. Patrick, J. A. Brown, L. E. Verhooven, E. J. Ripling, and S. Mostovoy, J. Adhes., 1, 136 (1969).
35. S. Mostovoy and E. J. Ripling, J. Appl. Polym. Sci., 15, 641 (1971).
36. S. Mostovoy and E. J. Ripling, J. Appl. Polym. Sci., 13, 1083 (1969).
37. S. Mostovoy and E. J. Ripling, J. Appl. Polym. Sci., 15, 661 (1971).
38. S. Mostovoy, P. B. Crosley, and E. J. Ripling, J. Mater., 2, 661 (1967).
39. R. L. Patrick, J. A. Brown, N. M. Cameron, and W. G. Gehman, Appl. Polym. Symp., 16, 87 (1971).
40. D. J. Falconer, N. C. MacDonald, and P. Walker, Chem. Ind., 1230–1231 (July 4, 1964).

References

41. L. H. Sharpe, Appl. Polym. Symp., 3, 353 (1966).
42. P. L. Northcott, H. G. M. Colebeck, and A. Kozak, ASTM STP 401, 3 (1966).
43. P. L. Northcott and W. V. Hancock, ASTM STP 401, 62 (1966).
44. J. A. Scott, ASTM STP 401, 16 (1966).
45. G. F. Carter, ASTM STP 401, 28 (1966).
46. R. M. Kell and C. W. Cooper, ASTM STP 401, 39 (1966).
47. R. F. Wegman, Appl. Polym. Symp., 19, 385 (1972).
48. R. F. Wegman, Appl. Polym. Symp., 32, 1 (1977).
49. R. F. Wegman, M. J. Bodnar, and M. C. Ross, Adhes. Age, 21(7), 38 (1978).
50. R. B. Krieger, Jr., Appl. Polym. Symp., 19, 409 (1972).
51. W. M. Scardino and J. A. Marceau, Appl. Polym. Symp., 32, 51 (1977).
52. J. D. Minford, Appl. Polym. Symp., 32, 91 (1977).
53. J. D. Minford, Adhes. Age, 21(3), 17 (1978).
54. N. J. DeLollis, Adhes. Age, 20(9), 41 (1977).
55. S. Mostovoy and E. J. Ripling, Appl. Polym. Symp., 19, 395 (1972).
56. D. K. Owens, J. Appl. Polym. Sci., 14, 1725 (1970).
57. A. N. Gent and J. Schultz, J. Adhes., 3, 281 (1972).
58. H. Schonhorn and H. L. Frisch, J. Polym. Sci., Polym. Phys. Ed., 11, 1005 (1973).
59. W. A. Zisman, Ind. Eng. Chem., 57(1), 26 (1965).
60. W. A. Zisman, Ind. Eng. Chem., Prod. Res. Dev., 8, 97 (1969).
61. W. D. Bascom, J. Colloid Interface Sci., 27, 789 (1968); Adv. Chem. Ser., 87, 38 (1968).
62. L. H. Lee, J. Colloid Interface Sci., 27, 751 (1968).
63. E. P. Plueddemann, in *Composite Materials*, L. J. Broutman and R. H. Krock, eds., Vol. 6, *Interfaces in Polymer Matrix Composites*, E. P. Plueddemann, ed., Academic Press, New York, 1974, pp. 173–216.
64. E. P. Plueddemann, J. Adhes., 2, 184 (1970).
65. E. G. Shafrin and W. A. Zisman, J. Am. Ceram. Soc., 50, 478 (1967).
66. E. P. Plueddemann, Proc. SPI Conf. Reinforced Plast. Div., 20th ANTEC, 19-A, 1965.
67. J. A. Laird and F. W. Nelson, SPE Trnas., 120 (April, 1964).
68. E. P. Plueddemann, J. Paint Technol., 42, 600 (1970).
69. E. P. Plueddemann, Proc. SPI Conf. Reinforced Plast. Div., 25th ANTEC, 13-D, 1970.
70. E. Lotz, D. Wood, and R. Barnes, Proc. SPI Conf. Reinforced Plast. Div., 26th ANTEC, 14-D, 1971.
71. R. C. Hooper, Proc. SPI Conf. Reinforced Plast. Div., 11th ANTEC, 8-B, 1956.

72. P. W. Erickson and A. A. Volpe, NOL Tech Rep. 63—10, 1963.
73. P. W. Erickson, A. A. Volpe, and E. R. Cooper, Proc. SPI Conf. Reinforced Plast. Div., 21-A, 1964.
74. C. A. Kumins and J. Roteman, J. Polym. Sci., *1A*, 527 (1963).
75. T. K. Kwei, J. Polym. Sci., *3A*, 3229 (1965).
76. R. C. Hartlein, Ind. Eng. Chem., Prod. Res. Dev., *10*(1), 92 (1971).
77. E. P. Plueddemann, Mod. Plast., *47*(3), 92 (1970).
78. J. C. Bolger and A. S. Michaels, in *Interface Conversion for Polymer Coatings*, P. Weiss and G. D. Cheever, eds., Elsevier, New York, 1968, pp. 3—60.
79. G. A. Parks, Chem. Rev., *65*, 177 (1965).
80. C. R. Noller, *Chemistry of Organic Compounds*, W. B. Saunders, Philadelphia, 1965, pp. 983—991.
81. J. A. Marceau, Y. Moji, and J. C. McMillan, Adhes. Age, *20*(10), 28 (1977).
82. S. Mostovoy and E. J. Ripling, Appl. Polym. Symp., *19*, 395 (1972).

Appendix I
CALCULATION OF SURFACE TENSION AND ITS NONPOLAR AND POLAR COMPONENTS FROM CONTACT ANGLES BY THE HARMONIC-MEAN AND THE GEOMETRIC-MEAN METHODS

I.1. HARMONIC-MEAN METHOD

A set of simultaneous quadratic equations is to be solved to obtain the nonpolar and polar components of the surface tension of a solid from the contact angles of two testing liquids by using the harmonic-mean equation, as discussed in Section 5.1.4.

The set of simultaneous equations is given below, following Eqs. (5.8) and (5.9).

$$(1 + \cos \theta_1)\gamma_1 = 4 \left[\frac{\gamma_1^d \gamma_S^d}{\gamma_1^d + \gamma_S^d} + \frac{\gamma_1^p \gamma_S^p}{\gamma_2^p + \gamma_S^p} \right] \quad (I.1)$$

$$(1 + \cos \theta_2)\gamma_2 = 4 \left[\frac{\gamma_2^d \gamma_S^d}{\gamma_2^d + \gamma_S^d} + \frac{\gamma_2^p \gamma_S^p}{\gamma_2^p + \gamma_S^p} \right] \quad (I.2)$$

where θ_1 is the contact angle of liquid 1 on the solid, θ_2 that of liquid 2 on the solid, γ the surface tension, and the superscripts d and p refer to nonpolar and polar compnents, respectively. To solve Eqs. (I.1) and (I.2) simultaneously, they are rearranged as

$$Q_1xy + R_1x + U_1y = V_1 \tag{I.3}$$

$$Q_2xy + R_2x + U_2y = V_2 \tag{I.4}$$

where

$$x = \gamma_S^d \text{ and } y = \gamma_S^p$$

$$Q_j = \frac{1}{4}(1 + \cos\theta_j)\gamma_j$$

$$D_j = \gamma_j^d$$

$$P_j = \gamma_j^p$$

$$S_j = D_j + P_j - Q_j$$

$$R_j = P_j(D_j - Q_j)$$

$$U_j = D_j(P_j - Q_j)$$

$$V_j = Q_j D_j P_j$$

where $j = 1$ or 2. Combining Eqs. (I.3) and (I.4) to eliminate y gives

$$x^2 + Mx + N = 0 \tag{I.5}$$

Let $A = 1/(S_1 S_2)$ and $W = (R_1/S_1) - (R_2/S_2)$, then we have

$$M = (1/W)[A(R_1 U_2 - R_2 U_1) + (V_2/S_2) - (V_1/S_1)]$$

$$N = (A/W)(U_1 V_2 - U_2 V_1)$$

Equation (I.5) is then solved to give

$$\gamma_S^d = x = -(M/2) \pm [(M^2/4) - N]^{1/2} \tag{I.6}$$

$$\gamma_S^p = y = (V_j - R_j x)/(S_j x + U_j) \tag{I.7}$$

Appendix I

where j = 1 or 2. The γ_S^d and γ_S^p have two roots; one is significant, the other insignificant. They are easily recognized.

1.2. GEOMETRIC-MEAN METHOD

The two simultaneous linear equations are given below; following Eqs. (5.10) and (5.11).

$$(1 + \cos \theta_1)\gamma_1 = 2[(\gamma_1^d \gamma_S^d)^{1/2} + (\gamma_1^p \gamma_S^p)^{1/2}] \tag{I.8}$$

$$(1 + \cos \theta_2)\gamma_2 = 2[(\gamma_2^d \gamma_S^d)^{1/2} + (\gamma_2^p \gamma_S^p)^{1/2}] \tag{I.9}$$

Let $e_j = (D_j)^{1/2} = (\gamma_j^d)^{1/2}$

$f_j = (P_j)^{1/2} = (\gamma_j^p)^{1/2}$

where j = 1 or 2. Then, Eqs. (I.8) and (I.9) can be solved to give

$$\gamma_S^d = x = 4\left[\frac{Q_1 f_2 - Q_2 f_1}{e_1 f_2 - e_2 f_1}\right]^2 \tag{I.10}$$

$$\gamma_S^p = y = 4\left[\frac{Q_2 e_1 - Q_1 e_2}{e_1 f_2 - e_2 f_1}\right]^2 \tag{I.11}$$

I.3. A FORTRAN PROGRAM FOR SURFACE TENSION CALCULATIONS

```
00100   *           SURF.FOR
00200   *           TO CALCULATE SURFACE TENSION AND COMPONENTS
00300   *           BY HARMONIC-MEAN (H) OR GEOMETRIC-MEAN (G) METHOD
00400               FG(X1,X2)=((X1*Q2-X2*Q1)/A)**2
00500               FQ(G,T)=0.5*G*(1.+COSD(T))
00600   1           TYPE 100
00700   100         FORMAT(/1X,'*'/1X,'POLYMER:   ',$)
00800               ACCEPT 101,I
00900   101         FORMAT(A1)
01000               IF(I.EQ.' ')STOP
```

Appendix I

```
01100       2           TYPE 201
01200     201           FORMAT('+METHOD(H OR G):  ',$)
01300                   ACCEPT 101,KM
01400                   IF(KM.NE.'H'.AND.KM.NE.'G')GO TO 1
01500                   TYPE 102
01600     102           FORMAT('+LIQUIDS(1 AND 2):  ',$)
01700                   ACCEPT 101,I
01800                   TYPE 103
01900     103           FORMAT('+CONTACT ANGLES(DEG):  ',$)
02000                   ACCEPT 104,T1,T2
02100     104           FORMAT(2G)
02200                   TYPE 105
02300     105           FORMAT('+SURFACE TENSIONS:  ',$)
02400                   ACCEPT 104,G1,G2
02500                   TYPE 106
02600     106           FORMAT('+DISPERSION COMPONENTS:  ',$)
02700                   ACCEPT 104,D1,D2
02800     *
02900                   P1=G1-D1
03000                   P2=G2-D2
03100     *
03200                   IF(KM.EQ.'H')GO TO 3
03300     *
03400     *   GEOMETRIC
03500                   Q1=FQ(G1,T1)
03600                   Q2=FQ(G2,T2)
03700                   IF(D1.OR.D2.OR.P1.OR.P2)GO TO 5
03800                   R1=SQRT(D1)
03900                   R2=SQRT(D2)
04000                   S1=SQRT(P1)
04100                   S2=SQRT(P2)
04200                   A=S1*R2-S2*R1
04300                   D3=FG(S1,S2)
04400                   P3=FG(R1,R2)
04500                   GO TO 4
04600     *
04700     *   HARMONIC
04800     3             Q1=0.5*FQ(G1,T1)
04900                   Q2=0.5*FQ(G2,T2)
05000                   S1=D1+P1-Q1
05100                   S2=D2+P2-Q2
05200                   R1=P1*(D1-Q1)
05300                   R2=P2*(D2-Q2)
05400                   U1=D1*(P1-Q1)
05500                   U2=D2*(P2-Q2)
05600                   V1=Q1*D1*P1
05700                   V2=Q2*D2*P2
05800                   A=1./S1/S2
05900                   W=R1/S1-R2/S2
06000                   X=A*(R1*U2-R2*U1)+V2/S2-V1/S1
06100                   Y=A*(U1*V2-U2*V1)
06200                   X=-0.5*X/W
06300                   Y=X**2-Y/W
06400                   IF(Y)GO TO 5
```

Appendix I

```
06500              Y=SQRT(Y)
06600              D3=X-Y
06700              P3=(V1-R1*D3)/(S1*D3+U1)
06800              G3=D3+P3
06900   *    OUTPUT
07000              TYPE 107,D3,P3,G3
07100              D3=X+Y
07200              P3=(V2-R2*D3)/(S2*D3+U2)

07300   4          G3=D3+P3
07400              TYPE 107,D3,P3,G3
07500              GO TO 1
07600   5          TYPE 109
07700   109        FORMAT(1X,'IMAGINARY ROOT'/)
07800              GO TO 1
07900   107        FORMAT(1X,'SOLID DISPERSION COMPONENT=',G14.6/)
08000       *      12X,'POLAR COMPONENT=',G14.6/22X,'TOTAL=',G14.6/)
08100              END
```

EXECUTE SURF.FOR

LINK: Loading
[LNKXCT SURF execution]
*
POLYMER: POLYTETRAFLUOROETHYLENE
METHOD(H OR G): H
LIQUIDS(1 AND 2): WATER(1), METHYLENE IODIDE(2)
CONTACT ANGLES(DEG): 108 77
SURFACE TENSIONS: 72.8 50.8
DISPERSION COMPONENTS: 22.1 44.1

SOLID DISPERSION COMPONENT=   20.5385
           POLAR COMPONENT=   2.00699
                    TOTAL=   22.5455   (significant root)

SOLID DISPERSION COMPONENT=   82.9956
           POLAR COMPONENT=   -4.44888
                    TOTAL=   78.5467   (insignificant root)
*
POLYMER: SAME AS ABOVE
METHOD(H OR G): G
LIQUIDS(1 AND 2): WATER(1), METHYLENE IODIDE(2)
CONTACT ANGLES(DEG): 108 77
SURFACE TENSIONS: 72.8 50.8
DISPERSION COMPONENTS: 21.8 49.5

SOLID DISPERSION COMPONENT=   18.5589
         POLAR COMPONENT=   0.497580
                  TOTAL=   19.0565
*
POLYMER:

STOP

REMARKS

Note that the preferred values for the nonpolar components of the testing liquids are different for the harmonic-mean and the geometric-mean methods. Also, note the difference in the calculated polar components of the surface tension of polytetrafluoroethylene by the two different methods.

Appendix II
UNIT CONVERSION TABLES

Table II.1. SI Metric Units

Quantity	Name	Symbol	Equivalent
Length	meter	m	—
Time	second	s	—
Mass	kilogram	kg	—
Electric current	ampere	A	—
Plane angle	radian	rad	—
Volume	liter	L	$10^{-3}\,m^3$
Frequency	hertz	Hz	s^{-1}
Force	newton	N	$m \cdot kg \cdot s^{-2}$
Pressure, stress	pascal	Pa	$N \cdot m^{-2}$
Energy, work, heat	joule	J	$N \cdot m$
Power	watt	W	$J \cdot s^{-1}$
Viscosity	pascal second	Pa·s	$N \cdot s \cdot m^{-2}$
Temperature	kelvin	K	—
	degree Celsius	°C	—

Table II.2. Multiple Units

Prefix	Symbol	Equivalent
giga	G	10^9
mega	M	10^6
kilo	k	10^3
deci	d	10^{-1}
centi	c	10^{-2}
milli	m	10^{-3}
micro	µ	10^{-6}
nano	n	10^{-9}

Table II.3. Conversion Factors

Length	1 in = 2.540 cm
	1 Å = 0.1 nm
Mass	1 lb = 453.59 g
Force	1 N = 10^5 dyne = 0.102 kgf = 0.2248 lbf
Pressure, stress	1 Pa = 10 dyne/cm^2 = 7.5 × 10^{-3} mmHg
	= 10^{-5} bar = 1.02 × 10^{-5} kg/cm^2
	= 1.45 × 10^{-4} psi
Energy	1 J = 0.2387 cal
Viscosity	1 Pa · s = 10 poise

REFERENCES

1. L. D. Pedde, W. E. Foote, L. F. Scott, D. L. King, and D. L. McGalliard, "Metric Manual," U.S. Government Printing Office, Washington, D.C., 1978.
2. F. S. Conant, Polym. Eng. Sci., 17, 222 (1977).
3. F. S. Conant, Rubber Chem. Technol., 48(1), 1 (1975).

Index

Absolute rate theory:
  in creep fracture, 577–578
  in fatigue fracture, 582
Acidity of surface, 600–603
Activation energy:
  for adhesive bonding, 375–379, 388–391, 404–405
  for diffusion, 388–391, 404–405
  for viscous flow, 377–379
  for wetting, 375–376
Adherend, 338
Adhesion (*see also* Adhesive bond strength):
  chemical, 377, 406–435, 600–603
  chemical bonding on, 406–435
  contact angle on, 360–380
  definition of, 337
  diffusion on, 380–406
  interfacial flaw on, 339–342
  interfacial structure on, 342–344
  interfacial tension on, 360–380

[Adhesion]
  internal stress on, 465–473
  mechanical, 280–284, 294, 337, 435–441
  nonreciprocal, 373–375
  optimum wettability for, 371–372
  reciprocal, 373–375
  shrinkage stress on, 465–473
  surface tension on, 360–380
  surface topography on, 435–441
  thermodynamic, 337
  wetting on, 360–380
  work of, 4, 98–106, 108–112, 338–340, 489–493, 589–593
Adhesion promotion (*see also* Chemical adhesion):
  with amino groups, 410, 423, 425, 602–603
  with carboxyl groups, 423–425, 602–603
  with chromium complexes, 427–434

[Adhesion promotion]
  with epoxide groups, 328, 425
  with hydroxyl groups, 423, 425, 602–603
  with isocyanates, 328, 425
  with methylol groups, 425
  with nitrogen-containing groups, 410, 423, 425, 602–603
  with phosphorus-containing groups, 423, 425
  with silanes, 410–420, 594–600
  with sulfonic groups, 423, 425
  with titanates, 430–435
Adhesive bond formation, 359–440
Adhesive bond strength (see also Adhesion, Fracture):
  blister tests for, 560–561
  of butt joints, 497–510
  cantilever beam tests for, 555–560
  coating adhesion tests for, 562
  cone tests for, 561–562
  in creep fracture, 572–578
  in fatigue fracture, 579–586
  fracture energy tests for, 501–503, 525–528, 534–554
  ideal (theoretical), 47–62, 338
  interfacial structure on, 342–344
  internal stress on, 465–473
  of lap joints, 510–529
  loading modes on, 480–482, 495–496, 509
  of peel joints, 497–510
  rate-temperature effects on, 491–494, 507–508, 539–551
  real (practical), 339–342
  shrinkage stress on, 465–473
  in stress corrosion cracking, 604–608
  stress distribution in, 498–501, 512–525, 530–533
  weak-boundary-layer, 344–351, 449–460
  in wet environment, 586–609

Adhesive joint strength (see Adhesive bond strength)
Adsorption:
  contact angle, effect on, 152–160
  Gibbs adsorption equation, 3, 152–155
  on high-energy surfaces, 215–219, 222–231
  spreading pressure, relation to, 9–11, 133–139, 152–160
  surface tension, effect on, 152–160
Aluminum:
  acid-etched, 437
  anodized, 439
Amino groups, 410–420, 423, 425, 430–435, 594–600
Antiplane shear mode, 480–482
Antonoff's rule, 96
Auger electron spectroscopy, 354
Autohesion, 312–321, 338, 381–382, 403–406
Autophobic liquids, 219
Aziridinyl group, 425

Bashforth and Adams equation, 5–7
Basicity of surface, 600–603
Benard cells, 161–164
Bikerman's doctrine, 345–351, 449–460
Blister test, 560–561
Bondability (see also Adhesion, Adhesive bond strength):
  of chemical-treated polymers, 280–296
  of flame-treated polymers, 296–298
  of photochemical-treated polymers, 312–321
  of plasma-treated polymers, 296–327
  of transcrystallized polymers, 323–327

Index

Breaking thread method, 273–274
Butt joints, 497–510

Cantilever beam tests, 555–561
Capillary height method, 263–264, 271–272
Captive bubble method, 257–260
Cassie and Baxter equation, 25
Cassie's equation, 21, 24
Chemical adhesion (*see also* Adhesion),
  406–435, 600–603
Chromium complexes, 427–434
Climbing drum peel test, 531
Coating adhesion tests, 562–563
Coupling agents:
  chromium complexes, 427–434
  silanes, 410–420, 427–428
  titanates, 430–435
Cohesion, 338
Cohesive energy, 51, 95–96, 104–106
Cohesive strength, ideal, 47–62, 338
Cone tests, 561–562
Contact angle
  advancing, 12
  Cassie's, 15
  dynamic, 235
  equilibrium, 15
  in forced motion, 235, 247–254
  heat of wetting relation, 148
  on high-energy surfaces, 222–231
  hysteresis, 15–26
  intrinsic, 15
  on liquids, 14
  measurements of, 257–274
  metastable, 11
  numerical values of, 108–112, 142–146, 162–163
  between polymer melts, 108–112
  prediction of, 152
  receding, 12
  on solid, 11

[Contact angle]
  on solid polymers, 133–165
  solute adsorption, 152–160
  in spontaneous motion, 235–243
  stable, 11
  static, 235
  temperature effect, 139–147
  thermodynamics of, 11–26
  transition (primary and secondary) effects, 147–148
  Wenzel's, 15
  Young's, 12, 15
Copper, electroformed, 440
Corresponding states theory, 85–86
Coupling agents (*see also* Chemical adhesion):
  chromium complexes, 427–434
  silanes, 410–420, 594–600
  titanates, 430–435
Crack extension force, 484, 557
Cratering, 164
Creep fracture, 571–578
Critical surface tension, 173, 181–193
Crystalline polymer surface, 201–209, 323–325
Curvature:
  surface tension, effect on, 8
  vapor pressure, effect on, 7–8

Diamagnetic sesceptibility, 38, 40
Dielectric constant, 40
  complex, 56–58
Diffusion:
  adhesion, effect on, 380–406
  coefficient, 388–396
  interfacial tension, effect on, 115–122
  kinetics of, 400–402
Dipole-dipole force, 33, 40–44
Dipole-induced dipole force, 35–44

Dipole moment, 37, 39
Dispersion force, 30–32, 40–44
Dugdale plastic zone, 485–486
Du Nouy ring method, 272–273

Electron microprobe, 354
Electron microscopy, 285–287, 291, 353, 437–439
Equation of state, 172–178
ESCA, 280–283, 287, 304–309, 352, 354
Excess surface free energy, 3

Failure (*see also* Fracture):
  adhesive (interfacial), 347–352
  cohesive, 347–352
  interfacial (adhesive), 347–352
  locus of, 344–354, 496, 583, 540–544
  mixed adhesive and cohesive, 496
Fatigue fracture, 579–586
  initiation and propagation, 579–586
  power law, 579, 583, 608–609
  theories of, 579–582
  time-dependent, 580
  time-independent, 580–581
  in wet environment, 608–609
Floating roller peel test, 531
Forces (*see also* Molecular forces):
  attractive, 47–52
  intermolecular, 29–62
  repulsive, 52–54
Fracture (*see also* Failure):
  adhesive (interfacial), 347–352
  cohesive, 347–352
  creep, *see* Creep fracture
  energy, *see* Fracture energy
  fatigue, *see* Fatigue fracture
  interfacial (adhesive), 347–352
  locus, 344–354, 496, 583, 540–544

[Fracture]
  mixed adhesive and cohesive, 496
  threshold, 571, 578, 583, 603, 608
  in wet environment, 586–609
Fracture energy:
  of butt joints, 501–503, 508–510
  chemical bond effect on, 491–492
  crack size effect on, 475–497, 509
  crack tip radius effect on, 479, 497
  definition of, 339
  equilibrium values of, 493
  in interfacial fracture, 489–491
  of lap joints, 509, 515–527
  loading mode effect on, 480–484, 496, 509, 560
  in mixed mode fracture, 496
  of peel joints, 492, 495, 509, 534–537, 550–551
  rate-temperature effects on, 509, 559
  tests for, 555–562
  threshold, 571, 578, 583, 603, 608
  wetting effect on, 370
  work of adhesion, relation to, 491, 493
  work of cohesion, relation to, 491
Fracture mechanics, 338–342, 475–497
Fracture toughness, 484

Geometric-mean approximation, 42–44, 53
Geometric-mean equation, 102
Geometric-mean method, 181
Gibbs adsorption equation, 3, 152–155
Gibbs-Duhem equation, 3

Index

Good and Girifalco theory, 96–98
Goland and Reissner theory, 516–523, 528–529
Greensmith crack theory, 572–573
Griffith-Irwin theory, 480–484
Griffith theory, 339–342, 475–480
Guggenheim equation, 68

Halpin-Polley theory, 229
Harmonic-mean approximation, 42–44
Harmonic-mean equation, 102
Harmonic-mean method, 178–181
Heat of wetting, 148
High-energy surfaces:
  adsorption on, 215–219, 222–231
  definition of, 215
  spreading on, 219–222
  surface tension of, 215–219
  wetting of, 215–231
Hydrogen bond, 35–36, 40, 42, 44
  adhesion, effect on, 313–322

Induction force, 35
Infrared spectroscopy, 283, 287, 297, 309, 316, 352–354
In-plane shear mode, 480–482
Interface (see also Interfacial tension, Surface):
  chemical bonding at, 409–420
  structure of, 342–344, 384–393
  thermodynamics of, 1–4
  thickness between polymers, 115–122, 384–393
Interfacial chemical bonding:
  adhesion, effect on, 406–435
  evidence of, 409–420
  fracture energy, effect on, 406–409
  wet adhesion, effect on, 586–609

Interfacial defect, 360, 375–376
Interfacial structure (see Interface)
Interfacial tension (see also Interface)
  additives, effect on, 122
  definition of, 1–4
  fractional polarity theory, 98–104
  geometry-mean equation, 103
  Good and Girifalco theory, 96–98
  harmonic-mean equation, 102
  lattice theories, 121–122
  mean-filed theory, 115–121
  molecular weight dependence of, 114–115
  numerical values of, 105, 108–111, 126–128
  polarity effect on, 106–108
  between polymers, 67–129
  between polymer solutions, 125–129
  prediction of, 106
  semicontinuum theory, 49–51
  temperature dependence of, 112–113
  thermodynamic theory of, 122
  thermodynamics of, 1–4
Interfacial thickness (see Interface)
Intermolecular forces (see Molecular forces)
Internal stress:
  adhesion, effect on, 465–473
  in adhesive joints, 465–473
  in coatings, 467–469
Ionic bonding at interface, 600–603
Ionization potential, 37, 39
Irwin plastic zone, 484–485
Isocyanates, 425
Isoelectric point, 600–603

Kelvin equation, 8

Kinetics:
  of adhesive bond formation, 375–384, 400–402
  of capillary penetration, 243–247
  of high-energy surface tension variation, 215–219, 229–231
  of diffusion, 400–402
  of silt penetration, 246–247
  of spontaneous spreading, 236–247
  of wetting, 236–254
Knauss theory, 576–577

Lap joints, 509, 510–529
Laplace equation, 5–7
Lennard-Jones potential, 38, 44–46
Lifshitz theory, 54–62
Lifshitz-van der Waals constant, 58–59
Loading modes, 480–482, 495–496, 509
London dispersion force, 33, 40, 43, 44
Long term durability limit, 605
Low-energy surfaces, 215

Macleod's equation, 70–71
Macleod's exponent, 70–71
Mechanical adhesion (see Adhesion)
Meniscus, 5–7
Microscopy, optical and electron, 353
Mixed adhesive and cohesive failure, 496
Mixed mode fracture, 480–484, 496, 509, 560
Mode I loading, 480–482
Mode II loading, 480–482
Mode III loading, 480–482
Modification of surfaces (see Surface modification)
Molecular forces:
  additivity of, 40–42

[Molecular forces]
  approximations for unlike molecules, 42–44
  attractive, between large bodies, 47–51
  classical theories of, 29–44
  dipole, 33, 40–44
  geometric-mean approximation, 42–44
  harmonic-mean approximation, 42–44
  hydrogen bonding (see Hydrogen bond)
  induction, 35
  Lennard-Jones, 44–46
  Lifshitz theory, 54–62
  London dispersion, 30–32, 40–44
  repulsive, between large bodies, 52–54, 58–59
Morpholine derivaties, 425
Morphology of surfaces:
  of aluminum, acid-etched, 437
  of aluminum, anodized, 439
  of copper, electroformed, 440
  of nickel, electroformed, 440
  of polymers, 201–209, 284–294
  of polymers, etched, 280, 284–294
  of steel, phosphated, 438

Neumann triangle, 14
Nickel, electroformed, 440
Nitrogen-containing groups:
  amino groups, 410, 425
  amino silanes, 410–420, 594–600
  amino titanates, 430–435
  aziridinyl group, 425
  isocyanates, 425
  morpholine derivatives, 425
  oxazine derivatives, 425
  oxazolidine derivatives, 425
  oxazoline derivatives, 425
  piperazine derivatives, 425

Index

[Nitrogen-containing groups]
piperidine derivatives, 425
pyridine derivatives, 425
urea derivatives, 425

Opening mode, 480–482
Oxazine derivatives, 425
Oxazolidine derivatives, 425
Oxazoline derivatives, 425

Peel joints, 509, 531–554
Pendent drop method, 68, 266–268
Photoelastic analysis, 500
Pi tensile test, 497
Pigments, 198–201
Piperazine derivatives, 425
Piperidine derivatives, 425
Plantema's stress analysis, 523–525
Plasma treatments (*see* Surface modifications)
Plastic zone models, 484–486
Polarity:
  of liquids, 148–151, 179, 184–189
  of pigments, 126–128, 178–181
  of polymers, 104–106, 180
Polarizability, 36–38
Power law:
  in creep, 572
  in fatigue, 579–580, 608–609
Pure shear specimen, 488, 501–503, 509–529
Pyridine derivatives, 425

Refractive index, 40
Repulsive force (*see* Molecular forces)
RFL dip, 328
Rivlin and Thomas theory, 486–488
Rotating drop method, 269–270

Roughness:
  adhesion, effect on, 435
  contact angle, effect on, 15–26, 435
  wetting, effect on, 15–26, 435

Scarfed joints, 513, 560
Secondary ion mass spectroscopy, 354
Sessile drop method, 257–260, 268–269
Shift factor, 494, 540
Shrinkage stress, 465–473
Significant structure theory, 94
Silane coupling agents (*see also* Adhesion promotion, Chemical adhesion, Coupling agents):
  chemical bonding of, 410–420
  chemical structures, 412–415
  deformable layer theory, 598
  dynamic equilibrium theory, 598
  mechanisms of coupling, 410–420, 594–600
  restrained layer theory, 598
  structure of adsorbed layer, 410–420
  surface tension of, 595–597
  wettability theory, 594–598
Simple tension specimen, 487–488, 497–510
Sliding mode of loading, 480–482
Slip-stick failure, 537–538
Spontaneous spreading, 236–247
Spontaneous wet failure, 589–593
Spontaneous wicking, 435
Spreading (*see also* Spreading coefficient, Spreading pressure):
  forced, 247–254
  on high-energy surface, 219–222
  liquid front configuration, 236–237
  of partially submerged drop, 164–165

[Spreading]
  shapes of drop in, 237–238
  spontaneous, 236–247
Spreading coefficient:
  definition of, 10
  final, 10
  initial, 10
  intermediate, 10
  of liquid on liquid, 139
  of polymer melts, 108–112
Spreading pressure:
  definition of, 9, 133
  equilibrium, 9, 133
  measurements of, 133–139
  numerical values of, 134–135, 137–138
Static endurance limit, 578
Steel, phosphated, 438
Strain energy density, 487–488
Strain energy release rate, 484
Stress concentration, 339, 360, 476, 498–499, 503, 510–529
Stress corrosion cracking, 560, 603–607
Stress distribution:
  in butt joints, 499
  in lap joints, 514–525
  in peel joints, 531–554
  in shrinkage, 465–473
Stress intensity factor, 482
Stress-strain curve, 547
Substrate, 338
Surface acidity, 600–603
Surface basicity, 600–603
Surface crystallinity, 76–79, 201–209, 286–295, 323–327, 453, 459–460
Surface energy (see also Surface tension), 3
Surface enthalpy, 3
Surface entropy, 3
Surface free energy (see also Surface tension), 3
Surface isoelectric point, 600–603
Surface modifications, 279–328
  by abrasion, 327

[Surface modifications]
  for aramid fiber, 328
  by chemical treatments, 280–296
  by colloidal deposition, 327
  with chromic acid, 284–291
  with chromic complexes, 427–434
  by crystallization, 323–327
  for EPDM rubber, 323
  by flame treatments, 296–297
  for fluoropolymers, 279–328
  by grafting on surface, 327
  for metal platings, 284–291, 436
  for nylons, 291–328
  by photochemical treatments, 322–323
  by plasma treatments, 298–321
  for polyacetals, 284–328
  for poly(ethylene terephthalate), 284–328
  for polyolefins, 284–328
  with RFL dip, 328
  with silanes, see Silane coupling agents
  with sodium complexes, 280–283
  with sulfuric acid, 293–294
  by thermal treatments, 296–298
  for tire cords, 328
  with titanates, 327, 430–435
Surface tension (see also Interfacial tension):
  additives, effect on, 84, 209–211
  adsorption of solute, effect on, 152–160
  of chemical-treated polymers, 280–296
  cohesive energy, relation to, 95–96
  conformational effect on, 209–211
  constitutive effect on, 71–72, 201
  of copolymers, 79–82, 209–211

Index

[Surface tension]
  corresponding states theory, 85–86
  critical, 181–193
  of crystalline surfaces, 201–209, 323–325
  curvature effect on, 8
  definition of, 1–4
  of flame-treated polymers, 296–298
  glass transition effect on, 76–79
  of high-energy surfaces, 215–219
  humidity effect on, 222–231
  kinetics of aging, 222–231
  Lifshitz theory, 85
  measurements of, 169–198, 265–274
  molecular-weight dependence, 72–76, 169–172
  morphological effect on, 201–209
  numerical values for liquids, 148–151, 179, 184–189
  numerical values of monolayers, 184–189
  numerical values for organic solids, 184–189
  numerical values for pigments, 198–201
  numerical values for polymer melts, 88–94
  numerical values for polymers, 88–94, 184–189
  numerical values for polymer solids, 88–94, 184–189
  of photochemical-treated polymers, 322–323
  of plasma-treated polymers, 309–312
  of polymer blends, 82–84, 209–211
  of polymer liquids, 67–129
  of polymer melts, 67–127
  of polymer solids, 88–94, 169–211

[Surface tension]
  of polymer solutions, 82–84
  Poser-Sanchez theory, 95
  semicontinuum theory, 49–51, 84–85, 169–211
  significant structure thoery, 94
  of silanes, 595–597
  tacticity effect on, 209–211
  temperature dependence, 3, 67–71, 172
  theories of, 84–96
  of thermal-treated polymers, 296–298
  thermodynamics of, 1–4
  of transcrystallized polymers, 323–327
  of UV-treated polymers, 322–323
Surface topography, 285–294, 436–441
Surface treatments (see Surface modifications)

Tearing mode of loading, 480–482
Tensile (simple) specimen, 487–488, 497–510
Tensiometric method, 260–262, 270–271
Threshold energy, 571
  for creep fracture, 578
  for fatigue fracture, 583, 608
  for stress corrosion cracking, 603–607
Time to break in creep, 572–578
Tire cord adhesion, 328
Titanates, 430–435
Topography of surfaces (see also Morphology of surfaces):
  adhesion, effect on, 280, 284–294, 435–441
  wetting, effect on, 15–26, 435
T-peel test, 530

Urea derivatives, 425

Van der Waals forces (*see* Molecular forces)
Vapor pressure, 7–8, 133–139
Volkersen's stress analysis, 514–516, 528–529

Washburn equation, 243
Weak boundary layer, 316–317, 327, 345–347, 351–352, 449–460
Wenzel's angle, 15
Wenzel's equation, 16
Wet adhesion, 586–609
Wet adhesive strength, 586–609
Wet static strength, 604
Wettability (*see also* Contact angle, Wetting)
  of high-energy surfaces, 215–219, 222–231
  optimum condition for adhesion, 371–372
  of polymers, chemical-treated, 208–296
  of polymers, chromic-acid-treated, 284–291
  of polymers, flame-treated, 296–298
  of polymers, photochemical-treated, 322–323
  of polymers, plasma-treated, 309–312
[Wettability]
  of polymers, thermal-treated, 296–298
  of polymers, transcrystallized, 201–209, 323–327
  of silanes, 594–598
Wetting (*see also* Contact angle, Wettability):
  adhesion, effect on, 375–380
  heat of, 148
  kinetics of, 235–254
Wicking, 435
Wilhelmy plate method, 260–262, 270–271
WLF equation, 494, 540
Work of adhesion:
  definition of, 4
  fracture energy, relation to, 338–340, 489–493, 589–593
  molecular theory of, 98–106
  between polymers, 108–112
Work of cohesion:
  definition of, 4
  fracture energy, relation to, 339–340, 489–493, 589–593
  molecular theory of, 98–106

X-ray fluorescence, 354

Young's angle, 12, 15
Young's equation, 12–14